THE GULAG ARCHIPELAGO

Aleksandr I. Solzhenitsyn

THE GULAG ARCHIPELAGO

1918–1956

An Experiment in Literary Investigation

Volume 3

Translated from the Russian by Harry Willetts
Foreword by Anne Applebaum

HARPER**PERENNIAL** ◉ MODERN**CLASSICS**

NEW YORK ● LONDON ● TORONTO ● SYDNEY

HARPER**PERENNIAL** ● MODERN**CLASSICS**

A hardcover edition of this book was published in 1978 by Harper & Row, Publishers, Inc.

P.S.™ is a trademark of HarperCollins Publishers.

THE GULAG ARCHIPELAGO 1918–1956 (VOLUME 3). Copyright © 1976 by The Russian Social Fund for Persecuted Persons and Their Families. English language translation copyright © 1978 by Harper & Row, Publishers, Inc. Foreword copyright © 2007 by Anne Applebaum. All rights reserved. Printed in the United States of America. No part of this book may be used or reproduced in any manner whatsoever without written permission except in the case of brief quotations embodied in critical articles and reviews. For information, address HarperCollins Publishers, 195 Broadway, New York, NY 10007.

HarperCollins books may be purchased for educational, business, or sales promotional use. For information, please e-mail the Special Markets Department at SPsales@harpercollins.com.

First HarperPerennial edition published 1992.
Reissued in Harper Perennial Modern Classics 2007.

Designed by Sidney Feinberg

The Library of Congress has catalogued the previous edition as follows:

Solzhenitsyn, Aleksandr Isaevich.
    The Gulag archipelago, 1918–1956: an experiment in
literary investigation, I-VII.
        Volumes I–IV translated from the Russian by Thomas P. Whitney.
        Volumes V–VII translated from the Russian by Harry Willetts.
        Translation of Arkhipelag Gulag, 1918–1956.
        Includes bibliographical references.
        Translated from the Russian.
        1. Political prisoners—Russia. I. Title.
HV9713.S6413   1974   365'.45'0947   73–22756
ISBN 0-06-013912-9

ISBN: 978-0-06-125373-7 (reissue)
ISBN-10: 0-06-125373-1 (reissue)

20 21 LSC 21

Acknowledgments

The translator wishes to express his warmest gratitude to Alexis Klimoff of Vassar College, who read the translation with minute care and whose suggestions were invariably helpful and frequently invaluable.

Contents

A section of photographs follows page 174.

Foreword

Although more than three decades have now passed since the winter of 1974, when unbound, hand-typed, samizdat manuscripts of Aleksandr Solzhenitsyn's *Gulag Archipelago* first began circulating around what was then the Soviet Union, the emotions the book stirred have left marks which remain today. Usually, readers were given only twenty-four hours to finish the lengthy manuscript before it had to be passed on to the next person. That meant spending an entire day and a whole night absorbed in Solzhenitsyn's prose—not an experience anyone was likely to forget. Members of that first generation of readers remember who gave the book to them, who else knew about it, whom they passed it on to next. They remember what the book felt like—the blurry, mimeographed text, the dog-eared paper, the dim glow of the lamp switched on late at night—and with whom they later discussed it.

In part, Russians responded so strongly because *The Gulag Archipelago*'s author was, at that time, simultaneously very famous and strictly taboo. Twelve year earlier, in 1962, Solzhenitsyn had attained an unusual distinction, becoming both the first authentic Gulag author to be published in the official press, as well as the last. In that year—the height of the post-Stalinist "Thaw"—the Soviet general secretary Nikita Khrushchev had personally permitted the publication of Solzhenitsyn's short novel *One Day in the Life of Ivan Denisovich*. The book, based on Solzhenitsyn's own camp experiences—like Ivan Denisovich, he too had been a camp bricklayer—described a single, ordinary day in the life of a Gulag prisoner.

Reading it now, it can be hard for contemporary readers to understand why Solzhenitsyn's only published work had created such a furor in the Soviet literary world. But in 1962, *Ivan Denisovich* came as a revelation. Instead of speaking vaguely about 'repressions,' as some other books did at the time, *Ivan Denisovich* was blunt and specific. The sufferings of its heroes were pointless. The work they did was boring and exhausting, and they tried to avoid it. They spoke using camp slang and were rude to one another. The Party did not triumph at the end of the story, and communism did not win out in the end. This honesty, unusual in an era of morality tales and social realism, won Solzhenitsyn admirers, particularly among camp survivors, who wrote him long letters of praise. Each new printing of the novel sold out instantly, and copies were eagerly shared among groups of friends.

Solzhenitsyn's honesty also quickly won him detractors. Within a month of publication, the novel had already been denounced at a meeting of the Soviet Writers' Union. Critics wrote that it was too bleak, too "amoral." Within a few more months, Solzhenitsyn himself was under personal attack, falsely accused of having surrendered to the Germans during the war, and of having been convicted on criminal charges. He fought back, but to no avail: thanks to the furor caused by this first published novel, none of his work would ever be officially published in the Soviet Union again.

Yet his name and his novels remained in circulation thanks to the world of underground publishing in Russia, which at that time was growing rapidly. In fact, in the years between his first burst of "official" fame and the appearance of *The Gulag*

Archipelago in samizdat form, Solzhenitsyn became if anything more notorious, and more celebrated, despite the official ban. The KGB began to follow him closely, and at one point stole his entire personal archive. His wife lost her job. Recently released archival documents show that his every move was closely analyzed at the highest levels of the Soviet security apparatus, and sometimes even by the Politburo itself. At the same time, his occasional lectures were wildly popular: six hundred people showed up for one of his first public readings in 1966. His books began to appear in foreign translations, to great acclaim, and were copied and re-copied in secret.

Then, in 1970, Solzhenitsyn won the Nobel Prize. Fearing he would be barred from returning to Russia, he decided not to travel to Stockholm to accept the award. But he issued a statement to be read out at the Nobel banquet, among other things noting the "remarkable fact that the day of the Nobel Prize presentation coincides with Human Rights day," and calling on all Nobel Prize winners to remember that fact: "Let none at this festive table forget that political prisoners are on hunger-strike this very day in defence of rights that have been curtailed or trampled underfoot."

The Swedish government was unnerved, and the Nobel Committee failed to read out that part of the statement. The Soviet authorities were furious, and boycotted the ceremony. The Soviet Writers' Union denounced Solzhenitsyn as the darling of "reactionary circles in the West," and reviewers described him as "a run-of-the-mill writer with an exaggerated idea of his own importance" whose literary gifts were "inferior to many of his Soviet contemporaries—writers the West chooses to ignore because it finds the impact of truth in their writing unbearable."

Still, millions of Russians learned of the prize through Western radio, as well as through the underground press (which circulated the statement that the Swedes had feared to read), and celebrated the award to their countryman. Thus when news that Solzhenitsyn had written a history of the Soviet Gulag began to filter out too, there was an enormous reading public—and a listening public, for excerpts were immediately read out on Radio Liberty—already waiting to receive it.

Yet the impression which *The Gulag Archipelago* made on its first Russian readers was not solely due to the author's notoriety, or to his Nobel Prize, or to the denunciations of him in the Soviet press. More importantly, the book's appearance also marked the first time that anyone inside Russia had ever tried to write a complete history of the Soviet concentration camps, using what information was then available, mostly the "reports, memoirs and letters by 227 witnesses," whom Solzhenitsyn cites in his introduction. Many knew fragments of the story, from the cousin who had been there or the neighbor's nephew who worked in the police. No one, however, had attempted to put it all together, to tell, in effect, an alternative history of the Soviet Union.

And the result was unique. Solzhenitsyn called *The Gulag Archipelago* an "experiment in literary investigation," and that remains the best description of a work which is otherwise impossible to categorize. The book is not quite a straight history— obviously, Solzhenitsyn did not have access to archives or historical records—and large sections are autobiographical. Solzhenitsyn describes in great detail his own arrest and interrogation, his first prison cell, and, courageously, his flirtation with camp police who asked him to serve as an informer. Other parts of the book rely heavily on the words and experiences of others, including some of Solzhenitsyn's camp friends, as well as many people he did not know but who wrote to him after the publication of *Ivan Denisovich*. Still other sections are based on Solzhenitsyn's research into what sources were available: legal tomes, official histories, and the Soviet press.

But all of the material was then filtered through Solzhenitsyn's unique sensibility, and retold in a style which was simultaneously angry, prophetic, ironic—and always opinionated. Thus *The Gulag Archipelago* is a history, but it is also an interpretation of history, and one which many at first found shocking. Up until the publication of *Gulag*, many in Russia and elsewhere were content to blame Stalin for Soviet terror, concentration camps, and mass arrests. Solzhenitsyn argued, and with real evidence, that Lenin, not Stalin, was responsible for creating the Gulag, and that the first Soviet concentration camps for political prisoners were built in the 1920s, not the 1930s. He also showed that the famous "great purge" of the 1930s, during which many leaders of the Bolshevik Revolution were put on public trial and then executed, was no aberration. In reality, it was only one of the many "waves which strained the murky, stinking pipes of our prison sewers to bursting," and not even the largest at that: far more people were killed during the era of mass collectivization, and the Gulag population actually reached its zenith a decade later, at the end of the 1940s and in the early 1950s.

Most importantly, Solzhenitsyn aimed to show that, contrary to what many believed, the Gulag was not an incidental phenomenon, something which the Soviet Union could eventually eliminate or outgrow. Rather, the prison system had been an essential part of the Soviet economic and political system from the very beginning. "We never did have empty prisons," he wrote, "merely prisons which were full or prisons which were very, very overcrowded." In fact, *The Gulag Archipelago* was intended to serve as a condemnation not just of the Soviet camp system, but of the Soviet Union itself. It succeeded—so much so that Soviet authorities decided they could no longer tolerate Solzhenitsyn's presence at all. As a result of its publication, not just in Russia but in multiple foreign countries, Solzhenitsyn was expelled from the country. He would not return until after the collapse of the Soviet Union in 1991.

With his expulsion, Solzhenitsyn became a true international celebrity, and the influence of *The Gulag Archipelago* began to spread rapidly outside Russia. The first German translation was received rapturously and in exactly the spirit its author had intended. One left-wing German newspaper wrote that *The Gulag Archipelago* constituted a "burning question mark over fifty years of Soviet power, over the whole Soviet experiment from 1918 on." The French and English translations appeared somewhat later, thanks to some misunderstandings over ownership of publication rights, but were equally influential. In the United States, where Solzhenitsyn ultimately chose to reside after his expulsion, the paperback edition of *The Gulag Archipelago*'s first volume sold more than two million copies. In France, it is no exaggeration to say that the book effectively ended the long-standing French intellectual flirtation with Soviet communism. So threatening was the book to the French status quo that Jean-Paul Sartre himself described Solzhenitsyn as a "dangerous element."

The West had heard of the Soviet camp system before, of course: credible witnesses had begun reporting on the growth of the Gulag as early as the 1920s. But what Solzhenitsyn produced was simply more thorough, more monumental, and more detailed than anything that had been produced previously. It could not be ignored, or dismissed as a single man's experience. No one who dealt with the Soviet Union, diplomatically or intellectually, could ignore it. Among other things, its horrific portrait of Soviet terror certainly contributed to the development, first under President Jimmy Carter and then under President Ronald Reagan, of an American foreign policy which recognized "human rights" as a legitimate element of international debate.

Since then, the stature of *The Gulag Archipelago*—now published in hundreds of editions, and in dozens of languages—has continued to grow. True, as open debate about the book and its subject has become possible in recent years, some legitimate criticisms of the work have been aired. Some camp survivors felt their memoirs, entrusted to Solzhenitsyn, were used in ways they didn't like, or to illustrate points they hadn't been making. Others objected to his almost fanatical insistence that any form of cooperation with the Gulag authorities had amounted to collaboration. The writer Lev Razgon, another Gulag memoirist, argued that for himself, as for many others, choosing to take an indoor accounting job was a matter of survival, not moral weakness: it was not immoral, Razgon wrote, to choose life.

It is also true that in the fifteen years that have passed since the Russian archives have opened, some errors have been found in Solzhenitsyn's work. His statistics are often wrong, and he sometimes garbles names and dates. Some of the stories he tells are impossible to verify. Some of the information he presents is partial or incomplete.

Nevertheless, what is most extraordinary about re-reading *The Gulag Archipelago* more than fifteen years after the collapse of the Soviet Union is how much it does get right. Although he did not have access to archival documents or government records, Solzhenitsyn's general outline of the history of the Gulag— from its origins in the Solovetsky Islands in the 1920s, through its expansion at the time of collectivization in the early 1930s, through the death of Stalin and the subsequent camp rebellions— has been proven correct. His description of the moral issues faced by the prisoners has never been disputed. His sociology of camp life, though presented in literary form, is unquestionably accurate. Among other things, the general reliability of the history presented in *The Gulag Archipelago* proves that "prisoners' gossip," so often dismissed by scholars as inaccurate, was often right. Indeed, part of the book's impact at the time of publication derived from the fact that both former victims and former perpetrators recognized Solzhenitsyn's descriptions and chronology as accurate, reflecting their own experiences.

This truthfulness continues to give the book a freshness and an importance which will never be challenged. For a contemporary reader, the book brilliantly evokes a mentality which no longer exists, and which is increasingly difficult to describe or explain: the atmosphere of constant fear; the constant temptation of betrayal; the ubiquity of secret police; the reversal of "normal" values; the generalized cruelty that permeated the culture of the Gulag, and of the Soviet Union itself.

And yet—no twenty-first century reader who picks up Solzhenitsyn's masterpiece for the first time should imagine that he or she is about to read a straightforward historical account. His book not only describes history: it is itself history. Thanks to Solzhenitsyn's obsessive attention to detail and his literary and polemical gifts, *The Gulag Archipelago* helped create the world that we live in today—a world in which Soviet communism is no longer held up as anybody's political ideal.

Anne Applebaum

Preface to the English Translation

To those readers who have found the moral strength to overcome the darkness and suffering of the first two volumes, the third volume will disclose a space of freedom and struggle. The secret of this struggle is kept by the Soviet regime even more zealously than that of the torments and annihilation it inflicted upon millions of its victims. More than anything else, the Communist regime fears the revelation of the fight which is conducted against it with a spiritual force unheard of and unknown to many countries in many periods of their history. The fighters' spiritual strength rises to the greatest height and to a supreme degree of tension when their situation is most helpless and the state system most ruthlessly destructive.

The Communist regime has not been overthrown in sixty years, not because there has not been any struggle against it from inside, not because people docilely surrendered to it, but because it is inhumanly strong, in a way as yet unimaginable to the West.

In the world of concentration camps, corrupt as everything within the Soviet system, the struggle began (alas, it could not begin otherwise) by terrorist actions. Terrorism is a condemnable tool, but in this case it was generated by forty years of unprecedented Soviet state terrorism, and this is a striking instance of evil generating evil. It shows that when evil assumes inhuman dimensions, it ends up by forcing people to use evil ways even to escape it. However, the concentration camp terrorism of the fifties, out of which heroic uprisings were born later on, was essentially different from the "left-wing" revolutionary terrorism which is shaking the Western world in our days, in that young Western terrorists, saturated with boundless freedom, play with innocent people's lives and kill innocent people for the sake of their unclear purposes or in order to gain material advantages. Soviet camp terrorists in the fifties killed proved traitors and informers in defense of their right to breathe.

However, there is no kind of terrorism that can be considered a pride of the twentieth century. On the contrary, terrorism has made it into one of the most shameful centuries of human history. And there is no guarantee that the darkest abyss of terrorism already lies behind us.

November 1977
Vermont

PART V

Katorga

■

"We shall turn the Siberia of *katorga,* Siberia in shackles, into a Soviet, socialist Siberia."

STALIN

THE DESTRUCTIVE-LABOR CAMPS

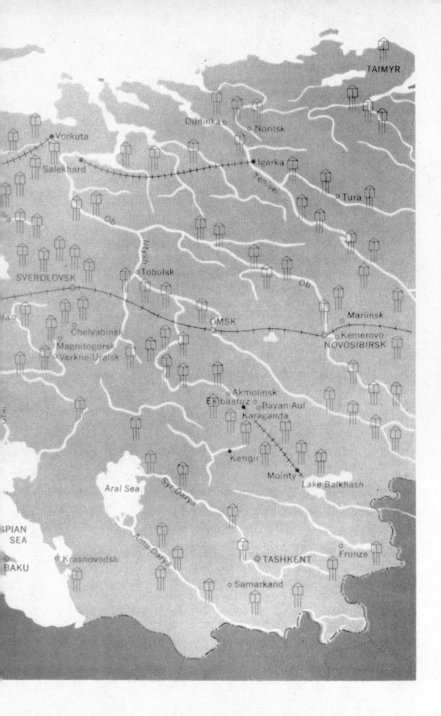

Krasnoyarsk
Taishet
Bratsk
Lena
Lena
Yana
Indigirka
Lake
Baikal
Irkutsk
Dzhida
BAIKAL-
AMUR
REGION
Nerchinsk

Labor
camps ●┼┼┼┼┼● Railroads
built by convicts ⌐⌐⌐⌐ Canals

0   300   600   900 km

© by Scherz Verlag

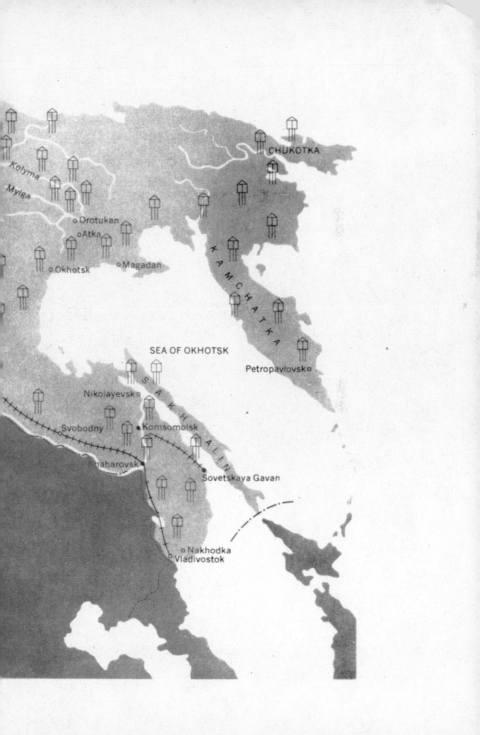

CHUKOTKA

Kolyma

Mylga

o Orotukan

o Atka

o Okhotsk   o Magadan

K A M C H A T K A

SEA OF OKHOTSK

Petropavlovsk

S A K H A L I N

Nikolayevsk o

Svobodny

Komsomolsk

Khabarovsk

Sovetskaya Gavan

o Nakhodka
o Vladivostok

Chapter 1

■

The Doomed

Revolution is often rash in its generosity. It is in such a hurry to disown so much. Take the word *katorga,** for instance. Now, *katorga* is a good word, a word with some weight in it, nothing like the runtish abortion DOPR or the pipsqueak ITL.* *Katorga* descends from the judicial bench like the blade of a guillotine, stops short of beheading the prisoner but breaks his spine, shatters all hope there and then in the courtroom. The word *katorzhane* holds such terror that other prisoners think to themselves: "These must be the real cutthroats!" (It is a cowardly but comfortable human failing to see yourself as not the worst of men, nor in the worst position. *Katorzhane* wear *numbers!* They are obviously beyond redemption! Nobody would pin a number on you or me! . . . They will, though—you'll see!)

Stalin was very fond of old words; he never forgot that they can cement a state together for centuries. It was not to meet some proletarian need that he grafted on again words too hastily lopped off: "officer," "general," "director," "supreme."* And twenty-six years after the February Revolution had abolished *"katorga,"* Stalin reintroduced it. This was in April, 1943, when he began to feel that he was no longer sliding downhill. For the home front the first fruits of the people's victory at Stalingrad were the decree on the militarization of the railroads (providing for trial of women and little boys by court-martial) and, on the following day (April 17), the decree introducing the *katorga* and the gallows. (The gallows is another fine old institution, much superior to a short

*See Notes, page 530.

sharp pistol shot: it makes death a leisurely process, which can be exhibited in detail to a large crowd of people.) Each subsequent victory drove fresh contingents of the doomed into *katorga* or to the gallows—first from the Kuban and the Don, then from the left-bank Ukraine and the Kursk, Orel, and Smolensk regions. On the heels of the army came the tribunals, publicly hanging some people on the spot, dispatching others to the newly created *katorga* forced-labor camps.

The first of them, of course, was Mine No. 17 at Vorkuta (those at Norilsk and Dzhezkazgan came soon after). Little attempt was made to conceal their purpose: the *katorzhane* were to be done to death. These were, undisguisedly, murder camps: but in the Gulag tradition murder was protracted, so that the doomed would suffer longer and put a little work in before they died.

They were housed in "tents," seven meters by twenty, of the kind common in the north. Surrounded with boards and sprinkled with sawdust, the tent became a sort of flimsy hut. It was meant to hold eighty people, if they were on bunk beds, or one hundred on sleeping platforms. But *katorzhane* were put into them two hundred at a time.

Yet there was no reduction of average living space—just a rational utilization of accommodation. The *katorzhane* were put on a twelve-hour working day with two shifts, and no rest days, so that there were always one hundred at work and one hundred in the hut.

At work they were cordoned off by guards with dogs, beaten whenever anybody felt like it, urged on to greater efforts by Tommy guns. On their way back to the living area their ranks might be raked with Tommy-gun fire for no good reason, and the soldiers would not have to answer for the casualties. Even at a distance a column of exhausted *katorzhane* was easily identified —no ordinary prisoners dragged themselves so hopelessly, so painfully along.

Their twelve working hours were measured out in full to the last tedious minute.

Those quarrying stone for roadmaking in the polar blizzards of Norilsk were allowed ten minutes for a warm-up once in the course of a twelve-hour shift. And then their twelve-hour rest was wasted in the silliest way imaginable. Part of these twelve hours went into moving them from one camp area to another, parading them, searching them. Once in the living area, they were immediately taken into a "tent" which was never ventilated—a window-less hut—and locked in. In winter a foul sour stench hung so heavy in the damp air that no one unused to it could endure it for two minutes. The living area was even less accessible to the *katorzhane* than the camp work area. They were never allowed to go to the latrine, nor to the mess hut, nor to the Medical Section. All their needs were served by the latrine bucket and the feeding hatch. Such was Stalin's *katorga* as it took shape in 1943–1944: a combination of all that was worst in the camps with all that was worst in the prisons.[1]

Their twelve hours of *rest* also included inspections, morning and evening—no mere counting of heads, as with ordinary zeks,* but a full and formal roll call at which each of a hundred *katorzhane* twice in every twenty-four hours had to reel off smartly his number, his abhorrent surname, forename, and patronymic, where and in what year he was born, under which article of the Criminal Code he was convicted and by whom, the length of his sentence and when it would expire: while the other ninety-nine, twice daily, listened to all this and suffered torments. Then again, food was distributed twice in the course of these twelve hours: mess tins were passed through the feeding hatch, and through the feeding hatch they were collected again. No *katorzhanin* was permitted to work in the kitchens, nor to take around the food pails. All the serving was done by the thieves—and the more brazenly, the more ruthlessly they cheated the accursed *katorzhane*, the better they lived themselves and the more the camp bosses liked it; as always when the 58's (politicals) were footing the bill, the interests of the NKVD and of the thieves coincided.

According to the camp records, which were not meant to preserve for history the fact that political prisoners were also starved to death, they were entitled to supplementary "miner's rations" and "bonus dishes," which were miserable enough even before three lots of thieves got at them. This was another lengthy procedure conducted through the feeding hatch—names were called

1. We have Chekhov's word for it that the Tsarist *katorga* was much less inventive. The *katorzhane* in the jail at Aleksandrovskaya (Sakhalin) could not only go out into the yard or to the latrine at all hours of the day and night (latrine buckets were not in use there at all), but at any time during the day could go into the town! Stalin, then, was the first to understand the word *katorga* in its original sense—a galley in which the rowers are shackled to their oars.

out one by one, and dishes exchanged for coupons. And when at last you were about to collapse onto the sleeping platform and fall asleep, the hatch would drop again, once again names were called, and they would start reissuing the same coupons for use the next day. (Ordinary zeks had none of this bother with coupons—the foreman took charge of them and handed them in to the kitchen.)

So that out of twelve leisure hours in the cell, barely four remained for undisturbed sleep.

Then again, *katorzhane* were of course paid no money, nor had they any right to receive parcels or letters (the memory of their former freedom must fade in their muddled, dully aching heads, till there was nothing left in the inscrutable polar night but work and barracks).

The *katorzhane* responded nicely to this treatment and quickly died.

The first *alphabet* at Vorkuta—twenty-eight letters,* with numbers from 1 to 1000 attached to each of them—the first 28,000 prisoners in Vorkuta all passed under the earth within a year.

We can only be surprised that it was not in a single month.[2]

A train was sent to Cobalt Mine No. 25 at Norilsk to pick up ore, and some *katorzhane* lay down in front of the locomotive to end it all quickly. A couple of dozen prisoners fled into the tundra in desperation. They were located by planes, shot, and their bodies stacked where the men lined up for work assignment would see them.

At No. 2 Mine, Vorkuta, there was a Women's Camp Division. The women wore numbers on their backs and on their head scarves. They were employed on all underground jobs, and—yes —they even overfulfilled the plan! . . .[3]

But I can already hear angry cries from my compatriots and contemporaries. Stop! *Who are* these people of whom you dare to speak? Yes! They were there to be destroyed—and rightly so! Why, these were traitors, Polizei,* burgomasters! They got what they asked for! Surely you are not sorry for them? (If you are, of

---

2. When Chekhov was there, the whole convict population of Sakhalin was—how many would you think?—5,905. Six letters of the alphabet would have been enough for all of them. Ekibastuz as we knew it was roughly as big, and Spassk very much bigger. The name Sakhalin strikes terror, yet it was really just one Camp Division! In Steplag alone there were twelve complexes as big as that of Sakhalin, and there were ten camps like Steplag. You can calculate how many Sakhalins we had.

3. On Sakhalin there was no hard labor for women (Chekhov).

course, further criticism is outside the competence of literature, and must be left to the *Organs.* *) And the women there were German bedstraw, I hear *women's* voices crying. (Am I exaggerating? It was our women who called other women "German bedstraw," wasn't it?)

I could most easily answer in what is now the conventional fashion—by *denouncing the cult.* I could talk about a few untypical cases of people sent to *katorga.* (The three Komsomol girl volunteers, for instance, who went up in a fighter bomber but were afraid to drop their bombs on the target, jettisoned them in open country, returned to base safely and reported that they had carried out their mission. Later on, her Komsomol conscience began troubling one of them, and she told the Komsomol organizer of her air squadron, also a girl, who of course went straight to the Special Section. The three girls collected twenty years of *katorga* each.) I could cry shame: to think that honest Soviet people like these were punished like criminals at the despot Stalin's whim! And I could wax indignant not so much at Stalin's high-handedness as about the fateful errors in the treatment of Komsomols and Communists, now happily corrected.

It would, however, be improper not to examine the question in depth.

First, a few words about our women, who, as everybody knows, are now emancipated. Not from working twice as hard, it's true, but from religious marriage, from the yoke of social contempt, and from cruel mothers-in-law. Just think, though—have we not wished upon them something worse than Kabanikha* if women who behave as though their bodies and their personal feelings are indeed their own are to be condemned as criminals and traitors? Did not the whole of world literature (before Stalin) rapturously proclaim that love could not be contained by national boundaries? By the will of generals and diplomats? But once again we have adopted Stalin's yardstick: except as decreed by the Supreme Soviet, thou shalt not mate! Your body is, first and foremost, the property of the Fatherland.

Before we go any further, how old were these women when they closed with the enemy in bed instead of in battle? Certainly under thirty, and often no more than twenty-five. Which means that from their first childhood impression onward they had been educated *after* the October Revolution, been brought up in Soviet schools and on Soviet ideology! So that our anger was for the work of our own hands? Some of these girls had taken to heart what we had tirelessly dinned into them for fifteen years on end—that there is no such thing as one's own country, that the Fatherland is a reactionary fiction. Others had grown a little bored with our puritanical Lenten fare of meetings, conferences, and demonstrations, of films without kisses and dancing at arm's length. Yet others were won by politeness, by gallantry, by male attention to the niceties of dress and appearance and to the ritual of courtship, in which no one had trained the young men of our Five-Year Plan epoch, or the officers of Frunze's army. Others again were simply hungry—yes, hungry in the most primitive sense: they had nothing to put in their bellies. And perhaps there was a fifth group, who saw no other way of saving themselves and their relatives, of avoiding separation from their families.

In the town of Starodub, in Bryansk Province, where I arrived hot on the heels of the retreating enemy, I was told that a Hungarian garrison had been stationed there for a long time, to protect the town from partisan raids. Orders came transferring them elsewhere, and dozens of local women, abandoning all shame, went to the station and wept as they said goodbye to the occupying troops—wept more loudly, added a sarcastic shoemaker, than "when they had seen their own husbands off to the war."

The military tribunal reached Starodub some days later. It would hardly fail to act on information received. It doubtless sent some of the weeping women of Starodub to Mine No. 2 at Vorkuta.

But who is really to blame for all this? Who? I ask you. Those women? Or—fellow countrymen, contemporaries—we ourselves, all of us? What was it in *us* that made the occupying troops much more attractive to our women? Was this not one of the innumerable penalties which we are continually paying, and will be paying for a long time yet, for the path we so hastily chose and have so stumblingly followed, with never a look back at our losses, never a cautious look ahead?

Perhaps all these women and girls deserved moral censure (though they, too, should have been given a hearing), perhaps they deserved searing ridicule—but to be sent to *katorga?* to the polar death house?

"Well, it was Stalin who sent them there! And Beria!"

I'm sorry, but it wasn't! Those who sent them there, kept them there, did them to death, now sit with other pensioners on social service councils, looking out for any lowering of moral standards. And the rest of us? We hear the words "German bedstraw" and nod in agreement. The fact that to this day we consider all these women guilty is much more dangerous for us than that they were once *inside*.

"All right, then, but the men at least were in for good reasons? They were traitors to their country, and to their class."

Here, too, we could prevaricate. We might recall (it would be quite true) that the worst criminals did not of course sit still and wait for our tribunals and the gallows. They made for the West as fast as they could, and many of them got away. While our punitive organs reached their target figures by including people innocent as lambs (denunciations by neighbors were a great help here). So-and-so had Germans billeted in his apartment—what made them take a liking to him? Somebody else carried hay for the Germans on his sledge—a straightforward case of collaboration with the enemy.[4]

We could then play the thing down, put all the blame on the *Stalin cult* again: there were excesses, now they have been corrected. All quite normal.

But since we have begun, let us go on.

What about the schoolteachers? Those whom our army in its panicky recoil abandoned with their schools, and pupils, for a year. For two years, or even for three. The quartermasters had been stupid, the generals no good—so what must the teachers do now? Teach their children or not teach them? And what were the kids to do—not kids of fifteen, who could earn a wage, or join the partisans, but the little kids? Learn their lessons, or live like sheep for two or three years to atone for the Supreme Commander's mistakes? If daddy doesn't give you a cap you let your ears freeze —is that it?

For some reason no such question ever arose either in Denmark or in Norway or in Belgium or in France. In those countries it was not felt that a people placed under German rule by its own foolish government or by force of overwhelming circumstances must thereupon stop living altogether. In those countries schools went on working, as did railways and local government.

4. To be fair, we should not forget that from 1946 such people were sometimes regraded and their twenty years of *katorga* commuted to ten years of corrective labor.

Somebody's brains (theirs, of course!) are 180 degrees out of true. Because in our country teachers received anonymous letters from the partisans: "Don't dare teach! You will be made to pay for it!" Working on the railways also became collaboration with the enemy. As for participation in local administration—that was treason, unprecedented in its enormity.

Everybody knows that a child who once drops out of school may never return to it. Just because the greatest strategic genius of all times and all nations had made a blooper, was the grass to wither till he righted it or could it keep growing? Should children be taught in the meantime, or shouldn't they?

Of course, a price would have to be paid. Pictures of the big mustache would have to be taken out of school, and pictures of the little mustache perhaps brought in. The children would gather round the tree at Christmas instead of New Year's, and at this ceremony (as also on some imperial anniversary substituted for that of the October Revolution) the headmaster would have to deliver a speech in praise of the splendid new life, however bad things really were. But similar speeches had been made in the past —and life had been just as bad then.

Or rather, you had to be more of a hypocrite before, had to tell the children many more lies—because the lies had had time to mature, and to permeate the syllabus in versions painstakingly elaborated by experts on teaching technique and by school inspectors. In every lesson, whether it was pertinent or not, whether you were studying the anatomy of worms or the use of conjunctions in complex sentences, you were required to take a kick at God (even if you yourself believed in Him); you could not omit singing the praises of our boundless freedom (even if you had lain awake expecting a knock in the night); whether you were reading Turgenev to the class or tracing the course of the Dnieper with your ruler, you had to anathematize the poverty-stricken past and hymn our present plenty (though long before the war you and the children had watched whole villages dying of hunger, and in the towns a child's ration had been 300 grams).

None of this was considered a sin against the truth, against the soul of the child, or against the Holy Ghost.

Whereas now, under the temporary and still unsettled occupation regime, far fewer lies had to be told—but they stood the old ones on their heads, that was the trouble! So it was that the voice of the Fatherland, and the pencil of the underground Party Committee, forbade you to teach children their native language, geography, arithmetic, and science. Twenty years of *katorga* for work of that sort!

Fellow countrymen, nod your heads in agreement! There they go, guards with dogs alongside, marching to the barracks with their night pails. Stone them—they taught your children.

But my fellow countrymen (particularly former members of specially privileged government departments, retired on pension at forty-five) advance on me with raised fists: Who is it that I am defending? Those who served the Germans as burgomasters? As village headmen? As Polizei? As interpreters? All kinds of filth and scum?

Well, let us go a little deeper. We have done far too much damage by looking at people as entries in a table. Whether we like it or not, the future will force us to reflect on the reasons for their behavior.

When they started playing and singing "Let Noble Rage"— what spine did not tingle? Our natural patriotism, long banned, howled down, under fire, anathematized, was suddenly permitted, encouraged, praised as *sacred*—what Russian heart did not leap up, swell with grateful longing for unity. How could we, with our natural magnanimity, help forgiving in spite of everything the native butchers as the foreign butchers drew near? Later, the need to drown half-conscious misgivings about our impulsive generosity made us all the more unanimous and violent in cursing the traitors—people plainly worse than ourselves, people incapable of forgiveness.

Russia has stood for eleven centuries, known many foes, waged many wars. But—have there been many traitors in Russia? Did traitors ever leave the country in *crowds?* I think not. I do not think that even their foes ever accused the Russians of being traitors, turncoats, renegades, though they lived under a regime inimical to ordinary working people.

Then came the most righteous war in our history, to a country with a supremely just social order—and tens and hundreds of thousands of our people stood revealed as traitors.

Where did they all come from? And why?

Perhaps the unextinguished embers of the Civil War had flared up again? Perhaps these were Whites who had not escaped extermination? No! I have mentioned before that many White émigrés (including the thrice-accursed Denikin) took sides with the Soviet

Union and against Hitler. They had freedom of choice—and that is what they chose.[5]

These tens and hundreds of thousands—Polizei and executioners, headmen and interpreters—were all ordinary Soviet citizens. And there were many young people among them, who had grown up since the Revolution.

What made them do it? . . . What sort of people were they?

For the most part, people who had fallen, themselves and their families, under the caterpillar tracks of the twenties and thirties. People who had lost parents, relatives, loved ones in the turbid streams of our sewage system. Or who themselves had time and again sunk and struggled to the surface in camps and places of banishment. People who knew well enough what it was to stand with feet numb and frostbitten in the queue at the parcels window. People who in those cruel decades had found themselves severed, brutally cut off from the most precious thing on earth, the land itself—though it had been promised to them, incidentally, by the great Decree of 1917, and though they had been called upon to shed their blood for it in the Civil War. (Quite another matter are the country residences bought and bequeathed by Soviet officers, the fenced-in manorial domains outside Moscow: that's ours, so it's all right.) Then some people had been seized for snipping ears of wheat or rye. And some deprived of the right to live where they wished. Or the right to follow a long-practiced and well-loved trade (no one now remembers how fanatically we persecuted craftsmen).

All such people are spoken of nowadays (especially by professional agitators and the proletarian vigilantes of *Oktyabr**) with a contemptuous compression of the lips: "people with a grudge against the Soviet state," "formerly repressed persons," "sons of the former kulak class," "people secretly harboring black resentment of the Soviet power."

One says it—and another nods his head. As though it explained anything. As though the people's state had the right to offend its citizens. As though this were the essential defect, the root of the evil: "people with a grudge," "secretly resentful". . .

And no one cries out: How can you! Damn your insolence! Do

5. They had not sipped with us the bitter cup of the thirties, and from a distance, from Europe, it was easy for them to be enthralled by the great patriotic feat of the Russian people, and overlook the twelve years of internal genocide.

you or do you not hold that being determines consciousness? Or only when it suits you? And when it doesn't suit you does it cease to be true?

Then again, some of us are very good at saying—and a shadow flits over our faces—"Well, yes, certain errors were committed." Always the same disingenuously innocent, impersonal form: "were committed"—only nobody knows by whom. You might almost think that it was by ordinary workers, by men who shift heavy loads, by collective farmers. Nobody has the courage to say: "The Party committed them! Our irremovable and irresponsible leaders committed them!" Yet by whom, except those who had power, could such errors be "committed"? Lump all the blame on Stalin? Have you no sense of humor? If Stalin committed all these errors—where were you at the time, you ruling millions?

In any case, even these mistakes have faded in our eyes to a dim, shapeless blur, and they are no longer regarded as the result of stupidity, fanaticism, and malice; they are all subsumed in the only mistake acknowledged—that Communists jailed Communists. If 15 to 17 million peasants were ruined, sent off for destruction, scattered about the country without the right to remember their parents or mention them by name—that was apparently no mistake. And all the tributary streams of the sewage system surveyed at the beginning of this book were also, it seems, no mistake. That they were utterly unprepared for war with Hitler, emptily vainglorious, that they retreated shamefully, changing their slogans as they ran, that only Ivan fighting for Holy Russia halted the Germans on the Volga—all this turns out to be not a silly blunder, but possibly Stalin's greatest achievement.

In the space of two months we abandoned very nearly one-third of our population to the enemy—including all those incompletely destroyed families; including camps with several thousand inmates, who scattered as soon as their guards ran for it; including prisons in the Ukraine and the Baltic States, where smoke still hung in the air after the mass shooting of political prisoners.

As long as we were strong, we smothered these unfortunates, hounded them, denied them work, drove them from their homes, hurried them into their graves. When our weakness was revealed, we immediately demanded that they should forget all the harm done them, forget the parents and children who had died of hunger in the tundra, forget the executions, forget how we ruined them, forget our ingratitude to them, forget interrogation and

torture at the hands of the NKVD, forget the starvation camps—
and immediately join the partisans, go underground to defend the
Homeland, with no thought for their lives. (There was no need for
us to change! And no one held out the hope that when we came
back we should treat them any differently, no longer hounding,
harassing, jailing, and shooting them.)

Given this state of affairs, should we be surprised that too many
people welcomed the arrival of the Germans? Or surprised that
there were so few who did? (The Germans could sometimes be the
instrument of justice: remember what happened to people who
had served in Soviet times as informers, the shooting of the deacon
at the Naberezhno-Nikolskaya Church in Kiev, for instance—and
there were scores of similar cases.)

And the believers? For twenty years on end, religious belief was
persecuted and churches closed down. The Germans came—and
churches began to open their doors. (Our masters lacked the nerve
to shut them again immediately after the German withdrawal.) In
Rostov-on-the-Don, for instance, the ceremonial opening of the
churches was an occasion for mass rejoicing and great crowds
gathered. Were they nonetheless supposed to curse the Germans
for this?

In Rostov again, in the first days of the war, Aleksandr Petrovich M——, an engineer, was arrested and died in a cell under
interrogation. For several anxious months his wife expected to be
arrested herself. Only when the Germans came could she go to bed
with a quiet mind. "Now at least I can get some sleep!" Should
she instead have prayed for the return of her tormentors?

In May, 1943, while the Germans were in Vinnitsa, men digging
in an orchard on Podlesnaya Street (which the city soviet had
surrounded with a high fence early in 1939 and declared a "restricted area under the People's Commissariat of Defense") found
themselves uncovering graves which had previously escaped notice because they were overgrown with luxuriant grass. They
found thirty-nine mass graves, 3.5 meters deep, 3 meters wide, 4
meters long. In each grave they found first a layer of outer garments belonging to the deceased, then bodies laid alternately head
first or feet first. The hands of all of them were tied with rope, and
they had all been shot by small-bore pistols in the back of the head.
They had evidently been executed in prison and carted out for
burial by night. Documents which had not decayed made it possible to identify people who had been sentenced to "20 years without

the right to correspond" in 1938. Plate No. 1 is one picture of the excavation site: inhabitants of Vinnitsa have come to view the bodies or identify their relatives. There was more to come. In June they began digging near the Orthodox cemetery, outside the Pirogov Hospital, and discovered another forty-two graves. Next the Gorky Park of Culture and Rest—where, under the swings and carrousel, the "funhouse," the games area, and the dance floor, fourteen more mass graves were found. Altogether, 9,439 corpses in ninety-five graves. This was in Vinnitsa alone, and the discoveries were accidental. How many lie successfully hidden in other towns? After viewing these corpses, were the population supposed to rush off and join the partisans?

Perhaps in fairness we should at least admit that if you and I suffer when we and all we hold dear are trodden underfoot, those *we* tread on feel no less pain. Perhaps in fairness we should at last admit that those whom we seek to destroy have a right to hate us. Or have they no such right? Are they supposed to die gratefully?

We attribute deep-seated if not indeed congenital malice to these Polizei, these burgomasters—but we ourselves planted their malice in them, they were "waste products" of our making. How does Krylenko's dictum go? "In our eyes every crime is the product of a particular social system!"[6] In this case—of your system, comrades! Don't forget your own doctrine!

Let us not forget either that among those of our fellow countrymen who took up the sword against us or attacked us in words, some were completely disinterested. No property had been taken from them (they had had none to begin with), they had never been imprisoned in the camps (nor yet had any of their kin), but they had long ago been sickened by our whole system: its contempt for the fate of the individual; the persecution of people for their beliefs; that cynical song "There's no land where men can breathe so freely"; the kowtowing of the devout to the Leader; the nervous twitching of pencils as everyone hurries to sign up for the state loan; the obligatory applause rising to an ovation! Cannot we realize that these perfectly normal people could not breathe our fetid air? (Father Fyodor Florya's accusers asked him how he had dared talk about Stalin's foul deeds when the Rumanians were on the spot. "How could I say anything different about you?" he answered. "I only told them what I knew. I only told them what

---

6. Krylenko, *Za Pyat Let (1918–1922)*, p. 337.

had happened." What we ask is something different: lie, go against your conscience, perish—just so long as it helps us! But this, unless I'm mistaken, is hardly materialism.)

In September, 1941, before I went into the army, my wife and I, young schoolteachers who had just started work in the settlement of Morozovsk (captured by the Germans in the following year), happened to rent lodgings on the same little yard as a childless couple, the Bronevitskys. Nikolai Gerasimovich Bronevitsky, a sixty-year-old engineer, was an intellectual of Chekhovian appearance, very likable, quiet, and clever. When I try now to recall his long face I imagine him with pince-nez, though he may not have worn them at all. His wife was even quieter and gentler than he was—a faded woman with flaxen hair close to her head, twenty-five years younger than her husband, but not at all young in her behavior. We were fond of them, and they probably liked us, particularly in contrast to our grasping landlord and his greedy family.

In the evenings the four of us would sit on the steps of the porch. They were quiet, warm, moonlit evenings, not as yet rent by the rumble of planes and by exploding bombs, but anxiety about the German advance was stealing over us like the invisible clouds stealing over the milky sky to smother the small and defenseless moon. Every day new trainloads of refugees stopped at the station, on their way to Stalingrad. Refugees filled the marketplace of the settlement with rumors, terrors, 100-ruble notes that seemed to burn holes in their pockets, then they continued their journey. They named towns which had surrendered, about which the Information Bureau, afraid to tell people the truth, would keep silent for a long time to come. (Bronevitsky spoke of these towns not as having "surrendered" but as having been "taken.")

We were sitting on the steps and talking. We younger people were full of ourselves, of anxiety for the future, but we really had nothing more intelligent to say about it than what was written in the newspapers. We were at ease with the Bronevitskys: we said whatever we thought without noticing the discrepancies between our way of looking at things and theirs.

For their part, they probably saw in us two surprising examples of naïvely enthusiastic youth. We had just lived through the thirties—and we might as well not have been alive in that decade at all. They asked what we remembered best about 1938 and 1939. What do you think we said? The university library, examinations, the fun we had on sporting trips, dances, amateur concerts, and of course love affairs—we were at the age for love. But hadn't any of our professors been put away at that time? Yes, we supposed that two or three of them had been. Their places were taken by senior lecturers. What about students—had any of them gone inside? We remembered that some senior students had indeed been jailed. And what did you make of it? Nothing; we carried on dancing. And no one near to you was—er—touched? No; no one.

It is a terrible thing, and I want to recall it with absolute precision. It is all the more terrible because I was not one of the young sporting and dancing set, nor one of those obsessive people buried in books and formulae. I was keenly interested in politics from the age of ten; even as a callow adolescent I did not believe Vyshinsky and was staggered by the fraudulence of the famous trials—but nothing led me to draw the line connecting those minute Moscow trials (which seemed so tremendous at the time) with the huge crushing wheel rolling through the land (the number of its victims somehow escaped notice). I had spent my childhood in queues—for bread, for milk, for meal (meat was a thing unknown at that time)—but I could not make the connection between the lack of bread and the ruin of the countryside, or understand *why* it had happened. We were provided with another formula: "temporary difficulties." Every night, in the large town where we lived, hour after hour after hour people were being hauled off to jail—but I did not walk the streets at night. And in the daytime the families of those arrested hung out no black flags, nor did my classmates say a word about their fathers being taken away.

According to the newspapers there wasn't a cloud in the sky.

And young men are so eager to believe that all is well.

I understand now how dangerous it was for the Bronevitskys to tell us anything. But he gave us just a peep into his past, this old engineer who had got in the way of one of the OGPU's* cruelest blows. He had lost his health in prison, been pulled in a time or two, got to know quite a few camps, but he talked with blazing passion only about Dzhezkazgan in its early days—about the water poisoned by copper; about the poisoned air; about the murders; about the futility of complaints to Moscow. The very word Dzhez-kaz-gan made your flesh creep—like steel wool rubbed on the skin, or like the tales of its pitiless ways. (And yet . . . did this Dzhezkazgan have the slightest effect on our way of looking at the world? Of course not. It was not very near. It was not happening to us. You have to experience it for yourself. It is better not to think about it. Better to forget.)

There in Dzhezkazgan, when Bronevitsky was allowed outside the guarded area, his present wife, then a mere girl, had come to him, and they had been married with the barbed wire for witness. When war broke out they were, by some miracle, at liberty in Morozovsk, with black marks in their passports, of course. He was working in some wretched construction agency, and she was a bookkeeper.

I went off to the army, and my wife left Morozovsk. The settlement came under German occupation. Then it was liberated. And one day my wife wrote to me at the front: "Can you imagine it —they say that Bronevitsky acted as burgomaster for the Germans while they were in Morozovsk. How disgusting!" I was just as shocked. "Filthy thing to do!" I thought.

But a few more years went by. Lying on the sleeping platform in some dark jail and turning things over in my mind, I remembered Bronevitsky. And I was no longer so schoolboyishly self-righteous. They had unjustly taken his job from him, given him work that was beneath him, locked him up, tortured him, beaten him, starved him, spat in his face—what was he supposed to do? He was supposed to believe that all this was the price of progress, and that his own life, physical and spiritual, the lives of those dear to him, the anguished lives of our whole people, were of no significance.

Through the smoke screen of the personality cult, thin and ineffectual as it is, through the intervening layers of time in which we have changed, each of which has its own sharp angle of refraction, we see neither ourselves nor the thirties in true perspective and true shape. Idolization of Stalin, boundless and unquestioning faith, were not characteristic of the whole people, but only of the Party and the Komsomol; of urban youth in schools and universities; of ersatz intellectuals (a surrogate for those who had been destroyed or dispersed); and to some extent of the urban petty bourgeoisie (the working class)[7]—their loudspeakers were never switched off from the morning chimes of the Spassky belfry to the playing of the Internationale at midnight, and for them the voice of the radio announcer Levitan* became the voice of conscience.

7. It was in the thirties that the working class merged completely with the petty bourgeoisie, and became its main constituent part.

(I say "to some extent" because labor legislation like the "twenty minutes late" decree and the tying of the workers to their factories enlisted no supporters.) All the same, there was an urban minority, and not such a small one, numbering at the least several millions, who pulled out the radio plug in disgust whenever they dared. On every page of every newspaper they saw merely a spreading stain of lies, and polling day for these millions was a day of suffering and humiliation. For this minority the dictatorship existing in our country was neither proletarian nor national in character, nor yet (for those who recalled the original sense of the word) Soviet, but the dictatorship of another minority, a usurping minority, which was very far from being a spiritual elite.

Mankind is almost incapable of dispassionate, unemotional thinking. In something which he has recognized as evil man can seldom force himself to see also what is good. Not everything in our lives was foul, not every word in the papers was false, but the minority, downtrodden, bullied, beset by stool pigeons, saw life in our country as an abomination from top to bottom, saw every page in the newspapers as one long lie. Let us recall that in those days there were no Western broadcasts in Russian (and the number of private radio sets was inconsiderable), so that a citizen could obtain information *only* from our newspapers and the official radio, in which Bronevitsky and his like expected from experience to find only cowardly suppression of facts or a vexatious tangle of lies. Everything that was written about other countries, about the inevitable collapse of the West in 1930, about the treachery of Western socialists, about the passionate hostility of all Spain to Franco (or in 1942 about Nehru's treasonable aspiration to freedom in India—which of course weakened our ally the British Empire), all this proved to be nothing but lies. The maddeningly monotonous, hate-filled propaganda conducted on the principle that "he who is not for us is against us" had never drawn distinctions between the attitudes of Mariya Spiridonova and Nicholas II, those of Léon Blum and Hitler, those of the British Parliament and the German Reichstag. So when Bronevitsky read apparently fantastic stories about bonfires of books in German squares, and the resurrection of some sort of ancient Teutonic savagery (we must not forget that Tsarist propaganda during the First World War had also told a few fibs about Teuton savagery), how could he be expected to distinguish them from all the rest, single them out as true, recognize in German Nazism (reviled in almost the same—inordinate—terms as Poincaré, Pilsudski, and the British conservatives earlier) a quadruped as dangerous as that which in reality and in the flesh had for a quarter of a century past been squeezing the life out of him, poisoning his existence, clawing him till he bled, and with him the whole Archipelago, the Russian town, the Russian village? Then the newspapers were forever changing their minds about the Hitlerites: at first it was friendly encounters between nice sentries in nasty Poland, and the newspapers were awash with sympathy for the valiant warriors standing up to French and English bankers, and Hitler's speeches, verbatim, filled a page of *Pravda* at a time; then one morning (the second morning of the war) an explosion of headlines—all Europe was piteously groaning under the Nazi heel. This only confirmed that newspaper lies changed as the wind shifted, and could do nothing to persuade Bronevitsky that other butchers on this earth were a match for ours, about whom he knew the truth. If someone had tried to convince him by putting BBC bulletins before him daily, he might at most have been made to believe that Hitler was a secondary danger to Russia but certainly not, while Stalin lived, the greatest. As it was, the BBC provided no bulletins; the Soviet Information Bureau from the day it was born commanded no more credit than Tass; the rumors carried by evacuees were not firsthand information (from Germany or from the occupied areas no living witness had yet appeared). What he did know at first hand was the camp at Dzhezkazgan, and 1937, and the famine of 1932, and "dekulakization," and the destruction of the churches. So that as the German army approached, Bronevitsky (and tens of thousands of lonely individuals like him) felt that their hour was drawing near—the hour which they had ceased to hope for twenty years ago, which is given to a man only once, then lost forever, since our lives are so short measured against the slow pace of historical change—the hour in which he can repudiate what has befallen, what has been visited upon, flogged, and trampled into his people, serve in some way still obscure his agonized country, help to revive some sort of public life in Russia. Yes, Bronevitsky had remembered everything and forgiven nothing. He could never accept as his own a regime which had thrashed Russia unmercifully, brought it to collectivized beggary, to moral degeneracy, and now to a stunning military defeat. He choked with anger as he looked at naïve creatures like me, like us, for it was beyond his power to convert us. He was waiting for *someone,* anyone, to take

power in place of Stalin! (The well-known psychological phenomenon of reaction to extremes: anything rather than the
nauseous reality we know! Can we imagine, anywhere in the
world, anyone worse than our rulers? Incidentally, this was in
the Don region—where half the population were just as eagerly awaiting the Germans.) So then Bronevitsky, who had
been an apolitical being all his life, resolved in his seventh decade to make a political move.

He consented to head the Morozovsk municipal authority. . . .

There, I think, he must quickly have seen what a silly situation
he had landed himself in, seen that for the new arrivals Russia was
even more insignificant and detestable than for those who had
gone away—that the vampire needed only Russia's vital fluids,
and that the body could wither and perish. The new burgomaster's
task was to be in charge not of public-spirited Russians, but of
auxiliaries to the German police. But he was fastened to the axle
and now, like it or not, he could only spin. Having freed himself
from one lot of butchers, he must help another. The patriotic idea,
which he had thought of as diametrically opposed to the Soviet
idea, he suddenly saw fused with it: in some incomprehensible
fashion patriotism had slipped away like water through a sieve
from the minority who had preserved it, and passed to the majority; it was forgotten how people had been shot for patriotism, how
it had been ridiculed, and now it was the main stem of someone
else's tree.

He, and others like him, must have felt trapped and terrified:
the crack had narrowed and the only way out led to death or to
katorga.

Of course, they were not all Bronevitskys. Of course, many
birds of prey greedy for power and blood had flocked to that brief
feast in time of plague. But their like will flock wherever there are
pickings. They were very much at home in the NKVD, too. Such
a one was Mamulov, or Antonov at Dudinsk, or Poisui-Shapka—
can anyone imagine fouler butchers? (See Volume II.) Yet they
lorded it for decades and bled the people dry a hundred times over.
We shall shortly meet Warder Tkach—one of those who managed
to fit into both contexts.

We have been talking about the towns, but we should not forget
the countryside. Liberals nowadays commonly reproach the village with its political obtuseness and conservatism. But before the
war the village to a man, or overwhelmingly, was sober, much

more sober than the town: it took no part at all in the deification of Daddy Stalin (and needless to say had no time for world revolution either). The village was, quite simply, sane and remembered clearly how it had been promised land, then robbed of it; how it had lived, eaten, and dressed before and after collectivization; how calves, ewes, and even hens had been taken away from the peasant's yard; how churches had been desecrated and defiled. Even in 1941 the radio's nasal bray was not yet heard in peasant huts, and not every village had even one person able to read the newspapers, so that to the Russian countryside all those Chang Tso-lins, MacDonalds, and Hitlers were indistinguishably strange and meaningless lay figures.

In a village in Ryazan Province on July 3, 1941, peasants gathered near the smithy were listening to Stalin's speech relayed by a loudspeaker. The man of iron, hitherto unmoved by the tears of Russian peasants, was now a bewildered old gaffer almost in tears himself, and as soon as he blurted out his humbugging "Brothers and Sisters," one of the peasants answered the black paper mouthpiece. "This is what you want, you bastard," and he made in the direction of the loudspeaker a rude gesture much favored by Russians: one hand grips the opposite elbow, and the forearm rises and falls in a pumping motion.

The peasants all roared with laughter.

If we questioned eyewitnesses in every village, we should learn of ten thousand such incidents, some still more pungent.

Such was the mood of the Russian village at the beginning of the war—the mood, then, of the reservists drinking the last half-liter and dancing in the dust with their kinsmen while they waited at some wayside halt for a train. On top of all this came a defeat without precedent in Russian memories, as vast rural areas stretching to the outskirts of both capitals and to the Volga, as many millions of peasants, slipped from under kolkhoz rule, and —why go on lying and prettifying history?—it turned out that the republics only wanted independence, the village only wanted freedom from the kolkhoz! The workers freedom from feudal decrees! If the newcomers had not been so hopelessly arrogant and stupid, if they had not preserved the bureaucratic kolkhoz administration for Great Germany's convenience, if they had not conceived the obscene idea of turning Russia into a colony, the patriotic cause would not have devolved on those who had always tried to smother it, and we should hardly have been called upon to celebrate the twenty-fifth anniversary of Russian Communism. (Somebody, someday, will have to tell us how the peasants in occupied areas never joined the partisan movement of their own free will, and how to begin with they took up arms against the partisans rather than hand over their grain and cattle.)

Do you remember the great exodus from the Northern Caucasus in January, 1943—and can you think of any analogy in world history? A civilian population, and a peasant population at that, leaving with a defeated enemy, with an alien army, rather than stay behind with the victors, their fellow countrymen—the wagon trains rolling as far as the eye could see through the fierce, icy January winds!

Here, too, lie the social roots of those hundreds of thousands of volunteers who, monster though Hitler was, were desperate enough to don enemy uniform. The time has come for us to give our views on the Vlasov movement* once again. In the first part of this book the reader was not yet prepared for the whole truth (nor am I in possession of the whole truth; special studies will be written on the subject, which is for me of secondary importance). There at the beginning, before the reader had traveled the high-roads and byroads of the camp world with me, he was merely alerted, invited to think. Now, after all those prison transports, transit jails, lumber gangs, and camp middens, perhaps the reader will be a little more open to persuasion. In Part I, I spoke of those Vlasovites who took up arms in desperation, because they were starving in camps, because their position seemed hopeless. (Yet even here there is room for reflection: the Germans began by using Russian prisoners of war only for nonmilitary tasks in the rear, in support of their own troops, and this, you might think, was the best solution for those who only wanted to save their skins—so why did they take up arms and confront the Red Army head on?) But now, since further postponement is impossible, should I not also talk about those who even before 1941 had only one dream —to take up arms and blaze away at those Red commissars, Chekists,* and collectivizers? Remember Lenin's words: "An oppressed class which did not aspire to possess arms and learn how to handle them would deserve only to be treated as slaves" (Fourth Edition, Volume 23, page 85). There is, then, reason to be proud if the Soviet-German war showed that we are not such slaves as all those studies by liberal historians contemptuously make us out to be. There was nothing slavish about those who reached for their sabers to cut off Daddy Stalin's head (nor about those on the other side, who straightened their backs for the first time when they put on Red Army greatcoats—in a strange brief interval of freedom which no student of society could have foreseen).

These people, who had experienced on their own hides twenty-four years of Communist happiness, knew by 1941 what as yet no one else in the world knew: that nowhere on the planet, nowhere in history, was there a regime more vicious, more bloodthirsty, and at the same time more cunning and ingenious than the Bolshevik, the self-styled Soviet regime. That no other regime on earth could compare with it either in the number of those it had done to death, in hardiness, in the range of its ambitions, in its thoroughgoing and unmitigated totalitarianism—no, not even the regime of its pupil Hitler, which at that time blinded Western eyes to all else. Came the time when weapons were put in the hands of these people, should they have curbed their passions, allowed Bolshevism to outlive itself, steeled themselves to cruel oppression again—and only then begun the struggle with it (a struggle which has still hardly started anywhere in the world)? No, the natural thing was to copy the methods of Bolshevism itself: it had eaten into the body of a Russia sapped by the First World War, and it must be defeated at a similar moment in the Second.

Our unwillingness to fight had already shown itself in the Soviet-Finnish war of 1939. V. G. Bazhanov, formerly Secretary of the Politburo and Orgburo of the CPSU(b)* and Stalin's close assistant, tried to exploit this mood: to turn captured Red Army men against the Soviet lines under the command of Russian émigré officers—not to fight their compatriots but to convert them. The attempt was abruptly terminated by the sudden capitulation of Finland.

When the Soviet-German war began, ten years after the slaughterous collectivization, eight years after the great Ukrainian famine (six million dead, unnoticed by neighboring Europe), four years after the devil's dance of the NKVD, one year after the workers were shackled to the new labor laws—and all this when there were 15 million in camps about the country, and while the older generation all clearly remembered what life was like before the Revolution—the natural impulse of the people was to take a deep breath and liberate itself, its natural feeling one of loathing for its rulers. "Caught us unawares"; "numerical superiority in aircraft and tanks" (in fact, all-round numerical superiority was enjoyed by the Red Army)—it was not this that enabled the enemy to close so easily those disastrous salients, taking 300,000 armed men at a time (Bialystok, Smolensk), or 650,000 (Bryansk, Kiev); not this that caused whole fronts to cave in, and rolled our armies back farther and faster than anything Russia had seen in all its one thousand years, or, probably, any other country in any other war —not this, but the instant paralysis of a paltry regime whose subjects recoiled from it as from a hanging corpse. (The raikoms and gorkoms* were blown away in five minutes, and Stalin was gasping for breath.) In 1941 this upheaval might have run its full course (by December, 60 million Soviet people out of a population of 150 millions were no longer in Stalin's power). The alarmist note in Stalin's Order No. 0019, July 16, 1941, was justified. "On all [!] fronts there are numerous [!] elements who even run to meet the enemy [!], and throw down their arms at the first contact with him." (In the Bialystok salient in July, 1941, among 340,000 prisoners there were 20,000 deserters.) Stalin thought the situation so desperate that in October, 1941, he sent a telegram to Churchill suggesting that twenty-five to thirty British divisions be landed on Soviet territory. What Communist has ever suffered a more complete moral collapse! This was the mood of the time: on August 22, 1941, the commanding officer of the 436th Light Infantry Regiment, Major Kononov, told his regiment to their faces that he was going over to the Germans, to join the "Liberation Army" for the overthrow of Stalin, and invited all who wished to go with him. Not only did he meet with no opposition—*the whole regiment* followed him! Only three weeks later Kononov had created a regiment of Cossack volunteers behind the enemy lines (he was a Don Cossack himself). When he arrived at the prisoner-of-war camp near Mogilev to enlist volunteers, 4,000 of the 5,000 prisoners there declared their readiness to join him, but he could not take them all. In the same year, half the Soviet prisoners of war in the camp near Tilsit—12,000 men—signed a declaration that the time had come to convert the war into a civil war. We have not forgotten how the whole population of Lokot-Bryansky, before the arrival of the Germans and independently of them, joined in the creation of an autonomous Russian local administration over a large and flourishing province, with eight districts, and more than a million inhabitants. The demands of the Lokot-Bryansky community were quite precise: a Russian national government to be established, Russians to administer themselves in all the occupied provinces, Russia to be declared independent within its 1938 frontiers, a "Liberation Army" to be formed under Russian command. Or again, a group of young people in Leningrad numbering more than 1,000 (led by the student Rutchenko) went out in the woods near Gatchina to await the Germans and fight against the Stalin regime. (The Germans, however, sent them behind the lines to work as drivers and kitchen orderlies.) The Germans were met with bread and salt in the villages on the Don. The pre-1941 population of the Soviet Union naturally imagined that the coming of a foreign army meant the overthrow of the Communist regime—otherwise it could have no meaning for us at all. People expected a political program which would liberate them from Bolshevism.

From where we were, separated from them by the wilderness of Soviet propaganda, by the dense mass of Hitler's army—how could we readily believe that the Western allies had entered this war not for the sake of freedom in general, but for their own Western European freedom, only against Nazism, intending to take full advantage of the Soviet armies and leave it at that? Was it not more natural for us to believe that our allies were true to the very *principle* of freedom and that they would not abandon us to a worse tyranny? . . . True, these were the same allies for whom Russians had died in the First World War, and who then, too, had abandoned our army in the moment of collapse, hastening back to their comforts. But this was a lesson too cruel for the heart to learn.

Having rightly taught ourselves to disbelieve Soviet propaganda, whatever it said, we naturally did not believe tall stories about the Nazis' wishing to make Russia a colony and ourselves German slaves; who would expect to find such foolishness in twentieth-century heads, unless he had experienced its effects for himself? Even in 1942 the Russian formation in Osintorf attracted more volunteers than a unit still not fully deployed could absorb, while in the Smolensk region and Byelorussia, a volunteer "people's militia" 100,000 strong was formed for purposes of self-defense against the partisans directed from Moscow (the Germans took fright and banned it). As late as spring, 1943, on his two propaganda tours in the Smolensk and Pskov regions, Vlasov was greeted with enthusiasm wherever he went. Even then, the population was still waiting and wondering when we should have our own independent government and our own army. I have testimony from the Pozherevitsky district of the Pskov oblast about the friendly attitude of the peasant population to the Vlasov unit there—which refrained from looting and brawling, wore the old Russian uniform, helped with the harvest, and was regarded as a Russian organ of authority opposed to kolkhozes. Volunteers from among the civilian population came to sign on (just as they did in Lokot-Bryansky with Voskoboinikov's unit)—and we are bound to wonder what made them do so. It was not as though they were getting out of a POW camp. In fact, the Germans several times forbade Vlasovites to take in reinforcements (let them sign on with the Polizei). As late as March, 1943, prisoners of war in a camp near Kharkov read leaflets about the Vlasov movement (so called) and 730 officers signed an application to join the "Russian Liberation Army"; they had the experience of two years of war behind them, many were heroes of the battle for Stalingrad, their number included divisional commanders and regimental commissars—moreover, the camp was very well fed, and it was not the desperation of hunger that induced them to sign. (The Germans, however, behaved with typical stupidity; of the 730 who signed, 722 had still not been released from the camp and given a chance to act when the war ended.) Even in 1943 tens of thousands of refugees from the Soviet provinces trailed along behind the retreating German army—anything was better than remaining under Communism.

I will go so far as to say that our folk would have been worth nothing at all, a nation of abject slaves, if it had gone through that war without brandishing a rifle at Stalin's government even from afar, if it had missed its chance to shake its fist and fling a ripe oath at the *Father of the Peoples.* The Germans had their generals' plot —but what did we have? Our generals were (and remain to this day) nonentities, corrupted by Party ideology and greed, and have not preserved in their own persons the spirit of the nation, as happens in other countries. So that those who raised their hands and struck were almost to a man from the lowest levels of society —the number of former gentry émigrés, former members of the wealthier strata, and intellectuals taking part was microscopically small. If this movement had been allowed to develop unhindered, to flow with the same force as in the first weeks of the war, it would have been like a second Pugachev rising*— resembling the first in the numbers and social level of those swept in its train, in the weight of popular support, in the part played by the Cossacks, in spirit (its determination to settle accounts with evildoers in high places), in the contrast between its elemental force and the weakness of its leadership. However this may be, it was very much more a movement of the people, the *common people,* than the whole "liberation movement" of the intelligentsia from the beginning of the twentieth century right up to February, 1917, with its pseudo-popular aims and its harvest in October. It was not, however, destined to run its course, but to perish ignominiously, stigmatized as "treason to our holy Motherland"!

We have lost the taste for social analysis of events—because such explanations are juggled around to suit the need of the moment. But what of our friendship pact with Ribbentrop and Hitler? The braggadocio of Molotov and Voroshilov before the war? And then, the staggering incompetence, the unpreparedness, the fumbling (and the craven flight of the government from Moscow), the armies abandoned, half a million at a time, in the salients—was this not betrayal of the Motherland? With more serious consequences? Why do we cherish *these* traitors so tenderly in their apartments on Granovsky Street?

Oh, the length of it! The length of the prisoners' bench with seats for *all* those who tormented and betrayed our people, if we could bring them all, from first to last, to account.

Awkward questions get no answers in our country. They are passed over in silence. Instead, this is the sort of thing they yell at us:

"It's the *principle!* The very principle of the thing! Does any Russian, to achieve his own political ends, however just they appear to him, have the right to lean on the strong right arm of German imperialism?! . . . And that at the moment of war to the death?"

True enough, this is the crucial question: Ought you, for what seem to you noble ends, to avail yourself of the support of German imperialists at war with Russia?

Today, everyone will join in a unanimous cry of "No!"

What, then, of the sealed German carriage from Switzerland to Sweden, calling on the way (as we have now learned) at Berlin? The whole Russian press, from the Mensheviks to the Cadets,* also cried "No!" but the Bolsheviks explained that it was permissible, that it was indeed ridiculous to reproach them with it. But this is not the only train journey worth mentioning. How many railroad cars did the Bolsheviks rush out of Russia in summer, 1918, some carrying foodstuffs, others gold—all of them into Wilhelm's capacious maw! Convert the war into a civil war! This was Lenin's proposal before the Vlasovites thought of it. .

—Yes, but his aims! Remember what his aims were!

Well, what were they? And what has become of them, those aims?

—Yes, but really—that was Wilhelm! The Kaiser! The little Emperor! A bit different from Hitler! And anyway, was there really any government in Russia at the time? The Provisional Government doesn't count. . . .

Well, there was a time when, inflamed with martial ardor, we never mentioned the Kaiser in print without the words "ferocious" or "bloodthirsty," and incautiously accused the Kaiser's soldiers of smashing the heads of babes against stones. But let's agree—the Kaiser was different from Hitler. The Provisional Government, though, was also different: it had no Cheka, shot no one in the back of the head, imprisoned no one in camps, herded no one into collective farms, poisoned no one's life: the Provisional Government was not Stalin's government.

We must keep things in proportion.

■

It was not that someone took fright as *katorga* killed off one "alphabet" after another, but simply that with the war drawing to an end there was no need for such a savage deterrent: no new Polizei units could be formed, working hands were needed, and in *katorga* people were dying off uselessly. So as early as 1945 huts in *katorga* ceased to be prison cells, doors were opened to let in the daylight, slop buckets were carried out to the latrines, prisoners were allowed to make their own way to the Medical Section and were trotted to the mess hall at the double to keep their spirits up. The thieves who used to filch other prisoners' rations were removed, and mess orderlies appointed from among the politicals themselves. Later on, prisoners were allowed to receive letters, two a year.

The line between *katorga* and the ordinary camps became blurred in the years 1946–1947. Unfastidious managing engineers did not let political distinctions stand in the way of plan fulfillment and began (in Vorkuta at least) to transfer political offenders with good qualifications to ordinary Camp Divisions, where nothing but the numbers on their backs reminded them of *katorga,* while rank-and-file manual laborers from Corrective Labor Camps were shoved into *katorga* to fill the gaps.

In this way the thoughtless managers might have thwarted Stalin's great idea of resurrecting *katorga*—except that in 1948 a new idea came to him just in time, that of dividing the natives of Gulag into distinct groups, separating the socially acceptable thieves and delinquents from the socially irredeemable 58's.

All this was part of a still greater concept, the Reinforcement of the Home Front (it is obvious from the choice of words that Stalin was preparing for war in the near future). Special Camps[8] were set up with a special regime, slightly milder than that of the *katorga* earlier on, but harsher than that of the ordinary camps.

To distinguish them from other camps, fantastic poetical titles were invented for them instead of ordinary geographical names. Such new creations included *Gorlag** at Norilsk, *Berlag* on the Kolyma, *Minlag* on the Inta, *Rechlag* on the Pechora, *Dubrovlag* at Potma, *Ozerlag* at Taishet, *Steplag, Peschanlag,* and *Luglag* in Kazakhstan, *Kamyshlag* in Kemerovo Province.

Dark rumors crept around the Corrective Labor Camps, that 58's would be sent to Special Extermination Camps. (It did not, of course, enter the heads either of those carrying out the orders or of the victims that any formal additional sentences might be necessary.)

The Registration and Distribution and the Security Operations sections worked furiously. Secret lists were made and driven away somewhere for approval. Long red prisoner-transport trains were moved in, companies of brisk red-tabbed guards* marched up with Tommy guns, dogs, and hammers, and the enemies of the people, as their names were called, meekly obeyed the inexorable summons to leave their cozy huts and begin the long transit.

But not all 58's were summoned. It was only later, comparing notes on their acquaintances, that the prisoners realized which of them had been left behind on Corrective Labor islands with the minor offenders. Among them were those convicted under Article 58, Section 10, with no further charges. This covered simple Anti-

8. Cf. the Special Purpose Camps set up in 1921.

Soviet Agitation, which meant that it was a gratuitous act, without accomplices, and not aimed at anyone in particular. (Though it may seem almost impossible to imagine such agitators, millions of them were on the books, and were left behind on the older islands of Gulag.) Agitators who had formed duets or trios, shown any inclination to listen to each other, to exchange views, or to grumble in chorus, had been burdened with an additional charge under Article 58, Section 11 (on hostile groups) and, as the leaven of anti-Soviet organizations, now went off to Special Camps. So, needless to say, did traitors to the Motherland (58-1a and 1b), bourgeois nationalists and separatists (58-2), agents of the international bourgeoisie (58-4), spies (58-6), subversives (58-7), terrorists (58-8), wreckers (58-9), and economic saboteurs (58-14). This was also the most convenient place to put those prisoners of war, German (in Minlag) or Japanese (in Ozerlag), whom it was intended to detain beyond 1948.

On the other hand, noninformers (58-12) and abettors of the enemy (58-3) in Corrective Labor Camps remained where they were. Whereas prisoners in *katorga* sentenced specifically for aiding and abetting the enemy now went to the Special Camps with all the rest.

The wisdom of the separators was even harder to fathom than appears from this description. Criteria still unexplained left in the Corrective Labor category female traitors serving twenty-five years (Unzhlag) and here and there whole Camp Divisions including nothing but 58's, Vlasovites and ex-Polizei among them. These were not Special Camps, the prisoners wore no numbers, but the regime was severe (Krasnaya Glinka, on the Samara bend of the Volga, the Tuim camp, in the Shirin district of Khakassiya, and the Southern Sakhalin camp were examples). These camps were so harsh that prisoners would have been no worse off in the Special Camps.

So that the Archipelago, once the Great Partition had been carried out, should never again lapse into confusion, it was provided from 1949 onward that every newly naturalized immigrant from the world outside should have written in his prison book, apart from his sentence, a ruling (of the State Security authorities and the Prosecutor's Office in the oblast) as to the type of cage in which this particular bird should always be kept.

Thus, like the seed that dies to produce a plant, Stalin's *katorga* grew into the Special Camp.

The red prisoner-transport trains traveled the length and breadth of the Motherland and the Archipelago carrying the new intake.

At Inta they had the sense simply to drive the herd out of one gate and in through another.

Chekhov complained that we had no "legal definition of *katorga,* or of its purpose."

But that was in the enlightened nineteenth century! In the middle of the twentieth, the cave man's century, we didn't even feel the need to understand and define. Old Man Stalin had decided that it would be so—and that was all the definition necessary.

We just nodded our heads in understanding.

Chapter 2

■

The First Whiff of Revolution

Dismayed by the hopeless length of my sentence, stunned by my first acquaintance with the world of Gulag, I could never have believed at the beginning of my time there that my spirit would recover by degrees from its dejection: that as the years went by, I should ascend, so gradually that I was hardly aware of it myself, to an invisible peak of the Archipelago, as though it were Mauna Loa on Hawaii, and from there gaze serenely over distant islands and even feel the lure of the treacherous shimmering sea between.

The middle part of my sentence I served on a golden isle, where prisoners were given enough to eat and drink and kept warm and clean. In return for all this not much was required of me: just twelve hours a day sitting at a desk and making myself agreeable to the bosses.

But clinging to these good things suddenly became distasteful. I was groping for some new way to make sense of prison life. Looking around me, I realized now how contemptible was the advice of the special-assignment prisoner from Krasnaya Presnya: "At all costs steer clear of general duties."* The price we were paying seemed disproportionately high.

Prison released in me the ability to write, and I now gave all my time to this passion, brazenly neglecting my boring office work. There was something I had come to value more than the butter and sugar they gave me—standing on my own feet again.

Well, they jerked a few of us to our feet—en route to a Special Camp.

They took a long time getting us there—three months. (It could be done more quickly with horses in the nineteenth century.) So long that this journey became, as it were, a distinct period in my life, and it even seems to me that my character and outlook changed in the course of it.

The journey was bracing, cheerful, full of good omens.

A freshening breeze buffeted our faces—the wind of *katorga* and of freedom. People and incidents pressed in on every hand to assure us that justice was on our side! on our side! on our side! not with our judges and jailers.

The Butyrki, our old home, greeted us with a heartrending female shriek from a window—probably that of a solitary-confinement cell. "Help! Save me! They're killing me! They're killing me!" Then the cries were choked in a warder's hands.

At the Butyrki "station" we were mixed up with raw recruits of the 1949 intake. They all had funny sentences—not the usual *tenners,* but *quarters.* When at each of the numerous roll calls they had to give dates of release, it sounded like a cruel joke: "October, 1974!" "February, 1975!"

No one, surely, could sit out such a sentence. A man must get hold of some pliers and cut the wire.

These twenty-five-year sentences were enough to transform the prisoners' world. The holders of power had bombarded us with all they had. Now it was the prisoners' turn to speak—to speak freely, uninhibitedly, undeterred by threats, the words we had never heard in our lives and which alone could enlighten and unite us.

We were sitting in a Stolypin car* at the Kazan station when we heard from the station loudspeaker that war had broken out in Korea. After penetrating a firm South Korean defense line to a depth of ten kilometers on the very first day, the North Koreans insisted that they had been attacked. Any imbecile who had been at the front understood that the aggressors were those who had advanced on the first day.

This war in Korea excited us even more. In our rebellious mood we longed for the storm. The storm must break, it must, it must, or else we were doomed to a lingering death! . . .

Somewhere past Ryazan the red rays of the rising sun struck with such force through the mole's-eye windows of the prison car that the young guard in the corridor near our grating screwed up his eyes. Our guards might have been worse: they had crammed us into compartments fifteen or so at a time, they fed us on herring, but, to be fair, they also brought us water and let us out morning and evening to relieve ourselves, so that we should have had no quarrel with them if this lad hadn't unthinkingly, not maliciously, tossed the words "enemies of the people" at us.

That started it! Our compartment and the next pitched into him.

"All right, we're enemies of the people—but why is there no grub on the kolkhoz?"

"You're a country boy yourself by the look of you, but I bet you'll sign on again—I bet you'd sooner be a dog on a chain than go back to the plow."

"If we're enemies of the people, why paint the prison vans different colors? Who are you hiding us from?"

"Listen, kid! I had two like you who never came back from the war—and you call *me* an enemy of the people?"

It was a very long time since words like this had flown through the bars of our cages! We shouted only the plainest of facts, too self-evident to be refuted.

A sergeant serving extra time came to the aid of the flustered youngster, but instead of hauling anyone off to the cooler, or taking names, he tried to help his subordinate to fight back.

Here, too, we saw a faint hint that times were changing—no, this was 1950, too soon to speak of better times; what we saw were signs of the new relationship between prisoners and jailers created by the new long sentences and the new political camps.

Our argument began to take on the character of a genuine debate. The young men took a good look at us, and could no longer bring themselves to call us, or those in the next compartment, enemies of the people. They tried trotting out bits from newspapers and from their elementary politics course, but their ears told them before their minds could that these set phrases rang false.

"Look for yourselves, lads! Look out the window," was the answer they got from us. "Look what you've brought Russia down to!"

Beyond the windows stretched a beggarly land of rotted thatch and rickety huts and ragged folk (we were on the Ruzayev line, by which foreigners never travel). If the Golden Horde had seen it so befouled, they would not have bothered to conquer it.

On the quiet station at Torbeyevo an old man walked along the platform in bast shoes. An old peasant woman stopped opposite the lowered window of our car and stood rooted to the spot for a long time, staring through the outer and inner bars at us prisoners tightly packed together on the top bed shelf. She stared at us with that look on her face which our people have kept for "unfortunates" throughout the ages. A few tears trickled down her cheeks. She stood there, work-coarsened and shabby, and she looked at us as though a son of hers lay among us. "You mustn't look in there, mamma," the guard told her, but not roughly. She didn't even turn her head. At her side stood a little girl of ten with white ribbons in her plaits. She looked at us very seriously, with a sadness strange in one of her years, her little eyes wide and unblinking. She looked at us so hard that she must have imprinted us on her memory forever. As the train eased forward, the old woman raised her blackened fingers and devoutly, unhurriedly made the sign of the cross over us.

Then at another station some girl in a spotted frock, anything but shy or timid, came right up to our window and started boldly asking us what we were in for and for how long. "Get away," bellowed the guard who was pacing the platform. "Why, what will you do? I'm the same as them! Here's a pack of cigarettes—give it to the lads," and she produced them from her handbag. (We had already realized that the girl had done time. So many of them, now roaming around free, had received their training on the Archipelago!) The deputy guard commander jumped out of the train. "Get away! I'll put you inside!" She stared scornfully at the old sweat's ugly mug. "You go and ——— yourself, you ———" "Give it to 'em, lads," she said to encourage us. And made a dignified departure.

So we rode on, and I don't think the guards felt that they were protecting the people from its enemies. On we went, more and more inflamed with the conviction that we were right, that all Russia was with us, that the time was at hand to abolish this institution.

At the Kuibyshev Transit Prison, where we *sunbathed* (i.e., loafed) for more than a month, more workers came our way. The air was suddenly rent by the sickening, hysterical yells of thieves (they even whine in a loathsome shrill way). "Help! Get us out of here! The Fascists are beating us! Fascists!"

Here was something new! "Fascists" beating thieves? It always used to be the other way around.

But shortly after, there was a reshuffle of prisoners, and we found that no miracles had happened yet. It was only the first swallow—Pavel Boronyuk. His chest was a millstone; his gnarled hands were ever ready for a friendly clasp or a blow; he was dark in complexion, aquiline, more like a Georgian than a Ukrainian. He had been an officer at the front, had prevailed in a machine-gun duel with three "Messerschmitts," had been recommended for the order of Hero of the Soviet Union and turned down by the Special Section, had been sent to a punitive battalion and returned with a decoration; and now he had a *tenner*, which as times now were, was hardly a "man's sentence."

He had sized up the thieves while he was still on his way from the jail at Novograd-Volynsk and had fought with them before. Now he was sitting in the next cell on the upper bed platform, quietly playing chess. The whole cell were 58's, but the administration had slipped two thieves in among them. On his way to clear his *rightful* sleeping space by the window, a Belomor cigarette dangling carelessly from his lip, Fiksaty said jokingly, "Might have known they'd put me with gangsters again!" The naïve Veliev, who didn't know much about thieves, hastened to reassure him: "No, we're all 58's here. What about you?" "I'm an embezzler, I'm an educated man!" The thieves chased two men away, slung their own sacks onto their "reserved" places, and walked through the cell examining other people's sacks and looking for trouble. The 58's—no, they hadn't changed yet; they put up no resistance. Sixty grown men waited tamely for their turn to be robbed. There is something hypnotically disarming about the impudence of thieves, who never for a moment expect to meet resistance. (Besides, they can always count on the support of authority.) Boronyuk went on pretending to move his chessmen, but by now he was rolling his eyes in fury and wondering how best to take care of them. When one of the thieves stopped in front of him, he swung his dangling foot and booted him in his ugly face, then jumped down, grabbed the stout wooden lid of the sanitary bucket, and brought it down in a stunning blow on the other thief's head. Then he began hitting them alternately with the lid until it fell to pieces, leaving its base, two solid bars joined crosswise, in his hands. The thieves changed their tune to a pathetic whine, but it must be admitted that there was a certain humor in their moans, that they seemed to see the funny side of it. "What do you think you're doing—hitting people *with a cross!*" "Just because you're strong you shouldn't bully others!" Boronyuk kept on hitting them till one of the thieves rushed to the window shouting, "Help! The Fascists are beating us!"

The thieves never forgot it, and threatened Boronyuk many times afterward. *"You smell like a dead man already! We'll take you with us!"* But they never attacked him again.

Soon afterward our cell also clashed with the bitches.* We were out in the yard to stretch our legs, and relieve ourselves while we were at it, when a woman prison officer sent a trusty to chase some of us out of the latrine. His arrogance (to the "politicals") outraged Volodya Gershuni, a high-strung youngish man, recently sentenced. Volodya pulled the trusty up short, and the trusty felled the lad with a blow. Previously the 58's would simply have swallowed this, but now Maxim the Azerbaijani (who had killed the chairman of his kolkhoz) threw a stone at the trusty, while Boronyuk laid one on his jaw. He slashed Boronyuk with his knife (the warders' assistants went around with knives; there was nothing unusual in this), and ran to the warders for protection, with Boronyuk chasing him. They quickly herded us all into the cell, and senior prison officers arrived to discover who was to blame and threaten us with additional sentences for gang fighting (the MVD man's heart always bleeds for his nearest and dearest, his trusties). Boronyuk's blood was up, and he stepped forward of his own accord. "I beat those bastards, and I'll go on beating them as long as I live!" The "godfather"* warned us that we Counter-Revolutionaries couldn't afford to put on airs and that it would be safer for us to hold our tongues. At this up jumped Volodya Gershuni. He was hardly more than a boy, a first-year university student when he was arrested, and not just a namesake but the nephew of that Gershuni who once commanded the SR* terrorist squad. He screamed at the godfather, as shrill as a fighting cock. "Don't dare call us Counter-Revolutionaries! That's all in the past. We're re-vo-lu-tion-aries again now! Against the Soviet state this time!"

How we enjoyed ourselves! This was the day we'd lived for! And the godfather just frowned and scowled and swallowed it all. Nobody was taken off to the lockup, and the prison officers beat an inglorious retreat. Was *this* how life in prison would be from now on? Could we then fight? Turn on our tormentors? Say out loud just what we thought? All that time we had endured it all like idiots! It's fun beating people who weep easily. We wept—so they beat us.

Now, in the legendary new camps to which they were taking us, where men wore number patches as in the Nazi camps, but where there would at last be only political prisoners, cleansed of the slimy criminal scum, perhaps the new life would begin. Volodya Gershuni, with his dark eyes and his peaked, dead-white face, said hopefully: "Once we get to the camp we shall soon know with whom we belong!" Silly lad! Did he seriously expect to find there a vigorous political life, with parties of many different shades feverishly contending, discussions, programs, underground meetings? "With whom we belong!" As though the choice had been left to us! As though those who drew up the target figures for arrests in each republic, and the bills of lading for camp-transport trains, had not decided it for us.

In our very long cell—once a stable, with two lines of two-tier bed platforms where the two rows of mangers used to stand, with pillars made of crooked tree trunks along the aisle propping up a decrepit roof, with typical stable windows in the long wall, shaped so that the hay could be forked straight into the mangers (and made narrower by "muzzles"*)—in our cell there were 120 men, of all sorts and conditions. More than half of them were from the Baltic States, uneducated people, simple peasants: the second purge was under way in that area, and all who would not voluntarily join collective farms, or who were suspected in advance of reluctance to join, were being imprisoned or deported. Then there were quite a few Western Ukrainians—members of the OUN,[1] together with anyone who had once given them a night's rest or a meal. Then there were prisoners from the Russian Soviet Federation—with fewer new boys among them, most of them "repeaters." And, of course, a certain number of foreigners.

We were all being taken to the same camp complex (we found out from the records clerk that it was the Steplag group). I looked carefully at those with whom fate had brought me together, and tried to see into their minds.

I found the Estonians and Lithuanians particularly congenial. Although I was no better off than they were, they made me feel ashamed, as though I were the one who had put them inside. Unspoiled, hard-working, true to their word, unassuming—what had they done to be ground in the same mill as ourselves? They had harmed no one, lived a quiet, orderly life, and a more moral life than ours—and now they were to blame because we were

1. Organization of Ukrainian Nationalists.

hungry, because they lived cheek by jowl with us and stood in our path to the sea.

"I am ashamed to be Russian!" cried Herzen when we were choking the life out of Poland. I felt doubly ashamed in the presence of these inoffensive and defenseless people.

My attitude to the Latvians was more complicated. There was a fatality in their plight. They had sown the seed themselves.*

And the Ukrainians? We have long ago stopped saying "Ukrainian nationalists"; we speak only of "Banderists," and this has become such a dirty word that no one thinks of inquiring into the reality. (We also call them "bandits," following our established rule that anyone, anywhere, who kills *for us* is a "partisan," whereas those who kill us are always "bandits," beginning with the Tambov peasants* in 1921.)

The reality is that although long ago in the Kiev period we and the Ukrainians constituted a single people, we have since then been torn asunder and our lives, our customs, our languages for centuries past have taken widely different paths. The so-called "re-union" was a very awkward though perhaps in some minds a sincere attempt to restore our former brotherhood. But we have not made good use of the three centuries since. No statesman in Russia ever gave much thought to the problem of binding the Ukrainians and Russians together in kinship, of smoothing out the lumpy seam. (Had the join been neater, the first Ukrainian Committees would not have been formed in spring, 1917, nor the Rada later on.)

The Bolsheviks before they came to power found the problem uncomplicated. In *Pravda* for June 7, 1917, Lenin wrote as follows: "We regard the Ukraine and other regions not inhabited by Great Russians as territories annexed by the Tsar and the capitalists." He wrote this when the Central Rada was already in existence. Then on November 2, 1917, the "Declaration of the Rights of the Peoples of Russia" was adopted. Was it just meant as a joke? Was it just a trick when they declared that the peoples of Russia did indeed have the right of self-determination, up to and including secession? Six months later the Soviet government *requested* the good offices of the Kaiser's Germany in helping Soviet Russia to conclude peace and define its boundaries with the Ukraine, and Lenin signed a treaty to this effect with Hetman Skoropadsky on June 14, 1918. By doing so he showed himself fully reconciled to the detachment of the Ukraine—even if it became a monarchy as a result!

But strangely enough, as soon as the Germans were defeated by the Entente (which could not affect in the least the principles governing our relations with the Ukraine), as soon as the Hetman had fallen, together with his patrons, as soon as we proved stronger than Petlyura (there's another word of abuse, "Petlyuro-vite": but these were merely Ukrainian townsfolk and peasants, who wanted to order their lives without our interference), we immediately crossed the border which we had recognized and imposed our rule on our blood brothers. True, for fifteen to twenty years afterward we made great play with the Ukrainian language, pushed it perhaps too hard, and impressed it on our brothers that they were completely independent and could break away whenever they pleased. Yet when they tried to do so at the end of the war we denounced them as "Banderists," and started hunting them down, torturing them, executing them, or dispatching them to the camps. (But "Banderists," like "Petlyurovites," are just Ukrainians who do not want to be ruled by others; once they discovered that Hitler would not bring them the freedom they had been promised, they fought against the Germans, as well as ourselves, throughout the war, but we kept quiet about this, since like the Warsaw rising of 1944 it shows us in an unfavorable light.)

Why are we so exasperated by Ukrainian nationalism, by the desire of our brothers to speak, educate their children, and write their shop signs in their own language? Even Mikhail Bulgakov (in *The White Guard*) let himself be misled on this subject. Given that we have not succeeded in fusing completely; that we are still different in some respects (and it is sufficient that *they,* the smaller nation, feel the difference); that however sad it may be, we have missed chance after chance, especially in the thirties and forties; that the problem became most acute not under the Tsar, but after the Tsar—why does their desire to secede annoy us so much? Can't we part with the Odessa beaches? Or the fruit of Circassia?

For me this is a painful subject. Russia and the Ukraine are united in my blood, my heart, my thoughts. But from friendly contact with Ukrainians in the camps over a long period I have learned how sore they feel. Our generation cannot avoid paying for the mistakes of generations before it.

Nothing is easier than stamping your foot and shouting: "That's mine!" It is immeasurably harder to proclaim: "You may live as you please." We cannot, in the latter end of the twentieth century, live in the imaginary world in which our last, not very bright Emperor came to grief. Surprising though it may be, the prophecy of our Vanguard Doctrine* that nationalism would fade has not come true. In the age of the atom and of cybernetics, it has for some reason blossomed afresh. Like it or not, the time is at hand when we must pay out on all our promissory notes guaranteeing self-determination and independence—pay up of our own accord, and not wait to be burned at the stake, drowned in rivers, or beheaded. We must prove our greatness as a nation not by the vastness of our territory, not by the number of peoples under our tutelage, but by the grandeur of our actions. And by the depth of our tilth in the lands that remain when those who do not wish to live with us are gone.

The Ukraine will be an extremely painful problem. But we must realize that the feelings of the whole people are now at white heat. Since the two peoples have not succeeded over the centuries in living harmoniously, it is up to us to show sense. We must leave the decision to the Ukrainians themselves—let federalists and separatists try their persuasions. Not to give way would be foolhardy and cruel. And the gentler, the more tolerant, the more careful to explain ourselves we are now, the more hope there will be of restoring unity in the future.

Let them live their own lives, let them see how it works. They will soon find that not all problems are solved by secession.[2]

For some reason the cell in the converted stables was our home for a long time, and it looked as though they would never send us on to Steplag. Not that we were in any hurry; we enjoyed life where we were, and the next place could only be worse.

We were not left without news—they brought us daily a sort of half-sized newspaper. I sometimes had the task of reading it aloud to the whole cell, and I read it with expression, for there were things there which demanded it.

The tenth anniversary of the "liberation" of Estonia, Latvia,

2. The fact that the ratio between those who consider themselves Russian and those who consider themselves Ukrainian varies from province to province of the Ukraine will cause many complications. A plebiscite in each province, and afterward a helpful and considerate attitude to those who wish to move, may be necessary. Not all of the Ukraine in its present official Soviet borders is really Ukrainian. Some of the left-bank provinces undoubtedly feel drawn to Russia.

and Lithuania came around just at this time. Some of those who understood Russian translated for the rest (I paused for them to do so), and what can only be called a howl went up from the bed platforms as they. heard about the freedom and prosperity introduced into their countries for the first time in history. Each of these Balts (and a good third of all those in the transit prison were Balts) had left behind a ruined home, and was lucky if his family was still there and not on its way to Siberia with another batch of prisoners.

But what of course most excited the transit prison were the reports from Korea. Stalin's blitzkrieg had miscarried. The United Nations volunteers had by now been assembled. We saw in Korea the precursor, the Spain, of the Third World War. (And Stalin probably intended it as a rehearsal.) Those U.N. soldiers were a special inspiration to us. What a flag to fight under! Whom would it not unite? Here was a prototype of the united mankind of the future!

We were wretched, and we could not rise above our wretchedness. Should this have been our dream—to perish so that those who looked unmoved on our destruction might survive? We could not accept it. No, we longed for the storm!

Some will be surprised. —What a desperate, what a cynical state of mind. Had you no thought for the hardships war would bring to those outside? —Well, the free never spared us a thought! —You mean, then, that you were capable of wishing for a world war? —When all those people were given sentences in 1950 lasting till the mid-1970s, what hope were they left with except that of world war?

I am appalled myself when I remember now the false and baneful hopes we cherished at the time. General nuclear destruction was no way out for anyone. And leaving aside the nuclear danger, a state of war only serves as an excuse for domestic tyranny and reinforces it. But my story will be distorted if I do not tell the truth about our feelings that summer.

Romain Rolland's generation in their youth were depressed by the constant expectation of war, but our generation of prisoners was depressed by its absence—and not to say so would be to tell less than the truth about the spirit of the Special Political Camps. This was what they had driven us to. World war might bring us either a speedier death (they might open fire from the watchtowers, poison our bread, or infect us with germs, German fashion), or it just might bring freedom. In either case, deliverance would be much nearer than the end of a twenty-five-year sentence.

This was what Petya P——v counted on. Among those in our cell Petya P——v was the last living soul to arrive from Europe. Immediately after the war, cells everywhere were packed with these Russkies returning from Europe. But the first arrivals were long ago in camps or in the ground, and the rest had vowed to stay away. Where, then, had Petya sprung from? He had come home of his own free will in November, 1949, when normal people were no longer returning.

The war had overtaken him just outside Kharkov, where he attended an industrial school in which he had been compulsorily enrolled. Just as unceremoniously the Germans carried these young lads off to Germany. There he remained as an "Ost-Arbeiter" to the end of the war, and there his philosophy of life was formed: a man must find an easy way of living, not work as he had been made to work from infancy. In the West, taking advantage of European credulity and lax frontier controls, he had smuggled French vehicles into Italy and Italian vehicles into France and sold them off cheaply. The French, however, had tracked him down and arrested him. He then wrote to the Soviet Embassy, saying that he wanted to return to his beloved Fatherland. P——v's reasoning was that in France he might get ten years, but would have to serve his sentence in full, whereas in the Soviet Union he would get twenty-five as a traitor—but then, the first drops of the coming storm, the Third World War, were already falling; the Union, he thought, wouldn't last even three years, so it would pay him to go to a Soviet prison. Instant friends arrived from the embassy and clasped Petya P——v to their bosoms. The French authorities were glad to hand over a thief.[3] Some thirty others just like Petya were assembled in the embassy. They were given a comfortable sea passage to Murmansk, let loose to wander freely about the town, and picked up again one by one in the course of the next twenty-four hours.

For his cellmates Petya now took the place of Western newspapers (he had followed the Kravchenko trial in detail), Western theatre (he skillfully performed Western tunes with

3. French statistics are said to show that between the First and Second World Wars the crime rate was lower among Russian émigrés than among any other ethnic group. After the Second World War the opposite was the case: of all the ethnic groups, the Russians —Soviet citizens who had fetched up in France—had the highest crime rate.

his cheeks and lips), and Western films (he told us the stories and mimed the action).

How free and easy things were in the Kuibyshev Transit Prison! The inmates of different cells occasionally met in the common yard. From under the muzzles we could exchange remarks with other transports as they were driven across the yard. On our way to the latrine we could approach the open windows (which were barred but unscreened) of the family barracks, where women with several children were held. (They, too, were on their way into exile from the Baltic States and the Western Ukraine.) And between the two converted stables there was a crack, known as the "telephone," where interested persons lay on either side of the wall discussing the news from morning to night.

All these freedoms excited us still more; we felt the ground firmer under our own feet and imagined that it was becoming uncomfortably warm under the feet of our jailers. When we walked about the yard we raised our faces to the sun-bleached July sky. We should not have been surprised, and not at all alarmed, if a V formation of foreign bombers had emerged from nowhere. Life as it was meant nothing to us.

Prisoners traveling in the other direction from the Karabas Transit Prison brought rumors of notices stuck on walls: "We won't take any more!" We worked ourselves up to white heat, and one sultry night in Omsk when we were being crammed and screwed into a prison van, like lumps of sweating, steaming meat through a mincer, we yelled out of the depths at our warders: "Just wait, you vermin! Truman will see you off! They'll drop the atom bomb on your heads!" And the cowards said nothing. They were uneasily aware that our resistance was growing stronger and —so we sensed—that justice was more and more clearly on our side. We were so sick with longing for justice that we should not have minded if we and our tormentors were incinerated by the same bomb. We were in that final stage at which there is nothing to lose.

If this is not brought into the open, the full story of the Archipelago in the fifties will not have been told.

The prison at Omsk, which had known Dostoyevsky, was not like any old Gulag transit prison, hastily knocked together from matchwood. It was a formidable jail from the time of Catherine II, and its dungeons were particularly terrible. You could never imagine a better film set than one of its underground cells. The small square window is at the top of an oblique shaft up to ground level. The depth of this opening—three meters—tells you what the walls are like. The cell has no ceiling, but massive, menacing vaults converge overhead. One wall is wet—water seeps through from the soil and leaks onto the floor. In the morning and in the evening it is dark, on the brightest afternoon half-dark. There are no rats, but you fancy that you can smell them. Although the vaulted roof dips so low that you can touch it in places, the jailers have contrived to erect two-tier bed platforms even here, with the lower level barely raised above the floor, ankle high.

You might think that this jail would stifle the vague mutinous anticipations which had grown in us in the slack Kuibyshev Transit Prison. But no! In the evening, by the light of a 15-watt bulb, no brighter than a candle, Drozdov, the bald, sharp-featured churchwarden of the cathedral church at Odessa, takes his stand near the mouth of the window shaft, and in a voice that is weak yet full of feeling, the voice of a man whose life is ending, sings an old revolutionary song.

> Black as the conscience of tyrant or traitor,
> The shades of the autumn night fall.
> Blacker than night, looming out of the darkness,
> Ghostlike—the grim prison wall.

He sings only for us, but in this place if you shouted aloud no one would hear. As he sings, his prominent Adam's apple runs up and down under the withered brown skin of his neck. He sings and shudders, he remembers, lets decades of Russian life flow through him, and we shudder in sympathy.

> Though all's silent within, it's a jail, not a graveyard—
> Sentry, ah, sentry, beware!

A song like that in a prison like that![4] Not a false note, not a false word! Every note, every word in tune with what awaited our generation of prisoners.

Then we settle down to sleep in the yellow gloom, the cold, the damp. Right, who's going to tell us a story?

A voice is heard—that of Ivan Alekseyevich Spassky, a sort of composite voice of all Dostoyevsky's heroes. A voice that falters,

---

4. It is a great pity that Shostakovich did not hear this song *in that place.* Either he wouldn't have touched it or he would have expressed its modern instead of its dead significance.

chokes, is never calm, seems about to break at any moment into weeping or a cry of pain. The most primitive tale by Breshko-Breshkovsky, "The Red Madonna," for instance, retold in such a voice, charged with faith, with suffering, with hatred, sounds like the *Chanson de Roland.* Whether it is true or pure fiction, the story of Victor Voronin, of how he raced 150 kilometers on foot to Toledo, and how the siege of Alcazar was raised, etches itself on our memories like an epic.

Spassky's own life would make a better novel than many. In his youth he took part in the Campaign on the Ice.* He fought throughout the Civil War. He emigrated to Italy. He graduated from a Russian ballet school abroad (Karsavina's, I think), and also learned cabinetmaking in the household of some Russian countess. (Later on, in the camps, he amazed us by making himself some miniature tools and fashioning for the bosses furniture of such exquisite workmanship, with such elegantly curving lines, that they were left speechless. True, it took him a month to make a little table.) He toured Europe with the ballet. He was a news cameraman for an Italian company during the Spanish Civil War. Under the slightly disguised name of Giovanni Paschi, he became a major commanding a battalion in the Italian army and in summer, 1942, arrived back on the Don. His battalion was promptly surrounded though the Russians were still retreating almost everywhere. Left to himself, Spassky would have fought to the death, but the Italians, mere boys, started weeping—they wanted to live! After some hesitation Major Paschi hung out the white flag. He could have committed suicide, but by now he was itching to take a look at some Soviet Russians. He might have gone through an ordinary prisoner-of-war camp and been back in Italy within four years, but his Russian soul was impatient of restraint and he got into conversation with the officers who had captured him. A fatal mistake! If you are unlucky enough to be Russian, conceal the fact like a shameful disease, or it will go hard with you! First they kept him for a year in the Lubyanka. Then for three years in the International Camp at Kharkov. (There was such a place—full of Spaniards, Italians, Japanese.) Then without taking into account the four years he had already served, they doled out another twenty-five. Twenty-five—what a hope! He was doomed to a speedy end in *katorga.*

The jails at Omsk, and then at Pavlodar, took us in because—and this was a serious oversight!—there was no specialized transit prison in either city. Indeed, in Pavlodar—what a disgrace!—there wasn't even a prison van and they marched us briskly from the station to the jail, many blocks away, without worrying about the local population—just like before the Revolution, or in the first decade after. In the parts of town we went through there were still neither pavements nor piped water, and the little one-story houses were sinking into the gray sand. The city proper began with the two-story white stone jail.

But by twentieth-century standards this was a jail to soothe rather than horrify, to inspire laughter rather than terror. A spacious, peaceful yard, with wretched grass growing here and there, divided by reassuringly low fences into little squares for exercise. There were widely spaced bars across the cell windows on the second floor, and no muzzles, so that you could stand on the window sill and examine the neighborhood. Directly below, under your feet, between the wall of the building and the outer prison wall, an enormous dog would run across the yard dragging his chain when something disturbed him, and give a couple of gruff barks. But he, too, was not a bit like a prison dog—not a terrifying German shepherd trained to attack people, but a shaggy yellow-white mongrel (they breed dogs like that in Kazakhstan), and already pretty old by the look of him. He was like one of those good-natured elderly wardens transferred to the camps from the army, who thought prison service a dog's life, and did not care who knew it.

Beyond the prison wall we could see a street, a beer stall, and people walking or standing there—people who had come to hand in parcels for the prisoners and were waiting to get their boxes and wrapping paper back. Farther on there were blocks and blocks of one-story houses, the great bend of the Irtysh and open country vanishing beyond the river into the distance.

A lively girl, who had just got back from the guardhouse with her empty basket, looked up and saw us waving to her from the window, but pretended not to notice. She walked unhurriedly, demurely past the beer stall, until she could not be seen from the guardhouse, and there her whole manner changed abruptly: she dropped the basket, frantically waved both arms in the air, and smiled at us. Then she signaled with nimble fingers: "Write notes!" then (an elliptical sweep of the arm): "Throw them to me, throw them to me!" then (pointing in the direction of the town): "I'll take them and pass them on for you." Then she opened both arms wide:

"What else do you need? What can I do for you? I'm a friend!"

Her behavior was so natural and straightforward, so unlike that of the harassed and hag-ridden "free population," our bullied and baffled free citizens. What could it mean? Were times changing? Or was this just Kazakhstan? Where half the population, remember, were exiles . . .

Sweet, fearless girl! How quickly and accurately you had learned the prison-gate skills! How happy it made me (and I felt a tear in my eye) to know that there are still people like you! Accept our homage, whoever you are! If our people had all been like you there would have been not a hope in hell of imprisoning them.

The infamous machine would have jammed!

We had, of course, bits of pencil lead in our jackets. And scraps of paper. And it would have been easy to pick off a lump of plaster, tie a note to it with thread, and throw it clear of the wall. But there was absolutely nothing we could ask her to do for us in Pavlodar! So we simply bowed to her and waved our greetings.

We were driven into the desert. Even the unprepossessing overgrown village of Pavlodar we should soon remember as a glittering metropolis.

We were now taken over by an escort party from Steplag (but not, fortunately, from the Dzhezkazgan Camp Division: throughout the journey we had kept our fingers crossed that we would not end up in the copper mines). The trucks sent to collect us had built-up sides and grilles were attached to the rear of their cabs, to protect the Tommy-gunners from us as though we were wild animals. They packed us in tightly, facing backward, with our legs twisted under us, and in this position jogged and jolted us over the potholes for eight hours on end. The Tommy-gunners sat on the roof of the cab, with the muzzles of their guns trained on our backs throughout the journey.

Up front rode lieutenants and sergeants, and in the cab of our truck there was an officer's wife with a little girl of six. When we stopped the little girl would jump down and run through the grass picking flowers and calling in a clear voice to her mother. She was not in the least put out by the Tommy guns, the dogs, the ugly shaven heads of the prisoners sticking up over the sides of the lorries; our strange world cast no shadow on the meadow and the flowers, and she didn't even spare us a curious glance. . . . I remembered the son of a sergeant major at the Special Prison in

Zagorsk. His favorite game was making two other little boys, the sons of neighbors, clasp their hands behind their backs (sometimes he tied their hands) and walk along the road while he walked alongside with a stick, escorting them.

As the fathers live, so the children play.

We crossed the Irtysh. We rode for a long time through water meadows, then over dead flat steppe. The breath of the Irtysh, the freshness of evening on the steppe, the scent of wormwood, enveloped us whenever we stopped for a few minutes and the swirling clouds of light-gray dust raised by the wheels sank to the ground. Thickly powdered with this dust, we looked at the road behind us (we were not allowed to turn our heads), kept silent (we were not allowed to talk), and thought about our destiny, the camp with the strange, difficult, un-Russian name. We had read the name on our case files, hanging upside down from the top shelf in the Stolypin—EKIBASTUZ. But nobody could imagine where it was on the map, and only Lieutenant Colonel Oleg Ivanov remembered that it was a coal-mining area. We even supposed that it might be somewhere quite close to the Chinese border (and this made some of us happy, since they had yet to learn that China was even worse than our own country). Captain Second Class Burkovsky (a new boy and a 25-er, he still looked askance at us, because he was a Communist imprisoned in error, while all around were enemies of the people: he acknowledged me only because I was a former Soviet officer and had not been a prisoner of war) reminded me of something I had learned at the university and forgotten: if we traced a meridian line on the ground at the autumnal equinox and subtracted the meridional altitude of the sun on September 23 from 90, we should find our latitude. This was reassuring—although there was no way of discovering our longitude.

On and on they drove. Darkness fell. The stars were big in the black sky and we saw clearly now that we were being carried south-southwest.

Dust danced in the beams of headlights behind us. Patches of the dust cloud whipped up over the whole road, but were visible only where the headlights picked them out. A strange mirage rose before me: the world was a heaving sea of blackness, except for those whirling luminous particles forming sinister pictures of things to come.

To what far corner of the earth, what godforsaken hole, were they taking us? Where were we fated to make our revolution?

Our legs, doubled under us, became so numb that they might not have been ours. It was very near midnight when we reached a camp surrounded by a high wooden fence and—out in the dark steppe, beside a dark sleeping settlement—bright with electric light, in the guardhouse and around the boundary fence.

After another roll call with full particulars—"March, nineteen hundred and seventy-five!"—they led us through the towering double gates for what was left of our quarter-century.

The camp was asleep, but all the windows of all the huts were brightly lit, as though the tide of life was running high. Lights on at night—that meant prison rules. The doors of the huts were fastened from outside by heavy padlocks. Bars stood out black against the brightly lit rectangles of the windows.

The orderly who came out to meet us had *number patches* stuck all over him.

You've read in the newspapers that in Nazi camps people had *numbers* sewn on their clothes, haven't you?

Chapter 3

∎

Chains, Chains ...

Our eager hopes, our leaping expectations, were soon crushed. The wind of change was blowing only in drafty corridors—in the transit prisons. Here, behind the tall fences of the Special Camps, its breath did not reach us. And although there were only political prisoners in these camps, no mutinous leaflets hung on posts.

They say that at Minlag the blacksmiths refused to forge bars for hut windows. All glory to those as yet nameless heroes! They were real people. They were put in the camp jail, and the bars for Minlag were forged at Kotlas. No one supported the smiths.

The Special Camps began with that uncomplaining, indeed eager submission to which prisoners had been trained by three generations of Corrective Labor Camps.

Prisoners brought in from the Polar North had no cause to be grateful for the Kazakh sunshine. At Novorudnoye station they jumped down from the red boxcars onto ground no less red. This was the famous Dzhezkazgan copper, and the lungs of those who mined it never held out more than four months. There and then the warders joyfully demonstrated their new weapon on the first prisoners to step out of line: handcuffs, which had not been used in the Corrective Labor Camps, gleaming nickel handcuffs, which went into mass production in the Soviet Union to mark the thirtieth anniversary of the October Revolution. (Somewhere there was a factory in which workers with graying mustaches, the model proletarians of Soviet literature, were making them—unless we suppose that Stalin and Beria did it themselves?) These handcuffs were remarkable in that they could be clamped on very tight.

Serrated metal plates were let into them, so that when a camp guard banged a man's handcuffed wrists against his knee, more of the teeth would slip into the lock, causing the prisoner greater pain. In this way the handcuffs became an instrument of torture instead of a mere device to inhibit activity: they crushed the wrists, causing constant acute pain, and prisoners were kept like that for hours, always with their hands behind their backs, palms outward. The warders also perfected the practice of trapping four fingers in the handcuffs, which caused acute pain in the finger joints.

In Berlag the handcuffs were used religiously: for every trifle, even for failure to take off your cap to a warder, they put on the handcuffs (hands behind the back) and stood you by the guardhouse. The hands became swollen and numb, and grown men wept: "I won't do it again, sir! Please take the cuffs off!" (Wondrous were the ways of Berlag: not only did prisoners enter the mess hall on command, they lined up at the tables on command, sat down on command, lowered their spoons into the gruel on command, rose and left the room on command.)

It was easy enough for someone to scribble the order: "Establish Special Camps! Submit draft regulations by such and such a date!" But somewhere hard-working penologists (and psychologists, and connoisseurs of camp life) had to think out the details: How could screws already galling be made yet tighter? How could burdens already backbreaking be made yet heavier? How could the lives of Gulag's denizens, already far from easy, be made harder yet? Transferred from Corrective Labor Camps to Special Camps, these animals must be aware at once of their strictness and harshness—but obviously someone must first devise a detailed program!

Naturally, the security measures were strengthened. In all Special Camps the perimeter was reinforced, additional strands of barbed wire were strung up, and coils of barbed wire were scattered about the camp's fringe area. On the path by which prisoners went to work, machine guns were set up in readiness at all main crossroads and turnings, and gunners crouched behind them.

Every Camp Division had its stone jailhouse—its Disciplinary Barracks (BUR).[1] Anyone put in the Disciplinary Barracks invari-

---

1. I shall continue to call it by this name, which prisoners remembered from the Corrective Labor Camps and went on using out of habit, although it is not quite accurate in this context: it was the camp jail, neither more nor less.

ably had his padded jacket taken from him: torture by cold was an important feature of the BUR. But every hut was just as much a jail, since all windows were barred, and latrine buckets were brought in for the night so that all doors could be locked. Moreover, there were one or two Disciplinary Barracks in each camp area, with intensified security, each a separate camplet within the camp; these were locked as soon as the prisoners got in from work —on the model of the earlier *katorga*. (They were BUR's really, but we called them "rezhimki.")

Then again, they quite blatantly borrowed from the Nazis a practice which had proved valuable to them—the substitution of a number for the prisoner's name, his "I," his human individuality, so that the difference between one man and another was a digit more or less in an otherwise identical row of figures. This measure, too, could be a great hardship, provided it was implemented consistently and fully. This they tried to do. Every new recruit, when he "played the piano" in the Special Section (i.e., had his fingerprints taken, as was the practice in ordinary prisons, but not in Corrective Labor Camps), had to hang around his neck a board suspended from a rope. His number—Shch 262 will do as an example—was set up on the board (in Ozerlag by now there were even numbers beginning with *yery:** the alphabet was too short!) and in this guise he had his picture taken by the Special Section's photographer. (All those photographs are still preserved somewhere! One of these days we shall see them!)

They took the board from around the prisoner's neck (he wasn't a dog, after all) and gave him instead four (or in some camps three) white patches measuring 8 centimeters by 15. These he had to sew onto his clothes, usually on the back, the breast, above the peak of his cap, and on one leg or arm (Plate No. 2)—but the regulations varied slightly from camp to camp. Quilted clothing was deliberately damaged in stipulated places before the patches were sewn on: in the camp workshops a separate team of tailors was detailed to damage new clothing: squares of fabric were cut out to expose the wadding underneath. This was done so that prisoners trying to escape could not unpick their number patches and pass as free workmen. In some other camps it was simpler still: the number was burned into the garments with bleaching fluid.

Warders were ordered to address prisoners by their numbers only, and to ignore and forget their names. It would have been pretty unpleasant if they had kept it up—but they couldn't. Russians aren't Germans. Even in the first year warders occasionally slipped up and called people by their names, and as time went by they did it more often. To make things easier for the warders, a plywood shingle was nailed onto each bunk, at every level, with the occupant's number on it. Thus the warder could call out the sleeper's number even when he could not see it on his garments, and if a man was missing the warder would know at once who was breaking the rules. Another useful field of activity opened up for warders: they could quietly turn the key in the lock and tiptoe into the hut before getting-up time, to take the numbers of those who had risen too soon, or they could burst into the hut exactly on time and take the numbers of those who were not yet up. In both cases you could be summarily awarded a spell in the hole, but in the Special Camp it was usually thought better to demand a *written explanation*—although pens and ink were forbidden and no paper was supplied. This tedious, long-winded, offensive procedure was rather a clever invention, especially as the camp administration had plenty of salaried idlers with leisure to scrutinize the explanations. Instead of simply punishing you out of hand, they required you to explain in writing why your bed was untidy, why the number plate at your bunk was askew and why you had done nothing about it, why a number patch on your jacket was soiled and why you had not put that right; why a cigarette had been found on you in the hut; why you had not taken your cap off to a warder.[2] Questions so profound that writing answers to them was even more of a torment to the literate than to the illiterate. But refusal to write meant that your punishment would be more severe! The note was written, with the neatness and precision which respect for the disciplinary staff demanded, delivered to the warder in charge of the hut, then examined by the assistant disciplinary officer or the disciplinary officer, who in turn wrote on it his decision about punishment.

In work rolls, too, it was the rule to write numbers before names. Why before and not instead of names? They were afraid to give up names altogether! However you look at it, a name is a reliable handle, a man is pegged to his name forever, whereas a number is blown away at a puff. If only the numbers were branded or picked out on the man himself, that would be something! But

2. Doroshevich was surprised to find prisoners taking their caps off to the prison governor on Sakhalin. But we had to uncover whenever we met an ordinary warder.

they never got around to it. Though they might easily have done so; they came close enough.

The oppressive number system tended to break down for yet another reason—because we were not in solitary confinement, because we heard each other's voices and not just those of the warders. The prisoners themselves not only did not use each other's numbers, they did not even notice them. (How, you may wonder, could anyone fail to notice those glaring white patches on a black background? When a lot of us were assembled—on work line-up, or for inspection—the bewildering array of figures gave you spots before your eyes. It was like staring at a logarithm table —but only while it was new to you.) So little did you notice them that you did not even know the numbers of your closest friends and teammates; your own was the only one you remembered. (Some dandified trusties carefully saw to it that their numbers were neatly, even jauntily, sewn on, with the edges tucked in, with minute stitching, to make them really pretty. Lackeys born and bred! My friends and I, on the contrary, took care that our numbers should look as ugly as possible.)

The Special Camp regime assumed a total lack of publicity, assumed that no one would ever complain, no one would ever be released, no one would ever break out. (Neither Auschwitz nor Katyn had taught our bosses anything.) And so the first Special Camps were Special Camps with truncheons. It was, as a rule, not the warders who carried them (they had the handcuffs!), but trusted prisoners—hut orderlies and foremen; they, however, could beat us to their hearts' content, with the full approval of authority. At Dzhezkazgan before work line-up the work assigners stood by the doors of the huts with clubs and shouted: "Out you come—and *no last man!!*" (The reader will have understood that if there should be a *last man,* it was immediately as though he had never been.)[3] For the same reason, the authorities were not greatly upset if, for instance, a winter transport from Karbas to Spassk—two hundred men—froze on the way, if all the wards and corridors of the Medical Section were packed with the survivors, rotting alive with a sickening stench, and Dr. Kolesnikov amputated dozens of arms, legs, and noses.[4] The wall of silence was

3. In Spassk in 1949 something snapped. The foremen were called to the staff hut, ordered to put away their clubs, and advised to do without them in future

4. This Dr. Kolesnikov was one of the "experts" who had shortly before signed the mendacious findings of the Katyn commission (to the effect that it was not we who had

so reliable that the celebrated disciplinary officer at Spassk, Captain Vorobyov, and his underlings first "punished" an imprisoned Hungarian ballerina by putting her in the black hole, then handcuffed her, then, while she was handcuffed, raped her.

The disciplinary regime envisaged patient attention to every detail. Thus prisoners were not allowed to keep photographs— either of themselves (which might help escapers!) or of their relatives. Should any be found they were confiscated and destroyed. A barracks representative in the women's division at Spassk, an elderly schoolteacher, put a small picture of Tchaikovsky on a table. The warder removed it and gave her three days in the black hole. "But it's a picture of Tchaikovsky!" "I don't care whose picture it is; in this camp women aren't allowed to have pictures of men." In Kengir prisoners were allowed to receive meal in their food parcels (why not?), but there was a rigorous prohibition against boiling it, and if a prisoner managed to make a fire between a couple of bricks the warder would kick over the pot and make the culprit smother the flames with his hands. (Later on, it is true, they built a little shed for cooking, but two months later the stove was demolished and the place was used to accommodate some pigs belonging to the officers, and security officer Belyaev's horse.)

While they were introducing various disciplinary novelties, our masters did not forget what was best in the practice of the Corrective Labor Camps. In Ozerlag Captain Mishin, head of a Camp Division, tied recalcitrants behind a sleigh and towed them to work.

By and large, the regime proved so satisfactory that prisoners from the former political camps *(katorga)* were now kept in the Special Camps on the same footing as the rest and in the same quarters, distinguished only by the serial letters on their number patches. (Though if there was a shortage of huts, as at Spassk, it was they who would be put to live in barns and stables.)

So that the Special Camp, though not officially called *katorga,* was its legitimate successor and merged with it.

For a prison regime to have a satisfactory effect on the prisoners, it must be grounded also on sound rules about work and diet.

The work chosen for the Special Camps was always the hardest

murdered the Polish officers). For this a just Providence had put him in this camp. But why did the powers of this world want him there? So that he would not talk too much. "Othello's occupation's gone."

in the locality. As Chekhov has truly remarked: "The established view of society, and with some qualifications of literature, is that no harder and more degrading form of hard labor can be found than that in the mines. If in Nekrasov's *Russian Women* the hero's job had been to catch fish for the jail or to fell trees, many readers would have felt unsatisfied." (Why speak so disparagingly of tree felling, Anton Pavlovich? Lumbering is not so bad; it will do the trick.) The first divisions of Steplag, those it began with, were all engaged in copper mining (the First and Second Divisions at Rudnik, the Third at Kengir, and the Fourth at Dzhezkazgan). They drilled dry, and the dust from the waste rock quickly brought on silicosis and tuberculosis.[5] Sick prisoners were sent to die in the celebrated Spassk camp (near Karaganda)—the "All-Union convalescent home" of the Special Camps.

Spassk deserves a special mention here.

It was to Spassk that they sent terminal cases for whom other camps could no longer find any use. But what a surprise! No sooner did the sick cross the salubrious boundary lines of Spassk than they turned into able-bodied workers. For Colonel Chechev, commandant of the whole Steplag complex, the Spassk Camp Division was one of his special favorites. This thick-set thug would fly in from Karaganda, have his boots cleaned in the guardhouse, and walk through the camp trying to spot prisoners not working. He liked to say: "I've only got one invalid in the whole Spassk camp—he's short of both legs. And even he's on light duties—he runs errands." All one-legged men were employed on sedentary work: breaking stones for road surfacing, or grading firewood. Neither crutches nor even a missing arm was any obstacle to work in Spassk. One of Chechev's ideas: putting four one-armed men to carry a stretcher (two of them left-armed, two of them right-armed). An idea thought up for Chechev: driving the machines in the engineering shop by hand when there was no electric power. Something Chechev liked: having his "own professor." So he allowed the biophysicist Chizhevsky to set up a *laboratory* at Spassk (with empty benches). But when Chizhevsky, using worthless waste materials, devised an antisilicosis mask for the Dzhezkazgan workers—Chechev would not put it into production. They've always worked without masks; why complicate things?

5. Under a law of 1886, no form of work which might be injurious to health was permitted even if it was the prisoner's own choice.

After all, there must be a regular turnover, to make room for the new intake.

At the end of 1948 there were about 15,000 prisoners, male and female, in Spassk. It was a huge camp area; the posts of the boundary fence went uphill in some places, and the corner watchtowers were out of sight of each other. The work of self-segregation gradually proceeded: the prisoners built inner walls to separate women, workers, complete invalids (this would hinder communications within the camp and make things easier for the bosses). Six thousand men building a dike had to walk 12 kilometers to work. Since they were sick men, it took them more than two hours each way. To this must be *added* an eleven-hour working day. (It was rare for anyone to last two months on that job.) The job next in importance was in the stone quarries—which were inside the camp itself, both in the men's and in the women's section (the island had its own minerals!). In the men's section the quarry was on a hillside. The stone was blasted loose with ammonal after the day's work was over, and next day the sick men broke the lumps up with hammers. In the women's zone they didn't use ammonal—instead, the women dug down to the rock layers with picks, then smashed the stone with sledge hammers. The hammer heads, of course, came away from the handles, and new ones sometimes broke. To replace a head, a hammer had to be sent to a different camp zone. Nonetheless, every woman had an output norm of 0.9 cubic meters a day, and since they could not meet it there was a long period during which they were put on short rations (400 grams)—until the men taught them to pinch stone from old piles before the daily accounting. Remember that all this work was done not only by sick people, not only without any mechanical aids at all, but in the harsh winter of the steppes (at temperatures as low as 30 to 35 degrees below freezing, and with a wind blowing), and what is more, in *summer clothing,* since there was no provision for the issue of warm clothing to *nonworkers,* i.e., to the unfit. P——r recalls how she wielded a huge hammer, practically naked, in frosts as severe as this. The value of this work to the Fatherland becomes very clear when we add that for some reason the stone from the women's quarry proved unsuitable as building material, and on a certain day a certain high official gave instructions that the women should dump all the stone they had quarried in a year back where it came from, cover it with soil, and lay out a park (they never, of course, got quite that far). In the men's zone the stone was good. The procedure for delivering it to the construction site was as follows: after inspection, the whole work force (around eight thousand men—all those who were alive on that particular day) was marched up the hill, and no one was allowed down again unless he was carrying stone. On holidays patients took their constitutional twice daily—morning and evening.

Then came such jobs as self-enclosure, building quarters for the camp administration and the guards (dwelling houses, a club, a bathhouse, a school), and work in the fields and gardens.

The produce from these gardens also went to the free personnel, while the prisoners got only beet tops: this stuff was brought in by the truckload and dumped near the kitchen, where it rotted until the cooks pitchforked it into their cauldrons. (A bit like feeding cattle, would you say?) The eternal broth was made from these beet tops, with the daily addition of one ladleful of mush. Here is a horticultural idyll from Spassk: about 150 prisoners made a concerted rush at one of the garden plots, lay on the ground, and gnawed vegetables pulled from the beds. The guards swarmed around, beating them with sticks, but they just lay there munching.

Nonworking invalids were given 550 grams of bread, working invalids 650. Medicines were as yet unknown in Spassk (where would you find enough for a mob like that! and they were there to peg out anyway), and so were proper beds. In some huts bunks were moved up together, and four men instead of two squeezed onto a double bed platform.

Oh, yes—there is one job I haven't yet mentioned! Every day 110 to 120 men went out to dig graves. Two Studebakers carried the corpses in slatted boxes, with their legs and arms sticking out. Even in the halcyon summer months of 1949, sixty or seventy people died every day, and in winter it was one hundred (the Estonians who worked in the morgue kept the count).

(In other Special Camps mortality was not so high; prisoners were better fed, but their work was harder, too, since they were not unfit: the reader can make the necessary adjustment himself.)

All this was in 1949 (the year one thousand nine hundred and forty-nine), the thirty-second year after the October Revolution, four years after the war, with its harsh imperatives, had ended, three years after the conclusion of the Nuremberg Trials, where mankind at large had learned about the horrors of the Nazi camps and said with a sigh of relief: "It can never happen again."[6]

Add to all this that on transfer to a Special Camp your links with the outside world, with the wife who waited for you and for your letters, with the children for whom you were becoming a mythical figure, were as good as severed. (Two letters a year—but even these were not posted, after you had put into them thoughts saved up for months. Who would venture to check the work of the women censors on the MGB staff? They often made their task lighter by burning some of the letters they were supposed to censor. If your letter did not get through, the post office could always be blamed. In Spassk some prisoners were once called in to repair a stove in the censors' office, and they found there hundreds of unposted letters which the censors had forgotten to burn. Conditions in the Special Camps were such that the stove menders were afraid to tell their friends—the State Security boys might make short work of them. . . . These women censors in the Ministry of State Security who burned the *souls* of prisoners to save themselves a little trouble—were they any more humane than the SS women who collected the skin and hair of murdered people?) As for family visits, they were unthinkable—the address of every Special Camp was classified and no outsider was allowed to go there.

Let us also add that the Hemingwayesque question *to have* or to *have not* hardly arose in the Special Camps, since it had been firmly resolved from the day of their creation in favor of *not having*. Not having money and receiving no wages (in Corrective Labor Camps it was still possible to earn a pittance, but here not a single kopeck). Not having a change of shoes or clothing, nor anything to put on underneath, to keep yourself warm or dry. Underwear (and what underwear—Hemingway's pauper would hardly have deigned to put it on) was changed twice a month; other clothes, and shoes, twice a year; it was all laid down with a crystalline clarity worthy of Arakcheyev. (Not in the first days of the camp, but later on, they fitted out a permanent storeroom,

---

6. Let me hasten to put the reader's mind at rest by assuring him that all these Chechevs, Mishins, and Vorobyovs, and also Warder Novgorodov, are flourishing: Chechev in Karaganda, retired with the rank of general. Not one of them has been brought to trial, or ever will be. And what could they be tried for? They were simply *carrying out orders.* They are not to be compared with those Nazis who were simply carrying out orders. If they in any way went beyond their orders, it was of course because of their ideological purity, with the sincerest intentions, out of simple unawareness that Beria, "Great Stalin's faithful comrade in arms," was also an agent of international imperialism.

where clothes were kept until the day of "release," and not handing in any article of wear among your personal belongings was considered a serious offense: it counted as preparation to escape, and meant the black hole and interrogation.) Not to keep food in your locker (you queued in the evening to hand it in at the food store, and in the morning to draw it out again—which effectively occupied those half hours in the morning and evening when you might have had time to think). Not to have anything in manuscript, not to have ink, indelible pencils, or colored pencils, not to have unused paper in excess of one school notebook. And finally, not to have books. (In Spassk they took away books belonging to a prisoner on admission. In our camp we were allowed to keep one or two at first, but one day a wise decree was issued: all books belonging to prisoners must be registered with the Culture and Education Section, where the words "Steplag, Camp Division No. ———" would be stamped on the title page. Henceforward all unstamped books would be confiscated as illegal, while stamped books would be considered the property of the library, not that of their former owners.)

Let us further remind ourselves that in Special Camps searches were more frequent and intensive than in Corrective Labor Camps. (Prisoners were carefully searched each day as they left for and returned from work [Plate No. 3]; huts were searched regularly—floors raised, fire bars levered out of stoves, boards pried up in porches; then there were prison-type personal searches, in which prisoners were stripped and probed, linings ripped away from clothes and soles from shoes.) That after a while they started weeding out every last blade of grass in the camp area "in case somebody hides a weapon there." That free days were taken up by chores about the camp.

If you remember all this, it may not surprise you to hear that making him wear numbers was not the most hurtful and effective way of damaging a prisoner's self-respect: when Ivan Denisovich says, "They weigh nothing, the numbers," it does not mean that he has lost all self-respect—as some haughty critics, who never themselves wore numbers or went hungry, have disapprovingly said—it is just common sense. The numbers were vexatious not because of their psychological or moral effects, as the bosses intended, but for a purely practical reason—that on pain of a spell in the hole we had to waste our leisure hours sewing up hems that had come unstitched, getting the figures touched up by the "artists," or searching for fresh rags to replace patches torn at work.

The people for whom the numbers were indeed the most diabolical of the camp's devices were the devout women members of certain religious sects. There were some of these in the Women's Camp Division near the Suslovo station (Kamyshlag)—about a third of the women there were imprisoned for their religion. Now, it is plainly foretold in the Book of Revelation (Chapter 13, Verse 16) that "it* causes all . . . to be marked on the right hand or the forehead."

These women refused, therefore, to wear numbers—the mark of Satan! Nor would they give signed receipts (to Satan, of course) in return for regulation dress. The camp authorities (Chief of Administration General Grigoryev, head of Separate Camp Site Major Bogush) showed laudable firmness! They gave orders that the women should be *stripped to their shifts,* and have their shoes taken from them (the job went to wardresses who were members of the Komsomol), thus enlisting winter's help in forcing these senseless fanatics to accept regulation dress and sew on their numbers. But even with the temperature below freezing, the women walked about the camp in their shifts and barefoot, refusing to surrender their souls to Satan!

Faced with this spirit (the spirit of reaction, needless to say; enlightened people like ourselves would never protest so strongly about such a thing!), the administration capitulated and gave their clothing back to the sectarians, who put it on without numbers! (Yelena Ivanovna Usova wore hers for the whole ten years; her outer garments and underwear rotted and fell to pieces on her body, but the accounts office could not authorize the issue of any government property without a receipt from her!)

Another annoying thing about the numbers was their size, which enabled the guards to read them from a long way off. They only ever saw us from a distance at which they would have time to bring their guns to the ready and fire, they knew none of us, of course, by name, and since we were dressed identically would have been unable to distinguish one from another but for our numbers. But now, if the guards noticed anybody talking on the march, or changing ranks, or not keeping his hands behind his back, or picking something up from the ground, the guard commander only had to report it to the camp and the culprit could expect the black hole.

The guards were yet another force which could crush a prisoner like a sparrow caught in a pulping machine. These "red tabs," regular soldiers, these little lads with Tommy guns, were a dark, unreasoning force, knowing nothing of us, never accepting explanations. Nothing could get through from us to them, and from them to us came only angry shouts, the barking of dogs, the grating of breechblocks, bullets. And they, not we, were always right.

In Ekibastuz, where they were adding gravel to a railroad bed, working without a boundary fence but cordoned by guards, a prisoner took a few steps, inside the permitted area, to get some bread from his coat, which he had thrown down—and one of the guards went for him and killed him. The guard, of course, was in the right. He would receive nothing but thanks. I'm sure he has no regrets to this day. Nor did we express our indignation. Needless to say, we wrote no letters about it (and if we had, our complaints would not have gone any further).

On January 19, 1951, our column of five hundred men had reached work site ARM. On one side of us was the boundary fence, with no soldiers between us and it. They were about to let us in through the gates. Suddenly a prisoner called Maloy ("Little," who was in fact a tall, broad-shouldered young man) broke ranks for no obvious reason and absent-mindedly walked toward the guard commander. We got the impression that he was not himself, that he did not know what he was doing. He did not raise his hand, he made no threatening gesture, he simply walked on, lost in thought. The officer in charge, a nasty-looking, foppish little fellow, took fright and started hastily backing away from Maloy, shouting shrilly, and try as he would, unable to draw his pistol. A sergeant Tommy-gunner advanced briskly on Maloy and when he was within a few paces gave him a short burst in the chest and the belly, slowly backing away in his turn. Maloy slowly advanced another two paces before he fell, and tufts of wadding sprang into sight in the back of his jacket, marking the path of the invisible bullets. Although Maloy was down, and the rest of the column had not stirred, the guard commander was so terrified that he rapped out an order to the soldiers and there was a rattle of Tommy guns on all sides, raking the air just above our heads; a machine gun, set up beforehand, began chattering, and many voices vying with each other in hysterical shrillness screamed: "Lie down! Lie down! Lie down!" While the bullets came lower and lower, to the level of the boundary wires. There were half a thousand of us, but we did not hurl ourselves on the men with the guns and trample on them; we prostrated ourselves and lay with our faces buried in the snow, in a humiliating and helpless position, lay like sheep for more than a quarter of an hour on that Epiphany morning. They could easily have shot every last one of us without having to answer for it: why, this was attempted mutiny!

This was what we were like in the first and second years of the Special Camps—pathetic, crushed slaves—but enough has been said about this period in *Ivan Denisovich.*

How did it come about? Why did so many thousands of these misused creatures, the 58's—damn it all, they were *political* offenders, and now that they were separated, segregated, concentrated, surely they would *behave* like *politicals*—why, then, did they behave so contemptibly, so submissively?

These camps could not have *begun* differently. Both the oppressed and their oppressors had come from Corrective Labor Camps, and both sides had decades of a master-and-slave tradition behind them. Their old way of life was transferred with them, they kept the old way of thinking alive and warm in each other's minds, because they traveled a hundred or so at a time from the same Camp Division. They brought with them to their new place the firm belief inculcated in all of them that men are rats, that man eats man, and that it can be no other way. Each of them brought with him a concern for his own fate alone, and a total indifference to the fate of others. He came prepared to give no quarter in the struggle for a foreman's job or a trusty's cozy spot in a warm kitchen, in the bread-cutting room, in the stores, in the accounts office, or in the Culture and Education Section.

When a man is being moved to a new place all by himself, he can base his hopes of *getting fixed up* there only on luck and his own unscrupulousness. But when men are transported together over great distances in the same boxcar for two or three or four weeks, are kept stewing in the same transit prisons, are marched along in the same columns, they have plenty of time to put their heads together, to judge which of them has a foreman's fist, which knows how to crawl to the bosses, to play dirty tricks, to feather his nest at the expense of the working prisoners—and a close-knit *family* of trusties naturally does not indulge in dreams of freedom but joins forces to uphold the cause of slavery, clubs together to seize the key posts in the new camp and keep out trusties from elsewhere. While the benighted workers, completely reconciled to their harsh and hopeless lot, get together to form good work teams and find themselves a decent foreman in the new place.

All these people had forgotten beyond recall not only that each of them was a man, that he carried the divine spark within him, that he was capable of higher things; they had forgotten, too, that they need not forever bend their backs, that freedom is as much man's right as air, that they were all so-called *politicals,* and that there were now no strangers in their midst.

True, there were still a very few thieves among them. The authorities had despaired of deterring their favorites from frequent attempts to break out (under Article 82 of the Criminal Code the penalty was not more than two years, and the thieves had already collected decades and centuries of extra time, so why should they not run away if there was no one to dissuade them?) and decided to pin charges under Article 58, Section 14 (economic sabotage), on would-be escapers.

Altogether not very many thieves went into Special Camps, just a handful in each transport, but in their code there were enough of them to bully and insult people, to act as hut wardens and walk around with sticks (like the two Azerbaijanis in Spassk who were subsequently hacked to death), and to help the trusties plant on these new islands of the Archipelago the flag (shit-colored, trimmed with black) of the foul and slavish Destructive-Labor Camps.

The camp at Ekibastuz had been set up a year before our arrival —in 1949—and everything had settled down in the old pattern brought there in the minds of prisoners and masters. Every hut had a warden, a deputy warden, and senior prisoners, some of whom relied on their fists and others on talebearing to keep their subjects down. There was a separate hut for the trusties, where they took tea reclining on their bunks and amicably settled the fate of whole work sites and whole work teams. Thanks to the peculiar design of the Finnish huts,* there were in each of them separate *cabins* occupied ex officio by one or two privileged prisoners. Work assigners rabbit-punched you, foremen smacked you in the kisser, warders laid on with the lash. The cooks were a mean and surly lot. All storerooms were taken over by freedom-loving Caucasians. Work-assignment duties were monopolized by a clique of scoundrels who were all supposed to be engineers. Stool pigeons carried their tales to the Security Section punctually and with impunity. The camp, which had started a year ago in tents, now had a stone jailhouse—which, however, was only half-built and so always badly overcrowded: prisoners sentenced to the hole had to wait in line for a month or even two. Law and order had broken down, no doubt about it! Queuing for the hole! (I was sentenced to the hole, and my turn never came.)

True, the thieves (or bitches, to be more precise, since they were not too grand to take posts in the camp) had lost a little of their shine in the course of the year. They felt themselves somehow cramped—they had no rising generation behind them, no reinforcements in sight, no one eagerly tiptoeing after them. Things somehow weren't working out for them. Hut warden Mageran, when the disciplinary officer introduced him to the lined-up prisoners, did his best to glower at them defiantly, but self-doubt soon took possession of him and his star sank ingloriously.

We, like every party of new arrivals, were put under pressure while we were still taking our bath on admission. The bathhouse attendants, barbers, and storemen were on edge and ganged up to attack anyone who tried to make the most diffident complaint about torn underwear or cold water or the heat-sterilization procedure. They were just waiting for such complaints. Several of them at once flew at the offender, like a pack of dogs, yelling in unnaturally loud voices—"You aren't in the Kuibyshev Transit Prison now"—and shoving their hamlike fists under his nose. (This was good psychology. A naked man is ten times more vulnerable than one in clothes. And if newly arrived prisoners are given a bit of a fright before they emerge from the inaugural bath, they will begin camp life with their wings clipped.)

That same Volodya Gershuni, the student who had imagined himself taking a good look around in the camp and deciding "whom to join," was detailed on his very first day to strengthen the camp—by digging a hole for one of the poles to which lights were strung. He was too weak to complete his stint. Orderly Baturin, one of the bitches, who was beginning to sing smaller but still had a bit of bluster in him, called him a *pirate,* and struck him in the face. Gershuni threw down his crowbar and walked right away from the hole. He went to the commandant's office, and made a declaration: "You can put me in the black hole if you like, but I won't go to work as long as your pirates hit people." (The word "pirate" had particularly upset him, because it was strange to him.) His request was not refused, and he spent two consecutive spells in the black hole, eighteen days in all. (This is how it's done: a prisoner is given five or ten days for a start, then when his time is up, instead of letting him out they wait for him to start protesting and cursing—whereupon they can legitimately *stick him* with a second spell.) After the black hole, they awarded him a further two months of Disciplinary Barracks—which meant that he stayed on in the jailhouse but would go out to work at the limekilns and get hot food and rations according to his output. Realizing that he was sinking deeper and deeper into the mire, Gershuni sought salvation through the Medical Section—he hadn't yet taken the measure of "Madame" Dubinskaya, who was in charge of it. He assumed that he could just present his flat feet for inspection and be excused from the long walk to and from the limekilns. But they wouldn't even take him to the Medical Section—the Ekibastuz Disciplinary Barracks had no use for the out patients' clinic. Gershuni was determined to get there, and he had heard a lot about methods of protest, so one morning when the prisoners were being lined up for work, he stayed on the bed platform, wearing only his underpants. Two warders, "Polundra" (a crack-brained ex-sailor) and Konentsov, dragged him off the bed platform by his feet and hauled him just as he was, in his underpants, to the line-up. As they dragged him he clutched at stones lying on the ground, ready for the builders, and tried to hang on to them. By now he was willing to go to the limekilns—"Just let me get my trousers on" he yelled—but they dragged him along just the same. At the guardhouse, while four thousand men were kept waiting for their work assignments, this puny boy struggled as they tried to handcuff him, shouting, "Gestapo! Fascists!" Polundra and Konentsov, however, forced his head to the ground, put the handcuffs on, and prodded him forward. For some reason, it was not they who were embarrassed, nor the disciplinary officer, Lieutenant Machekhovsky, but Gershuni himself. How could he walk through the whole settlement in his underpants? He refused to do it! A snub-nosed dog handler was standing nearby. Volodya remembered how he muttered: "Stop making such a fuss—fall in with the others. You can sit by the fire—you needn't work." And he held tightly onto his dog, which was struggling to break loose and get at Volodya's throat, because it could see that this lad was defying men with blue shoulder tabs! Volodya was removed from the work-assignment area and taken back to the Disciplinary

Barracks. The handcuffs cut more and more painfully into his wrists behind his back, and another warder, a Cossack, gripped him by the throat and winded him with his knee. Then they threw him on the floor, somebody said in a businesslike, professional voice, "Thrash him till he ———— himself," and they started kicking him with their jackboots about the temples and elsewhere, until he lost consciousness. The next day he was summoned to the Chief Security Officer, and they tried to pin a charge of *terrorist* intentions on him—when they were dragging him along he had clutched at stones! Why?

On another occasion Tverdokhleb tried refusing to report for work assignment. He also went on a hunger strike—he was not going to work for Satan! Treating his declarations with contempt, they forcibly dragged him out. This took place in an ordinary hut, so that he was able to reach the windowpanes and break them. The jangle of breaking glass could be heard by the whole line-up, a dismal accompaniment to the voices of work assigners and warders counting.

To the droning monotony of our days, weeks, months, years. And there was no ray of hope in sight. Rays of hope were not budgeted for in the MVD plan when these camps were set up.

Twenty-five of us newcomers, mostly Western Ukrainians, banded together in a work team and persuaded the work assigners to let us choose a foreman from our own number—Pavel Boronyuk, whom I have mentioned before. We made a well-behaved and hard-working team. (The Western Ukrainians, farm workers only yesterday and not in collectives, needed no urging on—at times they had to be reined in!) For some days we were regarded as general laborers, but then some of us turned out to be skilled bricklayers, others started learning from them, and so we became a building brigade. Our bricklaying went well. The bosses noticed it, took us off the housing project (building homes for free personnel), and kept us in the camp area. They showed our foreman the pile of stones by the Disciplinary Barracks—the same stones which Gershuni had tried to hang on to—and promised uninterrupted deliveries from the quarry. They explained that the Disciplinary Barracks as we saw it was only half a Disciplinary Barracks, that the other half must now be built onto it, and that this would be done by our team.

So, to our shame, we started building a prison for ourselves.

It was a long, dry autumn, not a drop of rain fell throughout

September and the first half of October. In the mornings it was calm, then the wind would rise, grow stronger by the middle of the day, and die away again toward evening. Sometimes this wind blew continuously, a thin, nagging wind which made you more painfully aware than ever of the heartbreaking flatness of the steppe, visible to us even from the scaffolding around the Disciplinary Barracks: neither the settlement with the first factory buildings, nor the hamlet where the guards lived, still less the wire fences around the camp, could conceal from us the endlessness, the boundlessness, the perfect flatness, and the hopelessness of that steppe, broken only by the first line of roughly barked telegraph poles running northeast to Pavlodar. Sometimes the wind freshened, and within an hour it would bring in cold weather from Siberia, forcing us to put on our padded jackets and whipping our faces unmercifully with the coarse sand and small stones which it swept along over the steppe. There is nothing for it; it will be simpler if I repeat the poem that I wrote at this time while I was helping to build the Disciplinary Barracks.

THE MASON

Like him of whom the poet sings, a mason, I
Tame the wild stones to make a jail. No city jail—
Here naught but fences, huts, and guard towers meets the eye,
And in the limpid sky the watchful buzzards sail.
None but the wind moves on the steppe—none to inquire
For whom I raise these walls . . . why dogs, machine guns, wire
Are still not jail enough. Trowel in hand, I too
Work thoughtlessly until—"The wall is out of true!
You'll be the first inside!" The major's easy jest
Adds naught to my fears. Informers have played their role.
My record is pocked like a face marked by black pest;
Neat brackets tie me to others bound for the hole.
Breaking, trimming, hammer to merry hammer calls.
Wall after gloomy wall springs up, walls within walls.
While we mix mortar we smoke, and await with delight—
Extra bread, extra slops in our basins tonight.
Back on our perch, we peer into cells walled with stones—
Black pits whose depths will muffle tortured comrades' groans.
Our jailers, like us, have no link with the world of men
But the endless road and the humming wires overhead. . . .
Oh, God, how lost we are, how impotent!
Was ever slave more abject, hope more dead!

Slaves! Not so much because, frightened by Major Mak-simenko's threats, we took care to lay the stones crisscross, with an honest layer of mortar between them, so that future prisoners would not easily be able to pull that wall down. But because even though we somewhat underfulfilled our norm, our team of prison builders was issued with supplementary rations, and instead of flinging them in the major's face we ate them. Our comrade Volo-dya Gershuni was sitting at that very time in the completed wing of the Disciplinary Barracks. And Ivan Spassky, for no known offense, but because of some mysterious black mark on his record, was already in the punishment cells. And for many of us the future held a spell in that same Disciplinary Barracks, in the very cells which we were building with such precision and efficiency. During working hours, when we were nimbly handling stones and mortar, shots suddenly rang out over the steppe. Shortly after, a prison van drove up to the guardhouse, where we were (it was assigned to the guard unit, a genuine prison van such as you see in towns—but they hadn't painted "Drink Soviet Champagne" on its sides for the benefit of the gophers).* Four men were bundled out of the van, all of them battered and covered with blood. Two of them stumbled, one was pulled out; only the first out, Ivan Vorobyov, walked proudly and angrily.

They led the runaways past us, right under our feet, under the catwalks we stood on, and turned with them into the already completed right wing of the Disciplinary Barracks. . . .

While we . . . went on laying our stones.

Escape! What desperate courage it took! Without civilian clothes, without food, with empty hands, to cross the fence under fire and run into the bare, waterless, endless open steppe! It wasn't a rational idea—it was an act of defiance, a proud means of sui-cide. A form of resistance of which only the strongest and boldest among us were capable!

But we . . . went on laying our stones.

And talking it over. This was the second escape attempt in a month. The first had also failed—but that had been rather a silly one. Vasily Bryukhin (nicknamed "Blyuker"), Mutyanov the en-gineer, and another former Polish officer had dug a hole, one cubic meter in capacity, under the room in which they worked in the engineering shop, settled down in it with a stock of food, and covered themselves over. They naïvely expected that in the eve-ning the guard would be taken off the working area as usual, and that they would then be able to climb out and leave. But when at knocking-off time three men were missed, with no breaks in the wire to account for it, guards were left on duty round the clock for days. During this time people walked about over their heads, and dogs were brought in—but the men in hiding held petrol-soaked wadding by a crack in the floor to throw the dogs off the scent. Three days and nights they sat there without talking or stirring, with their legs and arms contorted and entwined, three of them in a space of one cubic meter, until at last they could stand it no longer and came out.

Other teams came back into the camp area and told us how Vorobyov's group had tried to escape: they had burst through the fences in a lorry.

Another week. We were still laying stones. The layout of the second wing of the Disciplinary Barracks was now clearly discernible—here would be the cozy little punishment cells, here the solitary-confinement cells, here the "box rooms." We had by now erected a huge quantity of stone in a little space, and they kept bringing more and more of it from the quarries: the stone cost nothing, labor in the quarries or on the site cost nothing; only the cement was an expense to the state.

The week went by, time enough for the four thousand of Ekibastuz to reflect that trying to escape was insanity, that it led nowhere. And—on another equally sunny day—shots rang out again on the steppe. An escape!!! It was like an epidemic: again the guard troops' van sped into the camp, bringing two of them (the third had been killed on the spot). These two—Batanov and another, a small, quite young man—were led past us, all bloody, to the completed wing, there to be beaten, stripped, tossed onto the bare floor, and left without food or drink. What are your feelings, slave, as you look upon them, mangled and proud? Surely not a mean satisfaction that it is not you who have been caught, not you who have been beaten up, not you who have been doomed.

"Get on with it—we've got to finish the left wing soon!" yells Maksimenko, our potbellied major.

And we . . . lay our stones. We shall get extra kasha in the evening.

Captain Second Class Burkovsky carries the mortar. Whatever is built, he thinks, is for the good of the Motherland.

In the evening we were told that Batanov, too, had tried to break out in a lorry. It had been stopped by gunfire.

Surely you have understood by now, you slaves, that running away is suicide, that no one will ever succeed in running farther than one kilometer, that your lot is to work and to die.

Less than five days later, no shots were heard—but it was as though the sky were of metal and someone was banging on it with a huge iron bar when the news came. An escape! Another escape!!! And this time a successful one.

The escape on Sunday, September 17, was executed so neatly that the evening inspection went off without trouble—as far as the screws could see, the numbers tallied. It was only on the morning of the eighteenth that their sums wouldn't work out right—and work line-up was canceled for a general recount. There were several inspections on the central tract, then inspections by huts, inspections by work teams, then a roll call from filing cards—the dogs couldn't count anything except the money in the till. They arrived at a different answer every time! They still didn't know *how many* had run away, who exactly, when, where to, and whether on foot or with a vehicle.

By now it was Monday evening, but they gave us no dinner (the cooks, too, had been turned out onto the central tract to help with the counting!), but we didn't mind. We were only too happy! Every successful escape is a great joy to other prisoners! However brutally the guards behave afterward, however harsh discipline becomes, we don't mind a bit, we're only too happy! What d'you think of that, you dogs! Some of us have escaped! (We look our masters in the eye, all the time secretly thinking: Let them not be caught! Let them not be caught!)

What is more, they didn't lead us out to work, and Monday went by like a second day off. (A good thing the lads hadn't legged it on Saturday! They'd taken care not to spoil our Sunday for us!)

But who were they? Who were they?

On Monday evening the news went round: Georgi Tenno and Kolya Zhdanok.

We built the prison higher. We had already made the straight arches over the doors, built above the little window spaces, and we were now leaving sockets for the beams.

Three days since they had escaped. Seven. Ten. Fifteen.

Still no news!

They had got away!!

Chapter 4

∎

Why Did We Stand For It?

Among my readers there is a certain educated Marxist Historian. Sitting in his soft armchair, and leafing through this book to the passage about how we built the Disciplinary Barracks, he takes off his glasses, taps the page with something flat, a ruler perhaps, and nods his head repeatedly.

"Yes, yes . . . This bit I can believe. But all that stuff about the —er—whiff of revolution. I'll be damned if I do! You could not have a revolution, because revolutions take place in accordance with the laws of history. In your case all that had happened was that a few thousand so-called "politicals" were picked up—and did what? Deprived of human appearance, of dignity, family, freedom, clothing, food—what did you do? Why didn't you revolt?"

"We were earning our rations. I told you—building a prison."

"That's fine. Just what you should have been doing! It was for the good of the people. It was the only correct solution. But don't call yourselves revolutionaries, my friends! To make a revolution you must be linked with the one and only progressive class. . . ."

"Yes, but weren't we all workers by then?"

"That is neither here nor there. That is a philistine quibble. Have you any idea what historical necessity means?"

I rather think I have. I honestly have. I have an idea that when camps with millions of prisoners exist for forty years—that's where we can see historical necessity at work. So many millions, for so many years, cannot be explained by Stalin's vagaries or Beria's perfidy, by the naïve trustfulness of the ruling party, o'er which the light of the Vanguard Doctrine never ceased to shine. But I won't cast *this* example of historical necessity in my opponent's teeth. He would only smile sweetly and tell me that that was not the subject under discussion, that I was straying from the point.

He sees that I am at a loss, that I have no clear conception of historical necessity, and explains:

"Those were revolutionaries, who rose up and swept Tsarism away with their broom. Very simple. If Tsar Nicky had so much as tried to squeeze his revolutionaries so hard! If he had just tried to pin numbers on them! If he had even tried . . ."

"You are right. He didn't try. He didn't try, and that's the only reason why they survived to try it when he had gone."

"But he *couldn't* try it! He couldn't!"

Probably also correct. Not that he might not have liked to—but that he couldn't.

In the conventional Cadet (let alone socialist) interpretation, the whole of Russian history is a succession of tyrannies. The Tatar tyranny. The tyranny of the Moscow princes. Five centuries of indigenous tyranny on the Oriental model, and of a social order firmly and frankly rooted in slavery. (Forget about the Assemblies of the Land,* the village commune, the free Cossacks, the free peasantry of the North.) Whether it is Ivan the Terrible, Alexis the Gentle, heavy-handed Peter, or velvety Catherine, all the Tsars right up to the Crimean War knew one thing only—how to *crush*. To crush their subjects like beetles or caterpillars. If a man was sentenced to hard labor and deportation, they pricked on his body the letters "SK"* and chained him to his wheelbarrow. The state bore hard on its subjects; it was unflinchingly firm. Mutinies and uprisings were invariably crushed.

Only . . . only . . . Crushed, yes, but the word needs qualification. Not crushed in our modern technical sense. After the war with Napoleon, when our army came back from Europe, the first breath of freedom passed over Russian society. Faint as it was, the Tsar had to reckon with it. The common soldiers, for instance, who took part in the Decembrist rising*—was a single one of them strung up? Was a single one shot? And in our day would a single one of them have been left alive? Neither Pushkin nor Lermontov could be simply put inside for a *tenner*—roundabout ways of dealing with them had to be found. "Where would you have been in Petersburg on December 14?" Nicholas I asked Pushkin. Pushkin answered honestly, "On the Senate Square."* And by way of punishment . . . he was told to go home! Whereas all of us who have felt on our own hides the workings of a mechanized judicial system, and of course all our friends in public prosecutors' offices, know the proper price for Pushkin's answer: Article 58, Section 2 (armed insurrection), or—the mildest possible treatment—Article 19 (criminal intent)—and if not shooting, certainly nothing short of a *tenner. Our* Pushkins had heavy sentences slapped on them, went to the camps, and died. (Gumilyev never even got as far as a camp; they settled accounts with him in a cellar.)

Of all her wars, the Crimea was Russia's luckiest! It brought the emancipation of the peasants and Alexander's reforms, and what is more, the greatest of social forces—public opinion—appeared simultaneously in Russia.

On the face of it the Siberian *katorga* went on festering, and even spread: more transit prisons were brought into operation, prisoners were still transported in droves, courts were always in session. But what is this? The courts were in session but Vera Zasulich, who shot at the chief of police in the capital (!), was acquitted?!?

Seven attempts were made on the life of Alexander II himself (Karakozov's;[1] Solovyov's; one near Aleksandrovsk; one outside Kursk; Khalturin's explosion; Teterka's mine; Grinevitsky). Alexander II went around Petersburg with fear in his eyes (but, incidentally, without a bodyguard), "like a hunted animal" (according to Tolstoi, who met the Tsar on the staircase of a private house).[2] What did he do about it? Ruin and banish half Petersburg, as happened after Kirov's murder? You know very well that such a thing could never enter his head. Did he apply the methods of prophylactic mass terror? Total terror, as in 1918? Take *hostages?* The concept didn't exist. Imprison *dubious persons?* It simply wasn't possible. . . . Execute thousands? They executed . . . five. Fewer than three hundred were convicted by the courts in this period. (If just *one* such attempt had been made on Stalin,

1. Karakozov, incidentally, had a brother. Brother of the man who tried to shoot the Tsar! Measure that by our yardstick. What was his punishment? "He was ordered to change his name to Vladimirov." He suffered neither loss of property rights nor restrictions as to his place of residence.
2. *Lev Tolstoi v Vospominaniakh Sovremennikov (Lev Tolstoi Remembered by His Contemporaries),* Vol. 1, 1955, p. 180.

how many million lives would it have cost us?)

The Bolshevik Olminsky writes that in 1891 he was the only *political prisoner* in the whole Kresty Prison. Transferred to Moscow, he was the only one in the Taganka. It was only in the Butyrki, awaiting deportation, that a small party of them was assembled.

With every year of education and literary freedom the invisible but terrible power of public opinion grew, until the Tsars lost their grip on both reins and mane, and Nicholas II could only clutch at crupper and tail. It is true that the inertial undertow of dynastic tradition prevented him from understanding the demands of his age, and that he lacked the courage to act. In the age of airplanes and electricity he still lacked all social awareness, and thought of Russia as his own rich and richly variegated estate, in which to levy tribute, breed stallions, and raise armies for a bit of a war now and again with his imperial brother of the house of Hohenzollern. But neither he, nor any of those who governed for him, any longer had the will to fight for their power. They no longer crushed their enemies; they merely squeezed them gently and let them go. They were forever looking over their shoulders and straining their ears: what would public opinion say? They persecuted revolutionaries just sufficiently to broaden their circle of acquaintance in prisons, toughen them, and ring their heads with haloes. We now have an accurate yardstick to establish the scale of these phenomena—and we can safely say that the Tsarist government did not persecute revolutionaries but tenderly nurtured them, for its own destruction. The uncertainty, half-heartedness, and feebleness of the Tsarist government are obvious to all who have experienced an infallible judicial system.

Let us examine, for instance, some generally known biographical facts about Lenin. In spring, 1887, his brother was executed for an attempt on the life of Alexander III.[3] Like Karakozov's brother, Lenin was the brother of a would-be regicide. And what

3. It was incidentally established in the course of investigation that Anna Ulyanova had received a coded telegram from Vilna: "Sister dangerously ill," which meant "Weapons on the way." Anna was not surprised, although she had no sister in Vilna, and for some reason passed it on to Aleksandr; she was obviously his accomplice, and in our day she could have been sure of a *tenner*. But Anna was not even asked to account for it! In the same case it was established that another Anna (Serdyukova), a schoolteacher at Yekaterinodar, had direct knowledge of the planned attempt on the Tsar, and kept silent. What would have happened to her in our time? She would have been shot. And what did they give her? Two years . . .

happened to him? In the autumn of that very year Vladimir Ulyanov was admitted to the Imperial University at Kazan, and what
is more, to the Law Faculty! Surprising, isn't it?

True, Vladimir Ulyanov was expelled from the university in the
same academic year. But this was for organizing a student demonstration against the government. The younger brother of a would-
be regicide inciting students to insubordination? What would he
have got for that in our day? He would certainly have been shot!
(And of the rest, some would have got twenty-five and others ten
years.) Whereas he was merely expelled. Such cruelty! Yes, but he
was also banished. . . . To Sakhalin?[4] No, to the family estate of
Kokushkino, where he intended to spend the summer anyway. He
wanted to work—so they gave him an opportunity. . . . To fell
trees in the frozen north? No, to practice law in Samara, where
he was simultaneously active in illegal political circles. After this
he was allowed to take his examinations at St. Petersburg University as an external student. (With his curriculum vitae? What was
the Special Section thinking of?)

Then a few years later this same young revolutionary was arrested for founding in the capital a "League of Struggle for the
Liberation of the Working Class"—no less! He had repeatedly
made "seditious" speeches to workers, had written political leaflets. Was he tortured, starved? No, they created for him conditions
conducive to intellectual work. In the Petersburg investigation
prison, where he was held for a year, and where he was allowed
to receive the dozens of books he needed, he wrote the greater part
of *The Development of Capitalism in Russia,* and, moreover, for-
warded—legally, through the Prosecutor's Office—his *Economic
Essays* to the Marxist journal *Novoye Slovo.* While in prison, he
followed a prescribed diet, could have dinners sent in at his own
expense, buy milk, buy mineral water from a chemist's shop, and
receive parcels from home three times a week. (Trotsky, too, was
able to put the first draft of his theory of permanent revolution on
paper in the Peter and Paul Fortress.)

But then, of course, he was condemned by a three-man tribunal
and shot? No, he wasn't even jailed, only banished. To Yakutya,
then, for life? No, to a land of plenty, Minusinsk, and for three
years. He was taken there in handcuffs? In a prison train? Not at

4. There were, incidentally, political prisoners on Sakhalin. But, as it happened, not a
single notable Bolshevik (or for that matter Menshevik) was ever there.

all! He traveled like a free man, went around Petersburg for three days without interference, then did the same in Moscow—he had to leave instructions for clandestine correspondence, establish connections, hold a conference of revolutionaries still at large. He was even allowed to go into exile at his own expense—that is, to travel with free passengers. Lenin never sampled a single convict train or a single transit prison on his way out to Siberia or, of course, on the return journey. Then, in Krasnoyarsk, two more months' work in the library saw *The Development of Capitalism* finished, and this book, written by a political exile, appeared in print without obstruction from the censorship. (Measure that by our yardstick!) But what would he live on in that remote village, where he would obviously find no work? He asked for an allowance from the state, and they paid him more than he needed. It would have been impossible to create better conditions than Lenin enjoyed in his one and only period of banishment. A healthy diet, at extremely low prices, plenty of meat (a sheep every week), milk, vegetables; he could hunt to his heart's content (when he was dissatisfied with his dog, friends seriously considered sending him one from Petersburg; when mosquitoes bit him while he was out hunting, he ordered kid gloves); he was cured of his gastric disorders and the other illnesses of his youth, and rapidly put on weight. He had no obligations, no work to do, no duties, nor did his womenfolk exert themselves; for two and a half rubles a month, a fifteen-year-old peasant girl did all the rough work about the house. Lenin had no need to write for money, turned down offers of paid work from Petersburg, and wrote only things which could bring him literary fame.

He served his term of banishment (he could have "escaped" without difficulty, but was too circumspect for that). Was his sentence automatically extended? Converted to deportation for life? How could it be—that would have been illegal. He was given permission to reside in Pskov, on condition that he did not visit the capital. He did visit Riga and Smolensk. He was not under surveillance. Then he and his friend (Martov) took a basket of forbidden literature to the capital, traveling via Tsarskoye Selo, where there were particularly strict controls (they had been too clever by half). He was picked up in Petersburg. True, he no longer had the basket, but he did have a letter to Plekhanov in invisible ink with the whole plan for launching *Iskra.* * The police, though, could not put themselves to all that trouble; he was under arrest and in a cell for three weeks, the letter was in their hands—and it remained undeciphered.

What was the result of this unauthorized absence from Pskov? Twenty years' hard labor, as it would have been in our time? No, just those three weeks under arrest! After which he was freed completely, to travel around Russia setting up distribution centers for *Iskra,* then abroad, to arrange publication ("the police see no objection" to granting him a passport for foreign travel!).

But this was the least of it! As an émigré he would send home to Russia an article on Marx for the *Granat Encyclopedia!* And it would be printed.[5] Nor was it the only one!

Finally, he carried on subversive activity from a little town in Austrian Poland, near the Russian frontier, but no one sent undercover thugs to abduct him and bring him back alive. Though it would have been the easiest thing in the world.

Tsardom was always weak and irresolute in pursuit of its enemies—you can trace the same pattern in the story of any important Social Democrat (Stalin in particular—though here suspicion of other contributory factors insinuates itself). Thus, in 1904 Kamenev's room in Moscow was searched and "compromising correspondence" was seized. Under interrogation he refused all explanation. And that was that. He was banished—to his parents' place of residence.

The SR's, it is true, were persecuted more severely. But how severely? Would you say that Gershuni (arrested in 1903) had no serious crimes to answer for? Or Savinkov (arrested in 1906)? They organized the assassination of some of the highest-placed people in the empire. Yet they were not executed. Then Mariya Spiridonova was allowed to escape. She shot General Luzhenovsky, who put down the peasant rising in Tambov, shot him point-blank, and once again they could not bring themselves to execute a terrorist, and only sent her to forced labor.[6] Just imagine what would have happened if a seventeen-year-old schoolgirl had shot the suppressor of the 1921 peasant rising (also in Tambov!) —how many *thousands* of high school pupils and intellectuals

5. Just imagine the *Bolshaya Entsiklopedia* publishing an émigré article on Berdyayev.
6. She was released from forced labor by the February Revolution. But then in 1918 she was arrested on a number of occasions by the Cheka. Like other socialists, she was shuffled, redealt, and finally discarded in the Great Game of Patience. She spent some time in exile in Samarkand, Tashkent, and Ufa. Afterward her trail is lost in one of the political "isolators." Somewhere or other she was shot.

would have been summarily shot without trial in the wave of "retaliatory" red terror?

Were people shot for the naval mutiny at Sveaborg? No, just exiled.

Ivanov-Razumnik recalls how students were punished (for the great demonstration in Petersburg in 1901). The scene in the Petersburg prison was like a student picnic—roars of laughter, community singing, students walking around freely from cell to cell. Ivanov-Razumnik even had the impertinence to ask the prison governor to let him attend a performance by the touring company of the Moscow Art Theatre, so as not to waste his ticket! Later he was sentenced to banishment—to Simferopol, which was his own choice, and he hiked all over the Crimea with a rucksack.

Ariadna Tyrkova writes about this period as follows: "We stuck to our principles, and the prison regime was not strict." Gendarme officers offered them meals from the best restaurant, Dodon's. According to the indefatigably curious Burtsev, "The Petersburg prisons were much more humane than those of Western Europe."

For calling on the Moscow workers to rise up in arms (!) and overthrow (!) the autocracy, Leonid Andreyev was . . . kept in a cell for fifteen whole days. (He himself thought it was rather little, and added three weeks in his own account.) Here are some entries in his diary at the time:[7] "Solitary confinement! Never mind, it's not so bad. I make my bed, pull up my stool and my lamp, put some cigarettes and a pear nearby. . . . I read, eat my pear—just like home. I feel merry. That's the word, merry." "Sir! Excuse me, sir!" The warder calls to him through the feeding hatch. Several books have arrived. And notes from neighboring cells.

Summing it up, Andreyev acknowledged that as far as board and lodging were concerned, he lived better in his cell than he had as a student.

At this very time Gorky was writing *Children of the Sun* in the Trubetskoi Bastion.

The Bolshevik elite published a pretty shameless piece of self-advertisement in the shape of Volume 41 of the *Granat Encyclopedia:* "Prominent Personalities of the U.S.S.R. and the October Revolution—Autobiographies and Biographies." Read whichever of them you like; you will be astounded to find how lightly by our standards they got away with their revolutionary activity. And, in

---

7. Quoted from V. L. Andreyev, *Detstvo (Childhood)*.

particular, what favorable conditions they enjoyed in prison. Take Krasin, for instance: "he always remembered imprisonment in the Taganka with great pleasure. After the initial interrogations the gendarmes left him in peace" (why?), "and he devoted his involuntary leisure to unremitting toil: he learned German, read almost all the works of Schiller and Goethe in the original, acquainted himself with Schopenhauer and Kant, thoroughly studied Mill's *Logic* and Bundt's psychology," and so on. For his place of exile Krasin chose Irkutsk, the capital of Siberia and its most civilized town.

This is Radek, in prison in Warsaw in 1906. He "was in for half a year, of which he made splendid use, learning Russian, reading Lenin, Plekhanov, and Marx. While in prison, he wrote his first article [on the trade union movement] . . . and was terribly proud when he received [in jail] the issue of Kautsky's journal containing his contribution."

At the other extreme, Semashko's "imprisonment [Moscow, 1895] was unusually harsh": after three months in jail he was exiled for three years . . . to his native town, Yelets!

The reputation of the "terrible Russian Bastille" was created in the West by people demoralized by imprisonment, like Parvus, who wrote his highly colored, bombastic-sentimental reminiscences to avenge himself on Tsarism.

The same pattern can be traced in the experience of lesser personalities, in thousands of individual life stories.

I have on hand an encyclopedia, not the most obviously relevant, since it is the *Literary Encyclopedia,* and an old edition at that (1932), complete with "errors." Before someone eradicates these "errors" I will take the letter "K" at random.

Karpenko-Kary. While secretary to the city police (!) in Yelizavetgrad, provided passports for revolutionaries. (Translating as we go into our own language: an official in the passport section supplied an underground organization with passports!) Was he . . . hanged for it? No, banished for . . . 5 (five) years—to his own farm! In other words, to his country home. He became a writer.

Kirillov, V. T. Took part in the revolutionary movement of the Black Sea sailors. Shot? Hard labor for life? No; three years' banishment to Ust-Sysolsk. Became a writer.

Kasatkin, I. M. While in prison wrote stories which were published in newspapers! (In our time even ex-prisoners cannot get published.)

Karpov, Yevtikhi. After two (!) periods of banishment, was put in charge of the Imperial Aleksandrinsky Theatre and the Suvorin Theatre. (In our time he would never have obtained permission to reside in the capital, and in any case the Special Section would not have taken him on as a prompter.)

Krzhizhanovsky returned from banishment when the "Stolypin reaction"* was at its wildest and (while remaining a member of the underground Bolshevik Central Committee) took up his profession as an engineer without hindrance. (In our time he would have been lucky to find a job as mechanic in a Machine and Tractor Station!)

Although Krylenko hasn't got into the *Literary Encyclopedia,* it seems only right to mention him among the "K" 's. In all his years as a revolutionary hothead he three times "successfully avoided arrest"[8] and was six times arrested, but spent in all only fourteen months in prison. In 1907 (that year of reaction again) he was accused of agitation among the troops and participation in a military organization—and acquitted by the Military District Court (!). In 1915, for "evading military service" (he was an officer, and there was a war on), this future commander in chief (and murderer of his predecessor in that post) was punished by being . . . sent to a front-line (but not a punitive) unit. (This was how the Tsar's government proposed to damp down the fires of revolution while simultaneously defeating the Germans. . . .) And for fifteen years it was under the shadow of his unclipped procuratorial wing that the endless lines of those condemned in countless trials shuffled through the courts to receive their bullet in the back of the head.

During the Stolypin reaction again, V. A. Staroselsky, governor of Kutaisi, who unhesitatingly supplied revolutionaries with passports and arms, and betrayed the plans of the police and the government forces to them, got away with something like two weeks' imprisonment.[9]

Translate that into our language, if you have imagination enough!

During this same "reactionary" phase the Bolshevik philosophical and political journal *Mysl* was *legally* published. And the

---

8. This and what follows is taken from his autobiography in the *Granat Encyclopedia,* Vol. 41, pp. 237–245.

9. "Tovarishch Gubernator" ("Comrade Governor"), *Novy Mir,* 1966, No. 2.

"reactionary" *Vekhi* openly wrote about the "obsolete autocracy," "the evils of despotism and slavery"—fine, keep it up; we don't mind a bit!

The severity of those times was beyond human endurance. V. K. Yanovsky, an art photographer in Yalta, made a sketch showing the shooting of the Ochakov sailors and exhibited it in his shopwindow (much as if someone nowadays had exhibited episodes from the punitive operation at Novocherkassk* in a window on Kuznetsky Most). And what did the Yalta police chief do? Because Livadia* was so near he behaved with particular cruelty. He began by shouting at Yanovsky. And he went on to destroy . . . not Yanovsky's studio—oh, dear, no—not the sketch of the shooting, but . . . a copy of the sketch. (Some will explain this by Yanovsky's sleight of hand. But let us note that the governor did not order the window to be broken in his presence.) Thirdly, a very heavy penalty was inflicted on Yanovsky himself: as long as he continued to reside in Yalta he must not appear in the street . . . while the imperial family was passing through.

Burtsev, in an émigré journal, went so far as to cast aspersions on the private life of the Tsar. When he returned to the Motherland (on the flood tide of patriotism in 1914)—was he shot? He spent less than a year in prison, with permission to receive books and to carry on his literary pursuits.

No one stayed the axman's hand. And in the end the tree would fall.

When Tukhachevsky was "repressed," as they call it, not only was his immediate family broken up and imprisoned (it hardly needs to be said that his daughter was expelled from her institute), but his two brothers and their wives, his four sisters and their husbands, were all arrested, while all his nephews and nieces were scattered about various orphanages and their surnames changed to Tomashevich, Rostov, etc. His wife was shot in a camp in Kazakhstan, his mother begged for alms on the streets of Astrakhan and died there.[10] Similar stories can be told about the relatives of hundreds of other eminent victims. That is real persecution!

The most important special feature of persecution (if you can

10. I cite this example in sympathy for his innocent relatives. Tukhachevsky himself is becoming the object of a new cult, to which I do not intend to subscribe. What he reaped, he had sown when he directed the suppression of the Kronstadt rising and of the peasant rising in Tambov.

call it that) in Tsarist times was perhaps just this: that the revolutionary's relatives never suffered in the least. Natalya Sedova (Trotsky's wife) returned to Russia without hindrance in 1907, when Trotsky was a condemned criminal. Any member of the Ulyanov family (though nearly all of them were arrested at one time or another) could readily obtain permission to go abroad at any moment. When Lenin was on the "wanted" list for his exhortations to armed uprising, his sister Anna legally and regularly transferred money to his account with the Crédit Lyonnais in Paris. Both Lenin's mother and Krupskaya's mother as long as they lived received state pensions for their deceased husbands—one a high-ranking civil servant, the other an army officer—and it would have been unthinkable to make life hard for them.

Such were the circumstances in which Tolstoi came to believe that only moral self-improvement was necessary, not political freedom.

Of course, no one is in need of freedom if he already has it. We can agree with him that political freedom is not what matters in the end. The goal of human evolution is not freedom for the sake of freedom. Nor is it the building of an ideal polity. What matter, of course, are the moral foundations of society. But that is in the long run: what about the beginning? What about the first step? Yasnaya Polyana in those days was an open club for thinkers. But if it had been blockaded as Akhmatova's apartment was when every visitor was asked for his passport, if Tolstoi had been pressed as hard as we all were in Stalin's time, when three men feared to come together under one roof, even he would have demanded political freedom.

At the most dreadful moment of the Stolypin terror the liberal newspaper *Rus* was allowed to report, in bold type on its front page: "Five executions!" "Twenty executed at Kherson!" Tolstoi broke down and wept, said that he couldn't go on living, that it was *impossible* to imagine anything *more horrible.* [11]

Then there is the previously mentioned list in *Byloye:* 950 executions in six months. [12]

Let us take this issue of *Byloye.* Note that it appeared well within the eight-month period of Stolypin's "military justice" (August, 1906—April, 1907), and that the list was compiled from

11. *Tolstoy v Vospominaniakh Sovremennikov,* Vol. II, 1955, p. 232.
12. *Byloye,* No. 2/14, February, 1907.

data published by Russian news agencies. Much as though the Moscow papers in 1937 had given lists of those shot, which were then collated and republished, with the NKVD tamely turning a blind eye.

Secondly, this eight-month period of martial justice, which had no precedent and was not repeated in Tsarist Russia, could not be prolonged because the "impotent" and "docile" State Duma would not ratify such measures (indeed, Stolypin did not venture to submit them to the Duma for discussion).

Thirdly, the events during the previous six months invoked in justification of "military law" included the "murder of innumerable police officers for political motives," "many attacks on officials,"[13] and the explosion on Aptekarsky Island. If, it was argued, "the state does not put a stop to these terrorist acts then it will forfeit its right to exist." So Stolypin's ministry, impatient and angry with the jury courts and their leisurely inconsequences, their powerful and uninhibited bar (not a bit like our oblast courts or district tribunals, obedient to a telephone call), snatched at a chance to curb the revolutionaries (and also straightforward bandits, who shot at the windows of passenger trains and killed ordinary citizens for a few rubles) by means of the laconic court-martial. (Even so there were restrictions: a court-martial could be set up *only* in places in which martial law or a state of emergency had been declared; it convened only when the evidence of crime was fresh, not more than twenty-four hours after the event, and when a crime had manifestly been committed.)

If contemporaries were stunned and shocked, it was obviously because this was something new to Russia!

In the 1906–1907 situation we see that the revolutionaries must take their share of the blame for the "Stolypin terror," as well as the government. A hundred years after the birth of revolutionary terror, we can say without hesitation that the terrorist idea and terrorist actions were a hideous mistake on the part of the revolutionaries and a disaster for Russia, bringing her nothing but confusion, grief, and inordinate human losses.

Let us turn over a few more pages in the same number of *Byloye*.[14] Here is one of the earliest proclamations, dating from 1862, which were the start of it all.

---

13. The article in *Byloye* cited above does not deny these facts.
14. *Byloye,* 2/14, p. 82.

"What is it that we want? The good, the happiness of Russia. Achieving a new life, a better life, without casualties is impossible, because we cannot afford delay—we need speedy, immediate reform!"

What a false path! They, the zealots, could not afford to wait, and so they sanctioned human sacrifice (of others, not themselves) to bring universal happiness nearer! They could not afford to wait, and so we, their great-grandsons, are not at the same point as they were (when the peasants were freed), but much farther behind.

Let us admit that the terrorists were worthy partners of Stolypin's courts-martial.

What for us makes comparison between the Stolypin and the Stalin periods impossible is that in our time the barbarity was all on one side: heads were cut off for a sigh or for less than a sigh.[15]

"Nothing more horrible!" exclaimed Tolstoi. It is, however, very easy to imagine things more horrible. It is more horrible when executions take place not from time to time, and in one particular city of which everybody knows, but *everywhere* and *every day;* and not twenty but two hundred at a time, with the newspapers saying nothing about it in print big or small, but saying instead that "life has become better, life has become more cheerful."

They bash your face in, and say it was always ugly.

No, things weren't the same! Not at all the same, although the Russian state even then was considered the most oppressive in Europe.

The twenties and thirties of our century have deepened man's understanding of the possible degrees of *compression.* The terrestrial dust, the earth which seemed to our ancestors as compact as may be, is now seen by physicists as a sieve full of holes. An isolated speck in a hundred meters of emptiness—that is a model of the atom. They have discovered the nightmarish possibility of "atom packing"—forcing all the tiny nuclear specks from all those hundred-meter vacuums together. A thimbleful of such packing weighs as much as a normal locomotive. But even this packing is too much like fluff: the protons prevent you from compressing the nuclei as tight as you could wish. If you could compress neutrons

15. I state with confidence that our age has also surpassed that of the Tsars in the scale and technical level of its summary punitive operations. (Suppression of peasant revolt in 1918–1919; Tambov rising, 1921; Kuban and Kazakhstan, 1930.)

alone, a postage stamp made of such "neutron packing" would weigh five million tons!

And that is how tightly they squeezed us, without any help from pioneering physicists!

Through Stalin's lips our country was bidden henceforth to *renounce complacency.* But under the word used for complacency Dal* gives "kindness of heart, a loving state of mind, charity, a concern for the general good." That was what we were called upon to renounce, and we did so in a hurry—renounced all concern for the general good! Henceforth our own feeding trough was enough for us!

Russian public opinion by the beginning of the century constituted a marvelous force, was creating a climate of freedom. The defeat of Tsarism came not when Kolchak was routed, not when the February Revolution was raging, but much earlier! It was overthrown without hope of restoration once Russian literature adopted the convention that anyone who depicted a gendarme or policeman with any hint of sympathy was a lickspittle and a reactionary thug; when you didn't have to shake a policeman's hand, cultivate his acquaintance, nod to him in the street, but merely brush sleeves with him in passing to consider yourself disgraced.

Whereas we have butchers who—because they are now redundant and because their qualifications are right—are in charge of literature and culture. They order us to extol *them* as legendary heroes. And to do so is for some reason called . . . patriotism.

Public opinion! I don't know how sociologists define it, but it seems obvious to me that it can only consist of interacting individual opinions, freely expressed and independent of government or party opinion.

So long as there is no independent public opinion in our country, there is no guarantee that the extermination of millions and millions for no good reason will not happen again, that it will not begin any night—perhaps this very night.

The Vanguard Doctrine, as we have seen, gave us no protection against this plague.

But I can see my opponent pulling faces, winking at me, wagging his head. In the first place, *the enemy may overhear me.* And secondly, why such a broad treatment of the subject? The question was posed much more narrowly. It was not why were we jailed?

Nor why did those who remained *free* tolerate this lawlessness? Everyone knows that they didn't *realize* what was going on, that they simply *believed* (the party)[16] that if whole peoples are banished in the space of twenty-four hours, those peoples must be guilty. The question is a different one: Why did we in the camps, where we *did* realize what was going on, suffer hunger, bend our backs, put up with it all, instead of fighting back? The others, who had never marched under escort, who had the free use of their arms and legs, could be forgiven for not fighting—they couldn't, after all, sacrifice their families, their positions, their wages, their authors' fees. They're making up for it now by publishing critical reflections in which they reproach us for clinging to our rations instead of fighting, when we had nothing to lose.

But I have all along been leading up to my answer to this question. The reason why we put up with it all in the camps is that there was no public opinion *outside*.

What conceivable ways has the prisoner of resisting the regime to which he is subjected? Obviously, they are:

1. Protest.
2. Hunger strike.
3. Escape.
4. Mutiny.

So, then, it is *obvious to anybody*, as the Great Deceased liked to say (and if it isn't, we'll ram it into him), that if the first two have some force (and if the jailers fear them), it is *only* because of public opinion! Without that behind us we can protest and fast as much as we like and they will laugh in our faces!

It is a very dramatic way of obtaining your demands—standing before the prison authorities and tearing open your shirt, as Dzerzhinsky did. But only where public opinion exists. Without it—you'll be gagged with the tatters and pay for a government-issue shirt into the bargain!

Let me remind you of a celebrated event which took place in the Kara hard-labor prison at the end of the last century. Political prisoners were informed that in future they would be liable to corporal punishment. Nadezhda Sigida was due to be thrashed first (she had slapped the commandant's face . . . to force him to resign!). She took poison and died rather than

16. V. Yermilov's answer to I. Ehrenburg.

submit to the birch. Three other women then poisoned themselves—and also died! In the men's barracks fourteen prisoners
volunteered to commit suicide, though not all of them succeeded.[17] As a result, corporal punishment was abolished outright and forever! The prisoners had counted on frightening
the prison authorities. For news of the tragedy at Kara would
reach Russia, and the whole world.

But if we measure this case against our own experience, we shall
shed only tears of scorn. Smack the commandant's face? For an
injury inflicted on *someone else?* And what is so terrible about a
few thwacks across the backside? You'll go on living! And why did
her women friends take poison, too? And why fourteen men besides? We are only given one life! We must make the best of it! As
long as we get food and drink, why part with life? Besides, maybe
there will be an amnesty, maybe they'll start giving us good conduct marks.

You see from what a lofty plane prison behavior has declined.
And how low we have fallen. And how by the same token our
jailers have risen in the world! No, these are not the bumpkins of
Kara! Even if we had plucked up our courage and risen above
ourselves—four women and fourteen men—we should all have
been shot before we got at any poison. (Where, in any case, would
it come from in a Soviet prison?) If you did manage to poison
yourself, you would only make the task of the authorities easier.
And the rest would be treated to a dose of the birch for not
denouncing you. And needless to say, no word of the occurrence
would ever leak through the boundary wires.

This is the point, this is where their power lies: no news
could leak out. If some muffled rumor did, with no confirmation from newspapers, with informers busily nosing it out, it
would not get far enough to matter: there would be no outburst of public indignation. So what is there to fear? So why

17. We may note here some significant details from E. N. Kovalskaya, *Zhenskaya
Katorga (Women Political Prisoners),* Gosizdat, 1920, pp. 8–9; and G. F. Osmolovsky,
Kariiskaya Tragedia (The Tragedy at Kara), Moscow, 1920. Sigida struck and spat on an
officer for absolutely no reason, because of the "neurotic atmosphere" among political
prisoners. After this the gendarme officer (Masyukov) *asked a political prisoner* (Osmolovsky) *to interrogate him.* The governor of the prison (Bobrovsky) *died repentant,* and
would not even accept consolation from the priest. (If only we had had jailers with
consciences like these!) Sigida was beaten with her clothes on, and Kovalskaya's dress was
changed by other women, and not, as rumor had it, in the presence of men.

should they lend an ear to our *protests?* If you want to poison yourselves—get on with it.

The hopelessness of our hunger strikes has been sufficiently shown in Part I.

Escape, then? History has preserved for us accounts of some major escapes from Tsarist prisons. All of them, let us note, were engineered and directed *from outside*—by other revolutionaries, Party comrades of the escapers, with incidental help from many sympathizers. Many people were involved in the escape itself, in concealing the escapers afterward, and in slipping them across the frontier. ("Aha!" My Marxist Historian has caught me out here. "That was because the population sided with the revolutionaries, and because the future belonged to them!" "Perhaps also," I humbly reply, "because it was all a jolly game, and a legal one? Fluttering your handkerchief from a window, letting a runaway share your bedroom, helping him with his disguise? These were not indictable offenses. When Pyotr Lavrov ran away from his place of banishment, the governor of Vologda [Khominsky] gave his civil-law wife permission to leave and catch up with her man. . . . Even for forging passports you could just be rusticated to your own farm, as we saw. People *were not afraid*—do you know, from your own experience, what that means? While I think of it, how is it that *you* were never *inside?*" "Well, you know, it was all a *lottery. . . .*")

There is, however, evidence of another kind. We were all made to read Gorky's *Mother* at school, and some of you may remember the account of conditions in the Nizhni Novgorod jail: the warders had rusty pistols with which they would knock nails into the walls, and there was no difficulty at all in placing a ladder against the prison wall and calmly discharging yourself. The high police official Ratayev writes as follows: "Banishment existed only on paper. Prison didn't exist at all. Prison conditions at that time were such that a revolutionary who landed in prison could continue his former activities without hindrance. . . . The Kiev revolutionary committee were all in jail together, and while there directed a strike in the city and issued appeals."[18]

18. Letter from L. A. Ratayev to P. N. Zuyev in *Byloye*, No. 2/24, 1917. Ratayev goes on to speak of the general situation in Russia, *outside*. "Secret agents and free-lance detectives didn't exist anywhere [except in the two capitals—A.S.]. Surveillance was carried out if absolutely necessary by noncommissioned gendarme officers in disguise, who sometimes forgot to remove their spurs when they put on civilian clothes. . . . In these

I have at present no access to information about security at the principal locations of the Tsarist *katorga*; but if escape from them was ever as desperately difficult as it was from their Soviet counterparts, with one chance in 100,000 of success, I have never heard it. There was obviously no reason for prisoners to take great risks: they were not threatened with premature death from exhaustion by hard labor, nor with extensions of sentence which they had done nothing to deserve: the second half of their term they served not in prison but in places of banishment, and they usually put off escapes till then.

Laziness would seem to be the only reason for not escaping from Tsarist places of banishment. Evidently, exiles reported to the police infrequently, surveillance was poor, there were no secret police posts along the roads, you were not tied to your work day in and day out by police supervision; you had money (or it could be sent to you), and places of banishment were not remote from the great rivers and roads; again, no threat hung over anyone who helped a runaway, nor indeed was the runaway himself in danger of being shot by his pursuer, or savagely beaten, or sentenced to twenty-five years' hard labor, as in our day. A recaptured prisoner was usually reinstalled in his previous place, to complete his previous sentence. And that was all. You couldn't lose. Fastenko's departure abroad (Part I, Chapter 5) is typical of such ventures. But perhaps the Anarchist A. P. Ulanovsky's escape from the Turukhan region is even more so. In the course of his escape it was enough for him to look in at a student reading room and ask for Mikhailovsky's *What Is Progress?* and the students gave him a meal, a bed, and his fare. He escaped abroad by simply walking up the gangway of a foreign ship—no MVD patrol there, of course!—and finding a warm spot in the stokehold. More wonderful yet, during the 1914 war he voluntarily returned to Russia and to his place of banishment in Turukhan. Obviously a foreign spy! Shoot him! Come on, you reptile, tell us whose pay you're in! Well,

---

circumstances a revolutionary only had to transfer his activities outside the capitals . . . and they would remain an impenetrable secret for the department of police. In this way real nests of revolution and hotbeds of propaganda and agitation were created. . . ."

Our readers will readily grasp the difference between this and the Soviet period. Igor Sazonov, waiting his chance to kill Minister Plehve, disguised himself as a cabby and, with a bomb hidden in his droshky, stood outside the *main entrance of the police department* (!!) for a whole day—and no one took any notice of him or asked him what he was doing! Kalyaev, still inexperienced, spent a whole day on tenterhooks near Plehve's house on the Fontanka, fully expecting to be arrested—and no one touched him! . . . Golden days! . . . In such conditions revolution was easy.

no. For three years' absence abroad the magistrate ordered him to pay a fine of three rubles, or spend one day in the cells! Three rubles was a lot of money, and Ulanovsky preferred one day in detention.

Beginning with attempts to escape from Solovki by sea in some flimsy little boat, or in a hold among the timber, and ending with the insane, hopeless, suicidal breakouts from the camps in the late Stalin period (some later chapters are devoted to them), escape in our time has always been an enterprise for giants among men, but for doomed giants. Such daring, such ingenuity, such will power never went into prerevolutionary escape attempts—yet they were very often successful, and ours hardly ever.

"Because your attempts to escape were essentially reactionary in their class character! . . ."

Can a man's urge to stop being a slave and an animal ever be reactionary?

The reason for their failure was that success depends in the later stages of the attempt on the attitude of the population. And our population was *afraid* to help escapers, or even *betrayed* them, for mercenary or ideological reasons.

"So much for public opinion! . . ."

As for prison mutinies, involving as many as three, five, or eight thousand men—the history of our three revolutions knew nothing of them.

Yet we did.

But the same curse was upon them, and very great efforts, very great sacrifices, produced the most trivial results.

Because society was not ready. Because without a response from public opinion, a mutiny even in a huge camp has no scope for development.

So that when we are asked: "Why did you put up with it?" it is time to answer: "But we didn't!" Read on and you will see that we didn't put up with it at all.

In the Special Camps we raised the banner of the *politicals*—and politicals we became.

Chapter 5

■

Poetry Under a Tombstone, Truth Under a Stone

At the beginning of my camp career I was very anxious to avoid general duties, but did not know how. When I arrived at Ekibastuz in the sixth year of my imprisonment I had changed completely, and set out at once to cleanse my mind of the camp prejudices, intrigues, and schemes, which leave it no time for deeper matters. So that instead of resigning myself to the grueling existence of a general laborer until I was lucky enough to become a trusty, as educated people usually have to, I resolved to acquire a skill, there and then, in *katorga*. When we joined Boronyuk's team (Oleg Ivanov and I), a suitable trade (that of bricklayer) came our way. Later my fortunes took a different turn and I was for some time a smelter.

I was anxious and unsure of myself to begin with. Could I keep it up? We were unhandy cerebral creatures, and the same amount of work was harder for us than for our teammates. But the day when I deliberately let myself sink to the bottom and felt it firm under my feet—the hard, rocky bottom which is the same for all —was the beginning of the most important years in my life, the years which put the finishing touches to my character. From then onward there seem to have been no upheavals in my life, and I have been faithful to the views and habits acquired at that time.

I needed an unmuddled mind because I had been trying to write a poem for two years past. This was very rewarding, in that it helped me not to notice what was being done with my body. Sometimes in a sullen work party with Tommy-gunners barking about me, lines and images crowded in so urgently that I felt myself borne through the air, overleaping the column in my hurry to reach the work site and find a corner to write. At such moments I was both free and happy.[1]

But how could I *write* in a Special Camp? Korolenko tells us that he wrote in jail—but how different his conditions were! He wrote in pencil (why didn't they feel the seams of his clothes and take it away from him?), which he had carried in among his curls (and why wasn't his hair cropped?), wrote among all the noise (he ought to have been thankful that there was room to sit down and stretch his legs!). Indeed, he was so privileged that he could keep manuscripts or send them out (and that is hardest of all for our contemporaries to understand!).

You can't write like that nowadays, even in the camps! (Even saving names for a future novel was dangerous—the membership list of some organization, perhaps? I used to jot down only the etymological root, in the form of a common noun or an adjectival derivative.) Memory was the only hidey-hole in which you could keep what you had written and carry it through all the searches and journeys under escort. In the early days I had little confidence in the powers of memory and decided therefore to write in verse. It was of course an abuse of the genre. I discovered later that prose, too, can be quite satisfactorily tamped down into the deep hidden layers of what we carry in our head. No longer burdened with frivolous and superfluous knowledge, a prisoner's memory is astonishingly capacious, and can expand indefinitely. We have too little faith in memory!

But before you commit something to memory you feel a need to write it down and improve it on paper. In the camps you are allowed to have a pencil and clean paper but may not keep anything *in writing* (unless it is a poem about Stalin).[2] And unless you get a trusty's job in the Medical Section or sponge on the Culture and Education Section, you have to go through the morning and

---

1. Everything is relative! We read of Vasily Kurochkin that the nine years of his life after *Iskra* was closed down were "years of real agony": he was left without a press organ of his own! We who dare not even dream of an organ of our own find him incomprehensible: he had a room, quiet, a desk, ink, paper, there were no body searches, and nobody confiscated what he had written—why, then, the agony?

2. Dyakov describes one instance of such "artistic activity." Dmitrievsky and Chetverikov outlined a projected novel to the authorities and obtained their approval. The security officer saw to it that they were not put on general duties! Later on they were secretly taken out of the camp area ("in case the Banderists tore them to pieces") to continue their work: More poetry under the gravestone. But where *is* the novel?

evening searches at the guardhouse. I decided to write snatches of twelve to twenty lines at a time, polish them, learn them by heart, and burn them. I made it a firm rule not to content myself with tearing up the paper.

In prisons the composition and polishing of verses had to be done in my head. Then I started breaking matches into little pieces and arranging them on my cigarette case in two rows (of ten each, one representing units and the other tens). As I recited the verses to myself, I displaced one bit of broken match from the units row for every line. When I had shifted ten units I displaced one of the "tens." (Even this work had to be done circumspectly: such innocent match games, accompanied by whispering movements of the lips or an unusual facial expression, would have aroused the suspicion of stool pigeons. I tried to look as if I was switching the matches around quite absent-mindedly.) Every fiftieth and every hundredth line I memorized with special care, to help me keep count. Once a month I recited all that I had written. If the wrong line came out in place of one of the hundreds or fifties, I went over it all again and again until I caught the slippery fugitives.

In the Kuibyshev Transit Prison I saw Catholics (Lithuanians) busy making themselves rosaries for prison use. They made them by soaking bread, kneading beads from it, coloring them (black ones with burnt rubber, white ones with tooth powder, red ones with red germicide), stringing them while still moist on several strands of thread twisted together and thoroughly soaped, and letting them dry on the window ledge. I joined them and said that I, too, wanted to say my prayers with a rosary but that in my particular religion I needed one hundred beads in a ring (later, when I realized that twenty would suffice, and indeed be more convenient, I made them myself from cork), that every tenth bead must be cubic, not spherical, and that the fiftieth and the hundredth beads must be distinguishable at a touch. The Lithuanians were amazed by my religious zeal (the most devout among them had no more than forty beads), but with true brotherly love helped me to put together a rosary such as I had described, making the hundredth bead in the form of a dark red heart. I never afterward parted with this marvelous present of theirs; I fingered and counted my beads inside my wide mittens—at work line-up, on the march to and from work, at all waiting times; I could do it standing up, and freezing cold was no hindrance. I carried it safely through the search points, in the padding of my mittens, where it could not be felt. The warders found it on various occasions, but supposed that it was for praying and let me keep it. Until the end of my sentence (by which time I had accumulated 12,000 lines) and after that in my place of banishment, this necklace helped me to write and remember.

Even so, things were not so simple. The more you have written, the more days in each month are consumed by recitation. And the particularly harmful thing about these recitals is that you cease to see clearly what you have written, cease to notice the strong and weak points. The first draft, which in any case you approve in a hurry, so that you can burn it, remains the only one. You cannot allow yourself the luxury of putting it aside for years, forgetting it and then looking at it with a fresh critical eye. For this reason, you can never write really well.

Nor can you hang on to unburned scraps of paper for long. Three times I was caught, and was saved only because I never wrote the most dangerous names in full, but put dashes in their place. Once I was lying on the grass away from everyone else, too near the boundary wire (it was quieter there), and writing, concealing my scrap of paper in a book. Senior Warder Tatarin crept up behind me very quietly and saw that I was not reading, but writing.

"Right, let's have it!" I rose, in a cold sweat, and handed it over. These lines were written on it:

> All we have lost will be made good—
> None of our claims will be denied us.
> The Osterode-Brodnitsy route
> Was five weary days and nights on foot
> With an [escort] of K[azakhs] and T[atars] beside us. . . .

If the words "escort" and "Tatars" had been written in full, Tatarin would have hauled me before the security officer and they would have found me out. But the blanks told him nothing.

> With an ——— of K—— and T—— beside us. . . .

Our minds were running on different lines. I was afraid for the poem, and he had thought that I was making a sketch of the camp area and plotting escape. Still, even what he did find he read with a frown. Certain words seemed to him suggestive. But what really set his brain working furiously was the phrase "five weary days." I had overlooked its possible associations! "Five days" was a set formula in the camp, when prisoners were consigned to the hole.

"Who gets five days? Who's this all about?" he asked, looking black.

I barely managed to convince him (by pointing to the names Osterode and Brodnitsy) that I was trying to remember an army song someone had written, but couldn't recall all the words.

"Why do you want to do that? You aren't here to remember things!" was his surly warning. "If I catch you lying here again, you're in for it!"

When I talk about this incident now, it sounds trivial. But at the time, for a wretched slave like myself, it was an enormous event: I could never again lie on the grass away from all the noise, and if Tatarin caught me with any more verses, they might easily open a new file on me and put me under close surveillance.

But I could not stop writing now!

On another occasion I broke my usual rule. At the work site I wrote down sixty lines of a play[3] at one go, and failed to conceal this piece of paper at the camp entrance. True, I had again left a number of discreet blanks. The warder, a simple, flat-nosed young fellow, examined his catch with some surprise.

"A letter?" he asked.

(A letter taken to the work site had a whiff of the black hole about it. But it would seem a mighty strange letter if they passed it on to the security officer!)

"It's for a concert," I said, brazening it out. "I'm trying to write down a sketch from memory. Come and see it when we put it on."

The young man stared and stared at the paper, and at me, then said:

"You're a bigger fool than you look!"

And he ripped my page into two, four, eight pieces. I was terrified that he would throw the scraps, which were still large, on the ground there in front of the guardhouse, where they might catch the eye of a more vigilant staff member. Chief Disciplinary Officer Machekhovsky himself was only a few steps away, looking on while we were searched. But they evidently had orders not to leave litter by the guardhouse, or they would have to tidy it up themselves, and the warder put the torn-up pieces into my hand as though it were a refuse bin. I went through the gates and made haste to throw them into the stove.

3. *Pir Pobeditelei (Feast of the Victors).*

On a third occasion, while I still had a sizable piece of a poem unburned, I was working on the Disciplinary Barracks and the temptation to put "The Mason" on paper was too strong for me. At that time we never left the camp area, so that we did not undergo daily personal searches. When "The Mason" was three days old, I went out in the dark, before evening inspection, to go over it for the last time and then burn it at once. I was looking for a quiet, lonely spot, which meant somewhere toward the boundary fence, and I forgot entirely that I was near the place where Tenno had recently gone under the wire. A warder who had evidently been lying in ambush grabbed me immediately by the scruff of my neck and marched me through the darkness to the black hole. I took advantage of the darkness to crumple "The Mason" surreptitiously and toss it at random behind me. A breeze was beginning to blow and the warder did not hear the paper crackling and rustling.

I had quite forgotten that I still had another fragment of a poem on me. They found it when they searched me in the Disciplinary Barracks; fortunately it contained almost nothing that could incriminate me (it was a descriptive section from *Prussian Nights*).

The duty officer, a perfectly literate senior sergeant, read it through.

"What's this?"

"Tvardovsky," I answered unhesitatingly. *Vasily Tyorkin.* *

(This was where Tvardovsky's path and mine first crossed!)

"Tvardo-ovsky!" said the sergeant, nodding his head respectfully. "And what do you want it for?"

"Well, there aren't any books. I write down what I can remember and read it sometimes."

They took my weapon—half a razor blade—from me, but returned the poem, and they would have let me go (I wanted to run and find "The Mason"). But by then evening inspection was over and no one was allowed to move about the camp. The warder took me back to the hut and locked me in himself.

I slept badly that night. A gale-force wind had sprung up outside. Where would it carry the little ball of paper with "The Mason" on it? In spite of all the blanks, the sense of the poem remained obvious. And it was clear from the text that its author was in the team building the Disciplinary Barracks. Among all those Western Ukrainians it wouldn't be hard to find me.

So that the work of many years—that already done, and that I was planning—was a scrap of crumpled paper blown helplessly about the camp or over the steppe. I could only pray. When things are bad, we are not ashamed of our God. We are only ashamed of Him when things go well.

At five in the morning, as soon as we rose, I went to the spot, gasping for breath in the wind. It was so strong that it swept up small stones and hurled them in your face. It was a waste of time even looking! From where I was, the wind was blowing in the direction of the staff barracks, then the punishment cells (this place, too, was infested with warders, and there was a lot of tangled barbed wire), then beyond the camp limits, on to the street of the settlement. I prowled around, bent double, for an hour before dawn, and found nothing. By now I was in despair. Then when it got light . . . I saw something white three steps from the place where I had thrown it! The wind had rolled the ball of paper to one side and it had lodged among a pile of boards.

I still consider it a miracle.

So I went on writing. In winter in the warming-up shack, in spring and summer on the scaffolding at the building site: in the interval between two barrowloads of mortar I would put my bit of paper on the bricks and (without letting my neighbors see what I was doing) write down with a pencil stub the verses which had rushed into my head while I was slapping on the last hodful. I lived in a dream, sat in the mess hall over the ritual gruel sometimes not even noticing its taste, deaf to those around me—feeling my way about my verses and trimming them to fit like bricks in a wall. I was searched, and counted, and herded over the steppe —and all the time I saw the sets for my play, the color of the curtains, the placing of the furniture, the spotlights, every movement of the actors across the stage.

Some of the lads broke through the wire in a lorry, others crawled under it, others walked up a snowdrift and over it—but for me the wire might not have existed; all this time I was making my own long and distant escape journey, and this was something the warders could not discover when they counted heads.

I realized that I was not the only one, that I was party to a great secret, a secret maturing in other lonely breasts like mine on the scattered islands of the Archipelago, to reveal itself in years to come, perhaps when we were dead, and to merge into the Russian literature of the future.

(In 1956 I read the first small collection of Varlam Shalamov's poems in samizdat, which existed even then, and trembled as though I had met a long-lost brother. Here he declares his willingness to die like Archimedes during the siege of Syracuse:

> I know, none better, this is not a game—
> Or else a deadly game. But like the sage
> I'll welcome death rather than drop my pen,
> Rather than crumple my half-written page.

He, too, wrote in a camp. Keeping his secret from all around, like me expecting no answer to his lonely cry in the dark:

> A long, long row of lonely graves
> Are all I remember now.
> And I should have laid myself there,
> Laid my bare body down there,
> Had I not taken a vow:
> To sing and to weep to the very end
> And never to heed the pain,
> As though in the heart of a dead man
> Life yet could begin again.

How many of us were there? Many more, I think, than have come to the surface in the intervening years. Not all of them were to survive. Some buried manuscripts in bottles, without telling anyone where. Some put their work in careless or, on the contrary, in excessively cautious hands for safekeeping. Some could not write their work down in time.

Even on the isle of Ekibastuz, could we really get to know each other? encourage each other? support each other? Like wolves, we hid from everyone, and that meant from each other, too. Yet even so I was to discover a few others in Ekibastuz.

Meeting the religious poet Anatoly Vasilyevich Silin was a surprise which I owed to the Baptists. He was then over forty. There was nothing at all remarkable about his face. A reddish fuzz had grown in place of his cropped hair and beard, and his eyebrows were also reddish. Day in and day out he was meek and gentle with everyone, but reserved. Only when we began talking to each other freely, and strolling about the camp for hours at a stretch on our Sundays off, while he recited his very long religious poems to me (like me, he had written them right there in the camp), I was startled not for the first time or the last to realize what far from ordinary souls are concealed within deceptively ordinary exteriors.

A homeless child, brought up an atheist in a children's home, he had come across some religious books in a German prisoner-of-war camp, and had been carried away by them. From then on he was not only a believer, but a philosopher and theologian! "From then on" he had also been in prison or in camps without a break, and so had spent his whole theological career in isolation, rediscovering for himself things already discovered by others, perhaps going astray, since he had never had either books or advisers. Now he was working as a manual laborer and ditchdigger, struggling to fulfill an impossible norm, returning from work with bent knees and trembling hands—but night and day the poems, which he composed from end to end without writing a word down, in iambic tetrameters with an irregular rhyme scheme, went round and round in his head. He must have known some twenty thousand lines by that time. He, too, had a utilitarian attitude to them: they were a way of remembering and of transmitting thoughts.

His sensitive response to the riches of nature lent warmth and beauty to his view of the world. Bending over one of the rare blades of grass which grew illegally in our barren camp, he exclaimed:

"How beautiful are the grasses of the earth! But even these the Creator has given to man for a carpet under his feet. How much more beautiful, then, must we be than they!"

"But what about 'Love not this world and the things that are of this world'?" (A saying which the sectarians often repeated.)

He smiled apologetically. He could disarm anyone with that smile.

"Why, even earthly, carnal love is a manifestation of a lofty aspiration to Union!"

His theodicy, that is to say his justification of the existence of evil in the world, he formulated like this:

> Does God, who is Perfect Love, allow
> This imperfection in our lives?
> The soul must suffer first, to know
> The perfect bliss of paradise. . . .

> Harsh is the law, but to obey
> Is for weak men the only way
> To win eternal peace.

Christ's sufferings in the flesh he daringly explained not only by the need to atone for human sins, but also by God's desire to *feel* earthly suffering to the full.

"God always *knew* these sufferings, but never before had he *felt* them," Silin boldly asserted. Even of the Antichrist, who had

> Corrupted man's Free Will—perverted
> His yearning toward the One True Light

Silin found something fresh and humane to say:

> The bliss that God had given him
> That angel haughtily rejected:
> He nothing knew of human pain;
> He loved not with the love of men—
> By grief alone is love perfected.

Thinking so freely himself, Silin found a warm place in his generous heart for all shades of Christian belief.

> This is the crux:
> That though Christ's teaching is its theme
> Genius must ever speak with its own voice.

The atheist's impatient refusal to believe that spirit could beget matter only made Silin smile.

"Why don't they ask themselves how crude matter could beget spirit? That way round, it would surely be a miracle. Yes, a still greater miracle!"

My brain was full of my own verses, and these fragments are all that I have succeeded in preserving of the poems I heard from Silin—fearing perhaps that he himself would preserve nothing. In one of his poems, his favorite hero, whose ancient Greek name I have forgotten, delivered an imaginary speech at the General Assembly of the United Nations—a spiritual program for all mankind. A doomed and exhausted slave, with four number patches on his clothes, this poet had more in his heart to say to living human beings than the whole tribe of hacks firmly established in journals, in publishing houses, in radio—and of no use to anyone except themselves.

Before the war Anatoly Vasilyevich had graduated from a teachers' college, where he had specialized in literature. Like me, he now had about three years left before his "release" to a place of banishment. His only training was as a teacher of literature in schools. It seemed rather improbable that ex-prisoners like us would be allowed into schools. But if we were—what then?

"I won't put lies into children's heads! I shall tell the children the truth about God and the life of the Spirit."

"But they'll take you away after the first lesson."

Silin lowered his head and answered quietly: "Let them."

And it was obvious that he would not falter. He would not play the hypocrite just to go on handling a class register rather than a pickax.

I looked with pity and admiration at this unprepossessing man with ginger hair, who had known neither parents nor spiritual directors, for whom life was no harder in Ekibastuz, turning over the stony soil with a spade, than it always had been elsewhere.

Silin ate from the same pot as the Baptists, shared his bread and warm victuals with them. Of course, he needed appreciative listeners, people with whom he could join in reading and interpreting the Gospel, and in concealing the little book itself. But Orthodox Christians he either did not seek out (suspecting that they would reject him as a heretic), or did not find: there were few of them in our camp except for the Western Ukrainians, or else they were no more consistent in their conduct than others, and so inconspicuous. The Baptists, however, seemed to respect Silin, listened to him; they even considered him one of their own: but they, too, disliked all that was heretical in him, and hoped in time to bend him to their ways. Silin was subdued when he talked to me in their presence, and blossomed out when they were not there—it was difficult for him to force himself into their mold, though their faith was firm, pure, and ardent, helping them to endure *katorga* without wavering, and without spiritual collapse. They were all honest, free from anger, hard-working, quick to help others, devoted to Christ.

That is why they are being rooted out with such determination. In the years 1948–1950 several hundred of them were sentenced to twenty-five years' imprisonment and dispatched to Special

Camps *for no other reason* than that they belonged to Baptist communes (a commune is of course an *organization*).[4]

■

The camp is different from the Great Outside. Outside, everyone uninhibitedly tries to express and emphasize his personality in his outward behavior. In prison, on the contrary, all are depersonalized—identical haircuts, identical fuzz on their cheeks, identical caps, identical padded jackets. The face presents an image of the soul distorted by wind and sun and dirt and heavy toil. Discerning the light of the soul beneath this depersonalized and degraded exterior is an acquired skill.

But the sparks of the spirit cannot be kept from spreading, breaking through to each other. Like recognizes and is gathered to like in a manner none can explain.

You can understand a man better and more quickly if you know at least a fragment of his biography. Here are some trench diggers working side by side. Thick, soft snow has begun to fall. Perhaps because it is time for a break, the whole team goes into the dugout for shelter. But one man remains standing outside. At the edge of the trench, he leans on his spade and stands quite motionless, as though he found it comfortable, or as though he were a statue.

The snowflakes gathering on his head, shoulders, and arms make him still more like a statue. Doesn't he care? Does he even like it? He stares through the flurry of snowflakes—at the camp, at the white steppe. He has broad bones, broad shoulders, a broad face, with a growth of stiff blond bristles. He is always deliberate, slow-moving, very calm. He remains standing there—looking at the world and thinking. He is elsewhere.

I do not know him, but his friend Redkin tells me his story. This man is a Tolstoyan. He grew up with the antiquated notion that a man may not kill (even in the name of the Vanguard Doctrine!) and must not, therefore, take up arms. In 1941 he was called up. Near Kushka, to which he had been posted, he threw his gun away and crossed the Afghan border. There were no Germans around Kushka and none were expected, so that he could have had a quiet

4. The persecution became less severe in the Khrushchev period only in that shorter sentences were given: otherwise it was just as bad. (See Part VII.)

war, never shooting at a living thing, but even lugging that piece of iron around on his back went against his convictions. He supposed that the Afghans would respect his right not to kill people and let him go through to India, where there was religious tolerance. But the Afghan government turned out to be as cynically self-interested as governments always are. Fearing the wrath of its all-powerful neighbor, it put the runaway in the stocks. And kept him in prison, with his legs cramped in the stocks, unable to move, for three years, waiting to see which side would win. The Soviets did—and the Afghans obligingly returned the deserter to them. His present sentence was reckoned only from that date.

Now he stood motionless out in the snow, like part of the landscape. Had he been brought into the world by the state? Why, then, had the state usurped the right to decide how this man should live?

We don't mind having a fellow countryman called Lev Tolstoi. It's a good trademark. (Even makes a good postage stamp.) Foreigners can be taken on trips to Yasnaya Polyana. We are always ready to drool over his opposition to Tsarism and his excommunication (the announcer's voice will tremble at this point). But, my dear countrymen, if someone takes Tolstoi seriously, if a real live Tolstoyan springs up among us—hey, look out there! Mind you don't fall under our caterpillar tracks!

. . . Perhaps on the building site you run and ask the foreman, a prisoner, for his folding ruler, to measure how much wall you've laid. He sets great store by this ruler, and doesn't know you by sight—there are so many teams on the job—but for some reason he hands you his treasure without demur (sheer stupidity, to the camp-trained mind). And when you actually return the ruler to him, *he* will thank *you* very warmly. How can such a weird character be a foreman in a camp? He has an accent. Ah, yes— he turns out to be a Pole, his name is Jerzy Wegierski. You will hear more of him.

. . . Perhaps you are marching along in the column, and you ought to be telling your beads inside your mitten, or thinking about your next stanzas—but you find yourself very much interested by your neighbor in the ranks, a new face. (They have just sent a new brigade to your work site.) An elderly, likable Jewish intellectual with a mocking, intelligent expression. His name is Masamed, and he is a university graduate. From . . . ? Bucharest, Faculty of Biology and Psychology. He is, among other things, a physiognomist and graphologist. Moreover, he is a yogi, and will start you on a course in Hatha Yoga tomorrow. (That's the pity of it: our term in this university is so short! I can never get my breath! There's no time to take it all in.)

Later on I took a good look at him in the work zone and the living area. His fellow countrymen had offered to fix him up with an office job, but he wouldn't take it: it was important to him to show that Jews, too, make excellent general laborers. So at the age of fifty he fearlessly wields his pickax. Like a true yogi, he really is master of his own body: at minus 10 degrees Centigrade he strips and asks his workmates to hose him down from the fire hydrant. Unlike the rest of us, who shovel that wretched gruel into our mouths as fast as we can, he eats slowly, with concentration, looking away from his plate, swallowing a little at a time from a tiny spoon of his own.[5]

It does happen, and not so very rarely, that you make an interesting new acquaintance on the way to or from work. But you can't always get going in the column: with the guards shouting, and your neighbors hissing ("because of you . . . us as well . . ."), on the way to work you're too sluggish, and on the way back in too much of a hurry, and very likely there's a wind to shut your gob. Yet suddenly . . . well, of course these are untypical cases, as the socialist realists say. Quite exceptional cases.

In the outermost line walks a little man with a thick black beard (because he had it when he was last arrested, and was photographed with it, it has not been shaved off in the camp). He walks briskly, very much on his dignity, carrying a carefully tied roll of draftsman's paper under his arm. This is a "rationalization proposal" or invention of his, some new thing of which he is proud. He drew it at work, brought it to show to somebody in the camp, and is now taking it with him back to work. Suddenly a mischievous wind plucks the roll from under his arm and bowls it along away from the column. Arnold Rappoport (the reader already knows him) instinctively takes a step in pursuit of it, a second, a third . . . but the roll drifts on farther, between two guards, beyond the escort! This is where Rappoport should stop—"One step to right or left," remember, "and you get it without warning." But that's mine, my drawing, it's over there! And Arnold trots after it, bending forward, arms extended—an evil fate is carrying off his idea! Arnold reaches out eagerly, his hands rake the air. Barbar-

5. Nonetheless, he was to die soon, like an ordinary mortal, of an ordinary heart attack.

ian! Don't touch my blueprints! The column sees him, stumbles to a halt of its own accord. Guns are pointed, bolts click back! . . . Everything so far was typical, but now came something untypical: no one acted like a fool! No one fired! The barbarians realized that this was not an escape! Even to their befuddled brains this was an immediately comprehensible scene: the author pursuing his fleeing creation! Rappoport ran on another fifteen paces or so beyond the escort guards, caught his roll, straightened up, and returned to the ranks very pleased with himself. Returned . . . from the next world.

Although Rappoport had landed more than the average camp stint (after a kid's sentence and a *tenner* had come banishment, and now he was doing another *tenner*), he was full of life, agile, bright-eyed—and those eyes of his, although they were always merry, were made for suffering, were very expressive eyes. It was a matter of pride to him that years of prison had not aged and broken him. As an engineer, he had, however, always worked as a trusty on some production job, so that it was easy for him to keep his spirits up. He took a lively interest in his work, but over and beyond it he created for his soul's sake.

He was one of those versatile characters eager to embrace everything. At one time he was thinking of writing a book like this of mine, all about the camps, but he never got around to it. Another of his works made us, his friends, laugh: Arnold had for some years been patiently compiling a universal technical reference book, which would cover all the ramifications of modern science and technology (everything from types of radio valves to the average weight of elephants), and which was to be . . . pocket-size. The wiser for this laughter, Arnold showed me another of his favorite works in secret. Finding Stendhal's treatise *On Love* completely unsatisfactory, he had written a new one, in a glossy black exercise book. It consisted of unpolished, and for the present disconnected, remarks. For a man who had spent half his life in the camps, how chaste it all was! Here are some brief extracts.[6]

To possess a woman without love is the unhappy lot of the poor in body and spirit. Yet men boast of it as a "conquest."

Possession, without the preliminary organic development of feeling, brings not joy but shame and revulsion. The men of our age, who devote all their energy to making money, to their jobs, to the exercise of power,

6. This was all many years ago. Rappoport later abandoned his treatise, and I have his permission to quote it.

have lost the gene of higher love. On the other hand, woman's unerring instinct tells her that possession is only the first stage toward genuine intimacy. Only after it does a woman acknowledge a man as near and dear, and show it in her way of speaking to him. Even a woman who gives herself unintentionally feels an access of grateful tenderness.

Jealousy is injured self-esteem. Real love, unrequited, is not jealous but dies, ossifies.

Love, as much as science, art, and religion, is a mode of cognition.

Combining as he did such very different interests, Arnold Lvovich naturally knew the most various people. He introduced me to a man whom I should have passed by without noticing: at first sight he was just another of the walking dead, doomed to die of malnutrition, with collarbones sticking out from his unbuttoned camp jacket like those of a corpse. His lankiness made him even more astonishingly thin. He was naturally swarthy, and his shaven head had become still darker in the Kazakh sun. He still dragged himself out of the camp to work, still clung to his barrow to stay on his feet. He was a Greek and, once more, a poet! Yet another! A volume of his verse in modern Greek had been published in Athens. But since he was not an Athenian but a Soviet prisoner (and a Soviet citizen), our newspapers shed no tears for him.

He was only middle-aged, although so close to death. I made pitifully clumsy attempts to wave these thoughts away. He laughed wisely, and explained to me in imperfect Russian that not death itself, but only the moral preparation for it, holds terrors. He had finished with fear and grief and regret, shed all his tears, already lived through his inevitable death and was quite ready. It only remained for his body to finish dying.

So many people turn out to be poets! It's almost unbelievable. (Sometimes I was at a loss to understand it.) While the Greek waits to die, here are two young men waiting only for the end of their sentence and for future literary fame. They are poets— openly, without concealment. What they have in common is a certain radiant purity. Both are students who never graduated. Kolya Borovikov, an admirer of Pisarev (and therefore an enemy of Pushkin), works as a clinical orderly in the Medical Section. Yurochka Kireyev, from Tver, who admires Blok, and himself writes in the manner of Blok, goes out of camp to work in the office of the engineering shop. His friends (and what strange friends they are for him—twenty years older and fathers of families) tease him about the time in a Corrective Labor Camp up north when some generally available Rumanian woman offered herself to him, but he didn't understand and wrote sonnets to her. If you look at his innocent face you can very easily believe it. Now his virginity is a curse and a burden to be carried all through the camps!

. . . Sometimes people catch your eye, sometimes you catch theirs. In the big chaotic barracks where four hundred men live, moving restlessly around or lying on their bunks, I read after supper and during the tedious inspections the second volume of Dal's dictionary—the only book which I had brought as far as Ekibastuz, where I was forced to see it defaced with a stamp saying "Steplag, Culture and Education Section." I never flipped through the pages, because in the fag end of the evening I could hardly read through half a page. So I sat, or shuffled along for the inspection, engrossed in one particular passage. I was used to newcomers' asking what that fat book was, and wondering why the devil I wanted to read it. I used to answer with a joke. "It's the safest thing to read. No danger of catching a new sentence."[7]

But that book brought me a lot of interesting acquaintances, too. For instance, a small man like a bantam cock, with a fierce nose and a sharp mocking look, comes up to me and says in a singsong northern accent:

"May I inquire what book you have there?"

We exchange a few words and then as Sunday follows Sunday, as month follows month, in this one man a microcosm in which half a century of my country's history is densely packed opens out before me. Vasily Grigoryevich Vlasov (the one who was in the Kady trial,* and has now got through fourteen long years of his twenty) thinks of himself as an economist and politician, and has no idea that he is an artist in words—in the spoken word. Whether he tells me about haymaking or merchants' shops (he had worked in one as a boy), a Red Army unit or an old country house, an executioner from the Provincial Deserter Interception Organization or an insatiable woman in some small town, I see it all in the

---

7. But what is it *not* dangerous to read in a Special Camp? Aleksandr Stotik, an economist in the Dzhezkazgan Camp Division, used to read an adaptation of *The Gadfly* in the evening on the quiet. In spite of his secrecy he was denounced. The camp commander himself, and a pack of officers, came to join in the search. "Waiting for the Americans?" They made him read aloud in English. "How much longer are you in for now?" "Two years." "Make it twenty!" They also found some verses. "Interested in love, are you? . . . Make him so uncomfortable that not only his English but even his Russian will evaporate!" (What's more, the slavish trusties hissed at Stotik: "You'll land us in it, too! They'll drive us all back to work again.")

round and absorb it as thoroughly as if it had been part of my own experience. I wanted to write it all down at once—but it couldn't be done. I wish I could remember it word for word ten years later, but it is impossible.

I noticed that one man often stole glances at me and my book, but hesitated to start a conversation—a thin, fine-drawn, long-nosed young man whose politeness, diffidence even, seemed strange in those surroundings. I got to know him, too. He spoke in a quiet, shy voice, groping for words in Russian, and making hilarious mistakes, which he immediately redeemed with a smile. It emerged that he was Hungarian, and his name was Janos Rozsas. He nodded when I showed him Dal's dictionary, with my eyes on his shriveled, camp-worn face. "Yes, yes," he said, "a man must distract his attention to other things, not think about food all the time." He was only twenty-five, but there was no youthful flush in his cheeks; the dry, papery skin, made transparent by the winds, seemed to be stretched over the long, narrow bones of his skull with no flesh between. His joints ached—he had caught rheumatic fever, felling trees in the north.

There were two or three of his compatriots in the camp, but all day and every day they were obsessed with one thought only—how to survive and how to eat their fill. Whereas Janos ate whatever the foreman obtained for him, and even though he remained hungry, he made it a rule not to look for more. He was all eyes and ears; he wanted to understand. Understand what, you ask? . . . Us. He wanted to understand us Russians!

"My personal fate became very uninteresting when I got to know people here. I am most extremely surprised. They loved their own people, and for that they get *katorga*. But I think—this is wartime muddle, yes?" (He asks this in 1951! If it was still a wartime muddle then, maybe it dated from the First World War? . . .)

In 1944, when our troops captured him in Hungary, he was eighteen (and not in the army). "I had still had no time to do people either good or evil," he said, with a smile. "People had seen neither profit nor harm from me." Janos's interrogation went like this: the interrogator didn't understand a word of Hungarian, nor Janos a word of Russian. Occasionally some very bad interpreters came along, Carpathian Ukrainians. Janos signed a sixteen-page statement, with no idea what was in it. Nor, when an unknown officer read him something from a piece of paper, did he realize that this was the Special Board's sentence.[8] They sent him to the North, to fell trees, and there he *went under* and landed in the hospital.

Till then Russia had shown him only one side of herself (the part used for sitting on), but now she turned the other. In the camp hospital at the Symsk Separate Camp Site near Soli-kamsk there was a forty-five-year-old nurse called Dusya. She was a nonpolitical offender, not a part of the professional un-derworld, was in for five years, and had a pass. Her job as she saw it was not just to grab all she could and get through her sentence (a very common assumption in the camps, though Janos with his rosy view of things did not know it), but to look after these useless, dying people. What the hospital gave was not enough to save them. So Dusya used to exchange her morning ration of 300 grams for half a liter of milk in the vil-lage, and with this milk she nursed Janos back to life (as she had nursed others before him).[9] For this motherly woman's sake Janos came to love Russia and all Russians. He started diligently learning, there in the camp, the language of his warders and convoy guards, the great, the mighty Russian lan-guage. He spent nine years in our camps, and saw nothing of Russia except from prison trains, on little picture postcards, and in the camp. Yet he loved it.

Janos belonged to a breed which is steadily becoming rarer in our time: those whose only passion in childhood is reading. He kept this inclination as an adult, and even in the camps. In the north, and in the Ekibastuz Special Camp, he missed no opportu-nity to obtain and read new books. By the time I met him he already knew and loved Pushkin, Nekrasov, and Gogol. I ex-plained Griboyedov to him. But more than anyone else, more perhaps than even Petöfi and Arany, he came to love Lermontov, whom he had first read in captivity, not long before.[10] Janos iden-tified himself particularly with Mtsyri*—like himself a prisoner,

8. It is said that when Janos was rehabilitated after Stalin's death, curiosity prompted him to ask for a copy of the sentence in Hungarian, so that he could find out just why he had spent nine years in prison. But he was afraid to do so. "They may wonder what I want to do with it. And anyway, I don't really need it all that much. . . ." He had understood our way of thinking: why, indeed, did he need to know now? . . .

9. Can someone explain to me what ideology this behavior fits into? (Compare the Communist medical orderly in Dyakov: "Toothache, eh, you pig-faced Ukrainian bandit!")

10. I have been told more than once by foreigners that Lermontov is dearer to them than any other Russian poet. After all, they point out, Pushkin did write "To Russia's Slander-ers." Whereas Lermontov never did Tsarism the smallest service.

young and doomed. He had much of it by heart, and for years on end, as he dragged himself along with his hands behind his back, in a column of foreigners through an alien land, he would murmur to himself in the strangers' tongue:

> I knew then in my troubled mind
> My eager foot would no more find
> The native land I'd left behind.

Friendly, affectionate, with vulnerable pale-blue eyes—that was Janos Rozsas in our heartless camp. He would perch on my bunk, lightly, on the very edge, as though my sack of sawdust could be made any dirtier, or would lose its shape under his weight, and say in a low voice full of genuine feeling:

"With whom should I share my secret dreams?"

He never complained about anything.[11]

You move among the camp population as you would through a minefield, photographing each of them with the rays of intuition for fear of blowing yourself up. Yet in spite of this general rule of caution, how often I discovered a poetic personality under a zek's shaven skull and black jacket.

How many others kept their secret to themselves?

How many more—surely a thousand times as many—did not come my way at all?

11. All the Hungarians were allowed to go home after Stalin's death, so Janos escaped the fate of Mtsyri, for which he was fully prepared.

Twelve years have passed, 1956 among them. Janos is a bookkeeper in the little town of Nagy Kanizsa, where nobody knows Russian or reads Russian books. And what does he now write to me?

"After all that has happened, I can sincerely say that I would not give back my past. I learned in a harsh school what others can never know. . . . When I was freed I promised the comrades who stayed behind that I would never forget the Russian people, not for their sufferings but for their good hearts. Why do I follow newspaper reports about my former "motherland" with such interest? The works of the Russian classics are a whole shelf in my library, I have forty-one volumes in Russian and four in Ukrainian (Shevchenko). . . . Other people read the Russians as they read the English or the Germans, but I read them differently. To me Tolstoi is closer than Thomas Mann, and Lermontov much closer than Goethe.

"You cannot guess how much I miss Russia without talking about it. Sometimes people ask me what kind of crank I am, what good did the place ever do me, why do I feel drawn to Russians? How can I explain that all my youth went by there, and that life is an eternal farewell from the swiftly passing days. . . . How could I turn my back like a sulky child —for nine years my fate coincided with yours. How can I explain why my heart misses a beat when I hear a Russian folk song on the radio? I start singing to myself under my breath. 'See the reckless troika speeding . . .' But it is too painful for me to go on. My children ask me to teach them Russian. Wait awhile, children; for whom do you think I collect Russian books?"

And how many, in those decades, did you smother, infamous Leviathan?

■

Ekibastuz also had an official, though very dangerous, center for cultural intercourse—the Culture and Education Section, where they stamped black words on books and freshened up our numbers.

An important and very colorful figure in our CES was Vladimir Rudchuk, now an artist, but formerly an archdeacon, and possibly even personal secretary to the Patriarch. There is somewhere in the camp rules an unexpunged proviso that persons in holy orders shall not be shorn. Of course, this rule is never made public, and priests who do not know of it are shorn. But Rudchuk knew his rights, and he was left with wavy auburn locks, unusually long for a man. He took great care of them, and of his appearance generally. He was tall, well-made, attractive, with a pleasant bass voice, and it was easy to imagine him conducting a solemn service in a huge cathedral. Drozdov, the churchwarden who arrived at Ekibastuz with me, recognized the Archdeacon at once: he had formerly served in the cathedral church at Odessa.

Here in the camp he neither looked nor lived like a man of our convict world. He was one of those dubious personages who had insinuated themselves, or been insinuated, into the Orthodox Church as soon as they were no longer in official disfavor: They did much to bring the Church into disrepute. The story of how Rudchuk came to be in jail was also rather mysterious: for some reason he kept exhibiting a photograph (inexplicably left in his possession) of himself with Anastasi, the Metropolitan of the Russian Orthodox Church in Exile, on a New York street. He had a *cabin* to himself in the camp. When he got back from work line-up, where he disdainfully painted the numbers on our caps, jackets, and trousers, he would spend the day in idleness, occasionally making crude copies of tasteless pictures. He was allowed to keep a volume of reproductions from the Tretyakov Gallery, and it was because of these that I found my way to him: I wanted to take another look, perhaps my last. He had the *Bulletin of the Moscow Patriarchate* sent to him in the camp, and sometimes discoursed pompously on the great martyrs, or on details of the liturgy, but it was all affectation, all insincere. He also had a guitar, and the only thing he ever did sincerely was to sing, in a pleasant voice, to his own accompaniment. Even then his writhing was intended to suggest that he was haloed with a convict's crown of sorrow.

The better a man lives in the camp, the more exquisite his suffering. . . .

I was cautious to the nth degree in those days. I never called on Rudchuk again, I told him nothing about myself, and as a harmless insignificant worm escaped his sharp eye. Rudchuk's eye was the eye of the MGB.

Anyway, every old hand in the camps knew that any CES was riddled with informers, and the least suitable place imaginable to seek new acquaintances and company. In mixed Corrective Labor Camps the CES attracted prisoners as a meeting place for men and women. But in *katorga*, what reason could there be for going there?

It turned out that even a CES in *katorga* (full of informers though it was) could serve the cause of freedom. I learned this from Georgi Tenno, Pyotr Kishkin, and Zhenya Nikishin.

It was in the CES that I met Tenno, and that single brief meeting stayed clearly in my mind because Tenno himself was so memorable. He was a tall, slender man of athletic build. For some reason, they still hadn't stripped him of his naval tunic and breeches (some prisoners were wearing their own clothes for one last month). And although instead of the shoulder tabs of a captain second class he wore in various places the number SKh 520, he was every inch the naval officer, ready to step aboard at any moment. When his movements bared his arms above the wrists you saw little reddish hairs and tattoo marks: on one arm the word "Liberty" around an anchor, on the other "Do or die" [in English]. Tenno simply could not lower his lids or squint, to hide the sharpness of his eyes and the pride in them. And another thing he could not hide was his big, bright smile. (I did not yet know it, but that smile meant: My plan of escape is drawn up!)

That's the camp for you—a minefield! Tenno and I were both there, yet elsewhere: I was on the roads of East Prussia,* he on the route of his next escape. Each of us was charged with secret thoughts, but as we shook hands and exchanged casual words, not the smallest spark could leap from palm to palm or eye to eye! We said something unimportant, I buried myself in my newspaper, and he began discussing amateur performances with Tumarenko, a political in for fifteen years, but nonetheless in charge of the CES, a complicated man of many layers: I thought I knew the answer to him, but had no means of verifying it.

Ridiculous as it may seem, in *katorga* there was also a concert group attached to the CES, or rather in process of formation! Membership conferred none of the privileges of such groups in Corrective Labor Camps, exempted a man from nothing, so that only incurable enthusiasts would ever want to join. Among them, it appeared, was Tenno, though from his looks you would have thought better of him. Besides, he had been in the punishment cells ever since his arrival in Ekibastuz—it was from there that he had volunteered for the CES! The authorities interpreted this as an earnest of amendment and permitted him to attend. . . .

Petya Kishkin was no CES activist but the most famous man in the camp. All Ekibastuz knew of him. Any work site was proud to have him—with Kishkin around, no one could be bored. He appeared to be crazy, and was anything but. Though he acted the fool, everybody said, "He's cleverer than the lot of us." He was the same sort of fool as the "simple" youngest brother in a fairy tale! The Kishkins are a phenomenon of great antiquity in Russia: they loudly tell the truth to the wicked and powerful, they make the people see themselves as they are, and all this by foolery which involves no risk to themselves.

One of his favorite turns was dressing like a clown in a funny green waistcoat and collecting the dirty bowls from the tables. This in itself was a demonstration: the most popular man in the camp gathering dirty dishes so as not to die of hunger. A second reason for doing it was that while he was jigging and clowning around the tables, always the center of attention, he rubbed shoulders with the working convicts and sowed mutinous thoughts.

He would suddenly snatch a bowl from the table with the mush yet untouched—the prisoner was still only sipping his broth. The startled prisoner would grab at the bowl and Kishkin would dissolve in smiles (he had a moon face, but with a certain hardness in it): "Till somebody touches your mush, you never grasp anything."

And he dances lightly away with his mountain of dishes.

Before the day is out, Kishkin's latest joke will be going the rounds in other work teams, too.

Another time he leans out across the table, and the rest all look up at him from their plates. Rolling his eyes like a toy cat, with a buffoonish look on his face, Kishkin says:

"Listen, lads! If the father's a fool and the mother a whore, will the children be fed or go hungry?"

Without waiting for the obvious answer, he points at the litter of fish bones on the table, and says:

"Divide seven to eight billion poods per annum by two hundred million!"

And off he goes. It was so simple—why had none of us thought of it before? It had been reported long ago that we were harvesting eight billion poods of grain a year, which meant two kilograms of bread a day for everyone, including babes in arms. Right, we're grown men, making holes in the ground the whole day long—where's our share?

Kishkin varied his material. Sometimes he would put the same thought the other way around—with a lecture on anything and everything. When the column was waiting before the guardhouse, at the camp or the work site, and talking was allowed, he made use of the time to harangue us. One of his regular slogans was "Educate your faces!" "I walk around the camp and I look at you: you all have such uneducated faces. Can't think of anything except their barley cake."

Or he would suddenly, for no obvious reason, shout at a crowd of zeks: "Dardanelle! Tommyrot!" It seemed to make no sense. But after one or two shouts everybody clearly understood *who* "Dardanelle" was, and it seemed so apt and so funny that you almost saw the menacing mustache on his face. "Dardanelle!"

One of the bosses barks at Kishkin outside the guardhouse, trying to get a rise out of him for a change. "How come you're baldheaded, Kishkin, you so-and-so? Got dry rot, have you?" Without a minute's hesitation, Kishkin answered so that all the crowd could hear: "Is that what made Vladimir Ilyich bald, then?"

Or he might go around the mess hut announcing that after the dishes had been collected he would teach the goners* to do the Charleston.

Suddenly—a great surprise. A film had arrived. It was shown in the evening, in the same old mess hut, without a screen, on the whitewashed wall. The hut was filled to overflowing—there were people sitting on benches, on tables, between benches, on top of each other. But before it had run far, the film was stopped. A blank white beam played on the wall, and we saw that several warders had come in and were looking for the most comfortable places. They selected a bench and ordered the prisoners to vacate it. The prisoners decided to stay where they were—it was years since they'd seen a film! The warders' voices grew more threatening, and somebody said, "Right, take their numbers!" That was that; they had to give way. Then suddenly a familiar mocking voice, like the screech of a cat, was heard through the darkened hall:

"Now, lads, really, you know the warders can't see a film anywhere else—let's move."

They exploded in laughter. What a force is laughter! All the power belonged to the warders, but they beat an inglorious retreat, without taking numbers.

"Where's Kishkin?" they shouted.

Not another sound out of Kishkin; Kishkin was missing.

The warders went away, and the film continued.

Next day Kishkin was called before the disciplinary officer. He'd get five days, for sure! No, he came back smiling. He had given a written explanation as follows: "During the *argument* between warders and prisoners about seats for the film show, I called upon the prisoners to give up their seats, as they are supposed to, and to move away." What had he done to be put in the hole?

The prisoner's irrational passion for shows, his ability to forget himself, his grief, and his humiliation for a scrap of nonsense, on film or live, insultingly showing that all's right with the world, was another subject for Kishkin's skillful satire. Before a film show or concert, would-be spectators flock together like sheep. Time goes by and still the door stays shut, while they wait for the head warder to bring his lists and admit the best work teams. They wait half an hour, slavishly pressed together, crushing each other's ribs. Kishkin, at the back of the crowd, slips off his shoes, vaults with a hand from his neighbors onto the shoulders of those in front, and passes quickly, nimbly from shoulder to shoulder over the whole crowd—right up to the gate of paradise! He knocks, his short body writhes from head to foot in a pantomime of impatience to enter. Then he runs back just as quickly over all those shoulders, and hops down. At first the crowd just laughs. But then they feel deeply ashamed. Standing here like sheep! Right, let's give it a miss.

They disperse. When the warder comes with the lists, nobody is trying to break the door down; in fact, there's hardly anyone to let in. If he wants an audience he'll have to round them up.

Another time a concert was just beginning in the spacious mess hall. They were all in their seats. Kishkin wouldn't think of boycotting a concert. There he was in his green waistcoat, fetching and carrying chairs, helping to draw back the curtain. Each of his appearances drew applause and friendly shouts from the hall. Suddenly he runs across the front of the stage as though someone were pursuing him and, waving a warning hand, shouts, "Dardanelle! Tommyrot!" Roars of laughter. Then there was some sort of hitch. The curtain was up, but the stage was deserted. Kishkin dashed onto the stage. The audience started laughing, but immediately fell silent: he looked now no longer comic but insane, there was a wild glare in his eyes, he was terrifying. He declaimed a poem, trembling, gazing around unseeingly.

> I look and, ah, the sight I see:
> The police rain blows, the blood flows free—
> Streets littered with the dead and dying,
> Son beside father murdered lying. . . .

This was for the Ukrainians, who were half of the audience. They had only lately been brought from seething provinces, and it was like salt on a fresh wound. They howled. A warder was rushing at the stage and Kishkin. But Kishkin's tragic face relaxed in a clown's grin. He shouted, in Russian this time, "When I was in fourth grade we learned that poem for the Ninth of January!"*

And he left the stage, hobbling absurdly.

Zhenya Nikishin was a nice, simple, sociable lad, with an open freckled face. (There were many like him in the countryside, before its destruction. Nowadays you see mainly hostile expressions there.) Zhenya had a small voice, and liked singing for his friends, in the hut or from the stage.

One day "Wife of Mine, Little Wife" was announced. "Music by Mokrousov, words by Isakovsky. Performed by Zhenya Nikishin, with guitar accompaniment."

A sad, simple melody trickled from the guitar. And Zhenya faced that large hall and sang to each of us, showing us how much warmth and tenderness there still was in our hearts.

> Wife, oh, little wife,
> Dearer far than life—
> You and you alone live in my heart!

You and you alone! The long platitudinous slogan about output plans up above the stage grew dim. In the blue-gray half-light of the hall, the long years of camp life—years already lived through, years still remaining—faded. You and you alone! Not the crimes we were alleged to have committed, nor our reckoning with authority. Not our wolfish preoccupations. . . . You and you alone!

> None could be more dear,
> None could be so near,
> Whether we're together or apart.

It was a song about endless waiting for news that never comes. About loneliness and despair. How appropriate it was. Yet prison was never directly mentioned. It could all equally well refer to a lengthy war.

Though I was an underground poet, my instinct failed me: I did not realize at the time that the verses ringing from the stage were those of another underground poet (how many of them are there?!), but a more flexible one than I, better equipped to reach his public.

What could they do to him? Send for the music sheet, check whether it really was Isakovsky and Mokrousov? He had probably said that he could do it from memory.

In the blue-gray dimness sat or stood some two thousand men. They were so quiet and still that they might not have been there at all. Men calloused, brutalized, turned to stone—but now touched to the heart. Tears, it appeared, could still break through, still find a way.

> Wife, oh, little wife,
> Dearer far than life—
> You and you alone live in my heart!

Chapter 6

∎

The Committed Escaper

When Georgi Pavlovich Tenno talks nowadays about past escapes—his own, those of comrades, and those of which he knows only by repute—his words of praise for the most uncompromising and persistent heroes—Ivan Vorobyov, Mikhail Khaidarov, Grigory Kudla, Hafiz Hafizov—are these:

"There was a *committed* escaper!"

A committed escaper! One who never for a minute doubts that a man cannot live behind bars—not even as the most comfortable of trusties, in the accounts office, in the Culture and Education Section, or in charge of the bread ration. One who once he lands in prison spends every waking hour thinking about escape and dreams of escape at night. One who has vowed never to resign himself, and subordinates every action to his need to escape. One for whom a day in prison can never be just another day; there are only days of preparation for escape, days on the run, and days in the punishment cells after recapture and a beating.

A committed escaper! This means one who knows what he is undertaking. One who has seen the bullet-riddled bodies of other escapers on display along the central tract. He has also seen those brought back alive—like the man who was taken from hut to hut, black and blue and coughing blood, and made to shout: "Prisoners! Look what happened to me! It can happen to you, too!" He knows that a runaway's body is usually too heavy to be delivered to the camp. And that therefore the head alone is brought back in a duffel bag, sometimes (this is more reliable proof, according to the rulebook) together with the right arm, chopped off at the elbow, so that the Special Section can check the fingerprints and write the man off.

A committed escaper! It is for his benefit that window bars are set in cement, that the camp area is encircled with dozens of strands of barbed wire, towers, fences, reinforced barriers, that ambushes and booby traps are set, that red meat is fed to gray dogs.

The committed escaper is also one who refuses to be undermined by the reproaches of the average prisoner: You escapers make it worse for the rest! Discipline will be stiffer! Ten inspections a day! Thinner gruel! He ignores the whispered suggestions of other prisoners—not only those who urge resignation ("Life's not so bad even in a camp, especially if you get parcels"), but those who want him to join in protests or hunger strikes, because all that is not struggle but self-deception. Of all possible means of struggle, he has eyes only for one, believes only in one, devotes himself only to one—escape!

He cannot do otherwise! That is how he is made. A bird cannot renounce seasonal migration, and a committed escaper cannot help running away.

In the intervals between unsuccessful attempts, peaceful prisoners would ask Tenno: "Why can't you just sit still? Why do you keep running? What do you expect to find on the Outside—especially now?" Tenno was amazed. "What d'you mean—what do I expect to find? Freedom, of course! A whole day in the taiga without chains—that's what I call freedom!"

Gulag and the Organs had known no prisoners like him or Vorobyov in their *middle* period—the age of the chicken-hearted. Such prisoners came along only in the very early days or after the war.

That was Tenno for you. In each new camp (he was transferred frequently) he was depressed and miserable until his next escape plan matured. Once he had a plan, Tenno was radiant, and a smile of triumph never left his lips.

In fact, he recalls that when the general review of sentences and the rehabilitations began, he was dismayed: he felt the hope of rehabilitation sapping his will to escape.

There is no room in this book for his complicated life story. But the urge to escape had been with him from birth. As a small boy he had run away from boarding school in Bryansk to "America" —down the Desna in a rowboat. He had climbed the iron gates of the Pyatigorsk orphanage in his underwear in midwinter, and run away to his grandmother. He was a very unusual amalgam of sailor and circus performer. He had gone through a school for seamen, served before the mast on an icebreaker, as boatswain on a trawler, as navigation officer in the merchant navy. He had graduated from the army's Institute of Foreign Languages, spent the war with the Northern Fleet, sailed to Iceland and England as liaison officer with British convoys (Plate No. 4). But he had also, from his childhood on, practiced acrobatics; he had appeared in circuses during the NEP* period, and later in the intervals between voyages; had trained gymnasts on the beam, performed as a memory man (memorizing masses of words and figures) and as a mind reader. The circus, and living in seaports, had led to some slight contact with the criminal world: he had picked up something of their language, their adventurousness, their quick-wittedness, their daredeviltry. Later on, serving time with thieves in numerous Disciplinary Barracks, he had absorbed more and more from them. This, too, would come in handy for the committed escaper.

A man is the product of his whole experience—that is how we come to be what we are.

In 1948 he was suddenly demobilized. This was a signal from the other world (he knew languages, had sailed on an English vessel, and was, moreover, an Estonian, though it is true a Petersburg Estonian), but if we are to live we must hope against hope. On Christmas Eve that year in Riga, where Christmas still feels like Christmas, like a holiday, he was arrested and taken to a cellar on Amatu Street, next door to the conservatory. As he entered his first cell he couldn't resist the temptation to tell the apathetically silent warder, "My wife and I had tickets for *The Count of Monte Cristo* and should be watching it right now. He fought for freedom, and I shall never accept defeat."

But it was too early yet to start fighting. We are always at the mercy of our assumption that a *mistake* has been made. Prison? For what? It's impossible! *They'll soon get it sorted out.* Indeed, before his transfer to Moscow they deliberately reassured him (this is done as a safety measure when prisoners are in transit). Colonel Morshchinin, chief of counterespionage, even came to the station to see him off and shook hands with him. "Have a good journey!" There were four of them, Tenno and his special escort, and they traveled in a separate first-class compartment. When the major and the first lieutenant had finished talking about all the fun they would have in Moscow on New Year's Eve (perhaps special escort duties are merely an excuse for such trips?), they lay down on the upper bunks and appeared to be sleeping. On the other lower bunk lay a chief petty officer. He stirred whenever the prisoner opened his eyes. There was dim light from a blue bulb overhead. Under Tenno's pillow was his first, and last, hastily made parcel from his wife—a lock of her hair and a bar of chocolate. He lay and thought. The rhythm of the carriage wheels was soothing. We can fill their rattle with any meaning, any prophecy we please. It filled Tenno with hope that they would "get it sorted out." And so he had no serious intention of running away. He was only sizing up the best way to do it. (Later on he would often remember that night and cluck with annoyance. Never again would it be so easy to run, never again would freedom be so near!)

Twice in the course of the night Tenno went out along the deserted corridor, and the petty officer went with him. He had his pistol slung low, as sailors always do. He even squeezed into the lavatory together with the prisoner. For a master of judo and wrestling it would have been child's play to *pin him* there and then take his gun from him, order him to keep quiet, and calmly leave the train when it stopped.

The second time the petty officer was afraid to go into that narrow place, and waited outside the door. But the door was shut, and Tenno could have stayed there as long as he liked. He could have broken the window and jumped out onto the tracks. It was night! The train was not moving quickly—this was 1948—and it made frequent stops. True, it was winter, and Tenno had no overcoat and only five rubles on him, but his watch had not yet been taken away.

The luxury of a special escort came to an end at the station in Moscow. They waited for all the passengers to leave the train, and then the sergeant major with light-blue shoulder tabs who had brought the prison van came in and said, "Where is he?"

The admission routine, sleepless nights, solitary confinement, more solitary confinement. A naïve request to be called for interrogation soon. The warder yawned. "Don't be in such a hurry; you'll get more than you want shortly."

At last, the interrogator. "Right, tell me about your criminal activities." "I'm absolutely innocent!" "Only Pope Pius is absolutely innocent."

In his cell he was tête-à-tête with a stool pigeon. Trying to *box him in*. Come on, tell me what really happened. A few interrogations and it was all quite clear: they'd never straighten it out, never let him go. So he must escape!

The world fame of the Lefortovo Prison did not daunt Tenno. Perhaps he was like a soldier new to the front who has experienced nothing and therefore fears nothing? It was the interrogator, Anatoly Levshin, who inspired Tenno's escape plan. By turning mean and arousing his hatred.

People and peoples have different criteria. So many millions had endured beatings within those walls, without even calling it torture. But for Tenno the realization that he could be beaten with impunity was intolerable. It was an outrage, and he would sooner die than suffer it. So when Levshin, after verbal threats, first advanced on him and raised his fist, Tenno jumped up and answered with trembling fury: "Look, my life's worth nothing anyway! But I can gouge one or both of your eyes out right now! That much I can do!"

The interrogator retreated. One rotten prisoner's life in exchange for a good eye was not much of a bargain. Next he tried to wear Tenno down in the punishment cells, to sap his strength. Then he put on a show, pretending that a woman screaming with pain in the next office was Tenno's wife, and that if he did not confess she would undergo still worse tortures.

Again he had misjudged his man! If a blow from a fist was hard for Tenno to bear, the idea that his wife was being interrogated was no less so. It became increasingly obvious to the prisoner that the interrogator must be killed. This and his escape were combined in a single plan. Major Levshin, too, wore naval uniform, was tall and fair-haired. As far as the sentry on the interrogation block was concerned, Tenno could very easily pass for Levshin. True, Levshin's face was round and sleek whereas Tenno had grown thin. (It wasn't easy for a prisoner to get a look at himself in a mirror. Even if you asked to go to the lavatory when you were under interrogation, the mirror there was draped with a black curtain. Once he saw his chance, made one quick movement and twitched the curtain aside. God, how pale and worn out he looked! How sorry for himself he felt!)

In the meantime they had removed the useless stoolie from the cell. His bed was left there and Tenno examined it. A metal crosspiece was rusted through at the point where it was fixed to one of the legs. It was about 70 centimeters long. How could he break it off?

First he must . . . perfect his skill in counting seconds precisely. Then calculate for each warder the interval between two peeps through the spy hole. (You had to put yourself in the place of whichever warder was on duty, as he strolled at his own pace along the corridor.) The interval varied between 45 and 65 seconds.

During one such interval he tried his strength, and the metal bar cracked off at the rusted end. Breaking the other, solid end was harder. He would have to stand on it with both feet—but then it would crash onto the floor. So in the interval between two visits he must make time to put a pillow on the cement floor, stand on the bed frame, break it, replace the pillow, and hide the bar for the time being; say, in his bed. And all the time he must be counting seconds.

It broke. The trick was done!

But the problem was only half solved: if they came in and found it, he would be rotting in the punishment cells. Twenty days of that and he would lose the strength he needed to escape, or even to defend himself against the interrogator. Yes, that was it: he would tear the mattress with his fingernails. Extract a little of the flock. Wrap flock around the ends of the bar, and put it back where it had been. Counting the seconds! Right—it was there!

But this was still good only for a short time. Once every ten days you went to the bathhouse, and while you were away your cell was searched. They might discover the breakage. So he must act quickly. How was he to take the bar from the cell to the interrogation room? When they let you out of the cellblock there was no search. They only slapped your clothing when you came back from interrogation, and then only your sides and chest, where there were pockets. They were looking for a blade, to prevent suicide.

Under his naval jacket Tenno wore the traditional sailor's striped jersey—it warms body and soul alike. "The sailor leaves his troubles ashore." He asked a warder for a needle (they will give you one at certain fixed times), as if to sew on buttons made of bread. He undid his jacket, undid his trousers, pulled out the edge of his jersey, and turned it up and stitched it so that it formed a little pocket (for the lower end of the rod). He had previously snapped off a bit of tape from his underpants. Now he pretended to be sewing a button on his jacket and stitched this tape to the inside of his jersey at chest level, so that it formed a loop to hold the rod steady.

Next he put the jersey on back to front, and began practicing day after day. The rod was set in position down his back and under his jersey: it was pushed through the loop at the top until it rested in the pocket down below. The upper end of the rod came up to his neck, under his tunic collar. His training routine went like this: In the short time between two inspections he would have to fling his hand to the back of his neck, seize the end of the rod while bending his trunk backward, then with a reverse movement straighten like a released bowstring while simultaneously drawing the rod—and strike the investigating officer a smart blow on the head. Then he would put everything back in place. An eye at the spy hole. The prisoner would be leafing through a book.

The movement became quicker and quicker, until the rod fairly whistled through the air. If the blow was not fatal, the investigating officer would certainly be knocked out. If they had arrested his wife, too, he would show none of them mercy!

He also provided himself with two wads of flock, from the same mattress. These he could insert between his gums and his cheeks to make his face fuller.

He must also, of course, be clean-shaven on the day—and they scraped you with blunt razors only once a week. So that the day must be chosen carefully.

How was he to put some color into his cheeks? He would rub just a little blood on them. *That fellow's* blood.

An escaper cannot use his eyes and ears idly as other people do. He must look and listen for his own special purpose. He must let no trifle pass him by without comment. Wherever he is taken—to interrogation, to the exercise yard, to the lavatory— his feet count their steps, count stairs (not all of this will be useful, but they count anyway); his body notes the turns; he keeps his eyes on the ground as ordered, and they examine the floor (what is it made of? is the surface unbroken?), they search his surroundings as far as he can see, inspecting all doors (double or single? what sort of handles? what sort of locks? do they open inward or outward?); his mind assesses the function of every door; his ears listen and make comparisons (that's a sound

I've heard before from my cell; now I know what it means).

The famous K-shaped block of the Lefortovo Prison has one main stairway to all floors, metal galleries, a controller who sticks little flags on a chart. You cross into the interrogation block. The interrogators change rooms according to a roster. So much the better—you can study the layout of all the corridors and the position of the doors in the interrogation block. How do the interrogators get into the building? Through that door with the square window. The main document check is of course carried out not here but in the guardhouse outside, but here, too, they sign themselves in or are scrutinized. Listen. One man goes downstairs and shouts to somebody up above: "Right, I'm off to the ministry!" Splendid; that sentence could be useful to an escaper.

As to the rest of the route from here to the guardhouse, he would have to make a good guess and take the right way without hesitation. But no doubt a path had been worn in the snow. Or the asphalt would be darker and dirtier. How did they get past the guard? By showing an identification card? Or did they leave their cards on entry and give their names to reclaim them? Or perhaps they were all known by sight, and it would be a mistake to give a name instead of just holding your hand out?

You could find the answer to many things if you observed the interrogator closely, instead of attending to his silly questions. To sharpen his pencil he takes a razor blade from inside a little book, perhaps a personal document, which he keeps in his breast pocket. Questions immediately ask themselves.

"That's not his pass. Is his pass in the guardhouse?"

"That little book looks very much like a driver's license. So he comes by car? He must have his car key, then. Does he park outside the prison gates? I shall have to read the license number in his logbook before I leave the office or I may make for the wrong one."

They have no cloakroom. He hangs his overcoat and cap here in the office. So much the better.

Mustn't forget anything important, and must pack it all into four or five minutes. When he's lying there, knocked out, I must:

1. Slip off my jacket and put on his newer one with shoulder tabs.
2. Remove his shoelaces and lace my own floppy shoes up—that will take time.

3. Tuck his razor blade into a specially prepared place in the heel of my shoe (if they catch me and sling me into the nearest cell I can cut my veins).

4. Examine all his documents and take what I need.

5. Memorize the license plate number and find the ignition key.

6. Shove my own dossier into his bulky briefcase and take it with me.

7. Remove his watch.

8. Redden my cheeks with blood.

9. Drag his body behind the desk or screen, so that anybody coming in will think he's left and not raise a hue and cry.

10. Roll the flock into little balls and put them in my cheeks.

11. Put on his coat and cap.

12. Disconnect the wires to the light switch. If anybody comes soon afterward, finds it dark, and tries the switch, he will be sure to think that the bulb has burned out and that's why the interrogator has gone to another office. Even if they screw another bulb in, they won't immediately realize what has happened.

That makes twelve things to be done, and the escape itself will be number thirteen. . . . All this must be done during the night session. It won't be so good if the little book is not a driver's license. That will mean that he comes and goes by a special bus for interrogators (there must be special transport in the middle of the night), and the others will think it strange that Levshin couldn't wait till four or five o'clock but went off on foot in the middle of the night.

Something else to remember: when I go through the door with the square window I must raise my handkerchief to my face as though blowing my nose, and simultaneously turn my head to look at my watch. And to set the sentry's mind at rest I'll call upstairs, "Perov!" (That's his friend.) "I'm off to the ministry! We'll have a talk tomorrow!"

Of course, the odds were against him. For the moment, he gave himself a 3 to 5 percent chance of success: the outer guardroom was completely unknown and he had no real hope of getting past it. But he couldn't die there like a slave! Couldn't feebly submit to kicks! At least he would have the razor blade in the heel of his shoe!

So Tenno turned up, freshly shaven, for one nocturnal interrogation, with the iron bar behind his back. The interrogator questioned, abused, threatened, and all the time Tenno looked at him in surprise: couldn't he sense that his hours were numbered!

It was eleven o'clock. Tenno's plan was to sit tight till two in the morning. Interrogators sometimes wangled themselves a short night and began leaving about that time.

Now he must seize the right moment: either wait for the interrogator to bring some pages over for signature, as he always did, suddenly pretend to feel faint, let the pages slide to the floor, cause him to bend his head for a minute, and . . . Or else, without waiting for the papers, stand up, swaying, plead illness, ask for water. When he brought the enamel mug (the glass was reserved for his own use), Tenno would drain it, drop it, simultaneously raising his right hand to the back of his neck, which would seem quite natural since he was supposed to be dizzy; the interrogator would be bound to look down at the mug on the ground, and . . .

Tenno's heart thumped. A day of rejoicing was at hand. Or perhaps his last day.

Things turned out quite differently. Around midnight another interrogator hurried into the room and began whispering in Levshin's ear. This had never happened before. Levshin made hurried preparations to leave, pressed the button for the warder to come and remove the prisoner.

That was that. Tenno went back to his cell and replaced the iron bar.

Another time the interrogator sent for him when he was unshaven, and there was no point in taking the bar with him.

Then came a daytime interrogation. And it took a strange turn: the interrogator refrained from yelling, and weakened his resolve by predicting that he would get five to seven years, so that there was no need to be downhearted. Somehow Tenno no longer felt angry enough to split his head open. Tenno's wrath was not the sort that lasted.

The mood of high excitement had passed. It seemed to him now that the odds were too great, that it was too much of a gamble.

The escaper's moods are perhaps even more capricious than those of the artist.

All his lengthy preparations had gone for nothing. . . .

But the escaper must be ready for this, too. He had brandished his bar in the air a hundred times, killed a hundred interrogators.

A dozen times he had lived through every minute of his escape in detail—in the office, past the square window, along to the guard-room, beyond the guardroom. He had worn himself out with an escape which he would after all not be making.

Soon afterward they changed his interrogator and transferred him to the Lubyanka. There Tenno did not actively prepare to escape (his heart was not in it now that his interrogation seemed to have taken a more hopeful turn), but he was tirelessly observant, and he devised a training routine.

Escape from the Lubyanka? Is it even possible? If you think about it, it is perhaps easier than escaping from Lefortovo. You soon begin to know your way around those long, long corridors through which you are taken to interrogation. In the corridor you sometimes come across arrows on the wall: "To main entrance No. 2," "To main entrance No. 3." (You are sorry that you were so thoughtless when you were free—that you didn't walk around the outside of the Lubyanka to see where each of the entrances was.) It's easier here precisely because this is not just the territory of a prison but a ministry, where there are large numbers of interrogators and other officials whom the guards cannot know by sight. So that entry and exit is by pass, and the interrogator has his pass in his pocket. Again, if an interrogator is not known by sight, it is not so important to look exactly like him; a rough resemblance is good enough. The new interrogator wears khaki, not naval dress. So that it will be necessary to change into his uniform. No iron bar this time—but if the will is there you can manage. There are all sorts of suitable objects in his office—a marble paperweight, for instance. Anyhow, you needn't necessarily kill him—just stun him for ten minutes and you're away!

But vague hopes of clemency and reasonableness clouded Tenno's resolution. Only in the Butyrki Prison was he relieved of this burden: his sentence, read out from a piece of paper with a Special Board stamp, was confinement in camps for twenty-five years. He signed his name and felt relieved, found himself smiling, felt his legs carrying him easily to the cell for twenty-five-year prisoners. That sentence released him from humiliation, from the temptation to compromise, from humble submission, from truckling, from promises of five to seven years bestowed like alms on a beggar. Twenty-five is it, you bastards? Right; if that's all we can expect from you—we escape!!

Or die. But death was surely no worse than a quarter of a

century of slavery. Even the shaving of his head after his trial—
just an ordinary convict crop, never upset anybody—outraged
Tenno, as though they had spat in his face.

Now he must seek allies. And study the history of other escapes.
Tenno was a novice in this world. Someone must have tried to
escape before him.

How often we had all followed the warder through those iron
bulkheads which divide the corridors of the Butyrki into sections
—yet how many of us have noticed what Tenno saw at once: each
door had two locks, but swung open when the warder undid only
one of them. This meant that the second lock was for the moment
not in use: it consisted of three prongs which could emerge from
the wall and slide into the iron door.

Other people in the cell might talk about what they liked, but
Tenno wanted stories about escape attempts and those who took
part in them. There was even one prisoner—Manuel García—who
had been in a riot and seen the three prongs used. It had happened
a few months earlier. The prisoners in one of the cells had been
let out to relieve themselves, had seized the warder (although it
was against the rules that he was alone—there had been no trouble
for years, and they were used to submissiveness!), stripped him,
tied him up, and left him in the latrine, while one of the prisoners
put on his uniform. The lads took the keys and ran around opening all the doors on the corridor (they couldn't have chosen a
better place—some of the prisoners on that corridor were under
sentence of death!). They started shouting, whooping with joy,
calling to each other to go and liberate other corridors and take
over the whole prison. They were oblivious of the need for caution.
They should have stayed in their cells quietly preparing for flight,
allowing only the prisoner disguised as a warder to walk the
corridor, but instead they poured out in a noisy crowd. Hearing
all the noise, a warder from the next corridor looked through the
two-way spy hole in the iron door and pressed the alarm button.
When the alarm is given the second lock in each of the corridor
doors is turned from a central control point, and there is no key
to it on warders' key rings. The mutinous corridor was cut off. A
large body of prison guards was called out. They stood in facing
ranks, let the mutineers through one by one and beat them up,
identified the ringleaders and led them away. These men already
had a *quarter* each. Was the sentence doubled? Or were they shot?

In transit to the camp. The "watchman's cabin," so well known to prisoners, at the Kazan station—at a certain distance, of course, from busy public places. Here prisoners are brought in plain vans and the prison cars are loaded before they are coupled to the trains. Tense escort troops line the tracks on both sides. Dogs strain to be at someone's throat. The order is given: "Escort—at the ready!" and there is a deadly rattle of breechblocks. These people really mean it. The dogs go with them when they lead the prisoners along the tracks. Make a break for it? If you do a dog will catch you.

(But for the committed escaper, who is continually shunted from camp to camp, from jail to jail because of his attempts to escape, the future holds many such stations, many marches under escort along the tracks. And sometimes he will be marched along without dogs. Pretend to be lame and sick, scarcely able to drag your duffel bag and jacket behind you, and the escort will be more at ease. If there are several trains standing on the tracks—you might be able to get lost among them. That's it: drop your things, bend down, and hurl yourself under a railroad car. But as soon as you bend over you will see the boots of an extra guard striding along on the other side of the train. . . . Every contingency has been foreseen. All you can do is to pretend that weakness has caused you to fall and drop your things. Unless you are lucky enough to find a through train passing swiftly alongside! Run across the tracks right in front of the engine—no guard will run after you! You can risk your life for freedom, but why should he risk his? By the time the train flashes by you have gone! But for this you need two strokes of luck: the train must come just at the right time, and you must get past its wheels in one piece.)

From the Kuibyshev Transit Prison they were taking prisoners to the station in open trucks—making up a long train of red prison cars. In the transit prison Tenno obtained from a local sneak thief who "respected escapers" two local addresses to which he might go for initial support. He shared these addresses with two other would-be runaways and they concerted a plan; all three would try to sit in the back row and when the truck slowed down at a turning (Tenno had made the inward journey in a dark van and although his eyes would not recognize the turning, his sides had taken note of it) they would jump, all three of them at once—right, left, and rear—past the guards, knocking them over if necessary. The guards would open fire, but they would not hit all three. They might not shoot at all—there would be people in the streets.

Would they give chase? No, they couldn't abandon the other prisoners in the truck. So they would just shout and fire into the air. If the runaways were stopped, it would be by ordinary people, our Soviet people, passers-by. To frighten them off, the runaways must pretend to be holding knives! (They had no knives.)

The three of them maneuvered at the search point and hung back so that they would get onto the last truck and not leave before dusk. The last truck arrived, but . . . it was not a shallow three-tonner, like its predecessors, but a Studebaker with high sides. When he sat down even Tenno found that the top of his head was below the rim. The Studebaker moved quickly. Here was the turn! Tenno looked around at his comrades in arms. There was terror on their faces. No, they wouldn't jump. No, they were not committed escapers. ("But can you be sure of yourself?" he wondered.)

In the dark, with lanterns to light their way, to a confused accompaniment of barking, yelling, cursing, clanking, they were installed in cattle cars. Here Tenno let himself down—he was too slow to inspect the outside of the car (and your committed escaper must see everything while the seeing is good; he is not allowed to miss anything at all!).

At stops the guards anxiously sounded the cars with mallets. They sounded every single plank. They were afraid of something, then—but of what? Afraid that a plank might be sawn through. So that was the thing to do!

A small piece broken off a hacksaw and sharpened was produced (by the thieves). They decided to cut through a solid plank under the bottom bed shelf. Then when the train slowed down, to lower themselves through the gap, drop onto the line, and lie still until the cars had passed over them. True, the experts said that at the end of a cattle train carrying prisoners there was usually a *drag*—a metal scraper, with teeth which passed close to the ties, caught the body of anyone trying to escape, and dragged him over the ties to his death.

All night long they took turns slipping under the bed shelf and sawing away at a plank in the wall, gripping the blade, which was only a few centimeters long, with a piece of rag. It was hard going. Nonetheless, the first breach was made. The plank began to give a little. Loosening it, they saw in what was now the morning light white, unplaned boards outside their car. Why white? The reason was that an additional footboard for guards had been built onto their car. Right there, by the breach they had made, stood a sentry. It was impossible to go on sawing till the board came away.

Prison escapes, like all forms of human activity, have their own history, and their own theory. It's as well to know about them before you try your own hand.

The history is that of previous escapes. The security branch publishes no popular pamphlets on escape technology—it stores experience only for its own use. You can learn the history from others who once escaped and were recaptured. Their experience has been dearly bought—with blood, with suffering, almost at the cost of their lives. But to inquire in detail, step by step, about the attempts of one escaper, then a third, then a fifth, is no laughing matter; it can be very dangerous. It is not much less dangerous than asking whether anyone knows whom you should see about joining an underground organization. Stoolies may listen in to your long conversations. And worst of all, the narrators themselves, under torture after an attempted flight, forced to choose between life and death, may have lost their nerve and gone over to the other side, so that now they are live bait rather than fellow spirits. One of the godfathers' main tasks is to determine in good time who sympathizes with escape attempts or takes an interest in them—to forestall the lurking would-be escaper, make an entry in his dossier; from then on he'll be in a disciplinary squad and escape will be much more difficult.

Still, as he moved from prison to prison, camp to camp, Tenno eagerly interrogated escapers. He carried out escapes himself, he was caught, he had other escapers for cellmates in the camp jails —and that was his chance to question them. (Sometimes he made mistakes. The heroic escaper Stepan ──── sold him to the Kengir security officer Belyaev, who repeated to Tenno all the questions he had asked.)

As for the theory of escape—it is very simple. You do it any way you can. If you get away—that shows you know your theory. If you're caught—you haven't yet mastered it. The elementary principles are as follows. You can escape from a work site or you can escape from the living area. It is easier from work sites: there are many of them, the security measures are less rigid, and the escaper has tools to hand. You can run away alone—it is more difficult, but no one will betray you. Or you can run away in a group, which is easier, but then everything depends on whether you are a well-matched team. Theory further prescribes that you should know the geography as well as if you had an illuminated map in front of you. But you will never catch sight of a map in the camp. (The thieves, incidentally, are completely ignorant of geography: they take as north the transit prison where they felt cold last time through.) A further precept: you must know the people through whose region your escape route lies. Then there is the following general advice as to method: you must constantly prepare to escape according to plan, but be ready at any minute to do it quite differently, to seize a *chance*.

Here is an example of opportunism. At Kengir once, all the prisoners in the Disciplinary Barracks were marched out to make mud bricks. Suddenly they were hit by one of those dust storms which are so frequent in Kazakhstan: it grows darker and darker, the sun is hidden, handfuls of dust and small stones lash your face so painfully that you cannot keep your eyes open. Nobody was ready to run at such short notice, but Nikolai Krykov rushed to the boundary fence, flung his jerkin onto the barbed wire, scrambled across, scratching himself all over, and hid just outside the camp area. The storm passed. The jerkin on the wire told them that he had escaped. They sent out a mounted search party; the riders had dogs on leashes. But the cold storm had swept scent and tracks clean away. Krykov sat out the search in a pile of rubbish. Next day, however, he had to move on! And the motor vehicles sent to scour the steppe picked him up.

Tenno's first camp was Novorudnoye, near Dzhezkazgan. Now you're in the very place where they have doomed you to die. This is the place of all places from which you must escape! All around there is desert—salt flats and dunes, or firmer ground held together by tufted grass or prickly camel weed. In some parts of the plain Kazakhs roam with their herds, in others there is not a soul. There are no rivers, and you are very unlikely to come upon a well. The best time for flight is April or May, while melting snow still lingers here and there in puddles. But the camp guards are very well aware of this. At this time of year the search of prisoners going out to work becomes stricter, and they are not allowed to take with them a single bite or a single rag more than is necessary.

That autumn, in 1949, three runaways, Slobodyanyuk, Bazichenko, and Kozhin, risked a dash to the south: their idea was to walk along the river Sary Su to Kyzyl Orda. But the river had dried up completely. When they were caught they were nearly dead of thirst.

Taught by this experience, Tenno decided that he would not make his escape in autumn. He went along to the Culture and Education Section regularly—to show that he was no runaway, no rebel, but one of those rational prisoners who hope to mend their ways by the time their twenty-five-year sentence is up. He helped in every way he could, promised to perform his acrobatic stunts and his memory-man turn at camp concerts, and in the meantime went through every bit of paper in the Culture and Education Center until he found a rather poor map of Kazakhstan which the godfather had carelessly left around. Right. There was an old caravan route to Dzhusaly, 350 kilometers away, and there would quite probably be a well along the way. Then he could go northward 400 kilometers, toward Ishim; here there would perhaps be water meadows. Whereas in the direction of Lake Balkhash lay the Betpak-Dala—500 kilometers of unrelieved desert. But pursuit was unlikely in that direction.

Those were the distances. That was the choice.

The strangest ideas force their way into the mind of the inquisitive escaper. A sewage truck sometimes called at the camp—a tank with a suction pipe. The mouth of the pipe was wide. Tenno could easily crawl through it, stand up inside the tank, with his head bent, and then the driver could take in the liquid sewage as long as he didn't fill up to the top. You would be covered with filth, you might choke, drown, suffocate on the way—but this seemed less revolting to Tenno than slavishly serving out his sentence. He examined himself. Was he game? He was. What about the driver, though? He was a minor offender, serving a short sentence, and with an exit permit. Tenno had a smoke with him and looked him over. No, he was not the right man. He wouldn't risk his pass to help someone else. He had the mentality of the Corrective Labor Camp: only fools help other people.

In the course of that winter Tenno devised a plan and also picked himself four comrades. But one day while the plan was still being patiently worked out, as theory requires, he was unexpectedly marched out to a newly opened work site—a stone quarry. It was in a hilly spot, and invisible from the camp. As yet there were no watchtowers and no security fence: just stakes knocked into the ground and a few strands of wire. At one point there was a gap in the wire—which served as a gate. Six guards stood outside the wire, with nothing to raise them above ground level.

Beyond them was the April steppe, its grass still fresh and green, and a blaze of tulips as far as the eye could see! Those tulips, that April air, were more than the heart of an escaper could bear! Perhaps this was his *chance?*... While you're still not suspected, not yet in the Disciplinary Barracks—now's the time to run!

During his time there Tenno had got to know a lot of people in the camp and he now quickly assembled a team of four: Misha Khaidarov (he had been with the marines in North Korea, had crossed the 38th parallel to avoid a court-martial; not wishing to spoil the good relations firmly established in Korea, the Americans had handed him back and he had got a *quarter*); Jazdik, a Polish driver from the Anders army (he vividly summarized his life story with the help of his unmatching boots—"one from Hitler, one from Stalin"); and, lastly, Sergei, a railwayman from Kuibyshev.

Then a lorry arrived with real posts and rolls of barbed wire for a boundary fence—just as the dinner break was beginning. Tenno's team, loving forced labor as they did, especially when it was to make their prison more secure, volunteered to unload the lorry in the rest period. They scrambled onto the back. But since it was, after all, dinnertime, they took their time while they thought things over. The driver had moved away from his vehicle. The prisoners were lying all over the place, basking in the sun.

Should they run for it or not? They had nothing ready—no knife, no equipment, no food, no plan. But Tenno knew from his little map that if they were driving they must make a dash for Dzhezdy and then to Ulutau. The lads were eager to try it: this was their *chance!* Their lucky chance!

From where they were to the sentry at the "gate," the way was downhill. Just beyond the gate the road rounded a hill. If they drove out fast they'd soon be safe from marksmen. And the sentries could not leave their posts!

They finished unloading before the break was over. Jazdik was to drive. He jumped off, and puttered about the lorry while the other three lazily lay down in the rear, out of sight—hoping that some of the sentries hadn't seen where they had got to. Jazdik brought the driver over. We haven't kept you waiting—so let's have a smoke. They lit up. Right, wind her up! The driver got into the cab, but the engine obstinately refused to start. (The three in the back of the lorry didn't know Jazdik's plan and thought their attempt had misfired.) Jazdik began turning the crank. Still the engine would not start. Jazdik was tired and he suggested to the driver that they change places. Now Jazdik was in the cab. And the engine immediately let out a roar! The lorry rolled down the slope toward the sentry at the gate. (Jazdik told them later that he had tampered with the throttle while the driver was at the wheel, and quickly turned it on again before he himself took over.) The driver was in no hurry to jump in; he thought that Jazdik would stop the lorry. Instead it passed through the "gate" at speed.

Two shouts of "Halt!" The lorry went on. Sentries opened fire —shooting into the air at first, because it looked very much like a mistake. Perhaps some shots were aimed at the lorry—the runaways couldn't tell; they were lying flat. Around a bend. Once behind the hill they were safe from bullets. The three in the back kept their heads down. It was bumpy, they were traveling fast. Then—suddenly—they came to a stop and Jazdik cried out in despair: he had taken the wrong turn and they were pulled up short by the gates of a mine, with its own camp area and its own watchtowers.

More shooting. Guards ran toward them. The escapers tumbled out onto the ground face downward and covered their heads with their hands. Convoy guards kick, aiming particularly at the head, the ears, the temples, and, from above, at the spine.

The wholesome universal rule "Don't kick a man when he's down" did not apply in Stalin's *katorga!* If a man was down, that's just what they did—kicked him. And if he was on his feet, they shot him.

But the inquiry revealed that *there had been no breakout!* Yes! The lads said in unison that they'd been dozing in the back when the lorry started moving, then there was shooting and it was too late for them to jump off in case they were shot. And Jazdik? He was inexperienced, couldn't handle the lorry. But he'd steered for the mine next door, not for the steppe.

So they got off with a beating.[1]

---

1. Misha Khaidarov was to make many other attempts to escape. Even in the easiest years of the Khrushchev period, when habitual escapers were lying low and waiting to be released legally, he and his pals who had no hope of pardon would try to escape from the All-Union special prison Andzyoba-307: accomplices would throw homemade grenades under the watchtowers to distract the attention of the guards while the escapers tried to hack through the wire of the inner camp area with axes. But they would be kept back by machine-gun fire.

Preparations for *planned escape* take their own course. To make a compass: take a plastic container and mark the points on it. Break a bit off a spoke, magnetize it, and mount it on a wooden float. Then pour in water. And that's your compass. . . . Drinking water can conveniently be poured into an inner tube, which the escaper will carry like a greatcoat roll. All these things (together with food and clothing) are carried gradually to the woodworking plant, from which the escape is to be made, and hidden in a hole near the band saw. A free driver sells them an inner tube. Filled with water, this too now lies in the hole. Sometimes trains arrive by night and the loaders are left at the work site to deal with them. That's when they must run for it. One of the free employees, in return for a sheet brought out from the camp area (best prices paid!), has already cut the two lower strands of wire near the band saw, and the night for unloading timber is getting closer and closer! But one prisoner, a Kazakh, tracks them to the hole they use as a hiding place and denounces them.

Arrest, beatings, interrogations. In Tenno's case there were too many "coincidences" which looked like preparations to escape. They were sent off to the Kengir jail, and Tenno was standing face to the wall, hands behind his back, when the captain in charge of the Culture and Education Section went by, stopped near him, and exclaimed:

"Who'd have thought it of you! And you a member of the concert party!"

What most amazed him was that a peddler of prison-camp culture should want to escape. On concert days he was allowed an extra portion of mush—and yet he had tried to run away! Some people are never satisfied!!

On May 9, 1950, the fifth anniversary of victory in the Fatherland War, naval veteran Georgi Tenno entered a cell in the celebrated Kengir Prison. The cell was almost dark, with only one little window high up, and there was no air, but there were plenty of bugs—the walls were covered with splotches of bug blood. That summer a heat wave was raging, with temperatures between 40 and 50 degrees centigrade, and everyone lay around naked. It was a little cooler under the sleeping platform, but one night two prisoners shot out from there with a yell: poisonous spiders had perched on them.

It was a select company in the Kengir jail, brought together from various camps. In every cell there were experienced escapers, hand-picked champions. Tenno had found his committed escapers at last!

Among the prisoners was Captain Ivan Vorobyov, Hero of the Soviet Union. During the war he had been with the partisans in the Pskov oblast. He was a resolute man of indomitable courage. He had already made unsuccessful attempts to escape, and would make others. Unfortunately, he could not take on the jailbird coloring, the half-caste look which is so helpful to a runaway. He had preserved his soldier's straightforwardness; he had a chief of staff and they sat on the bed platform drawing a map of the locality and openly discussing plans. He could not adjust to the sly, furtive ways of the camps, and was invariably betrayed by stool pigeons.

A plan fermented in their heads: to overpower the warder supervising the issue of the evening meal if he came alone. Then open all the cells with his keys. Rush to the jailhouse exit and take control of it. Then open the jailhouse door and mob the camp guardroom. Take the guards along as prisoners and break out of the camp area as soon as darkness fell. Later they were taken out to work on a housing site, and a plan for escaping through the sewage system was born.

But these plans were never implemented. Before the summer was out this whole select company was manacled and transported for some reason to Spassk. There they were put into a hut with a separate security system. On the fourth night the committed escapers removed the bars from a window, got out into the service yard, noiselessly killed a dog, and tried to cross a roof to the huge main camp area. But the iron roof bent under their feet, and the noise in the quiet of night was like thunder. The warders gave the alarm. But when they arrived inside the hut, everyone was peacefully sleeping and the bars were back in place. The warders had simply imagined it all.

They were destined never, never to remain long in one place! The committed escapers, like Flying Dutchmen, were driven ever onward by their troubled destiny. If they didn't run away, they were transferred. This whole band of men in a hurry was switched, in handcuffs, to Ekibastuz camp jail. There the camp's own unsuccessful runaways—Bryukhin and Mutyanov—were added to their strength.

As part of their special punitive regime they were taken out to work at the limekilns. They unloaded quicklime from lorries with a wind blowing, and the lime was slaked in their eyes, mouths, windpipes. When they raked out the furnaces, their sweaty naked bodies were coated with slaked lime. This daily poisoning, intended to reform them, only forced them to hurry up with their escape.

The plan dictated itself. The lime was brought by lorries—they must make their break in a lorry. Break through the boundary fence, which still consisted of barbed wire in this place. Take a lorry with plenty of petrol in the tank. The ace driver among the escapers was Kolya Zhdanok, Tenno's partner in the unsuccessful breakout from the sawmill. It was agreed that he would drive the lorry. But agreement or no agreement, Vorobyov was too strong-willed, too much the man of action, to put himself in anyone else's hands. So that when they pinched the lorry (armed with knives, they climbed into the cab one on each side, and the white-faced driver could only sit there between them, an involuntary accomplice) it was Vorobyov who took over the steering wheel.

Every minute counted! They must all jump onto the lorry and break through the barrier. "Ivan, move over!" Tenno begged him. But that was something Ivan Vorobyov could not do! Having no faith in his skill as a driver, Tenno and Zhdanok stayed behind. There were now only three escapers: Vorobyov, Salopayev, and Martirosov. Suddenly, from nowhere, Redkin ran up—an intellectual, a mathematician, an eccentric, with no record at all as an escaper: he was in the Disciplinary Barracks for something quite different. But on this occasion he had been standing near, realized what was happening, and hopped onto the lorry, holding for some reason a lump not of bread but of soap.

"Freedom bound? I'm coming with you."

(Like somebody boarding a bus: "Is this right for Razgulyai?")

The lorry swung around and moved forward at low speed, so as to break through the strands gradually—the first of them with its bumpers, then it would be the turn of the engine, then of the cab. In the outer security zone the lorry could pass between the posts, but in the main boundary area it had to knock posts down, because they were staggered. In first gear, the lorry started pushing a post over.

The guards on the towers were taken aback: there had been an incident at another site a few days earlier, when a drunken driver had smashed a post in the maximum-security zone. Perhaps this was another drunk? The thought was with them for fifteen seconds. But by then the post was down, and the lorry had changed into second gear and driven over the barbed wire without a puncture. Shoot now! But there was nothing to shoot at: to protect the guards from the winds of Kazakhstan, their towers had been boarded up on three sides. They could only shoot into the enclosed area ahead of them. . . . By now the lorry was invisible to them, speeding over the steppe and raising dust. The watchtowers fired impotently into the air.

The roads were all free, the steppe was smooth, and in five minutes Vorobyov's lorry could have been on the horizon—but *purely by chance,* a prisoner-transport van belonging to the camp guards division drove up, on its way to the transport base for repairs. It quickly took some of the sentries aboard and gave chase to Vorobyov.

The breakout was over . . . within twenty minutes. The battered runaways, with Redkin the mathematician among them, his bloodied mouth full of the warm, salty taste of freedom, staggered their way to the camp jail.[2]

All the same, word went around the camp: the break had been a beautiful job! and they had been stopped only by accident! So ten days later the former air cadet Batanov and two of his friends repeated the maneuver: they broke through the barbed-wire barriers at another work site and raced off! Only in their haste they had taken the wrong road, and came under fire from a watchtower at the limekilns. A tire was punctured and the lorry came to a stop. Tommy-gunners surrounded it. "Out you come!" Should they get out? Or should they wait to be dragged out by the scruff of the neck? One of the three, Pasechnik, obeyed the order, got out of the lorry, and was immediately riddled with a furious burst of bullets.

In something like a month there had been three attempts to escape from Ekibastuz—and still Tenno was not on the run! He was pining away. A jealous longing to outdo them gnawed at him. From the sidelines, you see all the mistakes more clearly and always think that you could do better. If, for instance, Zhdanok

2. In November, 1951, Vorobyov again escaped from a work site, on a dump truck with five others. They were caught within a few days. Rumor has it that in 1953 Vorobyov was one of the mutinous "center" in the Norilsk rising, and was afterward in the Aleksandrovsk Central Prison. A biography of this remarkable man, beginning with his early years before the war and his wartime career as a partisan, would probably help us to understand our age much better than we do.

had been at the wheel instead of Vorobyov, they could, or so Tenno thought, have got away from the prison van. The minute Vorobyov's lorry was stopped, Tenno and Zhdanok sat down to discuss how they would make their own break.

Zhdanok was small, swarthy, very agile, and a "half-caste."* He was now twenty-six. He had been taken from his native Byelorussia to Germany and worked for the Germans as a driver. He, too, was serving a *quarter*. When he caught fire he was very energetic, he put everything he had into his work, into an impulse, a fight, an escape. Of course, he lacked discipline, but Tenno had plenty of that.

Everything pointed to the limekilns as the best place for their escape. If they couldn't make their break in a lorry, they must seize one outside the restricted area. But before the guards or the security officer could interfere with their plans, Tenno was called aside by the foreman of the punitive work gang, Lyoshka the Gypsy (Lyoshka Navruzov), a "bitch," and a puny creature who nonetheless struck terror into everybody, because in his time in the camps he had murdered dozens of people (he thought nothing of killing a man for a parcel or even a pack of cigarettes).

"I'm an escaper myself and I love escapers. Look at all these bullet scars; that's from when I ran away in the taiga. I know you meant to run away with Vorobyov. Just don't do it from the work site: I'm responsible there, and I'll get another stretch."

In other words, he loved escapers but loved himself more. Lyoshka the Gypsy was content with a "bitch's" life and wouldn't let anybody ruin it. That's how much your professional criminal "loves freedom."

But perhaps escape attempts at Ekibastuz really were becoming hackneyed? Everybody tried to escape from a work site, nobody from the living area. Dare he risk it? The living area also was at present surrounded only by wire; there was no solid fence.

One day at the limekilns they damaged the electric cable of a cement mixer. An electrician was called in from outside. While Tenno helped him with his repairs, Zhdanok stole some wire cutters from his pocket. The electrician missed them. Should he inform the guards? He couldn't—he would be punished himself for his carelessness. He begged the professional criminals to give him back his cutters, but they denied taking them.

While they were at the limekilns the would-be escapers made themselves two knives: they chiseled strips of metal from shovels, sharpened them at the blacksmith's shop, tempered them, and cast tin handles for them in clay molds. Tenno's was a "Turkish" knife; it would be a handy weapon to use, and what was more important, the flashing curve of its blade was terrifying. Their intention was to frighten people, not kill them. Wire cutters and knives they carried to the living area held to their ankles by the legs of their underpants, and stowed them away in the foundations of the hut.

Once again their escape plan hinged on the Culture and Education Section. While the weapons were being made and transferred, Tenno chose a suitable moment to announce that he and Zhdanok would like to take part in a camp concert. (This would be the first ever at Ekibastuz, and the camp command could not wait to get it rolling: they needed an extra item in their list of measures to take prisoners' minds off plotting, and besides, it would be fun to see them posturing on a stage after eleven hours of hard labor.) Sure enough, Tenno and Zhdanok were given permission to leave the punishment wing after it was locked for the night, and while the camp area as a whole was still alive and in motion for another two hours. They roamed the still unknown camp, noting how and when the guard was changed on the watchtowers, and which were the most convenient spots to crawl under the boundary fence. In the Culture and Education Section itself Tenno carefully read the Pavlodar provincial newspaper, trying to memorize the names of districts, state farms, collective farms, farm chairmen, Party secretaries, shock workers* of all kinds. Next he announced that he would put on a sketch, for which he must get hold of his ordinary clothes from the clothing store and borrow a briefcase. (A runaway with a briefcase—that was something out of the ordinary! It would help him to look important.) Permission was given. Tenno was still wearing his naval jacket, and now he took out his Icelandic gear, a souvenir of a Northern convoy. Zhdanok took from his pal's suitcase a gray Belgian suit, which looked incongruously elegant in camp surroundings. A Latvian prisoner had a briefcase among his belongings. This, too, was taken. Also real caps instead of the camp issue.

The sketch required so much rehearsing that the time left till lights out in the main camp area was too short. So there was one night, and later on another, when Tenno and Zhdanok did not return to the punishment wing at all, but spent the night in the hut which housed the Culture and Education Section, to accustom their own warders to their absence. (Escapers must have at least one night's head start!)

What would be the most propitious moment for escape? Evening roll call. When the lines formed outside the huts, the warders were all busy checking in prisoners, while the prisoners had eyes only for the doors, longing to get to their beds; no one was watching the rest of the camp area. The days were getting shorter, and they must hit on one when roll call would come after sundown, in the twilight, but *before* the dogs were stationed around the boundary fence. They must not let slip those five or ten uniquely precious minutes, because there would be no crawling out once the dogs were there. They chose Sunday, September 17. It would help that Sunday was a nonworking day, so that they could recruit their strength by evening, and take time over the final preparations.

The last night before escape! You can't expect much sleep. You think and think. . . . Shall I be alive this time tomorrow? Possibly not. And if I stay here in the camp? To die the lingering death of a goner by a cesspit? . . . No, you mustn't even begin to accept the idea that you are a prisoner.

The question is this: Are you prepared to die? You are? Then you are also prepared to escape.

A sunny Sunday. To rehearse their sketch, both of them were let out of the punishment wing for the whole day. To Tenno's surprise, there was a letter from his mother in the CES. On that day of all days. Prisoners can call to mind so many coincidences of this kind. . . . It was a sad letter, but perhaps it helped to steel his resolve: his wife was still in prison; she had not yet gone on to a camp. And his sister-in-law was demanding that his brother should break off relations with the traitor.

The runaways were very short of food: in the punishment wing they were on short rations, and hoarding bread would excite suspicion. They banked on seizing a lorry in the settlement and traveling quickly. However, that Sunday there was also a parcel from home—his mother's blessing on his escape. Glucose tablets, macaroni, oatmeal—these they could carry in the briefcase. The cigarettes they would exchange for makhorka.* Except for one packet, which they would give to the orderly in the sick bay—and Zhdanok would be on the list of those excused from duty for the day. The purpose of this was as follows: Tenno could go to the CES and say, "My Zhdanok is sick; we shan't be coming to rehearse this evening." While in the punishment barracks he would tell the warder and Lyoshka the Gypsy: "We shan't come back to the hut this evening; we're rehearsing." So no one could be expecting them in either place. They must also get hold of a "katyusha"—an improvised lighter consisting of a wick in a tube —and a steel and flint to light it. This was better than matches for a man on the run. Then they must pay their last visit to Hafiz in his hut. The old Tatar, an experienced escaper, was to have made the break with them. But then he had decided that he was too old and would only be a hindrance to their flight. Now he was the only man in the camp who knew about their plans. He was sitting on his stool with his legs tucked under him. He spoke in a whisper. "God give you good fortune! I shall pray for you!" He whispered a few more words in Tatar, running his hands over his face.

Also in Ekibastuz was Tenno's old cellmate in the Lubyanka, Ivan Koverchenko. He did not know about the escape plan, but he was a good comrade. He was a trusty, and lived in a cabin of his own: it was there that the runaways kept all the things *for the sketch.* It was the obvious place for them to boil the oatmeal which had arrived in the meager parcel from Tenno's mother. Some strong black tea was brewed at the same time. They were enjoying their miniature banquet, the guests overcome by the thought of what was before them, their host by the pleasure of a fine Sunday, when they suddenly saw through the window a coffin of rough boards being carried across the camp from the guardhouse to the morgue.

It was for Pasechnik, who had been shot a few days before.

"Yes," sighed Koverchenko. "It's useless trying to escape."

(If only he knew . . .)

Some devil prompted Koverchenko to rise, pick up their bulging briefcase, stride self-importantly about the cabin, and sternly declare:

"The investigating officer knows all about it! You are planning to escape!"

He was joking. He had taken it into his head to play the part of an interrogator. . . .

Some joke.

(Or perhaps it was a delicate hint? I can guess what you're up to, boys. But I advise against it!)

When Koverchenko went out, the runaways put on their suits under the clothes they were wearing and unpicked all their number patches, leaving them attached by the merest threads so that they would tear off with one pull. The caps without numbers went into the briefcase.

Sunday was coming to an end. A golden sun was setting. Tenno, tall and leisurely, and Zhdanok, small and vivacious, now draped padded jackets around their shoulders, took the briefcase (by now everyone in the camp was used to their eccentric appearance), and went to the prearranged departure point—on the grass between some huts, not far from the boundary fence and directly opposite a watchtower. The huts screened them from two other watchtowers. There was only this one sentry facing them. They opened out their padded jackets, lay down on them, and played chess, so that the sentry would get used to them.

The sky turned gray. There was the signal for roll call. The prisoners flocked to their huts. In the half-light, the sentry on his watchtower should not be able to make out that two men were still lying on the grass. His watch was nearly over, and he was less alert than he had been. A stale sentry always makes escape easier.

They intended to cut the wire, not in the open, but directly under the tower. The sentry certainly spent more time watching the boundary fence farther away than the ground under his feet.

Their heads were down near the grass, and besides, it was dusk, so they could not see the spot at which they would shortly crawl under. But it had been thoroughly inspected in advance. Immediately beyond the boundary fence a hole had been dug for a post, and it would be possible to hide there a minute. A little farther on there were mounds of slag: and a road running from the guards' hamlet to the settlement.

The plan was to take a lorry as soon as they reached the settlement. Stop one and say to the driver, "Do you want to earn something? We have to bring two cases of vodka up here from old Ekibastuz." What driver would refuse drink? They would bargain with him. "Half a liter all right? A liter? Right, step on it, but not a word to anybody." Then on the highway, sitting with the driver in his cab, they would overpower him, drive him out into the steppe, and leave him there tied up. While they tore off to reach the Irtysh in a single night, abandon the lorry, cross the river in a little boat, and move on toward Omsk.

It got a little darker still. Up in the towers searchlights were switched on. Their beams lit up the boundary fence, but the runaways for the time being were in a shadowy patch. The very time!

Soon the watch would be changed and the dogs would be brought along and posted for the night.

Now lights were switched on in the huts, and they could see the prisoners going in after roll call. Was it nice inside? It would be warm, comfortable. . . . Whereas here you could be riddled with Tommy-gun bullets, and it would be all the more humiliating because you were lying stretched on the ground.

Just so long as they didn't cough or sneeze under the tower.

Guard away, you guard dogs! Your job is to keep us here, ours is to run away!

But now let Tenno himself take up the story.

Chapter 7

■

The White Kitten
(Georgi Tenno's Tale)

I am the senior partner, so I must go first. Sheath knife at my belt, wire cutters in my hands. "Catch up with me when I cut the boundary wire!"

I crawl flat on my belly. Trying to press myself into the ground. Shall I look toward the sentry or not? If I do, I shall see what danger I'm in, and perhaps even draw his gaze upon me. How I'm tempted to look! But I won't.

Nearer to the watchtower. Nearer to death. I expect a burst of machine-gun fire to hit me. Any minute now I shall hear its chatter. Perhaps he can see me perfectly well, and is standing there laughing at me, letting me scrabble a bit further? . . .

Here's the boundary wire. I turn around and lie parallel with it. I cut the first strand. The severed wire twangs as it loses its tautness. Now for the machine-gun burst? . . . No. Perhaps no one else could hear that sound. Though it was very loud. I cut the second strand. And the third. I swing one leg over, then the other. My trousers catch on the barb of a trailing strand. I free myself.

I crawl over several meters of plowed land. There is a rustling behind me. It's Kolya—but why is he making so much noise? Of course, it's the briefcase dragging along the ground. Here are the abutments to the main fence. The wires are crisscrossed.

I cut a few of them. Now there is a spiral entanglement. I cut it twice and clear a way. Now I cut some strands of the main fence. Are we breathing at all? Probably not.

Still he doesn't shoot. Is he dreaming of home? Or thinking of the dance tonight?

I heave my body over the outer fence. There is yet another barbed-wire entanglement. I get caught in it. I cut my way out. I mustn't forget and mustn't get stuck: there must still be the outer sloping barriers ahead. Here they are. I cut them.

Now I am crawling toward the hole. Here it is, just where it should be. I lower myself into it. Kolya follows. We pause to get our breath. But we must hurry on! Any minute now the guard will be relieved and the dogs will be here.

We hoist ourselves out of the hole and crawl toward the slag heaps. We still can't bring ourselves to look around. In his eagerness to be out of the place, Kolya rises onto all fours. I push him down again.

We negotiate the first slag heap in a leopard crawl. I put the wire cutters under a stone.

Here's the road. A little way from it, we get to our feet.

No one opens fire.

We saunter along, without hurrying; the time has come to make ourselves look like inmates of the "open prison" nearby. We tear the numbers from our chests and knees, and suddenly two men come toward us out of the darkness. They are on their way from the garrison to the settlement. They are soldiers. And we still have numbers on our backs!

"Vanya!" I say loudly. "Maybe we can manage half a liter?"

We walk slowly, still not on the road itself, but toward it. We walk slowly, to let them go by first, but straight toward them, and without hiding our faces. They pass within two meters of us. To avoid turning our backs on them, we almost come to a standstill. They go by, talking about their own affairs, and we tear the numbers off each other's backs!

Have we escaped notice? . . . Are we free? Now to the settlement to find a lorry.

But what's that? A flare roars up over the camp! Another one! A third!

They have found us out! The pursuit will start right away! We must run!!

No more looking around, no more stopping to think, no more careful calculation—our magnificent plan is in ruins. We rush into the steppe, to get as far from the camp as we can! We gasp for breath, tumble over bumps in the ground, jump up again—while rocket after rocket shoots up into the sky! Remembering previous escape attempts, we imagine them shortly sending out mounted search parties with dogs on leashes, over the steppe in every direction. So we sprinkle all our precious makhorka in our tracks, and take big jumps.[1]

Now we shall have to make a wide detour around the settlement, and keep to the steppe. It takes a lot of time and trouble. Kolya begins to doubt whether I am leading him the right way. I am offended.

But here is the embankment of the Pavlodar railway. We are very glad to see it. From the embankment we are astounded by the widely scattered lights of Ekibastuz—it looks bigger than we have ever seen it.

We choose ourselves a stick. Holding on to it, we walk along, one on each rail. Once a train goes by, the dogs will be unable to pick up the scent from the rails.

We go on like that for about 300 meters, then take a few jumps —and into the steppe.

At last we can breathe freely! We want to sing and shout! We hug one another. We really are free! How we admire ourselves for resolving to escape, succeeding in doing so, and eluding the dogs.

Although the test of our will power is only just beginning, we feel as though the worst of it were already over.

The sky is clear. Dark and full of stars—you never see it like that from the camp because of all the lights. Guided by the polestar, we travel north-northeast. Later on we shall veer right and reach the Irtysh.

We must try to get as far away as possible this first night. In that way we shall increase by the power of two the area which our pursuers must keep under observation. All the brave cheerful

---

1. Pure chance! Like the prison van which those other escapers met! An unforeseeable accident! Chance events, favorable or hostile, lie in wait for us at every step. But it is only a runaway, on the razor edge of danger, who discovers how heavily they weigh in the balance. Quite accidentally, the lighting in the camp area failed a few minutes after Tenno and Zhdanok crawled out—and it was for this reason only that flares, of which there were still a great many in Ekibastuz, were so lavishly let off. If the runaways had crawled out five minutes later, the sentries, by then on the alert, might have noticed, and shot them. If the runaways had kept their self-possession under the brilliantly lit sky, coolly observed the camp area and seen that the lamps and the boundary searchlights were out, they could have calmly made their way to some motor vehicle, and their escape would have taken an entirely different course. But in their position—with flares over the camp just after they had crawled out—they could have no doubt that the hunt was on, and they must run for their lives.

A brief failure of the lighting system—and their whole escape plan was upset.

songs we can remember, in various languages, we sing as we go along, covering eight kilometers an hour. But because we have been confined for many months to our cells, we find that our legs have forgotten how to walk and are soon tired. (We had foreseen this, but had expected to be riding!) We start lying down to rest with our legs together in the air like the poles of a wigwam. Then on again. Then another lie-down.

Behind us the glow of Ekibastuz is a surprisingly long time fading. We have been walking for several hours, and still the glow is in the sky.

But now the night is ending, and the east grows pale. By day we cannot walk over the bare open steppe, nor indeed will it be easy for us to hide there: there are neither bushes nor long grass, and we know that they will be looking for us from the air, too.

So we dig ourselves a foxhole with our knives (the ground is hard and strong and digging is difficult), half a meter wide and thirty centimeters deep, and we lie there head to toe, covering ourselves with dry, prickly yellow steppe gorse. Now is the time to sleep and recover our strength! But sleep is impossible. This helpless lying around in the daytime, for more than twelve hours at a time, is much harder to bear than the nighttime walking. You cannot stop thinking. . . . The September sun is baking hot, there is nothing to drink, and there will be nothing. We have broken the rule for escapers in Kazakhstan—you must run away in the spring, not in the autumn. . . . But of course we had expected to be riding. . . . We suffer this misery from five in the morning until eight in the evening. Our bodies are painfully numb but we must not change position; if we raise ourselves or disturb the gorse, a man on horseback may see us from a distance. Wearing two suits each, we are dying from the heat. Grin and bear it!

Until at last the darkness comes—the only time for escapers. We rise. Our legs hurt, and standing is difficult. We walk slowly, trying to ease our cramped limbs. There is little strength in us. Except for chewing bits of dry macaroni and gulping down our glucose tablets, we have had nothing all day. We are thirsty.

Even in the dark we must beware of ambush tonight: they have of course broadcast the news far and wide, dispatched motor vehicles in all directions, and especially in the direction of Omsk. We wonder how and when they found our jerkins and the chessmen on the ground. They would realize at once from the numbers that we were the runaways—no need to consult the card index and call the roll.[2]

We move at not more than four kilometers an hour. Our legs ache. We often lie down to rest. Water, water! We cover no more than twenty kilometers that night. Then we have to look again for somewhere to hide, and lie down for our daytime torment.

We think we see buildings. We start crawling cautiously toward them. Unexpectedly, out there on the steppe, they turn out to be huge rocks.

Perhaps there will be water in their cavities? No . . . but there is a niche under one of the rocks. Scratched out by jackals, perhaps. Squeezing into it is difficult. And what if the rock topples over? It could flatten you like a pancake—and you might not die immediately. It is already rather cold. Morning finds us still awake. Nor can we sleep in the daytime. We take our knives and begin honing them on a stone: they lost their edge when we dug our foxhole at the last stopping place.

In the middle of the day we heard the rumble of wheels nearby. That was bad—we were near a road. A Kazakh rode right by us. Muttering to himself. Should we jump out and run after him? He might have water. But how could we tackle him without inspecting the area first? Perhaps we could be seen.

Might not the search party pass down this very road? We slid cautiously out of our hole and looked around from ground level. A hundred meters or so away there was a dilapidated structure. We crawled over. There was no one there. A well!! No; it was choked with rubbish.

There was some trampled straw in a corner. Should we lie down here for a bit? We lay down. Sleep would not come. Lord, how the fleas did bite! Fleas!! Such big ones, and so many of them! Kolya's light-gray Belgian jacket was black with fleas. We shook

---

2. What happened was this. In the morning some working prisoners found the padded jackets, so cold that they had obviously been out all night. They tore off the numbers and pinched them: a padded jacket is something worth having. The warders simply didn't see the things. Nor did they spot the cut wires till late on Monday afternoon. They had in fact to spend a whole day checking against the card index to discover who had escaped. The runaways could still have gone on walking or riding without concealment in the morning! You can see what a difference was made by their failure to investigate the flares.

Back in the camp, the picture of the escape on Sunday evening gradually became clear; people remembered that the lights had gone out, and exclaimed in admiration: "Aren't they crafty! Aren't they clever! However did they manage to put the lights out?" For a long time everybody thought that the light failure had been a help to them.

ourselves, cleaned ourselves. We crawled back again to the jackal's hole. Time was running out, our strength was running out, and we were not moving.

At dusk we got up. We were very weak. We were tortured by thirst. We decided to bear still more sharply to the right, so as to reach the Irtysh sooner. A clear night, a black sky with stars. The constellations of Pegasus and Perseus fuse as I look at them to form the outline of a bull, head down, pressing forward urging us on. And on we go.

Suddenly, rockets shoot into the sky before us! They're ahead of us now! We freeze in our tracks. We see an embankment. A railway line. No more rockets, but the beam of a searchlight travels along the rails, swinging from side to side. A handcar is reconnoitering the steppe. Any moment now they will spot us— and it will be all over. . . . We feel stupid and helpless: lying in range of the beam, waiting to be spotted.

It passed over us, and we were not seen. We jump up. We cannot run, but hurry as best we can away from the embankment. But the sky quickly clouds over and with our dashing from right to left we have lost our sense of direction. Now we are moving almost by guesswork. We cover only a few kilometers, and even these may be pointless zigzagging.

A wasted night! . . . It's getting light again. Once more we pluck steppe gorse. We must dig a hole, but I no longer have my curved Turkish knife. I lost it either when I was lying down or in my headlong dash away from the embankment. A disaster! How can an escaper manage without a knife? We dig our hole with Kolya's.

There's one good thing about it. A fortuneteller had told me that I should meet my end at the age of thirty-eight. Sailors can't help being superstitious. But the day now dawning is September 20—my birthday. I am thirty-nine today. The prophecy no longer affects me. I shall live!

Once more we lie in a hole motionless, without water. If only we could fall asleep—but we cannot. If only it would rain! Time drags by. Things are bad. We've been on the run for very nearly three days now, and still haven't had a single drop of water. We are swallowing five glucose tablets a day. And we haven't made much progress—perhaps a third of the way to the Irtysh. And our friends in the camp are feeling glad that we are enjoying freedom in the realm of the "green prosecutor."*

Twilight. Stars. Course—north by east. We struggle on. Suddenly we hear a shout in the distance: "Va-va-va-va!" What's this? We remember Kudla, an old hand at escaping: according to him, this is how the Kazakhs frighten wolves away from their sheep.

A sheep! Give us a sheep and we are saved! As free men we would never have dreamed of drinking blood. But here and now —just let us get at it.

We approach stealthily. Crawling. Buildings. We cannot see a well. Going into the house is too dangerous—if we meet people we shall leave a trail. We creep up to the adobe sheepfold. Yes, it was a Kazakh woman shouting, to scare off wolves. We heave ourselves over a low part of the wall into the enclosure. I have my knife between my teeth. We creep along the ground, sheep-hunting. I hear one of them breathing near me. But they shy away from us, again and again! Once more, we creep up on them from different directions. Can I somehow grab one by the leg? They run away! (Later on, my mistake will be explained to me. Because we are crawling, the sheep take us for wild animals. We should have approached them erect, as though we were their masters, and they would have submitted tamely.)

The Kazakh woman senses that something is wrong, comes close, and peers into the darkness. She has no light, but she picks up lumps of earth, starts throwing them, and hits Kolya. She's walking straight toward me; any minute now she will step on me! She either sees us or senses our presence, and screeches: "Shaitan! Shaitan!" She rushes away from us and we from her—over the wall, where we lie still. Men's voices. Calm voices. Probably saying, "Silly woman's seeing things."

A defeat. All right, let's wander on.

The silhouette of a horse. Ah, the beauty! Just what we need. We go up to it. It stands still. We pat its neck, slip a belt around it. I give Zhdanok a leg up, but cannot scramble on myself; I'm too weak. I cling on with my hands, press my belly against it, but cannot cock my leg over. The horse fidgets. Suddenly it breaks loose, bolts with Zhdanok and throws him. Luckily the belt remains in his hand; we've left no trace, and they can blame it all on Shaitan.

We have worn ourselves out with the horse. Walking is harder than ever. And now there is plowed land to cross. Our feet catch in the furrows and we have to drag them along. But this is not altogether a bad thing: where there are plowed fields, there are people, and where there are people, there is water.

We walk on, struggle on. Drag ourselves along. More silhouettes. Again we lie flat and crawl. Haystacks! Meadows? Fine! Are we near the Irtysh? (Alas, we are still a long, long way from it.) With one last effort we scramble onto the hay and burrow into it.

And this time we slept the whole day through. Counting the sleepless night before our escape, we had now missed five nights' sleep.

We wake at the end of the day and hear a tractor. Cautiously we part the straw, and poke our heads out an inch. Two tractors have arrived. There is a hut. Evening is drawing on.

A bright idea! There will be water in the tractors' cooling system! When the drivers go to bed we can drink it.

Darkness falls. We have reached the end of our *fourth* day on the run. We crawl up to the tractors.

Luckily there is no dog. Quietly, we make our way to the drainage cap and take a swig. No good—there is kerosene in the water. It's undrinkable. We spit it out.

These people have everything—they have food and they have water. Why don't we just knock on the door like beggars: "Brothers! Good people! Help us! We are convicts, escaped prisoners!" Just like it used to be in the nineteenth century—when people put pots of porridge, clothing, copper coins by the paths through the taiga.

> I had bread from the wives of the village
> And the lads saw me right for makhorka.

Like hell we will! Times have changed. Nowadays they turn you in. Either to salve their consciences, or to save their skins. Because for *aiding and abetting* you can have a *quarter* slapped on you. The nineteenth century failed to realize that a gift of bread and water could be a political crime.

So we drag ourselves farther. Drag ourselves on all through the night. We can't wait to reach the Irtysh; we look eagerly for signs of the river's proximity. There are none. We drive ourselves mercilessly on and on. Toward morning we come across another haystack. With even more difficulty than yesterday, we climb into it. We fall asleep. Something to be thankful for. When we wake up it is nearly evening.

How much can a man endure? *Five* days now we have been on the run. Not far away we see a yurt with an open shed near it.

Quietly we creep up to it. Coarse millet had been strewn on the ground. We stuff the briefcase with it, try to munch some, but we cannot swallow—our mouths are too dry. Suddenly we catch sight of a huge samovar near the yurt, big enough to hold several gallons. We crawl up to it. We turn the tap—the bloody thing is empty. When we tip it we get a couple of mouthfuls.

We stagger on again. Staggering and falling. Lying down, you breathe more easily. We can no longer get up off our backs. First we have to roll over onto our bellies. Then raise ourselves onto all fours. Then, swaying, onto our feet. Even this leaves us out of breath. We have grown so thin that our bellies seem stuck to our backbones. As dawn approaches we cover some 200 meters, no more. And lie down.

That morning no haystack came our way. There was some kind of burrow in a hill, dug by an animal. We lay in it through the day, but could not get to sleep. That day it got colder, and we felt a chill from the ground. Or perhaps our blood was no longer warming us? We tried to chew some macaroni.

Suddenly I see a line of soldiers advancing! With red shoulder tabs! They're surrounding us! Zhdanok gives me a shake: You're imagining things, it's a herd of horses.

Yes, it was a mirage. We lay down again. The day was endless. Suddenly a jackal arrived, coming home to his hole. We put some macaroni down for him and crawled away, hoping to lure him after us, stab him and eat him. But he wouldn't touch it. He went away.

To one side of us there was a slope, downhill a little the salt flats of a dried-up lake, and, on the other bank, a yurt and smoke drifting in the air.

Six days have gone by. We have reached the limit: we see red tabs in our hallucinations, our tongues are stuck to the roofs of our mouths. If we pass water at all there is blood in it. It's no good! Tonight we must get food and water at any price. We'll go over there, to the yurt. If they refuse us, we'll take it by force. I remember the old fugitive Grigory Kudla and his war cry: Makhmadera! (Meaning: "Stop asking—and grab!") Kolya and I come to an understanding: at the right moment I will say "Makhmadera!"

In the darkness we crept quietly toward the yurt. There was a well. But no bucket. Not far away a saddled horse stood at a hitching post. We glanced through the narrow opening. By the light of an oil lamp we saw a Kazakh couple and some children. We knocked and went in. I said, "Salaam!" And all the time there were big spots in front of my eyes and I was afraid of falling. Inside there was a low, round table (even lower than our modernist designers make them) for the beshbarmak.* Round the yurt were benches covered with felt. There was a big metal-bound chest.

The Kazakh muttered something in reply, and looked sullen— he was not a bit pleased to see us. To make myself look important (and anyway, I needed to conserve my strength), I sat down and put the briefcase on the table. "I'm in charge of a geological survey team, and this is my driver. We've left our transport back in the steppe with the others, five or maybe seven kilometers from here: the radiator leaks and the water's all run out. We ourselves haven't had anything to eat for three days; we're starving. Give us something to eat and drink, aksakal.* And tell us what you think we should do."

But the Kazakh screws up his eyes and offers no food and drink. "What you name, boss?" he asks.

I had it all ready once, but my head is buzzing, and I've forgotten.

"Ivanov," I answer. (Stupid, of course.) "Come on then, sell us some groceries, aksakal!" "No. Go to my neighbor." "Is it far?" "Two kilometers."

I sit there on my dignity, but Kolya can't hold out any longer, seizes a griddle cake and tries to chew it, though it is obviously hard work for him. Suddenly the Kazakh picks up a whip, with a short handle and a long leather lash, and threatens Zhdanok with it. I get to my feet. "So that's the sort of people you are! That's your famous hospitality!" Now the Kazakh is prodding Zhdanok in the back with the whip handle, trying to drive him out of the yurt. I give the command. "Makhmadera!" I take out my knife and tell the Kazakh, "In the corner! Lie down!" The Kazakh dives through a curtain. I am right behind him: he may have a gun there, may shoot us at any moment. But he flops on the bed, shouting, "Take it all! I won't say anything!" Oh, you miserable cur, you! What do I want with your "all"? Why couldn't you give me the little bit I asked for in the first place?

To Kolya I say, "Keep your eyes peeled!" I stand by the door with my knife. The Kazakh woman is screaming, the children start crying. "Tell your wife we won't hurt any of you. We just want to eat. Is there any meat?" He spread his hands. "Yok."* But

Kolya pokes about in the yurt and produces dried mutton out of a meat safe. "Why did you lie?" Kolya also grabs a basin with baursaki in it—lumps of dough, deep fried. I suddenly realize that there is kumiss in the jugs on the table. Kolya and I drink. With every swig life comes back to us! What a drink! My head begins to spin, but intoxication seems to help me, I feel stronger all the time. Kolya is beginning to enjoy himself. He passes some money to me. There turn out to be 28 rubles. There's probably more tucked away somewhere.

We drop the dried mutton into a sack, and scoop baursaki, griddlecakes, and sweets of some sort—dirty "satin cushions"—into another. Kolya also pinches a dish of roasted mutton scraps. A knife! That's something we really need. We try not to forget anything—wooden spoons, salt. . . . I carry the sack out. I come back and take a bucket of water. I take a blanket, a spare bridle, the whip. (He mutters to himself—he doesn't like this: how is he going to catch us?)

"Right, then," I tell the Kazakh. "Let that be a lesson to you, and mind you don't forget it. You should be more friendly to your guests! We would have gone on our knees to you for a bucket of water and a dozen baursaki. We do no harm to decent people. Here are your final instructions: lie where you are without stirring! We have friends outside."

I leave Kolya outside the door while I lug the rest of the loot over to the horse. I suppose we should be hurrying, but my mind works calmly. I take the horse to the well and let it drink its fill. It, too, has quite a job ahead of it; it will have to carry its excessive load all through the night. I drink from the well myself. So does Kolya. Just then some geese come up to us. Kolya has a weakness for poultry. "Shall we grab the geese?" he says. "Shall we wring their necks?" "It'll make too much noise. Don't waste time."

I let down the stirrups and tighten the girths. Zhdanok puts the blanket behind the saddle, and climbs onto it from the well wall. He takes the bucket of water in his hands. We have tied the two sacks together and slung them over the horse's back. I get into the saddle. And so by the light of the stars we ride off eastward, to throw our pursuers off the trail.

The horse objected to having two riders, strangers at that, and kept tossing its head, trying to turn back toward home. We mastered it somehow. It set off briskly. There were lights nearby. We made a detour. Kolya sang quietly in my ear.

Out on the range, where the wind blows free,
A cowboy's life is the life for me,
With a horse I can trust beneath me.

"I saw his passport, too," said Kolya. "Why didn't you take it? A passport always comes in handy. You can give them a peep at the cover—not too close."

On the road, we took frequent drinks of water and snacks without dismounting. Our mood was entirely different now! All we wanted now was to gallop as far away as we could before daylight!

We heard the cries of birds. A lake. It would be a long way around it, and we grudged the time. Kolya dismounted and led the horse along a slippery causeway. We got across. But we suddenly noticed—no blanket. It had slipped off. *We had left a trail.*

This was very bad. From the Kazakh's place many paths led in all directions, but if they found the blanket and drew a line from the yurt to that point, our route would be clear. Should we go back and look for it? There was no time. In any case, they must know we were going north.

We called a halt. I held the horse by a rein. We ate and drank, ate and drank, endlessly. The water was nearly down to the bottom of the bucket—we could hardly believe it.

Course—due north. The horse wouldn't break into a trot, but walked quickly, at 8 to 10 kilometers an hour. In six nights we had notched only 150 kilometers, but that night we did another 70. If we hadn't zigzagged about, we should be on the Irtysh by now.

Dawn. But nowhere to hide. We rode on. It was getting dangerous. Then we saw a deep hollow, almost a hole. We took the horse down into it, and ate and drank again. Suddenly, a motorcycle stuttered nearby. That was bad—there was a road. We must find a safer hiding place. We climbed out and looked around. Not too far off there was a dead and deserted Kazakh village.[3] We made our way there. We shed our load between the three walls of a ruined house, then hobbled the horse and turned it out to graze.

But there was no sleep that day: with the Kazakh and the blanket, we had *left a trail.*

3. The years 1930–1933 left many such ruined villages dotted about Kazakhstan. First Budenny passed through with his cavalry (to this day there is not a single kolkhoz called after him, not a single picture of him anywhere in Kazakhstan), then famine.

Evening. *Seven* days now. The horse was grazing some way off. We went after it—and it shied away, evaded us. Kolya grabbed it by the mane, but it dragged him along and he fell. It had freed its front legs—and now there was no holding it. We hunted it for three hours, wearing ourselves out, drove it in among the ruins, tried to slip a noose made from our belts around its neck, but still it wouldn't give in. We bit our lips, but we had to abandon it. All we had left was the bridle and the whip.

We ate, and drank our last water. We shouldered the sacks with the food, and the empty bucket. And off we went. Today we had the strength for it.

The following morning caught us in an awkward spot and we had to hide in some bushes not far from a road. Not the best of places—we could be spotted. A cart rattled by. We didn't sleep that day either.

As the *eighth* day ended we set off again. When we had gone a little way we suddenly felt soft earth underfoot: the plow had been here. We went on and saw headlights along the roads. Careful now!

There was a young moon up among the clouds. Yet another dead and ruined Kazakh hamlet. Farther on, the lights of a village, and the words of a song reached our ears:

"Hey, lads, unharness the horses. . . ."

We put the sacks down among the ruins, and made for the village with the briefcase and bucket. We had our knives in our pockets. Here's the first house—with a grunting piglet. If only we'd met him out on the steppe. A lad rode toward us on a bike. "Hey, pal, we've got a truck over there; we're moving grain. Where can we get some water for the radiator?" The boy got off, went ahead of us, and pointed. There was a tank on the edge of the village; probably the cattle drank from it. We dipped the bucket in and carried it away full, without taking a drink. We parted with the lad, then sat down and drank and drank. We half-emptied the bucket at one go (we were thirstier than ever today, because we had eaten our fill).

There seemed to be a slight chill in the air. And there was real grass under our feet. There must be a river near! We must look for it. We walk and walk. The grass is higher, there are bushes. A willow—where they are there is always water. Reeds! Water!!! No doubt a backwater of the Irtysh. Now we can splash around and wash ourselves. Reeds two meters high! Ducks start up from

under our feet. We can breathe freely here! We shan't come to grief
here!

And this was when for the first time in eight days the stomach
discovered that it was still working. After eight days out of action,
what torment it was! Birth pains are probably no worse.

Then back we went to the abandoned village. There we lit a fire
between the walls and boiled some dried mutton. We should have
used the night to move on, but all we wanted was to eat and eat
insatiably. We stuffed ourselves until we could hardly move. Then,
feeling pleased with life, we set off to look for the Irtysh. At a fork
in the road something happened for the first time in eight days—
we quarreled. I said, "Right," Zhdanok said, "Left." I felt sure
that it should be right, but he wouldn't listen. Another of the
dangers that lie in wait for escapers—falling out with one another.
When you are on the run, one of your number must be allowed
to have the last word, otherwise you are in trouble. Determined
to have my way, I went off to the right. I walked a hundred meters,
and still heard no footsteps behind me. My heart ached. We
couldn't just part like that. I sat down by a haystack and looked
back. . . . Kolya was coming! I hugged him. We walked on side
by side as though nothing had happened.

There are more bushes now and the air is chillier. We walk to
the edge of a sharp drop. Down below, the Irtysh splashes and
babbles and playfully breathes on us. We are overjoyed.

We find a haystack and burrow into it. What about it, tracker
dogs, still think you can find us? You haven't a hope! We fall into
a heavy sleep.

We were awakened by a shot! And dogs barking quite near! . . .
Was this it, then? Was our freedom to end so soon? We clung
together and stopped breathing. A man went by. With a dog. A
hunter! . . . We fell into an even deeper sleep . . . and slept the day
through. This was how we spent our *ninth* day.

When it got dark we set off along the river. Three days had gone
by since we left a trail. The dog handlers would only be looking
for us along the Irtysh by now. They would realize that we were
making for the water. If we went along the bank we might easily
stumble into an ambush. Besides, it was hard work—we had to go
around bends, creeks, reed beds. We needed a boat!

A light, a little house on the riverbank. The splash of oars, then
silence. We lay low and waited for some time. They put the light
out. We went quietly down to the water. There was the boat. And

a pair of oars. Splendid! (Their owner might have taken them with him.) "The sailor leaves his troubles ashore." My native element! Quietly, to begin with, without splashing. Once out in midstream, I rowed hard.

We move on down the Irtysh, and from around a bend a brightly lit steamer comes toward us. So many lights! The windows are all ablaze, the whole ship rings with dance music. Passengers, free and happy, stroll on the deck and sit in the restaurant, not realizing how happy they are, not even aware of their freedom. And how cozy it is in their cabins! . . .

In this way we traveled more than twenty kilometers downstream. Our provisions were running out. The sensible thing would be to stock up again while it was still night. We heard cocks crowing, put in to shore, and quietly climbed toward the sound. A little house. No dog. A cattle shed. A cow with a calf. Hens. Zhdanok is fond of poultry, but I say we'll take the calf. We untie it. Zhdanok leads it to the boat while I, in the most literal sense, wipe out our tracks, otherwise it will be obvious to the tracker dogs that we are traveling by boat.

The calf came quietly as far as the bank, but stubbornly refused to step into the boat. It was as much as the two of us could do to get him in and make him lie down. Zhdanok sat on him, to hold him down, while I rowed—once in the clear we would kill him. But that was our mistake—trying to carry him alive! The calf started getting to his feet, threw Zhdanok off, and heaved his forelegs into the water.

All hands on deck! Zhdanok hangs on to the calf's hindquarters, I hang on to Zhdanok, we all lean too far to one side, and water pours in on us. We are as near as need be to drowning in the Irtysh! Still, we drag the calf back in! But the boat is very low in the water and must be bailed out. Even that must wait, though, till we kill the calf. I take the knife and try to sever the tendon at the back of his neck—I know the place is there somewhere. But either I can't find it or the knife is too blunt; it won't go through. The calf trembles, struggles, gets more and more agitated—and I am agitated, too. I try to cut his throat—but this is no good either. He bellows, kicks, looks as if to jump clean out of the boat or sink us. He wants to live—but we have to live, too!

I saw away, but cannot cut deep enough. He rocks the boat, kicks its sides—the silly idiot will sink us any minute now! Because he is so nasty and so stubborn, a red hatred for him sweeps over me, as though he were my worst enemy, and I start savagely, randomly pricking and jabbing him with the knife.[4] His blood spurts forth and sprays us. He bellows loudly and kicks out desperately. Zhdanok clamps his hands around the calf's muzzle, the boat rocks, and I stab and stab again. To think that at one time I couldn't hurt a mouse or a fly! But this is no time for pity: it's him or us!

At last he lay still. We started hurriedly scooping out water with a bail and some tin cans, each of us using both hands. Then we rowed on.

The current drew us into a side channel. Ahead lay an island. This would be a good place to hide; it would soon be morning. We wedged the boat well into the reeds. We dragged the calf and all our goods onto the bank, and to make the boat still safer, strewed reeds over it. It wasn't easy to haul the calf by its legs up the steep overhang. But once there, there was grass waist high, and trees. Like a fairy tale! We had spent several years in the desert by now. We had forgotten what forests and grass and rivers were like. . . .

It was getting light. The calf looked aggrieved, we thought. But thanks to this little friend of ours we could now live for a while on the island. We sharpened the knife on a fragment of file made from our "katyusha." I had never before skinned a beast, but now I was learning. I slit the belly, pulled back the skin, and removed the entrails. In the depths of the wood we lit a fire and started stewing veal with oats. A whole bucketful of it.

A feast! And best of all, we feel at ease. At ease because we are on an island. The island segregates us from mean people. There are good people, too, but somehow runaways don't often come across them—only mean ones.

It is a hot, sunny day. No need for painful contortions to hide in a jackal's hole. The grass is thick and lush. Those who trample it every day don't know how precious it is, don't know what it means to plunge into it breast high, to bury your face in it.

We roam about the island. It is overrun by dog roses, and the hips are already ripe. We eat them endlessly. We eat more soup. And stew some more veal. We make kasha with kidneys.

We feel light-hearted. We look back at our difficult journey and find plenty to laugh at. We think of them waiting back there for

---

4. Is this not like the hatred our oppressors feel for us as they destroy us?

our sketch. Cursing us, explaining themselves to the administration. We make a play of it. We roar with laughter!

We tear bark from a thick trunk and burn in the following inscription with red-hot wire: "Here on their way to freedom in October, 1950, two innocent people sentenced to hard labor for life took refuge." Let this sign of our presence remain. Out in the wilds here it will not help our pursuers, and someday people will read it.

We decide not to hurry on. All that we ran away for we have: our freedom! (It can hardly be more complete when we reach Omsk or Moscow.) We also have warm, sunny days, clean air, green grass, leisure. And meat in plenty. Only we have no bread, and miss it greatly.

We lived on the island for nearly a week: from our *tenth* to the beginning of our *sixteenth* day. In the thickest part of the wood we built ourselves a shelter of dry boughs. It was cold at night even there, but we made up for lost sleep in the daytime. The sun shone on us all this time. We drank a lot, trying to store water as camels do. We sat serenely, looking for hours through the branches at life over yonder, on shore. Over there vehicles went by. The grass was being mown again, the second crop this summer. No one dropped in on us.

One afternoon while we were dozing in the grass, enjoying the last rays of the sun, we suddenly heard the sound of an ax at work on the island. Cautiously raising ourselves, we saw, not far away, a man lopping branches and moving gradually toward us.

In a fortnight, with no means of shaving, I had grown a beard, a terrible reddish bristling bush, and was now a typical escaped convict. But Zhdanok had no growth at all; he was like a smooth-faced boy. So I pretended to be asleep and sent him to head the man off, ask for a smoke, tell him that we were tourists from Omsk, and find out where he came from himself. If needs be, I was ready to act.

Kolya went over and had a chat with him. They lit up. He turned out to be a Kazakh from a nearby kolkhoz. Afterward we saw him walk along the bank, get into his boat, and row off without the branches he had cut.

What did this mean? Was he in a hurry to report us? (Or perhaps it was the other way round: perhaps he was afraid that we might inform on him; you can do time for wood-stealing, too. That was what our lives had come to—everybody feared everybody else.) "What did you say we were?" "Climbers." I didn't know whether to laugh or to cry—Zhdanok always made a muddle of things. "I told you to say hikers! What would climbers be doing on the flat steppe?!"

No, we couldn't stay there. Our life of bliss was over. We dragged everything back to the boat and cast off. Although it was daytime, we had to leave quickly. Kolya lay on the bottom, out of sight, so that from a little way off it would look as if there was just one man in the boat. I rowed, keeping to the middle of the Irtysh.

One problem was where to buy bread. Another was that we were now coming to inhabited places, and I could no longer go unshaven. We planned to sell one of our suits in Omsk, buy tickets several stations down the line, and get away by train.

Toward evening we reached a buoy keeper's hut and went up there. We found a woman, alone. She was frightened, and began rushing around. "I'll call my husband at once!" And off she went. With me following to keep an eye on her. Suddenly Zhdanok called out from the house in alarm: "Zhora!" Damn you and your big mouth! We had agreed that I would call myself Viktor Aleksandrovich. I went back. Two men, one with a hunting rifle. "Who are you?" "Tourists, from Omsk. We want to buy some groceries." And, to lull their suspicions: "Let's go into the house—why are you so inhospitable?" It worked, and they relaxed. "We've got nothing here. Maybe at the sovkhoz.* Two kilometers farther down."

We went to the boat and traveled another twenty kilometers downstream. It was a moonlit night. We climbed the steep bank: a little house. No light burning. We knocked. A Kazakh came out. And this first man we saw sold us half a loaf and a quarter of a sack of potatoes. We also bought a needle and thread (probably rather rash of us). We asked for a razor, too, but he was beardless and had no use for one. Still, he was the first kind person we had met. We got ambitious and asked whether there was any fish. His wife rose and brought us two little fishes and said, "Besh denga." "No money." This was more than we had hoped for—she was giving them to us free! These really were kind people! I started stowing the fish in my sack, but she pulled them back again. "Besh denga—five rubles," the man of the house explained.[5] Ah, so that's

_____

5. The narrator misunderstands the Kazakh words "besh denga" ("five rubles") as Russian "bez deneg" ("without money").

it! No, we won't take them; too dear. We rowed on for the rest of the night. Next day, our *seventeenth* on the run, we hid the boat in the bushes and slept in some hay. We spent the *eighteenth* and *nineteenth* in the same way, trying not to meet people. We had all we needed: water, fire, meat, potatoes, salt, a bucket. On the precipitous right bank there were leafy woods, on the left bank meadows, and a lot of hay. In the daytime we lit a fire among the bushes, made a stew, and slept.

But Omsk was not far off, and we should be compelled to mix with people, which meant that I must have a razor. I felt completely helpless: with neither razor nor scissors, I couldn't imagine how I was going to rid myself of all that hair. Pluck it out a hair at a time?

On a moonlit night we saw a mound high over the Irtysh. Was it, we wondered, a lookout post? From the times of Yermak? We climbed up to look. In the moonlight we saw a mysterious dead township of adobe houses. Probably also from the early thirties . . . What would burn they had burned, the mudbrick walls they had knocked down, some of the people they had tied to the tails of their horses. Here was a place the tourists never visited. . . .

It had not rained once in those two weeks. But the nights were already very cold. To speed things up, I did most of the rowing, while Zhdanok sat at the tiller, freezing. And sure enough, on the twentieth night he started asking for a fire, and hot water to warm himself. I put him at the oars, but he shivered feverishly and could think of nothing but a fire.

His comrade in flight could not deny him a fire—Kolya should have known that and denied himself. But that was the way with Zhdanok—he could never control his desires: remember how he had snatched the griddlecake from the table, and what a temptation the poultry was to him.

He kept shivering and begging for a fire. But they would be keeping their eyes skinned for us all along the Irtysh. It was surprising that no search party had crossed our path so far. That we had not been spotted on a moonlit night in the middle of the Irtysh and stopped.

Then we saw a light on the higher bank. Kolya stopped begging for a fire and wanted to go inside for a warm-up. That would be even more dangerous. I should never have agreed. We had gone through so much, suffered so many hardships—and for what? But how could I refuse him—perhaps he was seriously ill. And he could refuse himself nothing.

In the light of an oil lamp two Kazakhs, a man and a woman, were sleeping on the floor. They jumped up in fright. "I have a sick man here," I explained. "Let him get warm. We are on official business from the Grain Procurement Agency. They ferried us over from the other side." "Lie down," the Kazakh said. Kolya lay down on a heap of felt, and I thought it would look better if I lay down a bit, too. It was the first roof we had had over our heads since our escape, but I was on hot bricks. I couldn't even lie still, let alone go to sleep. I felt as though we had betrayed ourselves, stepped into a trap with our eyes open.

The old man went out, wearing nothing but his underwear (otherwise I would have gone after him), and was away a long time. I heard whispering in Kazakh behind the curtain. Young men. "Who are you?" I asked. "Buoy keepers?" "No, we're from the Abai State Livestock Farm, number one in the republic." We couldn't have chosen a worse place. Where there was a state farm there was officialdom and police. And the best farm in the republic, at that! They must be really keen. . . .

I pressed Kolya's hand. "I'm off to the boat—come after me. With the briefcase." Out loud I said, "We shouldn't have left the provisions on the bank." I went through to the entranceway and tried the outer door; it was locked. That's it, then. I went back in, alerted Kolya by pulling his sleeve, and returned to the door. The carpenters had made a botched job of it, and one of the lower planks was shorter than the rest. I shoved my hand through, stretched my arm as far as I could and felt around. . . . Ah, there we were—it was held by a peg outside. I dislodged it.

I went out. Hurried down to the bank. The boat was where it had been. I stood waiting in broad moonlight. But there was no sign of Kolya. This was dreadful! Evidently he couldn't make himself get up. He was enjoying an extra minute in the warmth. Or else they had seized him. I should have to go and rescue him.

I climbed the cliff again. Four people were coming toward me from the house, Zhdanok among them. "Zhora!" he shouted. ("Zhora" again!) "Come here! They want to see our papers." He wasn't carrying the briefcase, as I had told him to.

I go up to them. A new arrival with a Kazakh accent says, "Your papers!" I behave as calmly as I can. "Who are you, then?" "I'm the commandant." "All right, then," I say reassuringly.

"Let's go. You can check our papers anytime. There's more light in the house there." We go into the house.

I slowly lifted the briefcase from the floor, and went over to the lamp, looking for an opportunity to side-step them and dash out of the house, and talking all the time to distract them: "Welcome to see our papers anytime, of course. Papers must always be checked in such circumstances. You can't be too careful. We had a case once in the Procurement Agency . . ." My hand was on the lock now, ready to undo the briefcase. They crowded around me. Then . . . I butted the commandant with my shoulder, he bumped into the old man, and they both fell. I gave the young man on my right a straight punch on the jaw. They yelled, they howled. "Makhmadera," I shouted, and bounded with the briefcase through the inner then the outer door. Then Kolya shouted after me from the entrance way: "Zhora! They've got me!" He was clinging to the doorpost, while they tried to pull him back inside. I tugged at his arm, but couldn't free him. Then I braced myself against the doorpost with my foot and gave such a heave that Kolya flew over my head as I fell to the ground. Two of them flung themselves on top of me. I don't know how I wriggled out from under them. Our precious briefcase was left behind. I ran to the cliff, and bounded down it! Behind me I hear someone say in Russian: "Use the ax on him! The ax!" Probably trying to scare us—otherwise they would be speaking Kazakh. I can almost feel their outstretched hands on me. I stumble, and almost fall! Kolya is in the boat already. "Good thing they didn't have a gun," I shout. I pushed the boat out and was up to my knees in water before I jumped into it. The Kazakhs were reluctant to get wet. They ran along the bank, yelling, "Gir-gir-gir!" I shouted back at them: "Thought you had us, didn't you, you bastards?"

Yes, it was lucky they had no gun. I made the boat race with the current. They bayed after us, running along the bank, until a creek barred their way. I took off my two pairs of trousers—naval and civilian—and wrung them out. My teeth were chattering. "Well, Kolya," I said, "we got warmed up, all right." He was silent.

It was obviously time to say goodbye to the Irtysh. At daybreak we must go ashore and thumb rides the rest of the way to Omsk. It wasn't so very far now.

The "katyusha" and the salt had been left behind in the briefcase. And where could we get a razor? It wasn't worth

1. Relatives identifying the corpses of those executed at Vinnitsa.

I went right through Ekibastuz with the number Shch #232 until my last few months, when I was ordered to change it to Shch #262. I smuggled patches with this number on them out of Ekibastuz, and I have kept them to this day.

2. The author in 1953, right after his release from the special camp.

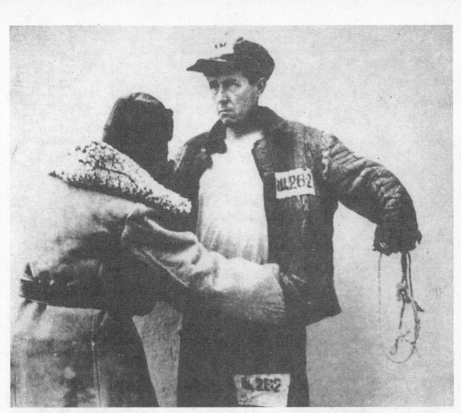

3. Body search.

4. Georgi Pavlovich Tenno.

5. The door to the Ekibastuzsky BUR (Disciplinary Barracks).

6. N. V. Surovtseva beside the hut.

7. VGS Barracks.

8. Kadatskaya and her husband
in their youth.

9. Kadatskaya in 1968.

10. Sculpture by Nedov.

asking myself how we could dry our clothes. Look—on the bank there: a boat and a hut. Obviously a buoy keeper. We went ashore and knocked. No light came on. A deep male voice: "Who is it?" "Let us in for a warm-up! Our boat capsized and we nearly drowned." There was a lot of fumbling, then the door opened. In the dimly lit entrance a sturdy old man, Russian, stood to one side of the door, arms raised, threatening us with an ax. He could bring it down on one of us, and there would be no stopping him. I tried to reassure him. "Don't be afraid. We're from Omsk. We've been on business to the Abai State Farm. We intended to go by boat to the district center downstream, but there were nets in the shallows a bit higher up; we fouled them and turned over." He still looked suspicious and didn't lower his ax. Where had I seen him before, in what picture? An old man out of a folk tale, with his gray mane, his gray beard. At last he decided to answer: "You were going to Zhelezyanka, you mean?" Fine, now we know where we are. "That's right, Zhelezyanka. The worst of it is my briefcase sank and there's a hundred and fifty rubles in it. We bought some meat at the state farm, but we have no use for it now. Perhaps you'll buy it from us?" Zhdanok went to get the meat. The old man let me into the inner room, where there was a kerosene lamp, and a sporting gun on the wall. "Now we'll check your papers." I tried to speak as confidently as I could. "I always keep my documents on me; it's lucky they were in my top pocket or they would be soaked. I'm Stolyarov, Viktor Aleksandrovich, representing the Provincial Livestock Administration." Now I must quickly seize the initiative. "What about you?" "I'm a buoy keeper." "Name and patronymic?" Just then Kolya arrived and the old man didn't mention papers again. He said that he couldn't afford to buy meat but that he could give us a drink of tea.

We sat with him about an hour. He warmed up some tea for us on a fire of wood chips, gave us bread, and even cut off a piece of fat bacon. We talked about the navigation channels of the Irtysh, how much we had paid for our boat, where to sell it. He did most of the talking. He looked at us with compassion in his wise old eyes, and it seemed to me that he knew all about us, that he was a real human being. I even felt like confiding in him. But it wouldn't have helped us: he obviously had no razor—he was as shaggy as everything else in the forest. Besides, it was less dangerous for him not to know: otherwise it would be "You knew and didn't tell."

We left him our veal, and he gave us some matches. He came out to see us off, and explained which side we should keep to at which points. We pushed off and rowed quickly to get as far away as possible in our last night. They would be looking for us on the right bank, so now we hugged the left most of the time. The moon was hidden by our bank but the sky was clear, and we saw a boat following the steep wooded right bank, going downstream like ourselves, but not as fast.

Could it be an MVD operations group? . . . We were following parallel courses. I decided to brazen it out, rowed strongly, and came closer to them. "Hey, pal! Where are you headed?" "Omsk." "And where are you from?" "Pavlodar." "Why so far?" "We're moving for good." His voice, with its peasant "o" 's, was too uneducated for an operations officer, he answered unhesitatingly, and . . . he even seemed glad to see us. His wife was sleeping in the boat, while he spent the night at the oars. I looked in: it was more like an ox wagon, crammed with goods and chattels, heaped high with packages.

I did a bit of quick thinking. A meeting like this—on our last night, in our last hours on the river! If he's pulling up his roots they must be carrying provisions, and money, and passports, and clothing, and even a razor. And no one, anywhere, will wonder where they are. He's alone and there are two of us—his wife doesn't count. I'll travel on his passport. Kolya can dress up like a woman: he's small, has a smooth face, we'll mold him a figure. They must surely have a suitcase, to help us look like genuine travelers. Any driver we meet will drop us in Omsk this very morning.

Who ever heard of a Russian river without pirates? Fate is cruel, but what else can we do? Now that we have left a trail on the river, this is our last, our only chance. It's a pity to rob a workingman of his belongings—but who ever took pity on us? And who ever would?

All this flashed through my mind, and through Zhdanok's, too, in a moment. I only had to ask quietly, "Uh-huh?" And he quietly replied, "Makhmadera."

I get steadily nearer and am now forcing their boat toward the steep bank, toward the dark forest. I must be quick to prevent them from reaching the next bend in the river, in case the forest ends there. I change my voice to one of authority and give my orders.

"Attention! We are an MVD operations group. Put in to shore. I want to inspect your papers!"

The rower threw down his oars: had he lost his head, or was he perhaps overjoyed to find that we were policemen and not robbers?

"Of course," he said. "You can inspect them here, or on the river."

"I said put in to the shore, and that's what you'll do. And be quick about it."

We got close to them. Our sides were almost touching. We jumped across, he scrambled with difficulty over his bundles, and we saw that he had a limp. His wife woke up. "Is it far now?" The young man handed over his passport. "What about your draft card?" "I was invalided out, wounded; I'm exempt. Here's the certificate." I saw a gleam of metal in the prow of their boat—an ax. I signaled to Kolya to remove it. He rushed too abruptly and seized the ax. The woman felt that something was wrong and set up a howl. "What's all that noise?" I said sternly. "Cut it out. We're looking for runaways. Criminals. And an ax is as good a weapon as any." She calmed down a little.

I give Kolya his orders.

"Lieutenant! Slip down to the observation post. Captain Vorobyov should be there."

(The name and rank came to me automatically—I'll tell you why: we had left a pal of ours, Captain Vorobyov, behind in Ekibastuz, confined to the cells for trying to escape.) Kolya understood: he was to see whether there was anyone around up top, or whether we could act. Up he ran. In the meantime I carried on questioning and inspecting. My suspect obligingly struck matches for me. I ran through their passports and certificates. His age was just right, too—the veteran was under forty. He had worked as a buoy keeper. Now they had sold their home and their cow. (He would have all the money with him, of course.) They were going to seek their fortune. They couldn't get there in a day, so they had set off by night.

A rare chance, an extraordinary chance, above all because no one would miss them. But what did we need from them? Did we need their lives? No, I had never murdered, and I didn't want to now. An interrogator, or an operations officer, when he was tormenting me—yes; but I couldn't raise my hand against ordinary working people. Should I take their money? All right, but just a little. How little, though? Enough for two tickets to Moscow, and some food. And some of their gear. That wouldn't ruin them. What if we left them their papers and the boat, and made a deal with them not to report us? It wouldn't be easy to trust them. And how could we manage without papers?

If we took their papers, they would have no choice but to report us. To prevent their doing so we must tie them up here and now. Tie them up well and truly so that we should have two or three days' head start.

But in that case wouldn't it be better simply to . . . ?

Kolya came back and signaled that everything was all right up above. He was waiting for me to say "Makhmadera!" What was I to do?

The slave camp of Ekibastuz rose before my eyes. Could I go back to that? Surely we had the right . . .

And suddenly—suddenly something very light touched my legs. I looked down: something small and white. I bent over; it was a white kitten. It had jumped out of the boat, and with its tail stiff as a stalk in the air, it purred and rubbed itself against my legs.

It didn't know what I was thinking. I felt as though the touch of this kitten had sapped my will power. Stretched taut for twenty days, ever since I had slipped under the wire, it suddenly seemed to snap. I felt that, whatever Kolya might say to me now, I could never take their lives nor even the money they had earned in the sweat of their brows.

Still keeping a stern face, I said, "Right, wait here; we'll soon see what's what."

We climbed the cliff. I had the papers in my hand. I told Kolya what I was thinking.

He said nothing. He disagreed, but he said nothing.

That's how the world is arranged: *they* can take anyone's freedom from him, without a qualm. If we want to take back the freedom which is our birthright—they make us pay with our lives and the lives of all whom we meet on the way.

They can do anything, but we cannot. That's why *they* are stronger than we. Without coming to an agreement, we went down again. Only the lame man was by the boat. "Where's your wife?"

"She was frightened; she ran off into the forest."

"Here are your papers. You can go on your way."

He thanks me, and shouts into the forest:

"Ma-ria! Come back! They're honest people!"

We push off. I row quickly. The ordinary workingman, the man with the bad leg, suddenly remembers and shouts after me:

"Comrade officer! We saw two chaps yesterday—looked just like bandits. If we'd known, we'd have held the rotters!"

"Still feel sorry for him?" asks Kolya.

I say nothing.

■

From that night—from the moment we went indoors for a warm-up, or perhaps when we met the white kitten—our escape began to go wrong. We had lost something: our confidence? our tenacity? our ability to think straight? the instinctive understanding between us? Now that we were nearly in Omsk we started making mistakes, pulling different ways. When runaways behave like that, they do not run much farther.

Toward morning we abandoned the boat. We slept through the day in a haystack, but uneasily. Darkness fell. We were hungry. It was time to stew some meat, but we had lost our bucket in the retreat. I decided to fry it. We found a tractor seat—that would do for a frying pan; the potatoes we could bake.

Nearby stood a tall hut, left behind by haymakers. In the mental blackout which had come upon me that day, I thought it a good idea to light my fire inside the hut: it would be invisible from all sides. Kolya didn't want any supper at all. "Let's move on!" Once again we couldn't see eye to eye.

I did light a fire in the hut, but I put too much wood on. The whole hut went up in flames, and I barely managed to crawl out. Then the fire jumped to the stack—the one in which we had spent the day—and it blazed up. Suddenly I felt sorry for that hay—so sweet-scented, and so kind to us. I started scattering it, and rolling on the ground in an attempt to put it out, to prevent the fire from spreading. Kolya sat aloof, sulked, and offered no help.

What a trail I'd left now! What a conflagration it was; the glow could be seen many kilometers away. What's more, this was an act of *sabotage.* For running away they would only give us the same *quarter* we already had. But for malicious destruction of kolkhoz hay they could "put us under" if they wished.

The worst of it is that each mistake increases the likelihood of further mistakes; you lose your self-confidence, your feel for the situation.

The hut had burned down, but the potatoes were baked. The cinders took the place of salt. We ate some of them.

We walked on in the night. Skirted a big village. Found a shovel. Picked it up in case it might be useful. We moved in closer to the Irtysh. And were brought to a halt by a creek. Should we make another detour? It was a nuisance. We looked around a bit and found a boat without oars. Never mind; the shovel would do for an oar. We crossed the creek. Then I strapped the shovel to my back, so that the handle would stick up like the barrel of a gun. In the dark we might pass for hunters.

Soon afterward someone came toward us and we stepped aside. "Petro!" he said. "You've got the wrong man; I'm not Petro."

We walked all night. Slept in a haystack again. We were awakened by a steamer whistle. We stuck our heads out, and saw a wharf quite near. Lorries were carrying melons onto it. Omsk is near, Omsk is near, Omsk is near. Time to shave and get hold of some money.

Kolya keeps on nagging me. "We shan't make it now. What was the good of running away in the first place if you're going to feel sorry for people? Our fate was in the balance, and you had to feel sorry for them. We shan't make it now."

He was right. It seemed so senseless now: we had neither razor nor money; both had been in our hands and we didn't take them. To think that after all those years longing to escape, after showing so much cunning, after crawling under the wire, expecting a bullet in the back any moment, after six days without water, after two weeks crossing the desert—we had not taken what was ours for the taking! How could I go into Omsk unshaven? How were we going to pay for the journey on from Omsk?

We lay through the day in a haystack. Couldn't sleep, of course. About five o'clock Zhdanok says, "Let's go right now and take a look around while it's light." "Certainly not," I say. He says, "It's nearly a month now! You're overdoing the caution! I'm getting out of this and going by myself." I threaten him: "Watch you don't get a knife in you." But of course I would never stab him.

He quieted down and lay still. Then suddenly he rolled out of the stack and walked off. What should I do? Let him go, just like that? I jumped down, too, and went after him. We walked on in broad daylight, following the road along the Irtysh. We sat behind a haystack to talk things over: if we met anyone now we couldn't let him go in case he reported us before it was dark. Kolya carelessly ran out to see whether the road was clear, and a young fellow immediately spotted him. We had to call him over. "Come on over here, pal, and let's watch our troubles go up in smoke." "What troubles have you got?" "Me and my brother-in-law are on holiday, taking a trip on the river. I'm from Omsk and he's a fitter in the ship-repair yards at Pavlodar, and, well, our boat slipped its moorings in the night and got away, all we've got left is what was on the bank. Who are you, then?" "I'm a buoy keeper." "Haven't seen our boat anywhere, have you? In the reeds, maybe?" "No." "Where's your post?" "Over there"—he pointed to a little house. "So let's go to your place, and we'll stew some meat. And have a shave."

Off we go. But the house we'd seen turned out to be that of his neighbor, another buoy keeper, and our man's house was 300 meters farther on. More company—no sooner had we entered the house than the neighbor cycled over to see us, with his sporting gun. He eyed my stubble and questioned me about life in Omsk. Some good, asking a jailbird like me about life outside. I babbled something vague, the gist of which was that the housing situation was bad, the food situation was bad, and the consumer-goods situation was bad—couldn't go far wrong there, I thought. He looked sour and contradicted me—it appeared that he was a Party member. Kolya made soup—we must eat our fill while we could; we might not have another chance till Omsk.

It was a wearisome wait for darkness. We couldn't let either of them leave us. And what if a third came along? At last they both got ready to go and attend to their lights. We offered our help. The Party man refused. "I shall just set two lights and then I have to go to the village. I'm taking my family a load of brushwood. I'll look in again later." I signal to Kolya not to take his eyes off the Party man and at the slightest hint of anything wrong to dive into the bushes. I show him where to meet me. I go with our man. From his boat I inspect the lie of the land and question him about distances. We return at the same time as the neighbor. That sets my mind at rest: he hasn't had time to turn us in yet. Shortly afterward he drives up, as he had said he would, with a load of brushwood on his cart. But instead of driving on, he sits down to sample Kolya's soup. He won't go away. What are we to do? Tie the pair of them up? Shut one in the cellar, tie the other to a bed? . . . They both have papers, and the neighbor has a bicycle and a gun. That's what life on the run does to you—simple hospitality isn't enough; you have to take more by force. . . .

Suddenly—the creak of rowlocks. I look through the window: three men in a boat, which makes it five to two now. My host goes out, and immediately returns for jerry cans. "Foreman's brought kerosene," he says. "Funny he's come himself; it's Sunday today."

Sunday! We had stopped reckoning by the day of the week— it wasn't the name that made one of our days different from another. It had been Sunday evening when we escaped. So that we had been on the run exactly three weeks! What was going on in the camp? The dog pack would have despaired of catching us by now. In three weeks, if we had torn off in a lorry, we could long ago have fixed ourselves up somewhere in Karelia or Byelorussia, got passports and jobs. Or, with a bit of luck, even farther west . . . How galling it would be to have to give in now, after three weeks!

"Right, Kolya—now we're stoked up, what do you say to a hearty crap?" We go out into the bushes and watch what is going on: our host is taking kerosene from the newly arrived boat, and the neighbor with the Party card has also joined them. They are talking about something, but we can't hear what.

They've gone. I send Kolya back to the house on the double. I don't want to leave the buoy keepers alone to talk about us. I myself go quietly to our host's boat. So as not to rattle the chain, I make an effort and pull up the post to which it is attached. I calculate how much time we have: if the foreman buoy keeper has gone to report us, he is seven kilometers, which means about forty minutes, from the village. If there are "red tabs" in the village, it will take them another fifteen minutes or so to get ready and drive over here.

I go into the house. The neighbor is still not ready to leave. He's entertaining them with his conversation. Very strange. So we shall have to take both of them at once. "What about it, Kolya—shall we go and have a wash before bedtime?" (We must agree on a plan.) The moment we go out we hear the tramp of boots in the darkness. Stooping, we can see against the pale sky (the moon hasn't risen yet) men running in line past the bushes to surround the house.

"To the boat," I whisper to Kolya. I run toward the river, slide down the steep bank, fall, and reach the boat. Every second may mean the difference between life and death. But Kolya is missing! Where, oh, where can he have got to? I can't desert him.

At last he comes running along the bank in the darkness, straight toward me. "That you, Kolya?" A flash! A shot, point-blank! I do a swan dive—arms outstretched into the boat. Bursts of submachine-gun fire from the steep bank. Shouts: "We got one of them." They bend over me. "Wounded?" I groan. They drag me out and lead me off. I limp (if I'm injured they will beat me less). In the darkness I surreptitiously throw the two knives into the grass.

Up top, the red tabs ask my name. "Stolyarov." (Maybe I can still wriggle out of it somehow. I am reluctant to give my name —if I do, that's the end of my freedom.) They hit me in the face. "Name!" "Stolyarov." They drag me into the hut, strip me to the waist, tie my hands tightly behind my back with wire that cuts into me. They press the points of their bayonets against my belly. A trickle of blood runs from under one of them. The militiaman who captured me, Senior Lieutenant Sabotazhnikov, jabs his revolver in my face and I can see that it is cocked. "Name!" Resistance is useless. I tell them. "Where's the other one?" He wags his revolver, the bayonets bite deeper. "Where's the other?" I feel happy for Kolya. "We were together," I tell them. "Most likely he was killed."

A security officer with bright blue facings arrived, a Kazakh. He shoved me onto the bed with my hands tied and as I half-sat, half-lay there, began rhythmically striking me in the face—left, right, left, right, as though he were swimming. With every blow my head banged against the wall. "Where's your weapon?" "What weapon?" "You were seen in the night with a gun." So the night hunter we had seen had also betrayed us. "That was a shovel, not a rifle." He didn't believe me and went on hitting me. Suddenly there was no more pain—I had lost consciousness. When I came around someone was saying: "Don't forget, if any one of us is wounded, we'll finish you off on the spot!"

(They must somehow have sensed it: Kolya really did have a gun. It all became clear to me later: when I said "To the boat," Kolya had run the other way, into the bushes. His explanation was that he hadn't understood . . . but there was more to it: he had been itching to go his own way all day, and now he did so. Besides, he had remembered the bicycle. Taking his direction from the shots, he rushed away from the river and crawled back the way he had come. By now it was really dark, and while the whole pack of them were crowding around me, he rose to his feet and ran. Ran and wept as he went, thinking they'd killed me. He ran as far as the second little house, the neighbor's. He kicked in a window and started searching for the gun. He fumbled around until he found it on the wall, and with it a pouch of cartridges. He loaded it. What he was thinking of, so he said, was whether he should avenge me. "Shall I go and take a few shots at them for Zhora?" But he thought better of it. He found the bicycle, and he found an ax. He chopped the door down from inside, put salt into a bag—I don't know whether this seemed the most important thing to him or whether he simply had no time to think—and rode off, first by a dirt lane, then through the village, straight past the soldiers. They thought nothing of it.)

Meanwhile I was put in a cart, still tied up, with two soldiers sitting on top of me, and taken to a state farm two kilometers away. It had a telephone, and it was from there that the forest ranger (he had been in the boat with the foreman buoy keeper) had summoned the red tabs. That's why they had arrived so quickly —because they had been phoned. I hadn't allowed for that.

A scene was enacted by myself and this forester which may seem unpleasant to relate but is typical of what a recaptured prisoner can expect. I wanted to relieve myself—standing up— and someone had to help me, in the most intimate way, since my hands were twisted behind my back. The Tommy-gunners felt that this was beneath them and ordered the forester to go outside with me. In the darkness we walked a little way from where the soldiers stood and as he was assisting me he asked my forgiveness for betraying me. "It's my job. I had no choice."

I didn't answer. How can anybody pass judgment? We had been betrayed by people with duties and people without. Everybody we met had betrayed us, except that old, old man with the gray mane.

I sit in a hut by the highroad, stripped to the waist and bound. I am very thirsty but they give me nothing to drink. The red tabs glare at me like wild beasts, and every one of them looks for an excuse to prod me with the butt of his gun. But here they can't very well kill me: they can kill you when there are only a few of them, and no witnesses. (Their rage is understandable. For so many days now they have been wading among the reeds, with no pause for rest, and eating from cans, with never a hot meal.)

The whole family is in the cottage. The little children look at me curiously but are afraid to come nearer; they even tremble with fear. The militia lieutenant sits drinking vodka with his host, well pleased with his success and the reward it will bring. "Know who he is?" he says boastfully to his host. "He's a colonel, a famous American spy, a major criminal. He was running away to the American Embassy. They've murdered people and eaten them on their way here."

He may even believe it himself. The MVD will have disseminated rumors of this sort to catch us more easily, to make everyone denounce us. They're not satisfied with the advantages of power, weapons, speed of movement—they need the help of slander as well.

(Meanwhile Kolya rides his bicycle along the road past the cottage, with the rifle slung over his shoulder, as though he hadn't a care in the world. He sees a brilliantly lighted cottage, soldiers smoking and noisily talking on the veranda, and through the window, me, half-naked. And he pedals hard for Omsk. Soldiers will lie in wait all night around the bushes where I was caught, and comb them in the morning. Nobody knows yet that the neighboring buoy keeper's bicycle and gun have disappeared—he, too, has probably sloped off to brag over a few drinks.)

When he has reveled long enough in his success—an unheard-of success by local standards—the militia lieutenant gives orders for me to be delivered to the village. Once again they throw me on the cart, and take me to the lockup. (There's always one handy! Every village soviet has one.) Two Tommy-gunners stand guard in the corridor, two more outside the window! An American espionage colonel! They untie my hands but order me to lie on the floor in the middle of the room and not edge toward any of the walls. That is how I spend an October night: lying on the floor, the upper half of my body bare.

In the morning a captain arrives, and bores through me with his eyes. He tosses me my tunic (they'd already sold the rest of my things for drink). Quietly, and with one eye on the door, he asks me a strange question.

"How do you come to know me?"

"I don't know you."

"Then how did you know that the officer in charge of the search was Captain Vorobyov? Do you know what sort of position you've put me in, you swine?"

His name was Vorobyov! And he was a captain! In the night, when we were posing as security troops, I had mentioned Captain Vorobyov, and the workingman whose life I had spared had reported it all carefully. And now the captain was having trouble! If the commander of the pursuit has connections with an escaped prisoner, it's not surprising that three weeks go by and still they can't catch him! . . .

Another pack of officers arrive, shout at me, and among other things ask about Vorobyov. I say that it's a coincidence.

They tied my hands with wire again, removed my shoelaces, and led me through the village in broad daylight. There must have been twenty Tommy-gunners in the escort party. The whole village poured out, women shook their heads, kids ran after me, shouting: "The bandit! They're taking him off to shoot him!"

The wire was cutting into my arms, my shoes fell off at every step, but I held my head high and looked openly and proudly at the villagers: letting them see that I was an honest man.

They were taking me this way as an object lesson, something for these women and children to remember (legendary tales would be told of it twenty years from now). On the edge of the village they bundled me into the back of a truck, bare and seatless, with splintering old boards. Five Tommy-gunners sat with their backs to the cab, so as not to take their eyes off me.

Now I must rewind all those kilometers in which we so rejoiced, all those kilometers which took us farther from the camp. By the roundabout motor road it came to half a thousand. They put handcuffs on my wrists, tightened them to the limit. My hands were behind my back, and I had no means of protecting my face. I lay there more like a block of wood than a man. But this is how they punish our kind.

And then the road became very bad. It rained and rained, and the lorry bumped over the potholes. At every bump the bottom of the lorry scraped my head and face, scratched me, drove splinters into me. Not only could my hands not protect my face, but they themselves were cut more severely than ever over the bumps, and it felt as though the handcuffs were sawing through my wrists. I tried to crawl to the side and sit there with my back propped against it. No good! There was nothing to hold on to, and at the first big bump I was hurled across the floor, and found myself sprawling helplessly. Sometimes I was tossed and hit the boards so violently that I thought my insides would jump out. I couldn't stay on my back: it would tear my hands off at the wrist. I turn onto my side—no good. I roll over on my belly—no good. I try arching my neck so as to raise my head and save it from these blows. But my neck gets tired, my head droops, and my face strikes the boards.

The five guards watch my torment, unconcerned.

This trip will form part of their psychological training.

Lieutenant Yakovlev, who is riding in the cab, looks into the back at every stop and says with a grin: "Haven't escaped, then?" I ask permission to relieve myself, and he guffaws. "Go on, do it in your trousers; we don't mind!" I ask him to take off the handcuffs, and he laughs. "Lucky you weren't caught by the lad who was on duty when you went under the wire. You wouldn't be alive now."

The day before I had been glad that the beatings so far were "less than I had earned." But why damage your fists, when the back of a lorry will do it all for you? Every inch of my body was bruised and lacerated. My hands were being sawn off. My head was splitting with pain. My face was battered, full of splinters from the boards; my skin was in ribbons.[6]

We traveled the whole day and almost all night.

When I stopped struggling with the lorry, and ceased to feel my head banging against the boards, one of the sentries couldn't stand it any more, put a sack under my head, eased the handcuffs while no one was looking, and bending over me, whispered, "It's all right, hold on, we'll soon be there." (What prompted the lad to do it? Who was responsible for his upbringing? Not Maxim Gorky, and not his company political officer, that's for sure.)

Ekibastuz. A cordon. "Get out!" I couldn't stand up. (And if I had, they would have made me run the gauntlet to celebrate.) They let down the side, and yanked me out onto the ground. The camp guards, too, came out to have a look and a laugh at me: "Ooh, you *aggressor,* you!" somebody yelled.

They dragged me through the guardhouse and into the Disciplinary Barracks. They didn't shove me in solitary but straight into

6. Moreover, Tenno had hemophilia. He shrank from none of the risks of escaping, but a single scratch could have cost him his life.

the common cell so that anyone who fancied a bid for freedom could take a look at me.

In the cell I was lifted by gentle hands and placed on the upper bed platform. But they had no food to give me until the morning rations came.

That same night, Kolya was riding on toward Omsk. He avoided the traffic, and whenever he saw headlights, rode into the steppe and lay down. Then in some lonely homestead he squeezed into a henhouse, and gratified the urge which had haunted him all the time he was on the run—by wringing the necks of three hens and tucking them in his sack. The others started squawking, so he hurried away.

The irresolution which had made us so unsteady after our first mistakes tightened its hold on Kolya now that I was captured. Easily swayed and impressionable, he was fleeing now in desperation, unable to think clearly what to do next. He was incapable of realizing the most obvious of facts: that the disappearance of the bicycle and the gun would of course have been discovered by now, so that they no longer camouflaged him, and he ought to throw them away first thing in the morning as too conspicuous; and also that he should not approach Omsk from that side and by the highway, but after a wide detour, by wasteland and back ways. The gun and the bicycle should be sold quickly, and he would have the money he needed. But he sat for half a day in the bushes near the Irtysh, then yet again lost patience before nightfall and set off by footpaths along the river. Very probably his description had already been broadcast by the local radio station—they have fewer inhibitions about this in Siberia than in the European part of the country.

He rode up to a little house and went in. Inside were an old woman and her thirty-year-old daughter. There was also a radio. By an extraordinary coincidence, a voice was singing:

> From Sakhalin a convict fleeing
> By narrow tracks and hidden trails . . .

Kolya went to pieces and began shedding tears. "What are you so unhappy about?" the woman asked. Their sympathy caused Kolya to weep unashamedly. They began comforting him. "I'm all alone, abandoned by everybody," he explained. "Get married, then," the old woman said, whether in jest or in earnest. "My girl's single, too." Kolya, more maudlin than ever, started taking peeps at the would-be bride. She gave the matter a businesslike twist: "Got any money for vodka?" Kolya dug out his last few rubles, but there wasn't enough. "Never mind, I'll give you more later." She went off. "Oh, yes," Kolya remembered, "I've shot some partridges. Cook the marriage feast, mother-in-law." The old woman took them. "Hey, these are hens!" "So it was dark when I shot them; couldn't tell the difference." "Yes, but why have their necks been wrung?"

Kolya asked for a smoke, and the old woman asked her daughter's suitor for money in return for some makhorka. Kolya took his cap off, and the old woman became agitated. "You're a convict, aren't you, with your head shaven like that? Go away while you're safe. Or else when my daughter comes back we'll turn you in!"

All the time the thought was going round and round in Kolya's head: Why did we take pity on free people there on the Irtysh, when free people have no pity for us? He took a Moscow jacket down from the wall (it was getting cold out of doors, and he was wearing only a suit) and put it on: just his size. The old woman yelled, "I'll hand you over to the militia!" But Kolya looked through the window and saw the daughter approaching with someone on a bicycle. She had already informed on him!

Only one thing for it—"Makhmadera!" He seized the gun. "Into the corner! Lie down!" he said to the old woman. He stood against the wall, let the other two come through the door, and ordered them to lie down. To the man he said: "And you give me your shoes, for a wedding present! Take them off one at a time!" With the gun trained on him, the man took them off and Kolya put them on, throwing away his wornout camp shoes, and threatened that he would wing anyone who followed him out.

Off he rode. But the other man dashed after him on his own bicycle. Kolya dismounted, and put the gun to his shoulder. "Stop! Leave the bike there! Get away from it!" He drove the man away, went over to the bike, broke its spokes, slit a tire with his knife, and rode on.

Soon he came out onto the highway. Omsk was ahead. So he just rode straight on. There was a bus stop. Women were digging potatoes in their gardens. A motorcycle carrying three workmen in jerkins tagged behind him. It drove on steadily for some time, then suddenly went straight for Kolya so that the sidecar struck him and knocked him off. They all jumped off, piled on top of Zhdanok, and hit him on the head with a pistol.

The women in the vegetable garden shrieked: "What are you doing that for? What's he done to you?"

What, indeed, had he *done to them?* . . .

But who has done, and will yet do, what to whom is beyond the understanding of the common people. Under their jerkins all three turned out to be wearing uniforms (the operations group had been on duty round the clock day after day at the entrance to the city). The women got their answer: "He's a murderer." That was the simplest thing to say. And the women, trusting the Law, went back to digging potatoes.

The first thing the operations group did was to ask the penniless runaway if he had any money. Kolya said quite honestly that he hadn't. They started searching him, and in one pocket of his new Moscow jacket found fifty rubles. They confiscated it, drove to an eating house, and spent the lot on food and drink. They did, however, feed Kolya, too.

So we came to anchor in jail for a long time. We were not tried until July of the following year. For nine months we festered in the Disciplinary Barracks, except when we were dragged out occasionally for interrogation. This was conducted by Chief Prison Officer Machekhovsky and Security Lieutenant Weinstein. What the interrogators wanted to know was which prisoners had helped us. Who among the civilian personnel had "conspired with us" to switch off the lights at the moment of our escape? (We didn't, of course, explain that our plan had been quite different, and that the lights' going out had only been a hindrance to us.) Where was our rendezvous in Omsk? Which frontier were we intending to cross eventually? (They found it incredible that people might want to stay in their native land.) "We were running away to Moscow to the Central Committee, to tell them about illegal arrests, and that's all there is to it!" They didn't believe us.

Having failed to get anything "interesting" out of us, they pinned on us the usual escaper's posy: Article 58-14 (Counter-Revolutionary sabotage); Article 59-3 (banditry); the "Four-sixths" decree, Article "One-two" (robbery carried out by a gang); the same decree, Article "Two-two" (armed robbery with violence endangering life); Article 182 (making and carrying an offensive weapon other than a firearm).

But this daunting array of charges threatened us with chains no heavier than those we wore. The penal practice of the courts had long exceeded all reasonable bounds and, on these charges, promised nothing worse than the twenty-five years which a Baptist could be given for saying his prayers, and which we were serving before we tried to escape. The only difference now was that when the roll was called we should have to say "end of sentence—1975" instead of "1973."

Hardly a palpable difference for us in 1951!

Only once did the interrogation take a menacing turn—when they promised to try us as economic *disrupters.* This innocent word was more dangerous than the hackneyed "saboteur," "bandit," "robber," "thief." This word opened up the possibility of capital punishment, which had been introduced about a year before.

We were "disrupters" because we had brought disorder into the economy of the people's state. As the interrogators explained it, 102,000 rubles had been spent on our recapture: some work sites had been at a standstill for several days (the prisoners were not marched out to work because their guards had been called off to join in the hunt); twenty-three vehicles had carried soldiers day and night about the steppe, and had spent their annual allotment of petrol in three weeks; operations groups had been dispatched to all neighboring towns and settlements; a nationwide search had been ordered, and four hundred pictures of myself and four hundred of Kolya distributed throughout the country.

We listened to this inventory with pride. . . .

Well, they sentenced us to twenty-five each.

When the reader picks up this book our sentences will probably still not be at an end.[7]

7. Before the reader could pick up this book, Georgi Pavlovich Tenno, athlete and theorist of athletics into the bargain, died on October 22, 1967, of a cancer which ran its course very quickly. Already bedridden, he lived barely long enough to read through these chapters and amend them with fingers which were beginning to lose their feeling. This was not the way he had promised himself and his friends to die! Once it had been his plan of escape that fired his passion; later it was the thought of death in battle. He used to say that he was determined to take with him a dozen murderers, first among them Vyachik Karzuby (i.e., Molotov), and, at all costs, Khvat (investigator in the Vavilov case). This would not be murder, but judicial execution, given that the law of the state protected murderers. "With your first shots you have already justified your existence," said Tenno, "and you

Another consequence of Tenno's escape attempt was that the concert party at the CES was disbanded for a year (on account of the ill-starred sketch).

Because culture is a good thing. But culture must serve oppression, not freedom.

---

gladly overfulfill the plan." But his illness came upon him suddenly, instantly robbing him of his strength and making it impossible for him to find a weapon. When he was already sick, Tenno went around posting copies of my letter to the Writers' Congress in different parts of Moscow. It was his wish to be buried in Estonia. The pastor was also an ex-prisoner —both in Hitler's and in Stalin's camps.

Molotov lives on in safety, leafing through old newspapers and writing the memoirs of a public executioner, while Khvat, too, is still peacefully spending his pension at 41 Gorky Street.

Chapter 8

■

Escapes—Morale and Mechanics

Escapes from Corrective Labor Camps, provided they were not to somewhere like Vienna or across the Bering Strait, were apparently viewed by Gulag's rulers and by Gulag's regulations with resignation. They saw them as only natural, a manifestation of the waste which is unavoidable in any overextended economic enterprise—a phenomenon of the same sort as cattle losses from disease or starvation, the logs that sink instead of floating, the gap in a wall where a half-brick was used instead of a whole one.

It was different in the Special Camps. In accordance with the particular wish of the Father of the Peoples, these camps were equipped with greatly reinforced defenses and with greatly reinforced armament, at the modern motorized infantry level (these are the units which will never lay down their arms even under the most general disarmament agreement). In these camps they did not keep any "class ally" type of prisoners,* whose escape could cause no great damage. Here it was no longer possible to plead that the men under arms were too few or their weapons too old. At the moment of their foundation it was laid down in the instructions for Special Camps that there *could be no escape* from them, because if one of these prisoners escaped it was just as though a major spy had crossed the frontier, and a blot on the political record of the camp administration and of the officers commanding the convoy troops.

But from that very moment 58's to a man started getting, not *tenners* as before, but *quarters*—i.e., the limit allowed by the Criminal Code. This senseless, across-the-board increase in severity carried with it one disadvantage: just as murderers were undeterred from fresh murders (each time their *tenner* was merely slightly updated), so now political prisoners were no longer deterred by the Criminal Code from trying to escape.

Besides, the people herded into these camps were not the sort who try to rationalize and justify the arbitrary behavior of camp authorities in the light of the One and Only True Theory, but sturdy, healthy lads who had crawled on their bellies all through the war, whose fingers were still cramped from clutching hand grenades. Georgi Tenno, Ivan Vorobyov, Vasily Bryukhin, their comrades, and many like them in other camps even without arms proved a match for the motorized infantry equipment of the new regular army guard.

And although there were fewer escapes from the Special Camps than from the Corrective Labor Camps (and anyway, the Special Camps did not exist for so many years), they were rougher, grimmer, more ruthless, more desperate, and therefore more glorious.

Stories told about them can help us to make up our minds whether our people really was so long-suffering, really was so humbly submissive in those years.

Here are just a few of them.

One attempt was made a year before that of Tenno, and served as the model for it. In September, 1949, two convicts escaped from the First Division of Steplag (Rudnik, Dzhezkazgan)—Grigory Kudla, a tough, steady, level-headed old man, a Ukrainian (but when his dander was up he had the temper of a Zaporozhian Cossack, and even the hardened criminals were afraid of him), and Ivan Dushechkin, a quiet Byelorussian some thirty-five years old. In the pit where they worked they found a prospecting shaft in an old workings, with a grating at its upper end. When they were on night shift they gradually loosened this grating, and at the same time they took into the shaft dried crusts, knives, and a hot-water bottle stolen from the Medical Section. On the night of their escape attempt, once down the pit each of them separately informed the foreman that he felt unwell, couldn't work, and would lie down a bit. At night there were no warders underground; the foreman was the sole representative of authority and he had to bully discreetly or else he might be found with his head smashed in. The escapers filled the hot-water bottle, took their provisions, and went into the prospecting shaft. They forced the grating and crawled out. The exit turned out to be near the watchtowers but outside the camp boundary. They walked off unnoticed.

From Dzhezkazgan they bore northwest through the desert. They lay down in the daytime and walked at night. Not once did they come across water, and after a week Dushechkin no longer felt like standing up. Kudla got him on his feet with the hope that there might be water in the hills ahead. They dragged themselves that far, but the hollows held no water, only mud. Then Dushechkin said, "I can't go on anyway. *Cut my throat* and drink my blood."

You moralists! What was the right thing to do? Kudla, too, could no longer see straight. Dushechkin was going to die—why should Kudla perish, too? But if he found water soon afterward, *how* could he live with the thought of Dushechkin for the rest of his days? I'll go on a bit, Kudla decided, and if in the morning I come back without water I'll put him out of his misery, and we needn't both perish. Kudla staggered to a hillock, saw a cleft in it and—just as in the most improbable of novels—in the cleft there was water! Kudla slithered to it, fell flat on his face, and drank and drank. (Only in the morning had he eyes for the tadpoles and waterweed in it.) He went back to Dushechkin with the hot-water bottle full. "I've brought you some water—yes, water." Dushechkin couldn't believe it, drank, and still didn't believe it (for hours he had been imagining that he was drinking). They dragged themselves as far as the cleft and stayed there drinking.

When they had drunk, hunger set in. But the following night they climbed over a ridge and went down into a valley like the promised land: with a river, grass, bushes, horses, life. When it got dark Kudla crept up to the horses and killed one of them. They drank its blood straight from the wounds. (Partisans of peace! That very year you were loudly in session in Vienna or Stockholm, and sipping cocktails through straws. Did it occur to you that compatriots of the versifier Tikhonov and the journalist Ehrenburg were sucking the blood of dead horses? Did they explain to you in their speeches that that was the meaning of *peace,* Soviet style?)

They roasted the horse's flesh on fires, ate lengthily, and walked on. They by-passed Amangeldy on the Turgai, but on the high-road Kazakhs in a lorry going their way asked to see their papers and threatened to hand them over to the militia.

Farther on they frequently came across streams and pools. Kudla also caught and killed a ram. By now they had been a *month* on the run! October was nearing its end; it was getting cold.

In the first wood they reached they found a dugout and set up house in it. They couldn't bring themselves to leave this land of plenty. That they settled in such surroundings, that their native places did not call to them or promise them a more peaceful life, meant that their escape lacked a goal and was doomed to fail.

At night they would raid the village nearby, filch a pot or break into a pantry for flour, salt, an ax, some crockery. (Inevitably the escaper, like the partisan, soon becomes a thief, preying on the peaceful folk all around him.) Another time they took a cow from the village and slaughtered it in the forest. But then the first snow came, and to avoid leaving tracks they had to sit tight in their dugout. Kudla went out just once for brushwood and the forester immediately opened fire on him. "So you're the thieves, are you! You're the ones who stole the cow." Sure enough, traces of blood were found around the dugout. They were taken to the village and locked up. The people shouted that they should be shot out of hand and no mercy shown to them. But an investigating officer arrived from the district center with the picture sent around to assist the nationwide search, and addressed the villagers. "Well done!" he said. "These aren't thieves you've caught, but dangerous political criminals."

Suddenly there was a complete change of attitude. The owner of the cow, a Chechen as it turned out, brought the prisoners bread, mutton, and even some money, collected by the Chechens. "What a pity," he said. "You should have come and told me who you were and I'd have given you everything you wanted!" (There is no reason to doubt it; that's how the Chechens are.) Kudla burst into tears. After so many years of savagery, he couldn't stand sympathy.

The prisoners were removed to Kustanai and put in the railroad jail, where their captors not only took away the Chechen's offering (and pocketed it), but gave them no food at all. (Didn't Korney-chuk tell you about it at the Peace Congress?) Before they were put on the train out of Kustanai, they were made to kneel on the station platform with their hands handcuffed behind their backs. They were kept like that for some time, for the whole world to see.

If it had been on a station platform in Moscow, Leningrad, or Kiev, or any other flourishing city, everybody would have passed by the gray-headed old man, kneeling and manacled, like a figure in a Repin picture, without noticing him or turning around to look —publishing executives, progressive film producers, lecturers on humanism, army officers, not to mention trade union and Party officials. And all ordinary, undistinguished citizens occupying no position worth mentioning would also have tried to go by without noticing, in case the guard asked their names and made a note of them—because if you have a residence permit for Moscow, where the shops are so good, you must not take risks. . . . (Easy enough to understand in 1949—but would it have been any different in 1965? Would our educated youngsters have stopped to intercede with his escort for the gray-haired old man in handcuffs and on his knees?)

The people of Kustanai, however, had little to lose. They were all either "sworn enemies," or persons with black marks against them, or simply exiles. They started crowding around the prisoners, and tossing them makhorka, cigarettes, bread. Kudla's wrists were shackled behind his back, so he bent over to pick up a piece of bread with his teeth—but the guard *kicked it out of his mouth.* Kudla rolled over, and again groveled to pick it up—and the guard kicked the bread farther away. (You progressive film makers, when you are taking shots of inoffensive "senior citizens"—perhaps you will remember this scene and this old man?) The people began pressing forward and making a noise. "Let them go! Let them go!" A militia squad appeared. The policemen had the advantage and dispersed the people.

The train pulled in, and the prisoners were loaded for transport to the Kengir jail.

Escape attempts in Kazakhstan are as monotonous as the steppe itself—but perhaps this monotony makes it easier to understand the most important thing?

There was another escape from a mine, also in Dzhezkazgan, but in 1951 this time: three men climbed an old shaft to the surface at night and walked for three nights. Thirst made them desperate, and when they saw some yurts two of them suggested that they should go over and get a drink from the Kazakhs. But Stepan ———— refused and watched them from a hill. He saw his comrades enter a yurt, and come out running, pursued by a number of Kazakhs, who quickly caught them. Stepan, a puny little man, went away, keeping to the low places, and continued his flight alone, with a knife as his sole possession. He tried to make for the northwest, but was continually changing course to avoid people—he preferred wild animals. He cut himself a stick to hunt gophers and jerboas: he would fling it at them from some distance while they were sitting up on their hind legs by their burrows and squeaking, and he killed some of them in this way. He sucked their blood as best he could and roasted their flesh on a fire of dry steppe gorse.

It was a fire that gave him away. One day Stepan saw a Kazakh horseman in a big red-brown fur hat galloping toward him, and he barely had time to hide his shashlik under some gorse so that the Kazakh would not see what choice food he was eating. The Kazakh rode up and asked who he was and where from. Stepan explained that he worked in the manganese mine at Dzhezdy (free men as well as convicts were employed there), and was on his way to a state farm 150 kilometers away to see his wife. The Kazakh asked the name of the farm. Stepan chose the most plausible—"The Stalin State Farm."

Son of the steppes! Why couldn't you gallop on your way! What harm had the poor wretch done you? But no! The Kazakh said menacingly: "You sit prison! You go with me!" Stepan cursed him and walked on. The Kazakh rode alongside, ordering him to come quietly. Then he galloped off a little way, waving his arm and calling to his fellows. But the steppe was deserted. Son of the steppes! Why, oh, why could you not just leave him? You could see that he had hundreds of versts of steppe to cross, with nothing but a bare stick in his hands and without food, so that he would perish anyway. Did you need a kilogram of tea so badly?

In the course of that week, living on equal terms with the wild animals, Stepan had grown used to the rustling and hissing sounds of the desert: suddenly his ear caught a new whistling sound in the air, and he was not mentally aware of his danger but like an animal sensed it in the pit of his stomach, and leaped to one side. This saved him! The Kazakh, he realized, had tried to lasso him, but he had dodged the noose.

Hunting bipeds! A man's life for a kilogram of tea! The Kazakh swore and hauled in his lasso, and Stepan went on, warily taking care not to let his enemy out of his sight. The other rode up closer, coiled his rope, and flung it again. As soon as he had made his throw, Stepan rushed at him, struck him on the head with the stick, and knocked him off his horse. (He had barely strength enough for it, but he was fighting for his life.) "Here's your reward, friend!" Relentlessly, with all the savagery of one beast goring another, Stepan began beating him. But when he saw blood he stopped. He took both lasso and whip from the Kazakh and scrambled onto the horse. There was a saddlebag with provisions.

Stepan was on the run for quite a while longer—another two weeks or so—but everywhere he went he religiously avoided his worst enemies: people, his fellow countrymen. He parted with the horse, tried to swim some river or other (although he couldn't swim!), tried making a raft of rushes (and of course couldn't do that either); he hunted, and got away from some large animal, perhaps a bear, in the dark. Then one day, tormented beyond endurance by hunger, thirst, and fatigue, and the longing for hot food, he made up his mind to enter a lonely yurt and beg something. There was a small enclosure in front of the yurt, with an adobe wall, and when he was already close to it Stepan belatedly saw two saddled horses standing there, and a young Kazakh, in a bemedaled tunic and breeches, coming toward him. He had missed his chance to run and thought he was done for. The Kazakh had stepped outside for a breather. He was very drunk, was delighted to see Stepan, and seemed not to notice his tattered and scarcely human appearance. "Come in, come in, be our guest!" Inside the yurt sat an identical young Kazakh with medals, and an old man: the two brothers, who had seen service at the front, were now both important people in Alma-Ata, and had borrowed horses from a kolkhoz to gallop over and pay their respects to their father in his yurt. These two young fellows had tasted war, and it had made human beings of them. Besides, they were drunk and bursting with drunken good nature (that good nature which, though he made it his business to do so, the Great Stalin never fully succeeded in eradicating). They were happy that another guest had joined the feast, though he was only a simple mine-worker on his way to Orsk, where his wife was expecting a baby at any moment. They did not ask to see his papers, but gave him food and drink and a place to sleep. That sort of thing sometimes happens, too. . . . (Is drink always man's enemy? Does it not sometimes bring out the best in him?)

Stepan woke up before his hosts, and fearing a trap in spite of everything, he went out. No, both horses were just where they had been, and he could have galloped off on one of them immediately. But he was not the man to harm kind people—and he left on foot.

After a few more days' walking, he started meeting cars and lorries. He was always quick enough to get out of their way. At last he reached a railway line and followed it until that same night he found himself near Orsk station. All he had to do was board a train! He had won! He had performed a miracle—crossed a vast expanse of desert all alone, with nothing but a homemade knife and a stick—and now he had reached his goal.

Suddenly by the light of the station lamps he saw soldiers pacing along the tracks. So he continued on foot along a cart track parallel with the railway line. He no longer troubled to hide, even when morning came: because he was in Russia now, his native land! A cloud of dust came toward him and for the first time Stepan did not run away from a car. Out of this first real Russian car jumped a real Russian militiaman. "Who are you? Show me your papers." Stepan explained that he was a tractor driver, looking for work. As it happened, a kolkhoz chairman was with the militiaman. "Let him be! I desperately need tractor drivers! How many people have papers down on the farm?"

They traveled all day, haggling as they went, stopping for drinks and snacks, but just before nightfall Stepan couldn't stand any more and ran for the woods, which were some 200 meters away. The militiaman rose to the occasion and fired!

Fired again! Stepan had to stop. They tied him up.

It may well be that his trail was cold, that they had given him up for dead, that the soldiers at Orsk had been lying in wait for somebody else, because the militiaman was for releasing him, and at the district MVD station they made a great fuss over him to begin with—gave him tea and sandwiches and Kazbek cigarettes, and the commandant questioned him in person, addressing him politely (you never know with these spies—he'll be taken to Moscow tomorrow, and might easily lodge a complaint). "Where's your transmitter? Been dropped here to make a map or two, have you—which service are you with?" Stepan was puzzled. "I've never worked in the geological service; I'm more of a miner."

This escape ended with something worse than sandwiches or even physical capture. When he got back to the camp he was beaten lengthily and unmercifully. Worn out and broken by all his sufferings, Stepan fell lower than ever before: he *signed on* with the Kengir security officer Belyaev to help him flush out would-be escapers. He became a sort of decoy duck. He gave one or two cellmates in the Kengir jail a detailed account of his escape, and watched their reactions. If there was any response, any obvious hankering to repeat the attempt, Stepan ———— reported it to the godfather.

The incidental cruelties which mark any difficult escape attempt were seen vividly enlarged in a bloody and confused breakout, also from Dzhezkazgan and also in summer, 1951.

Six prisoners escaping by night from a pit began by killing a seventh, whom they believed to be a stool pigeon. Then they climbed an old prospecting shaft out onto the steppe. The six prisoners included people of very different stripe and they immediately decided to separate. This would have been the right thing to do, if only they had had a sensible plan.

But one of them went straight to the settlement where the free workers lived, right next to the camp, and knocked on the window of his woman friend. His intention was not to hide, to wait awhile under the floorboards or in the attic (that would have been very sensible), but simply to have a good time with her while it lasted (we recognize at once the characteristics of the professional criminal). He whooped it up for a day and a night, and then the following evening put on her former husband's suit and took her to a film show in the club. Some of the jailers from the camp were there, recognized him, and *collared* him immediately.

Two of the others, Georgians, thoughtlessly sure of themselves, walked to the station and got on the train to Karaganda. But from Dzhezkazgan, apart from cattle trails and escapers' trails, there is no other way through to the outside world except this one—toward Karaganda and by train. Along the line there are camps, and at every station there is a security post, so that they were both *collared* before they reached Karaganda.

The other three took the most difficult road—to the southwest. There were no people, but there was no water either. The elderly Ukrainian, Prokopenko, who had seen active service, had a map and persuaded them to choose this route, telling them that *he* would find them water. His companions were a Crimean Tatar turned into a criminal by the camps, and a foul, "bitching" thief. They went on for four days and nights without food or water. When they could stand it no longer, the Tatar and the thief told Prokopenko: "We've decided to finish with you." He didn't understand. "What do you mean, pals? Do you want to go your own way?" "No, to *finish* you. We can't all get through." Prokopenko started pleading with them. He slit the lining of his cap and took out a photograph of his wife and children, hoping to stir their pity. "Brothers! Brothers! I thought we were all on our way to freedom together! I'll get you through! There should be a well soon! There's bound to be water! Hang on a bit! Have some mercy!"

But they stabbed him to death, hoping to quench their thirst with his blood. They cut his veins, but the blood wouldn't flow—it had curdled immediately! . . .

Another striking scene: two men bending over another on the steppe, wondering why he wouldn't bleed . . .

Eyeing each other like wolves, because now one of them must die, they went on in the direction which the "old boy" had pointed out to them, and *two hours later* found a well!

The very next day they were sighted from a plane and captured.

They admitted it all under interrogation, the camp got to know about it—and decided to avenge Prokopenko by *knocking off* the pair of them. But they were kept in a separate cell and taken elsewhere for trial.

You may believe, if you wish, that everything depends on the stars under which an escape begins. Your plans have been laid oh, so carefully, so very long in advance, but then at the crucial moment the lights go out in the compound and your chance of seizing a lorry goes up in smoke. Whereas sometimes when an attempt is made on the spur of the moment, circumstances fall into place as though made to order.

In summer, 1948, in Dzhezkazgan again, First Division (not yet a Special Camp), a dump truck was detailed one morning to take on a load of sand at a quarry some distance away and deliver it to the cement mixers. The sandpit was not a *work site*, which meant that it was not guarded and that the loaders—three long-sentence prisoners, one serving a *tenner*, the other two *quarters*—had to be taken on the lorry. Their escort consisted of a lance corporal and two soldiers, and the driver was a nonpolitical offender, a trusty. Here was a chance! But chances must be seized as quickly as they arrive. A decision had to be taken and a plan concerted—all in sight and hearing of the guards, who stood by while the sand was loaded. The biographies of all three were identical—and like those of millions at that time: first the front, then German prisoner-of-war camps, escape, recapture, punitive concentration camps, liberation when the war ended, and by way of thanks for it all—imprisonment by their own side. They hadn't been afraid to flee across Germany; what could stop them from trying it at home? They finished loading. The corporal took his seat in the cab. The two soldiers sat to the front of the lorry, backs to the cab, with their Tommy guns trained on the convicts, who sat on the sand to the rear. As soon as they drove out of the quarry the prisoners exchanged signals, threw sand in the eyes of the guards, and piled on top of them. They took away their Tommy guns, and stunned the corporal with a blow from a gun butt through the cab window. The lorry stopped; the driver was almost dead with fright. "Don't be afraid," they told him. "We won't hurt you—you aren't one of those dogs! Dump your load!" The engine raced and the sand, that precious sand, worth more than its weight in gold because it had brought them freedom, poured onto the ground.

Here, too, as in almost all escapes—let history not forget it! —the slaves showed themselves more generous than their guards: didn't kill them, didn't beat them up, merely ordered them to remove their clothes and their boots, and released them in their underwear. "What about you, driver, are you with us or with them?" "You, of course—what do you think?" the driver decided.

To confuse the barefooted guards (this was the price of their clemency), they drove first to the west (on the flat steppe you can drive where you like), then one of them changed into the corporal's clothes and the other two into those of the soldiers, and they sped northward; they were all armed. The driver had a pass; no one could suspect them. All the same, whenever their path crossed a telegraph route they broke the wires to disrupt communications (they tied a stone to a rope, slung it over the wires to weigh them down, then tugged at them with a hook). This took time, but gained them more in the end. They tore on at full speed all day long until the odometer had clocked up 300 kilometers and the petrol gauge registered zero. They began sizing up passing cars. A Pobeda came along. They stopped it. "Sorry, comrade, we're only doing our duty. Please let me see your papers." VIP's, they turned out to be. District Party bosses visiting their kolkhozes, to inspect, inspire, or maybe just to eat beshbarmak. "Right, out you get! Strip!" Don't shoot us, the big boys implored. The escapers led them out on the steppe, tied them up, took their documents, and rolled off in the Pobeda. It was not till evening that the soldiers whom they had stripped earlier in the day reached the nearest pit, only to hear from the watchtower: "Don't come any nearer!" "We're soldiers like you." "Oh, no you aren't—not while you're walking around in your underpants!"

As it happened, the Pobeda's tank wasn't full. When they had driven about 200 kilometers, the petrol ran out and there was nothing in the jerry can either. It was getting dark by then. They saw some horses grazing, managed to catch them although they had no bridles, and galloped bareback. The driver fell from his horse and hurt his leg. They suggested that he get up behind another rider. He refused. "Don't be afraid, lads, I won't rat on you!" They gave him some money, and the papers from the Pobeda, and galloped away. After the driver, no one ever saw them again. They were never taken back to their camp. And so the lads had left their *quarters* and their "ten-no-change" behind in the security officer's safe. The "green prosecutor" favors the bold!

The driver kept his word and did not give them away. He fixed himself up in a kolkhoz near Petropavlovsk and lived in peace for four years. Love of art was his undoing. A good accordionist, he performed in the kolkhoz club, then competed at amateur festivals, first in the district center, then in the provincial capital. He himself had practically forgotten his former life, but one of the Dzhezkazgan jailers was in the audience and recognized him. He was arrested as soon as he left the stage—and this time they slapped twenty-five years on him under Article 58. He was sent back to Dzhezkazgan.

■

In a category of their own we must put those escapes which originate not in a despairing impulse but in technical calculations and the love of fine workmanship.

A celebrated scheme for escaping by rail was conceived in Kengir. Freight trains carrying cement or asbestos were regularly pulled in at one of the work sites for unloading. They were unloaded within the restricted area, and left empty. And five convicts planned their escape as follows. They made a false end wall for a heavy boxcar, and what is more, hinged it like a folding screen, so that when they dragged it up to the car it looked like nothing more than a wide ramp, convenient for wheelbarrows. The plan was this: while the boxcar was being unloaded, the convicts were in charge of it; they would haul their contraption inside and open it out; clamp it to the solid side of the boxcar; stand, all five of them, with their backs to the wall and raise the false wall into position with ropes. The boxcar was completely covered with asbestos dust, and so was the board. A casual eye would not see the difference in the boxcar's depth. But the timing was tricky. They had to finish unloading the train ready for its departure while the convicts were still on the work site, but they couldn't board it too soon: they must be sure that they would be moved immediately. They were making their last-minute rush, complete with knives and provisions—and suddenly one of the escapers caught his foot in a switch and broke a leg. This held them up, and they didn't have time to complete the installation before the guards checked the train. So they were discovered. A full-dress investigation and trial followed.[1]

The same idea was adopted by trainee-pilot Batanov, in a single-handed attempt. Doorframes were made at the Ekibastuz wood-works for delivery to building sites. Work at the plant went on all around the clock and the guards never left the towers. But at the building sites a guard was mounted only in the daytime. With the help of friends, Batanov was boarded up with a frame, loaded onto a lorry, and unloaded at a building site. His friends back at the woodworks muddled the count when the next shift came on, so that he was not missed that evening; at the building site, he released himself from his box and walked away. However, he was seized that same night on the road to Pavlodar. (This was a year after his other attempt, when they punctured a tire as he was escaping in a car.)

Escapes successful and escapes frustrated at the start; events which were already making the ground hot underfoot;[2] the deeply-thought-out disciplinary decisions of prison officers; sit-down strikes and other forms of defiance—these were the reasons why the ranks of the Disciplinary Brigade at Ekibastuz swelled steadily. The two stone wings of the prison and the Disciplinary Barracks (hut No. 2, near the staff building) could no longer hold them. Another Disciplinary Barracks was established (hut No. 8), specially for the Banderists.

After every fresh escape and every fresh disturbance, the regime

1. However, my wardmate in the Tashkent cancer clinic, an Uzbek camp guard, told me that this attempt—of which he spoke with reluctant admiration—had in fact succeeded.
2. See Chapter 10.

in all three Special Sections became more and more severe. (For the historian of the criminal world we may note that the "bitches" in the Ekibastuz camp jail grumbled about it. "Bastards! Time you gave up this escaping. Because you keep trying to escape, they won't let us breathe any more. You can get your mug bashed in for this sort of thing in an ordinary camp." In other words, they said what the bosses wanted them to say.)

In summer, 1951, Disciplinary Barracks No. 8 took it into their heads to escape all together. They were about thirty meters from the camp boundary and made up their minds to tunnel. But too many tongues wagged and the Ukrainian lads discussed it almost openly among themselves, never thinking that any member of the Bandera army could be an informer—but some of them were. Their tunnel was only a few meters long when they were sold.

The leaders of Disciplinary Barracks No. 2 were greatly vexed by this noisy stunt—not because they feared reprisals, as the "bitches" did, but because they, too, were only thirty meters from the fence, and had planned and made a start on a high-class tunnel before hut No. 8. Now they were afraid that if the same thought had occurred to both Disciplinary Barracks, the dog pack would realize it and check. But the Ekibastuz bosses were more afraid of escapes in vehicles, and made it their chief concern to dig trenches a meter deep around all the work sites and the living area, so that any vehicle trying to leave would plunge into them. As in the Middle Ages, walls were not enough, and moats, too, were needed. An excavator neatly and accurately scooped out trenches of this kind, one after another, around all the sites.

Disciplinary Barracks No. 2 was a small compound, hemmed in with barbed wire, inside the big Ekibastuz compound. Its gate was always locked. Apart from the time spent at the limekilns, the disciplinary regime prisoners were allowed outside only for twenty minutes' exercise in their little yard. For the rest of the time they were locked up in their barracks, and crossed the main compound only to and from work line-up. They were not allowed into the general mess hall at all; the cooks brought food to them in mess tins.

As far as they were concerned, the limekilns were a chance to enjoy the sun and rest a bit, and they took care not to overstrain themselves shoveling noxious lime. When on top of this a murder took place at the end of August, 1951 (the criminal Aspanov killed the escaper Anikin with a crowbar: Anikin had crossed the wire with the help of a snowdrift during a blizzard, but had been recaptured a day later, which was why he was in the Disciplinary Barracks [see also Part III, Chapter 14]), the Mining Trust refused to have any more to do with such "workers," and throughout September the disciplinary regime prisoners were not marched out at all, but lived as if in a regular prison.

There were many "committed escapers" among them, and in summer twelve well-matched men had gradually come together in a safe escape team. (Mohammed Gadzhiev, leader of the Moslems in Ekibastuz; Vasily Kustarnikov; Vasily Bryukhin; Valentin Ryzhkov; Mutyanov; a Polish officer who made a hobby of tunneling; and others.) They were equals, but Stepan Konovalov, a Cossack from the Kuban, was nonetheless the leading spirit. They sealed their compact with an oath: if any one of them blabbed to a living soul, his number was up—either he would finish himself off or the others would stick their knives in him.

By then the Ekibastuz compound was already surrounded by a solid board fence four meters high. A belt of plowed land four meters wide followed the fence, and outside it a fifteen-meter forbidden area had been marked off, ending in a trench one meter deep. They resolved to dig a tunnel under this whole defense zone, so carefully concealed that it could not possibly be discovered before their escape.

A preliminary inspection showed that the barracks had shallow foundations, and that with so little space under the floorboards there would be nowhere to put the excavated soil. The problem seemed insurmountable. Should they give up the idea of escaping? . . . Then someone remarked that there was plenty of room in the loft, and suggested hoisting the soil up into it! At first thought it wasn't worth considering. Raise dozens and dozens of cubic meters of earth into the rafters, without attracting attention, through the living space of the barracks, which was constantly under observation, regularly inspected—raise it daily, hourly, and of course without spilling the merest pinch for fear of giving themselves away?

But when they thought of a way to do it, they were overjoyed, and their decision to escape was final. What helped them to decide was their choice of a "section," meaning a room. Their Finnish hut, originally intended for free workers, had been erected in the compound by mistake and there wasn't another like it anywhere in the camp: it had small rooms with three bunk beds wedged into each of them (not seven, as everywhere else), so that each held twelve men. They selected a section in which some of their number were already living. By various means—voluntary exchanges, forcing out unwanted roommates, teasing and ridiculing them ("you snore too loud, and you————too much")—they shunted outsiders to other sections and brought their own team together.

The more strictly the Disciplinary Barracks prisoners were segregated from the camp at large, the more severely they were punished and bullied, the higher their moral standing in the camps became. Any request from the prisoners in the Disciplinary Barracks held the force of law for all in the camps. They now started ordering whatever technical aids they needed, and these were made on one or another of the work sites, carried, at some risk, past the "frisking" points, and passed with further risk on to the Disciplinary Barracks—in the dishwash soup, in a loaf of bread, or with the medicines.

The first things ordered and delivered were knives and whetstones. Then nails, screws, putty, cement, whitening, electric wire, casters. They neatly sawed through the grooved ends of three floorboards with their knives, removed the skirting board which held them down, drew the nails from the butt ends of these boards near the wall, and the nails which fastened them to a joist in the middle of the room. The three loosened planks they fastened widthwise to a lath underneath them, so that they formed a single slab. The main nail was driven into this lath from above. Its broad head was smeared with putty the same color as the floor, and dusted over. The slab fitted snugly into the floor, and there was no way of getting a hold on it, but they never once pried it up by inserting an ax between the cracks. The way to raise the slab was to remove the skirting board, slip a piece of wire into the little gap around the broad head of the nail—and pull. Every time the tunnelers changed shifts, the skirting board was replaced and taken out all over again. Every day they "washed the floor"— soaking the floorboards to make them swell and close up chinks or cracks. This *entrance problem* was one of the most important. The tunneling room was always kept particularly clean and in exemplary order. Nobody lay on a bunk with his shoes on, nobody smoked, objects were not scattered around, there were no crumbs in the lockers. Nowhere was there less reason for an inspecting officer to linger. "Very civilized." And on he went.

The second problem was that of the *lift*—from the ground to the loft. The tunneling section, like all the others, had a stove. A narrow space into which a man could just about squeeze had been left between stove and wall: the idea they had was to *block up* this space—transfer it from the living area to the tunneling area. In an unoccupied section they dismantled completely one of the bunks. They used these planks to board up the gap, immediately covered them with lath and plaster, which they then whitewashed to match the stove. Could the warders remember in which of twenty little rooms in the barracks the stove was flush with the wall, and in which it stood out a little? In fact, they didn't even spot the disappearance of the bunk bed. The one thing the jailers might have noticed in the first day or two was the wet plaster, but to do so they would have had to go around the stove and crouch over a bunk—and there was no point in this, because the section was in exemplary condition! Even if they had been caught out in this, it need not have meant that the tunneling operation had failed: this was merely work they had done to beautify that section: the gap was always so dusty, and spoiled its appearance!

Only when the plaster and whitewash had dried out were the floor and ceiling of the now enclosed niche sawn through with knives, and a stepladder knocked together from the same cannibalized bunk bed placed there—so that the shallow space under the floor was connected with the spacious bower in the roof. It was a *mine shaft,* cut off from the gaze of their jailers, and the only mine for years in which these strong young men had felt like working themselves into a lather!

Can any work in a prison camp merge with your dreams, absorb your whole soul, rob you of sleep? It can—but only the work you do to escape!

The next problem was that of digging. They must dig with knives, and keep them sharp—so much was obvious—but this still left many other problems. For one thing, the *"mine surveyor's calculations"* (Engineer Mutyanov): you must go down deep enough for safety, but no farther than was necessary, must take the shortest route, must determine the optimum cross section for the tunnel, must always know where you are, and fix the right point of exit. Then again, the *organization of shifts:* you must dig as many hours as possible in every twenty-four, but without too many changes of shift, and taking care to be all present and impeccably correct at morning and evening inspections. Then there were *working clothes* to be thought of, and washing arrangements—you couldn't come up to the surface bedaubed with clay! Then the question of *lighting*—how could you drive a tunnel sixty meters long in the dark? They ran a wire under the floor and along the tunnel (and still had to find a way to connect it up inconspicuously!). Then there must be a *signal system:* how did you recall the man digging at the far end of the tunnel and out of hearing, if someone entered the barracks unexpectedly? And how could the tunnelers safely signal that they must come out immediately?

The strictness of the Disciplinary Barracks routine was also its weakness. The jailers could not creep up on the barracks and descend unexpectedly—they always had to follow the same path between the barbed-wire entanglements to the gate, undo the lock, then walk to the barracks and undo another lock, rattle the bolt. All their movements could easily be observed—not, it is true, from the window of the tunneling room, but from that of the empty "cabin," by the entrance—all they had to do was to keep an observer there. Signals to the working force were given by lamps: two blinks meant Caution, prepare to evacuate; repeated blinks meant Danger! Red alert! Hop out quickly!

Before they went under the floor they stripped naked, putting all their clothes under pillows or under a mattress as they took them off. Once past the trap door, they slipped through a narrow aperture beyond which no one would have supposed that there could be a wider *chamber,* where a light bulb burned continually and working jackets and trousers were laid out. Four others, naked and dirty (the retiring shift), scrambled up above and washed themselves thoroughly (the clay hardened in little balls on their body hair, and had to be soaked off or pulled away with the hair).

All this work was already in progress when the careless tunneling operation from Disciplinary Barracks No. 8 was discovered. You can easily understand the annoyance, or rather the outrage, of the artists whose ingenious conception had been so insulted. But nothing disastrous happened.

At the beginning of September, after nearly a year in prison, Tenno and Zhdanok were transferred (returned) to this Disciplinary Barracks. As soon as he got his breath back, Tenno became restless—he must plan his escape! But however he reproached them with letting the best season for escape slip by, with sitting around and doing nothing, no one in the Disciplinary Barracks, not even the most committed and desperate escapers, reacted.

(The tunnelers had three four-man shifts, and a thirteenth man, whoever he might be, was of no use to them.) Then Tenno, without beating about the bush, suggested tunneling—and they replied that they had already thought of it, but that the foundations were too shallow. (This was heartless, of course: looking into the eager face of a proven escaper and apathetically shaking your head is like forbidding a clever hunting dog to sniff out game.) But Tenno knew these lads too well to believe in this epidemic of indifference. They couldn't *all* have gone rotten at once!

So he and Zhdanok kept them under keen and expert observation—something of which the warders were incapable. Tenno noticed that the lads often went for a smoke to the same "cabin" near the entrance, always one at a time, never in company. That the door of their section was always on the hook in the daytime, that if you knocked they didn't open immediately, and that some of them would be sound asleep, as though the nights weren't long enough. Or else Vaska Bryukhin would come out all wet from the room where the close-stool was kept. "What are you up to?" "Just thought I'd have a wash."

They were digging, no doubt about it! But where? Why wouldn't they tell him? Tenno went first to one, then to another, and tried bluff. "You ought to be more careful about it, lads! It doesn't matter if I spot you digging, but what if an informer does?"

In the end, they had a confab and agreed to take in Tenno and three suitable companions. They invited him to inspect the room and see if he could find any traces. Tenno felt and sniffed every inch of wall and floorboard, and found nothing!—to his delight and the delight of all the lads. Trembling with joy, he slid under the floor to work *for himself.*

The shift underground was deployed as follows: One man, lying down, picked away at the face; another, crouching behind him, stuffed the loose earth into small canvas bags specially made for the purpose; a third, on all fours, dragged the bags backward along the tunnel, by means of straps over his shoulders, then went under the floor to the shaft and attached them one by one to a hook lowered from the loft. The fourth man was up there. He threw down the empty bags, hauled up the full ones, carried them, treading lightly, to different parts of the attic, scattered their contents in a thin layer over the whole area, and at the end of the shift covered this soil with cinders, of which there were a lot in the loft. The team might change jobs in the course of the shift, but did not always do so, because not everyone could perform the heaviest, utterly exhausting jobs: digging and dragging the soil away.

To begin with they dragged two bags at a time, but then they made it four. For this purpose they finagled a wooden tray from the cooks, which they towed along with the sacks on it. The strap went around the neck from behind, then passed under the armpits. Necks were rubbed raw, shoulders ached, knees were bruised; after a single trip a man was in a lather, and after a whole shift, you could pour him into a jug.

The digging had to be done in a very uncomfortable position. There was a short-handled shovel, which they sharpened every time. With this they had to cut vertical grooves to the depth of a bayonet thrust; then, semi-recumbent, with their backs against the fresh-dug earth, they had to gouge lumps of earth and throw them over their shoulders. The soil was sometimes stony, sometimes stiff clay. They had to go around the biggest stones, making a bend in the tunnel. An eight-to-ten-hour shift added not more than two meters lengthwise, and sometimes less than a meter.

The worst of it was the lack of air in the tunnel: your head swam, you fainted, you felt sick. So yet another problem had to be solved—that of *ventilation*. Air holes could be bored only upward—into that most dangerous area near the boundary fence, which was under constant surveillance. But without them breathing was impossible. They ordered a steel "propeller" blade, fixed a stick to it crosswise, so that they had something like a carpenter's brace, and bored their first hole to the wide world outside. A draft was created, and breathing was easier. (When the tunnel went beyond the fence and outside the camp, they made another hole.)

They were forever sharing their experience, advising each other how best to do things. And reckoning up how far they had gone.

The escape hole, or tunnel, plunged under the foundation wall, and from then on diverged from the straight only to avoid stones or because of inaccurate digging. It was half a meter wide and ninety centimeters high, and it had a vaulted roof. Its top, according to their calculations, was about 1.3 or 1.4 meters from the surface. The sides of the tunnel were reinforced with planks, and as they advanced the electric cable was extended and more and more electric bulbs were hung up.

If you looked along it, it was like an underground railway, a Gulag metro!

By now the tunnel had gone some dozens of meters, and they were digging beyond the boundary fence. Overhead the tramp of guards on their rounds, and the barking and whining of dogs, could be clearly heard.

When suddenly . . . one day suddenly, after morning inspection, before the day shift had gone below and (according to the team's strict rules) while nothing incriminating was visible, they saw a pack of jailers walking toward the barracks, with the short and short-tempered chief prison officer Lieutenant Machekhovsky at their head. The escapers' hearts sank. Had they been detected? Had they been sold? Or was it just a random check?

An order rang out: "Collect your personal belongings! Out, out of the barracks, every last man!"

The order was obeyed. All the prisoners were driven out and squatted on their haunches in the little exercise yard. From inside dull thuds were heard—they were stripping the planks from the bunks. "Get your tools in here!" Machekhovsky yelled. And the jailers dragged in their crowbars and axes. They could hear boards creaking and groaning as they were ripped up.

Such is the fate of the escaper! Not only would all their cleverness, their work, their hopes, their excitement go for nothing, but it would mean the punishment cells again, beatings, interrogations, fresh sentences. . . .

However . . . neither Machekhovsky nor any of the jailers ran out waving their arms in savage glee. They came out perspiring, brushing off dirt and dust, panting, not a bit pleased to have done all that donkey work in vain. A half-hearted order: "Step up one at a time." A search of personal belongings began. The prisoners returned to the barracks. What a shambles! The floor had been taken up in several places (where the boards had been badly nailed down or there were obvious cracks). In the sections everything had been tossed around, and even the bunks had been spitefully tipped over. Only in the *civilized* section had nothing been disturbed.

Prisoners not in on the secret were outraged: "Why can't they sit still, the dogs?! What do they think they're looking for?"

The escapers realized now how clever they had been not to stack piles of soil under the floor: they would have been spotted at once through the gaps in the boards. Whereas no one had climbed into the loft—you could only escape from the loft with wings! In any case, everything up there was carefully covered with cinders.

The dogs hadn't nosed out the game! What joy! If you work stubbornly, and keep a close watch on yourself, you are bound to reap your reward. Now we know we shall finish the job! There were six or eight meters to go to the circular trench. (The last few meters had to be dug with special precision, so that the tunnel would come out at the bottom of the trench—neither higher nor lower.)

What then? Konovalov, Mutyanov, Gadzhiev, and Tenno had by now worked out a plan, to which the other twelve agreed. The escape would be in the evening at about ten o'clock, when the evening inspection had been carried out throughout the camp, the jailers had dispersed to their homes or gone off to the staff hut, the sentries had been relieved on the watchtowers, and the changing of the guard was over.

They would go down into the underground passage one by one. The last man would watch the camp grounds from the cabin till the others were down; then he and the last but one would nail the removable part of the skirting board firmly to the boards of the trap door so that when they let it fall behind them the skirting board also would fit into place. The broad-headed nail would be drawn downward till it would go no farther, and in addition to all this, bolts would be ready under the floor to fasten the trap door, so that it would not budge even if anyone tried to wrench it upward.

Another thing—before escaping they would remove the grating from one of the corridor windows. The warders, finding sixteen men missing at morning inspection, would not immediately conclude that they had tunneled their way out, but would rush to search the compound, thinking that the Disciplinary Barracks prisoners had gone to settle accounts with informers. They would also look in the other camp area, in case the missing men had climbed over the wall into it. It was a beautiful job! They'd never find the tunnel, there were no tracks under the window, and sixteen men had gone—carried by angels up to heaven!

They would crawl out into the peripheral trench, then creep along the bottom away from the watchtower (the exit from the tunnel was too close to it); they would go out to the road one by one; they would leave intervals between groups of four, to avoid arousing suspicion and to give themselves time to look around (the last man would take another precaution: he would close the escape hole from *outside* with a wooden manhole cover smeared with clay, pressing it home with the weight of his body, and then covering it with earth; so that in the morning it would be impossible to discover any trace of a tunnel from the trench either.

They would go through the settlement in groups, making loud carefree jokes. If there should be any attempt to stop them, they would all join in resisting it, with knives if need be.

The general assembly point was a railway crossing much used by motor vehicles. There was a hump in the road where it crossed the tracks, and they would all lie on the ground nearby without being seen. The crossing was rough (they had seen it and walked over it on their way to work), the boards had been laid any old way, and lorries, whether carrying coal or empty, struggled across it slowly. Two men would raise their hands to stop a lorry as it was leaving the crossing, and approach the driver's cab from both sides. They would ask for a lift. At night drivers were more often than not alone. If this one was, they would immediately draw their knives, overpower him, and make him sit in the middle, while Valka Ryzhkov took the wheel, all the others hopped into the back, and—full speed ahead to Pavlodar! They could certainly put 130 or 140 kilometers behind them in a few hours. They would turn off upstream before they reached the ferry (their eyes had not been idle when they were first brought to these parts), tie the driver up and leave him lying in the bushes, abandon the lorry, row across the Irtysh, split up into groups—and go their different ways! The grain crop was being taken into storage just then, and all the roads were full of lorries.

They were due to finish work on October 6. Two days earlier, on the fourth, two members of the group were transferred: Tenno and Volodya Krivoshein, the thief. They tried reporting sick in order to stay behind at any price, but the operations officer promised to transport them in handcuffs, whatever their condition. They decided that excessive resistance would arouse suspicion. They sacrificed themselves for their friends and submitted.

So Tenno did not benefit from his insistence on joining the tunneling party. Instead of him, the thirteenth man would be his recruit and protégé, Zhdanok, who was by now far too unstable and erratic. It was a bad day for Stepan Konovalov and his friends when they gave in and told Tenno their secret.

They finished digging and ended in the right place. Mutyanov had made no mistake. But there was snow on the ground, and they postponed their flight until it was drier.

On October 9 they did everything precisely as planned. The first four emerged successfully—Konovalov, Ryzhkov, Mutyanov, and the Pole, who was his regular partner in escapes involving engineering.

But the next to crawl into the trench was the hapless little Kolya Zhdanok. It was no fault of his, of course, that footsteps were suddenly heard up above and not far away. But he should have been patient, lain down, made himself inconspicuous, and crawled on when the footsteps had passed. Instead, in his usual smart-aleck way, he stuck his head out. He had to *see* who was walking up there.

The quickest louse is first on the nit comb. But as well as himself, *this* stupid louse destroyed an escape team equally remarkable for the cogency of their plan and for their ability to work smoothly together—ruined fourteen long and difficult lives whose paths had intersected in this escape. In each of these lives this attempt had a special importance; without it past and future were meaningless. On each of them people somewhere still depended—women, children, and children still unborn. But a louse raised its head, and the whole thing went up in smoke.

The passer-by proved to be the deputy guard commander. He saw the louse, shouted, fired. And so the guards, although the plan had been far too clever for them and they had no inkling of it, became heroes. And my reader the Marxist Historian, tapping the page with his ruler, asks me in his patronizing drawl:

"Ye-e-es . . . but why didn't you run away? Why didn't you revolt? . . ."

The escapers had levered back the grating, crawled into the hole, and nailed the skirting board to the trap—and now they had to crawl slowly, painfully back.

Who has ever plumbed such depths of disillusionment and despair? Seen such efforts so insultingly derided?

They went back, switched off the lights in the tunnel, fitted the grating in its socket.

Very soon the whole Disciplinary Barracks was swarming with camp officers, division officers, escort troops, jailers. All prisoners were checked against the records, and driven into the stone jail.

But they did not find the tunnel! (How long would they have searched if everything had gone according to plan?) Near the place where Zhdanok had *broken cover,* they found a half-filled-in hole. But even when they followed the tunnel under the barracks, they could not understand how the men had got down there or where they had put the earth.

But the *civilized* section was four men short, and they gave the remaining eight a merciless *going over:* the easiest way for bone-heads to get at the truth.

What point was there in further concealment? . . .

Later on, guided tours of the tunnel were arranged for the whole garrison and all the jailers. Major Maksimenko, the big-bellied commandant of the Ekibastuz camp, used to boast to other area commandants at Gulag HQ: "You should see the tunnel at my place. Like the metro! But we . . . our vigilance . . ."

No; just one little louse . . .

The hue and cry also prevented the first four out from reaching the railway crossing. The plan had collapsed! They climbed the fence of an empty work site on the other side of the road, walked through it, climbed the fence again, and headed out onto the steppe. They decided against waiting in the settlement to pick up a lorry because it was already full of patrols.

Like Tenno the year before, they quickly lost speed and faith.

They set off southwest, toward Semipalatinsk. They had neither provisions nor strength for a long journey on foot—in the past day or two they had strained themselves to the breaking point finishing the tunnel.

On the fifth day of their run they looked in at a yurt and asked the Kazakhs for something to eat. They, needless to say, refused and shot at these hungry people with a hunting rifle. (Is this tradition native to these shepherds of the steppe? And if not, where does it come from?)

Stepan Konovalov attacked a Kazakh, knife against gun, and took the gun and some provisions from him. They went on farther. But the Kazakhs tracked them down on horseback, came upon them near the Irtysh, and summoned an operations group.

After that they were surrounded, beaten to a bloody pulp, and after that . . . There's no need to continue. . . .

If anyone can show me any attempt to escape by nineteenth- or twentieth-century Russian revolutionaries attended with such difficulties, such an absence of support from outside, such a hostile attitude on the part of the surrounding population, such lawless reprisals against the recaptured—let him do so!

And after that let him say, if he can, that we did not put up a fight.

Chapter 9

■

The Kids with Tommy Guns

The camps were guarded by men in long greatcoats with black
cuffs. They were guarded by Red Army men. They were guarded
by prisoner guards. They were guarded by elderly reservists. Last
came the robust youngsters born during the First Five-Year Plan,
who had seen no war service when they took their nice new
Tommy guns and set about guarding us.

Twice every day, for an hour at a time, we and they shuffled
along, tied together in silent and deadly brotherhood: any one
of them was at liberty to kill any one of us. Every morning we
dawdled listlessly along, we on the road and they on the shoulder, to a place where neither they nor we wished to go. Every
evening we stepped out briskly—we to our pen, they to theirs.
Since neither we nor they had any real home, these pens were
home to us.

We walked with never a glance at their sheepskin coats and
their Tommy guns—what were they to us? They walked watching
our dark ranks all the time. It was there in the regulations that
they must watch us all the time. Orders were orders. Duty was
duty. Any wrong movement, any false step, they must cut short
with a bullet.

How did they think of us, in our dark jackets, our gray caps of
Stalin fur,* our grotesque felt boots that had served three sen-
tences and shed four soles, our crazy quilt of number patches?
Decent people would obviously never be treated like that.

Was it surprising that our appearance inspired disgust? It was
intended to do just that. From their narrow footpaths the free
inhabitants of the settlement, especially the schoolchildren and

219

their women teachers, darted terrified glances at our columns as we were led down the broad street. They showed how much they dreaded that we, the devil's brood of fascism, might suddenly go berserk, overcome the convoy guards, and rush around looting, raping, burning, and killing. Obviously such bestial creatures could conceive no other desires. And inhabitants of the settlement were protected from these wild beasts by . . . convoy troops. The noble convoy troops. A sergeant of convoy guards, inviting the schoolteacher to dance in the club we had built for them, could feel like a knight in shining armor.

These kids watched us the whole time, from the cordon and from the watchtowers, but they were not allowed to know anything about us; they were allowed only the right to shoot without warning!

If they had just visited us in our huts of an evening, sat on our bunks and heard why this old man, or the other old fellow over there, was inside . . . those towers would have been unmanned and those Tommy guns would never have fired.

But the whole cunning and strength of the system was in the fact that our deadly bond was forged from ignorance. Any sympathy they showed for us was punishable as treason; any wish to speak to us, as a breach of a solemn oath. And what was the point of talking to us when the political instructor would come at fixed intervals to lead a discussion on the political and moral character of the enemies of the people whom they were guarding. He would explain in detail and with much repetition how dangerous these scarecrows were and what a burden to the state. (Which made it all the more tempting to try them out as living targets.) He would bring some files under his arm and say that the Special Section had let him take a few cases just for the evening. He would read out some typewritten pages about evil doings for which all the ovens of Auschwitz would not be punishment enough—attributing them to the electrician who was mending the light on a post nearby, or to the carpenter from whom certain soldiers had imprudently thought of ordering lockers.

The political instructor will never contradict himself, never make a slip. He will never tell the boys that some people are imprisoned here simply for believing in God, or simply for desiring truth, or simply for love of justice. Or indeed for nothing at all.

The whole strength of the system is in the fact that one man cannot speak directly to another, but only through an officer or a political instructor.

The whole strength of these boys lies in their ignorance.

The whole strength of the camps is in these boys. The boys with the red shoulder tabs. The murderers from watchtowers, the hunters of fugitives.

Here is one such political lesson, as remembered by a former convoy guard (at Nyroblag). "Lieutenant Samutin was a lanky, narrow-shouldered man, and his head was flat above the temples. He looked like a snake. He was towheaded and almost without eyebrows. We knew that in the past he had shot people with his own hands. Now he was a political instructor, reciting in a monotonous voice: 'The enemies of the people, over whom you stand guard, are the same as the Fascists, filthy scum. We embody the power and the punitive sword of the Motherland and we must be firm. No sentimentality, no pity.'"

That is how they mold the boys who make a point of kicking a runaway's head when he is down. The boys who can boot a piece of bread out of the mouth of a gray-haired old man in handcuffs. Who can look with indifference at a shackled runaway jounced on the splintery boards of a lorry; his face is bloodied, his head is battered, but they look on unmoved. For they are the Motherland's punitive sword, and he, so they say, is an American colonel.

After Stalin's death, then living in eternal banishment, I was a patient in an ordinary *free* Tashkent clinic. Suddenly I heard a patient, a young Uzbek, telling his neighbors about his service *in the army*. His unit had, he said, kept guard on beasts and butchers. The Uzbek admitted that the convoy guards were also somewhat underfed, and it enraged them that prisoners, like miners, got rations (if, of course, they fulfilled the norm by 120 percent) not much smaller than those of honest soldiers. It also enraged them that they, the convoy guards, had to freeze on top of watchtowers in the winter (in sheepskin coats down to their heels, it's true), while the enemies of the people, once they entered the working area, scattered about the warming-up shacks (even from the watchtower he could have seen that it was not so) and slept there all day (he seriously imagined that the state's treatment of its enemies was philanthropic).

Here was an interesting opportunity—to look at a Special Camp through the eyes of a convoy guard! I began asking what kind of reptiles they were and whether my Uzbek friend had talked to them personally. And of course he told me that he had learned all this from political officers, that they had even had "cases" read out to them in their political indoctrination sessions. And his malicious misconceptions about prisoners sleeping all day had of course been reinforced in him by the approving nods of officers.

Woe unto you that cause these little ones to stumble. Better for you had you never been born! . . .

The Uzbek also told us that a private in the MVD troops received 230 rubles a month. (Twelve times more than he would in the army! Why such generosity? Perhaps his service was twelve times more difficult?) And in the polar regions as much as 400 rubles for a fixed period of service, and with all expenses covered.

He further told us of a number of incidents. Once, for instance, one of his comrades was marching in convoy and fancied that somebody *was about to* run out of the column. He pressed his trigger and killed *five* prisoners with a single burst. Since all the other guards later testified that the column had been moving quietly along, this soldier incurred a terrible punishment: for killing five people he was put in detention for fifteen days (in a warm guardhouse, of course).

But which of the Archipelago's inhabitants has no stories of this sort to tell! . . . We knew so many of them in the Corrective Labor Camps. On a work site which had no fence but an invisible boundary line, a shot rang out and a prisoner fell dead: he had stepped over the line, they said. He may have done nothing of the sort— the line was invisible, remember—but no one else would hurry over to check, for fear of lying beside him. Nor would a commission come to verify where the dead man's feet were lying. Perhaps, though, he had overstepped the line—after all, the guards could keep their eyes on it, whereas the prisoners had to work. The most conscientious prisoner, the prisoner with thoughts for nothing but his work, is the one most likely to be shot in this way. At Novochunka station (Ozerlag), haymaking, a prisoner sees a little more hay two or three steps away and like a thrifty peasant decides to rake it in. A bullet! And the soldier gets . . . a month's leave!

Sometimes a particular guard is angry with a particular prisoner (for refusing to do him a favor, or failing to make something ordered), and avenges himself with a bullet. Sometimes it is done treacherously: the guard orders the prisoner to fetch something from beyond the boundary line. The prisoner goes off unsuspectingly—and the guard shoots. Or perhaps he tosses him a cigarette

—here, have a smoke! Even a cigarette will serve as bait for a prisoner, contemptible creature that he is.

Why do they shoot? Sometimes there's no answer. In Kengir, for instance, in a tightly organized camp, in broad daylight when there was not the faintest possibility of an escape, a girl called Lida, a Western Ukrainian, contrived to wash her stockings in working hours and put them out to dry on the sloping ground of the camp fringe area. The man on the tower took aim and killed her outright. (There was some vague rumor that he tried to commit suicide afterward.)

Why? Because the man has a gun! Because one man has the arbitrary power to kill or not to kill another.

What's more—it pays! The bosses are always on your side. They'll never punish you for killing somebody. On the contrary, they'll commend you, reward you, and the quicker you are on the trigger—bring him down when he's only put half a foot wrong—the more vigilant you are seen to be, the higher your reward! A month's pay. A month's leave. (Put yourself in the position of HQ: what sort of division is it that can point to no display of vigilance? What is wrong with its officers? Or perhaps the zeks are so docile that the guard can be reduced? A security system once created *needs deaths!*)

Indeed, a sort of competition springs up between sharpshooters: you killed somebody and bought butter with your prize money, so I'll kill somebody and buy butter, too. Want to pop home and paw your girl a bit? Just plug that gray thing over there and away you go for a month.

These were all cases of a kind familiar to us in the Corrective Labor Camps. But in the Special Camps there were novelties: shooting straight into a marching column, for instance, as the Uzbek's comrade did. Or as they did at the guardhouse in Ozerlag on September 8, 1952. Or shooting from the towers into the camp area.

It meant that they'd been trained to it. This was the work of the political officers.

In Kengir in May, 1953, these kids with Tommy guns suddenly and without provocation opened fire on a column which had halted outside the camp and was waiting to be searched on entry. Sixteen were wounded—but these were no ordinary wounds. The guards were using dumdum bullets, which capitalists and socialists had long ago joined in banning. The bullets left craters in their bodies, tore holes in their entrails and their jaws, mangled their extremities.

Now, why should convoy guards at a Special Camp be armed with *dumdum* bullets? On *whose* authority? We shall never know. . . .

All the same, they were very hurt when they read in my story that prisoners called their escort troops "screws," saw the word repeated for all the world to hear. No, no, the prisoners should have loved them and called them guardian angels!

One of these youngsters—one of the better ones, it's true, who didn't take offense but wants to defend the truth as he sees it— is Vladilen Zadorny, born 1933, served in the MVD troops (Infantry Guard Unit) at Nyroblag from the age of eighteen to the age of twenty. He has written me several letters.

Boys did not join of their own wish—they were called up by the Commissariat of War. The Commissariat passed them on to the Ministry of Internal Affairs. The boys were taught to shoot and stand guard. The boys cried with the cold at night. What the hell did they want with places like Nyroblag and all that they contained. You mustn't blame the lads —they were soldiers, serving their Motherland, and although there were things they didn't understand about this absurd and dreadful form of service [*How much, then, did they understand?* . . . It's either all or nothing—A.S.], they had taken their oath. Their service was not easy.

That is frank and honest. There is something to think about here. These lads were shut up inside a picket fence of words: Remember your oath! Serving your Motherland! You're soldiers now!

Yes, but their underlying common humanity must have been weak, or altogether lacking, if it was not proof against an oath and a few political discussion periods. Not every generation, and not every people, is of the stuff from which such boys are fashioned.

This is surely the main problem of the twentieth century: is it permissible merely to carry out orders and commit one's conscience to someone else's keeping? Can a man do without ideas of his own about good and evil, and merely derive them from the printed instructions and verbal orders of his superiors? Oaths! Those solemn pledges pronounced with a tremor in the voice and intended to defend the people against evildoers: see how easily they can be misdirected to the service of evildoers and against the people!

Let us remember what Vasily Vlasov meant to say to his executioner back in 1937. "It is your fault! You alone are to blame that they kill people! My death is on your head *alone,* and you must live with that! If there were no executioners, there would be no executions!"

If there were no convoy troops, there would be no camps.

Of course, neither our contemporaries nor history will ignore the hierarchy of guilt. Of course, it is obvious to all that their officers were more guilty; the security officers more guilty still; those who drew up the orders and instructions even more guilty, and those who ordered them to be drawn up most guilty of all.[1]

But shots were fired, camps were guarded, Tommy guns were held at the ready, not by them, but by those boys! Men were kicked about the head as they lay on the ground—by no one else, by those boys!

Vladilen writes also:

They dinned into us, they forced us to learn by heart USO 43SS—the 1943 Regulations of the Infantry Prison Guard (Top Secret)[2]—a cruel and frightening document. Then there was the oath. Then we had to heed the security officers and political officers. There was slander and denunciations. Cases were trumped up against some of the soldiers themselves. . . . Divided as they were by the stockade and the barbed wire, the men in the overall jackets and the men in greatcoats were equally prisoners —the former for twenty-five years, the latter for three.

It is putting it rather strongly to say that the infantrymen were also *imprisoned,* only not by a military tribunal, but by the Commissariat of War. *Equally* prisoners they were not—because the people in greatcoats could freely cut down with their Tommy guns people in work jackets, mow down a crowd of them, indeed, as we shall soon see.

Vladilen further explains that:

---

1. This does not mean that they will be brought to trial. Someone should ascertain whether they are satisfied with their pensions and dachas.

2. I wonder, by the way, whether we are fully aware of the part which this sinister double sibilant plays in our lives, in one abbreviation after another, beginning with KPSS [CPSU —Communist Party of the Soviet Union—Trans.], and KPSS-ovtsy [Party members]. Here we find that the regulations, too, were SS [*sovershenno sekretny*—top secret—Trans.], as is everything excessively secret. Obviously, those who drafted these regulations were conscious of their foulness, but drafted them nonetheless—and at what a time: we had just driven the Germans back from Stalingrad! One more fruit of the people's victory.

The lads were of all sorts. There were blinkered old sweats who hated zeks. Incidentally, the new recruits from some ethnic minorities—Bashkir, Buryat, Yakut—were very keen. Then there were those who just didn't care—the majority. They performed their service quietly and uncomplainingly. What they liked best was the tear-off calendar and post time. Finally there were decent chaps who felt sorry for the zeks as victims of misfortune. And most of us realized that our service was unpopular with the people. When we went on leave we did not wear uniforms.

Vladilen's own life story is his best argument in support of his ideas; though we must remember that there were very few like him.

He was allowed into the convoy guard service by an oversight on the part of a lazy Special Section. His stepfather, the old trade union official Voinino, had been arrested in 1937, and his mother expelled from the Party as a result. His father, who had commanded a Cheka squad, and had been a Party member since 1917, made haste to disown both his former wife and his son (and so he kept his Party card, but he was nevertheless demoted).[3] His mother tried to wash away her stain as a blood donor during the war. (It was all right; Party members and non-Party members alike took her blood.) The boy "had hated the bluecaps* from childhood, and now they put one on my own head. . . . The terrible night when people in my father's uniform roughly searched my cot had left too deep a mark in my young memory."

I was not a good convoy guard: I used to get into conversation with the zeks and do them favors. I would leave my rifle by the campfire while I went to buy things for them at the stall or post their letters. I daresay that there were people at the Intermediate, Mysakort, and Parma Separate Camp Sites who still remembered Private Volodya. A prisoner foreman told me: "If you look at people and hear their troubles you will understand. . . ." I saw my grandfather, my uncle, my aunt in every one of the political prisoners. . . . I simply hated my officers. I grumbled and complained and told the other soldiers: "There are the real enemies of the people!" For this, for open insubordination ("sabotage"), and for consorting with a zek, I was sent for interrogation. . . . Lanky Samutin . . . whipped me across the face, rapped my knuckles with a paperweight . . . because I would not give him a signed confession that I had posted

3. Although we have long ago learned to expect anything, there are still a few surprises left: because his abandoned wife's second husband is arrested, must a man disown his four-year-old son? And not just any man, but a squad commander in the Cheka?

letters for zeks. I would have squelched that tapeworm—I'm a silver medalist at boxing; I could cross myself with an eighty-pound weight—but there were two warders hanging on to me. . . . Still, the interrogators had other things on their minds—the MVD in 1953 didn't know whether it was coming or going. They didn't put me inside—they just gave me a dishonorable discharge under Article 47-G: "Discharged from the organs of the MVD for extreme insubordination and gross breaches of MVD regulations." Then they threw me out of the divisional guardhouse, beaten up and frozen, to make my way home. . . . Arsen, a foreman, released at this time, looked after me on the way.

Let us imagine that a convoy guard *officer* wanted to show the prisoners some leniency. He could, of course, only do it through his soldiers and in their presence. Which means that, given the general savagery, it would be "embarrassing" and in fact impossible. Besides, someone would immediately inform on him.

That's the system for you!

Chapter 10

■

Behind the Wire the Ground Is Burning

No, the surprising thing is not that mutinies and risings did not occur in the camps, but that in spite of everything they *did*.

Like all embarrassing events in our history—which means three-quarters of what really happened—these mutinies have been neatly cut out, and the gap hidden with an invisible join. Those who took part in them have been destroyed, and even remote witnesses frightened into silence; the reports of those who suppressed them have been burned or hidden in safes within safes within safes—so that the risings have already become a myth, although some of them happened only fifteen and others only ten years ago. (No wonder some say that there was no Christ, no Buddha, no Mohammed. There you're dealing in thousands of years. . . .)

When it can no longer disturb any living person, historians will be given access to what is left of the documents, archaeologists will do a little digging, heat something in a laboratory, and the dates, locations, contours of these risings, with the names of their leaders, will come to light.

We shall see the first outbreaks, like the Retyunin affair of January, 1942, in the Osh-Kurye Separate Camp Site near Ust-Usy. Retyunin is said to have been a free employee, perhaps even in charge of this detachment. He sounded the call to the 58's and the socially harmful (Article 7, Section 35) and rallied a few hundred volunteers. They disarmed the convoy (which consisted of short-sentence prisoner guards) and escaped with some horses into the forest, to live as partisans. They were killed off gradually. In spring, 1945, people who had no connection with it at all were still being jailed for the Retyunin affair.

Perhaps we (no, not we ourselves) shall learn at the same time about the legendary rising in 1948 at public works site No. 501, where the Sivaya Maska–Salekhard railway was under construction. It was legendary because everybody in the camps talked about it in whispers, but no one really knew anything. Legendary also because it broke out not in the Special Camp system, where the mood and the grounds for it by now existed, but in a Corrective Labor Camp, where people were isolated from each other by fear of informers and trampled under foot by thieves, where even their right to be "politicals" was spat upon, and where a prison mutiny therefore seemed inconceivable.

According to the rumors, it was all the work of ex-soldiers (recent ex-soldiers!). It could not have been otherwise. Without them the 58's lacked stamina, spirit, and leadership. But these young men (hardly any of them over thirty) were officers and enlisted men from our fighting armies, or their fellows who had been prisoners of war; among these some had been with Vlasov or Krasnov, or in the nationalist units. There were men who had fought against each other, but here oppression united them. These young men who had served on all fronts in a world war, who were expert in modern infantry warfare, camouflage, and picking off patrols—these young men, except those who were scattered singly, still retained in 1948 their wartime élan and belief in themselves, and they could not accept the idea that men like themselves, whole battalions of them, should meekly die. Even escape seemed to them a contemptible half-measure, rather like deserting one by one instead of facing the enemy together.

It was all planned and begun in one particular team. An ex-colonel called Voronin or Voronov, a one-eyed man, is said to have been the leader. A first lieutenant of armored troops, Sakurenko, is also mentioned. The team killed their convoy guards (in those days convoy guards, often unlike their charges, were not real soldiers, but reservists). Then they went and freed a second team, and a third. They attacked the convoy guards' hamlet, then the camp from outside, removed the sentries from the towers, and opened up the camp area. (The inevitable schism now took place: though the gates were wide open, most of the zeks would not go through them. These included prisoners with short sentences who had nothing to gain from mutiny. There were others with *tenners,* or even with fifteen years under the "Seven-eighths" and "Four-sixths" decrees, who would be worse off under Article 58. Finally, there were even 58's of the sort who preferred to die kneeling loyally rather than on their feet. Nor were all those who poured out through the gate necessarily on their way to help the mutineers: some felons happily broke bounds to plunder the free settlements.)

Arming themselves with weapons taken from the guards (who were later buried at the Kochmas cemetery), the rebels went on to capture the neighboring Camp Division. With their combined forces they decided to advance on Vorkuta! It was only sixty kilometers away. But this was not to be. Parachute troops were dropped to bar their way. Then low-flying fighter planes raked them with machine-gun fire and dispersed them.

They were tried, more of them were shot, and others given twenty-five or ten years. (At the same time, many of those who had not joined in the operation but remained in the camp had their sentences "refreshed.")

The hopelessness of this rising as a military operation is obvious. But would you say that dying quietly by inches was more "hopeful"?

The Special Camps were set up soon after this, and most of the 58's were raked out. What was the result?

In 1949 in the Nizhni Aturyakh division of Berlag, a mutiny started in much the same way: they disarmed their guards, took six or eight Tommy guns, attacked the camp from outside, knocked the sentries off the towers, cut the telephone wires, and opened up the camp. This time there was no one in the camp who was not numbered, branded, doomed, beyond all hope.

And what do you think happened?

Most prisoners would not pass through the gates. . . .

Those who had started it all and now had nothing to lose turned the mutiny into a breakout, and a group of them headed for Mylgi. At Elgena Toskana their way was barred by troops and light tanks. (General Semyonov was in command of the operation.)

They were all killed.[1]

---

1. I do not claim that my account of these risings is entirely accurate. I shall be grateful to anyone who can correct me.

Riddle: What is the quickest thing in the world? Answer: Thought.

It is and it isn't. It can be slow, too—oh, how slow! Only slowly and laboriously do men, people, society, realize what has happened to them. Realize the truth about their position.

In herding the 58's into Special Camps, Stalin was exerting his strength mainly for his own amusement. He already had them as securely confined as they could be, but he thought he would be craftier than ever and improve on his best. He thought he knew how to make it still more frightening. The results were quite the opposite.

The whole system of oppression elaborated in his reign was based on keeping malcontents apart, preventing them from reading each other's eyes and discovering how many of them there were; instilling it into all of them, even into the most dissatisfied, that no one was dissatisfied except for a few doomed individuals, blindly vicious and spiritually bankrupt.

In the Special Camps, however, there were malcontents by the thousands. They knew their numerical strength. And they realized that they were not spiritual paupers, that they had a nobler conception of what life should be than their jailers, than their betrayers, than the theorists who tried to explain why they must rot in camps.

This novel aspect of the Special Camp went almost unnoticed at first. On the face of it, things went on as though it was a continuation of the Corrective Labor Camp. Except that the thieves, those pillars of camp discipline and of authority, soon lost heart. Still, crueler warders and an enlarged Disciplinary Barracks seemed to compensate for this loss.

But mark this: when the thieves lost heart, there was no more pilfering in the camp. You could now leave your rations in your locker. You could drop your shoes on the floor for the night instead of putting them under your pillow, and they would be there in the morning. You could leave your pouch on your locker for the night, instead of lying with it in your pocket and rubbing the tobacco to dust.

Trivialities, you say? No, all enormously important! Once there was no pilfering, people began to look at their neighbors kindly and without suspicion. Listen, lads, maybe we really are . . . *politicals?*

And if we're politicals we can speak a bit more freely, between two bunks or by the fire on the work site. Better look around, of course, in case somebody's listening. Come to think of it, it doesn't matter a damn if they frame us—if you've got a *quarter* already, what else can they do to you?

The old camp mentality—you die first, I'll wait a bit; there is no justice, so forget it; that's the way it was, and that's the way it will be—also began to disappear.

Why forget it? Why must it always be so?

Teammates begin quietly talking to each other, not about rations, not about gruel, but about things which you never hear mentioned Outside—and talking more and more freely all the time. Until the foreman's fist suddenly ceases to throb with self-importance. Some foremen stop raising their fists altogether, others use them less often and less heavily. The foreman himself, instead of towering over his men, sits down to listen and to chat. And his team begin to look on him as a comrade—after all, he's *one of us.*

The foremen come to the Production Planning Section or the accounts office, and from discussing with them dozens of small questions—whose rations should be cut and whose shouldn't, whom to assign to which job—the trusties, too, are affected by this breath of fresh air, this small cloud of seriousness, responsibility, and purpose.

Not all of them at once. They had come to these camps greedily intent on grabbing the best jobs, and they had succeeded. They saw no reason why they shouldn't live as well as in the Corrective Labor Camps, shutting themselves up in their private rooms, roasting their potatoes with pork fat; live their own lives, separately from the workers. But no! It turned out that there was something more important than all this. And what might that be? . . . It became indecent to boast of being a bloodsucker, as people did in the Corrective Labor Camps, to boast that you lived at the expense of others. So trusties would make friends among the workers, spread their nice new jerkins beside the workers' grubby ones, and lie on the ground happily chatting their Sundays away.

People could no longer be divided into such crude categories as in the Corrective Labor Camps: trusties and workers, nonpoliticals and 58's. The breakdown was much more complex and interesting: people from the same region, religious groups, men of practical experience, men of learning.

It would take the authorities a long, long time to notice or understand. But the work assigners no longer carried clubs, and didn't even bellow as loudly as before. They addressed the foremen in a *friendly* way: for example, "Must be about time to get your men out to work, Komov." (Not because they were overcome by remorse, but because there was something new and disquieting in the air.)

But all this happened *slowly*. These changes took months and months and months. They did not affect every foreman and every trusty, only those in whom some remnants of conscience and fraternal feeling still glowed under the ashes. Those who preferred to go on being bastards found no difficulty in doing just that. As yet there was no real shift of consciousness, no heroic shift, no spiritual upheaval. The camp remained a camp as before, we were as much oppressed and as helpless as ever, there was still no hope for us but to crawl away under the wire and run out into the steppe, to be showered with bullets and hunted down with dogs.

A bold thought, a desperate thought, a thought to raise a man up: how could things be changed so that instead of *us* running from *them, they* would run from *us?*

Once the question was put, once a certain number of people had thought of it and put it into words, and a certain number had listened to them, the age of escapes was over. The age of rebellion had begun.

But how to begin? Where to begin? We were shackled, we were wrapped about by tentacles, we were deprived of freedom of movement—where could we begin?

Far from simple in this life are the simplest things of all. Even in the Corrective Labor Camps some people seem to have got around to the idea that stool pigeons should be killed. Even there accidents were sometimes arranged: a log would roll off a pile and knock a stoolie into the swollen river. So that it shouldn't have been difficult to figure out in the Special Camps which tentacles to hack off first. You would expect everybody to see that, yet nobody did.

Suddenly—a suicide. In the Disciplinary Barracks, hut No. 2, a man was found hanging. (I am going through the stages of the process as they occurred in Ekibastuz. But note that the stages were just the same in other Special Camps!) The bosses were not greatly upset; they cut him down and wheeled him off to the scrap heap.

A rumor went around the work team. The man was an informer. He hadn't hanged himself. He had been hanged.

As a lesson to the rest.

There were a lot of filthy swine in the camp, but none so hoggish, so crude, so brazen as Timofey S——,[2] who was in charge of the mess hut. His bodyguards were the fat, pig-faced cooks, and he had also hand-reared a retinue of thuggish orderlies. He and his retainers beat the zeks with fists and sticks. On one occasion, quite unjustly, he struck a short, swarthy prisoner whom everybody called "the kid." It was not his habit to notice whom he was beating. But there were no mere "kids" as things now were in the Special Camps, and this kid was also a Moslem. There were quite a few Moslems in the camp. They were not just common criminals. You could see them praying at sunset at the western end of the camp area (in a Corrective Labor Camp, people would have laughed, but we did not), throwing up their arms and pressing their foreheads to the ground. They had elders, and true to the spirit of the times, they met in council. The motion was to avenge themselves!

Early one Sunday morning, the victim, together with an adult companion, an Ingush, slipped into the trusties' hut while they were still wallowing in bed, and each of them quickly stuck a knife into the fat hog.

But how green we still were. They did not try to hide their faces or to run away. At peace with themselves, now that duty was done, they went with their bloody knives straight from the corpse to the warders' room, and gave themselves up. They would be put on trial.

All these were tentative fumblings. All this could perhaps have happened in a Corrective Labor Camp. But civic thought made further strides: perhaps this was the crucial link at which the chain must be broken?

"Kill the stoolie!" That was it, the vital link! A knife in the heart of the stoolie! Make knives and cut stoolies' throats—that was it!

Now, as I write this chapter, rows of humane books frown down at me from the walls, the tarnished gilt on their well-worn spines glinting reproachfully like stars through cloud. Nothing in the world should be sought through violence! By taking up the sword, the knife, the rifle, we quickly put ourselves on the level of our

---

2. I am not concealing his surname; I just don't remember it.

tormentors and persecutors. And there will be no end to it. . . .

There will be no end. . . . Here, at my desk, in a warm place, I agree completely.

If you ever get twenty-five years for nothing, if you find yourself wearing four number patches on your clothes, holding your hands permanently behind your back, submitting to searches morning and evening, working until you are utterly exhausted, dragged into the cooler whenever someone denounces you, trodden deeper and deeper into the ground—from the hole you're in, the fine words of the great humanists will sound like the chatter of the well-fed and free.

There will be no end of it! . . . But will there be a beginning? Will there be a ray of hope in our lives or not?

The oppressed at least concluded that evil cannot be cast out by good.

Stoolies, you say, are human beings, too? . . . Warders went around the barracks and to intimidate us read out an order addressed to the whole Peschany camp complex. At one of the women's divisions, two girls (their dates of birth were given; they were very young) had indulged in anti-Soviet talk. "A tribunal consisting of . . ." ". . . death by shooting!"

Those girls, whispering together on their bunks, had ten years hanging around their necks already. What foul creature, with a burden of its own to carry, had turned them in? How can you say that stoolies are human beings?

There were no misgivings. But the first blows were still not easy.

I do not know about other places (they started killing in *all* the Special Camps, even the Spassk camp for the sick and disabled), but in our camp it began with the arrival of the Dubovka transport —mainly Western Ukrainians, OUN members. The movement everywhere owed a lot to these people, and indeed it was they who set the wheels in motion. The Dubovka transport brought us the bacillus of rebellion.

These sturdy young fellows, fresh from the guerrilla trails, looked around themselves in Dubovka, were horrified by the apathy and slavery they saw, and reached for their knives.

In Dubovka it had quickly ended in mutiny, arson, and disbandment. But the camp bosses were so blindly sure of themselves (for thirty years they had met no opposition, and had grown unused to it) that they did not take the trouble even to keep the rebels separated from us. They were scattered throughout the camp in various work teams. This was a Corrective Labor Camp practice: there, dispersal muffled protest. But in our purer air, dispersal only helped the flames to engulf the whole mass more rapidly.

The newcomers went to work with their teams, but never lifted a finger, or just made a show of it; instead they lay in the sun (it was summer), conversing quietly. At such times, to the casual eye they looked very much like thieves *making law*, especially as they, too, were well-nourished, broad-shouldered young men.

A law indeed emerged, but it was a new and surprising law: "You whose conscience is unclean—this night you die!"

Murders now followed one another in quicker succession than escapes in the best period. They were carried out confidently and anonymously: no one went with a bloodstained knife to give himself up; they saved themselves and their knives for another deed. At their favorite time—when a single warder was unlocking huts one after another, and while nearly all the prisoners were still sleeping—the masked avengers entered a particular section, went up to a particular bunk, and unhesitatingly killed the traitor, who might be awake and howling in terror or might be still asleep. When they had made sure that he was dead, they walked swiftly away.

They wore masks, and their numbers could not be seen—they were either picked off or covered. But if the victim's neighbors should recognize them by their general appearance, so far from hurrying to volunteer information, they would not now give in even under interrogation, even under threat from the godfathers, but would repeat over and over again: "No, no, I don't know anything, I didn't see anything." And this was not simply in recognition of a hoary truth known to all the oppressed: "What you don't know can't hurt you"; it was self-preservation! Because anyone who *gave names* would have been killed next 5 A.M., and the security officer's good will would have been no help to him at all.

And so murder (although as yet there had been fewer than a dozen) became *the rule*, became a normal occurrence. "Anybody been killed today?" prisoners would ask each other when they went to wash or collect their morning rations. In this cruel sport the prisoner's ear heard the subterranean gong of justice.

It was done in a strictly conspiratorial fashion. Somewhere, someone (of recognized authority) simply gave a name to someone else: he's the one! It was not his concern who would do the killing, or on what date, or where the knives would come from. And the hit men, who were concerned with all these things, did not know the judge whose sentence they must carry out.

Bearing in mind the impossibility of documentary confirmation that a man was an informer, we are bound to acknowledge that this improperly constituted, illegal, and invisible court was much more acute in its judgments, much less often mistaken, than any of the tribunals, panels of three, courts-martial, or Special Boards with which we are familiar.

The *chopping,* as we called it, went so smoothly that it began to encroach on the daytime and become almost public. A small, blotchy-faced "barracks elder," once a big NKVD man in Rostov, and a notorious louse, was killed one Sunday afternoon in the "bucket room." Prisoners had become so hardened that they crowded in to see the corpse lying in a pool of blood.

Next the avengers ran through the camp with knives in broad daylight, chasing the informer who had *betrayed* a tunnel under the camp area from the Disciplinary Barracks, hut No. 8 (the camp command had woken up and herded the Dubovka ringleaders into it, but by now the *chopping* went on just as well without them). The informer fled from them into the staff barracks, they followed, he rushed into the office of the divisional commander, fat Major Maksimenko, and they were still behind him. At that moment the camp barber was shaving the major in his armchair. The major was unarmed, in accordance with camp regulations—they are not supposed to carry firearms into the camp area. When he saw the murderers armed with knives, the terrified major jumped from under the razor and begged for mercy, thinking that they were about to knife him. He was relieved to see them cut up the stoolie before his very eyes. (Nobody was even after the major. The nascent movement had issued a directive: only stoolies to be killed, warders and officers not to be touched.) All the same, the major jumped out of the window half-shaven, with a white smock wrapped around him, and ran for the guardhouse, shouting, in panic: "You in the watchtower, shoot! Shoot, I say." But the watchtower did not open fire.

On one occasion a stoolie broke away before they could finish the job and rushed wounded into the hospital. There he was operated on and bandaged up. But if the major had been frightened out of his wits by knives, could the hospital save a stoolie? Two or three days later they finished him off on a hospital bed. . . .

Out of five thousand men about a dozen were killed, but with every stroke of the knife more and more of the clinging, twining tentacles fell away. A remarkable fresh breeze was blowing! On the surface, we were prisoners living in a camp just as before, but in reality we had become free—free because for the very first time in our lives, we had started saying openly and aloud all that we thought! No one who has not experienced this transition can imagine what it is like!

And the informers . . . stopped informing.

Until then a security officer could make anyone he liked stay behind in camp in the daytime, talk to him for hours on end—whether to collect denunciations, give new instructions, or elicit the names of prisoners who looked out of the ordinary, who had done nothing so far but were capable of it, who were suspected of being possible nuclei of future resistance.

When his work team came back in the evening they would question their mate. "Why did they send for you?" And he would always reply, whether it was the truth or impudent bluff: "They wanted to show me some photographs."

True enough, many prisoners were shown photographs and asked to identify people whom they might have known during the war. But the security officers couldn't show them to everybody, and anyway it would have been pointless. Yet everybody—friends and traitors alike—always mentioned them. Suspicion moved in with us, and forced us into our shells.

But now the air was being cleansed of suspicion! Now even if a security officer ordered somebody to stay away from work line-up—he *would not!* Incredible! Unheard of in all the years of the Cheka-GPU-MVD's existence! They summoned a man, and instead of dragging himself there with his heart missing beats, instead of trotting in with a servile look on his silly face, he preserved his dignity (his teammates were watching) and refused to go! An invisible balance hung in the air over the work line-up. In one of its scales all the familiar phantoms were heaped: interrogation officers, punches, beatings, sleepless standing, "boxes" (cells too small to sit or lie down in), cold, damp punishment cells, rats, bedbugs, tribunals, second and third sentences. But this could not all happen at once, this was a slow-grinding bone mill, it could not devour all of us at once and process us in a single day. And even when they had been through it—as every one of us had—men still went on existing.

While in the other scale lay nothing but a single knife—but that knife was meant for you, if you gave in! It was meant for you alone, in the breast, and not sometime or other, but at dawn tomorrow, and all the forces of the Cheka-MGB could not save you from it! It was not a long one, but just right for neat insertion between your ribs. It didn't even have a proper handle, just a piece of old insulation tape wound around the blunt end of the blade—but this gave a very good grip, so that the knife would not slip out of the hand!

And this bracing threat weighed heavier! It gave the weak strength to tear off the leeches, to pass by and follow their mates. (It also gave them a good excuse later on: We would have stayed behind, citizen officer, but we were afraid of the knife. . . . You aren't threatened by it; you can't imagine what it's like.)

This was not all. Not only did they stop answering the summonses of the security officers and other camp authorities; they were now chary of dropping an envelope or any bit of paper with writing on it into the mailbox, which hung in the camp area, or the boxes for complaints to higher authority. Before mailing a letter or putting in a complaint, they would ask someone to look at it. "Go on, read it, it isn't a denunciation. Come with me while I mail it."

So that now the bosses were suddenly blind and deaf. To all appearances, the tubby major, his equally tubby second in command, Captain Prokofiev, and all the warders walked freely about the camp, where nothing threatened them; moved among us, watched us—and yet saw nothing! Because a man in uniform sees and hears nothing without stoolies: prisoners stop talking, turn their backs, hide things, move away at his approach. . . . A few yards off, faithful informants are swooning with desire to sell their comrades—but not one of them even makes a secret sign.

The information machine on which alone the fame of the omnipotent and omniscient Organs had been based in decades past had broken down.

On the face of it, the same teams still went to work at the same sites. (We had, however, agreed among ourselves to resist the convoy guards, too, not to let them rearrange the ranks of five, or to re-count us on the march—and we succeeded! There were no stoolies among us—and the Tommy-gunners also weakened!) The teams worked to fill their norms satisfactorily. On their return they allowed the warders to search them as before. (Their knives were never found, though.) But in reality other forms of human association now bound people more closely than the work teams artificially put together by the administration. Most important were national ties. National groups—Ukrainians, United Moslems, Estonians, Lithuanians—which informers could not penetrate, were born and flourished. No one elected the leadership, but its composition so justly satisfied the claims of seniority, wisdom, and suffering that no one disputed its authority over its own nation. A consultative and coordinating body evidently came into being as well—a "Council of Nationalities," as it were.[3]

The teams were the same as before, and there were no more of them, but here was something strange: a sudden *shortage of foremen* in the camp—an unheard-of phenomenon in Gulag. At first the wastage looked natural: one went into the hospital, another went to work in the service yard, a third was due for release. But the work assigners had always had crowds of candidates eager to buy a foreman's job with a piece of fatback or a sweater. Now instead of candidates there were some foremen who hung about the Production Planning Section daily, asking to be relieved of their jobs as soon as possible.

As things now were, the old methods used by foremen to drive a worker into a wooden overcoat were hopelessly out of date, and not everyone had the wit to devise new ones. There was soon such a shortage of foremen that work assigners would come into a squad's quarters for a smoke and a chat, and simply beg for their help. "Come on, lads, you've got to have a foreman; this is a ridiculous state of affairs. Come on now, choose somebody for yourselves and we'll promote him right away."

---

3. Here some qualification is necessary. It was not all as clean and smooth as it looks from this description of the main trend. There were rival groups—the "moderates" and the "ultras." Personal predilections and dislikes and the clash of ambitions among men eager to be "leaders" also crept in. The "hit men," the young bulls of the herd, were far from being men of broad political vision; some were apt to demand extra rations for their "work" and to try to get them by threatening the cook in the hospital kitchen. They demanded, in other words, to be fed at the expense of the sick, and if the cook refused, they would kill him without any formal court of morals: they had the knack of it, masks, knives in their hands. In a word, corruption and decay—old, invariable feature of revolutionary movements throughout history—were already burrowing into the healthy core.

There was one case of simple error: a crafty informer induced a good-natured working prisoner to change beds with him—and the man was murdered in the morning.

But in spite of these lapses, the movement as a whole kept strictly on course. We knew where we were going. The required social effect was achieved.

This happened more and more often when foremen started *escaping* into the Disciplinary Barracks—hiding behind stone walls! Not only they, but bloodsucking work assigners like Adaskin, stoolies on the brink of exposure or, something told them, next on the list, suddenly took fright and *ran for it!* Only yesterday they had put a brave face on it, behaving and speaking as though they approved of what was afoot (just try telling the zeks otherwise in their present mood!); only last night they had gone to bed in the common hut (whether to sleep or to lie there tense and ready to fight for their lives, vowing that there would be no more nights like this); but today they have vanished. An orderly receives instructions: carry so-and-so's things over to the Disciplinary Barracks.

This was a new period, a heady and spine-tingling period in the life of the Special Camp. It wasn't we who had taken to our heels —they had, ridding us of their presence! A time such as we had never experienced or thought possible on this earth: when a man with an unclean conscience could not go quietly to bed! Retribution was at hand—not in the next world, not before the court of history, but retribution live and palpable, raising a knife over you in the light of dawn. It was like a fairy tale: the ground is soft and warm under the feet of honest men, but under the feet of traitors it prickles and burns. If only our Great Outside were as lucky, the Land of the Free, which never has seen and perhaps never will see such a time.

The grim stone jailhouse, by now enlarged and completed, with its tiny muzzled windows, cold, damp, and dark, surrounded with a fence of overlapping two-inch boards—the Disciplinary Barracks so lovingly prepared by the masters of the camps for recalcitrants, runaways, awkward customers, protesters, people of courage—has suddenly become a rest home for retired stoolies, bloodsuckers, and bully boys.

It was, surely, a witty fellow who first had the idea of running to the Chekists and begging them to let their faithful servant take sanctuary from the people's wrath in the stone sack.

They themselves had begged to be locked up more securely, they had run not away from but into jail; they had voluntarily agreed never again to breathe clean air or see sunlight. I don't think history records anything like it.

The prison chiefs and security officers had pity on the first of them and took them under their wings; they had to look after their own. The best cell in the barracks was assigned to them (the camp wits called it the safe deposit), mattresses were sent in, better heating was ordered, a one-hour exercise period was prescribed.

But behind the first smart alecks came a long line of others less smart but no less eager to live. (Some tried to save their faces even in flight: who knows, they might someday have to go back and live among zeks again. Archdeacon Rudchuk's flight to the Disciplinary Barracks was carefully staged: warders came into the hut after locking-up time, executed a ruthless search, even shaking out the contents of his mattress, "arrested" Rudchuk, and took him away. The camp, however, was soon reliably informed that the haughty archdeacon, amateur of the paintbrush and the guitar, was with the others in the "safe deposit.") Their numbers shortly topped a dozen, fifteen, twenty! ("Machekhovsky's squad," people started calling them, after the chief disciplinary officer.) A second cell had to be brought into use, which further reduced the Disciplinary Barracks' productive area.

But a stoolie is only wanted, only useful, so long as he can rub shoulders with the crowd and pass undetected. Once detected, he is worthless, and cannot go on serving in the same camp. He eats the bread of idleness in the Disciplinary Barracks, doesn't go out to work, isn't worth his salt. Even the MVD's philanthropy must have some limits!

So the flow of stoolies begging to be saved was stemmed. Latecomers had to remain in their sheep's clothing and await the knife.

An informer is like a ferryman: once he's served his purpose, nobody wants to know him.

The camp authorities were more concerned with countermeasures to stop the menacing movement and break its back. They clutched first at the method with which they were most familiar —issuing written orders.

The masters of our bodies and souls were particularly anxious not to admit that our movement was political in character. In their menacing orders (warders went around the huts reading them out) the new trend was declared to be nothing but *gangsterism*. This made it all simpler, more comprehensible, somehow cozier. It seemed only yesterday that they had sent us gangsters labeled "politicals." Well, politicals—real politicals for the first time— had now become "gangsters." It was announced, not very confidently, that these gangsters would soon be discovered (so far not one of them had been), and still less confidently, that they would be shot. The orders further appealed to the prisoner mass to *condemn* the gangsters and *struggle* against them!

The prisoners listened and went away chuckling. Seeing the disciplinary officers afraid to call "political" behavior by its name (although the purpose of all investigations for thirty years past had been the imputation of political motives), we were aware of their weakness.

And weakness it was! Calling the movement "gangsterism" was a ruse to relieve the camp administration of responsibility for allowing a political movement to develop in the camp! This pretense was just as convenient, just as necessary higher up: in the provincial and camp administrations of the MVD, the central offices of Gulag, the ministry itself. A system which lives in constant dread of publicity loves to deceive itself. If the victims had been warders or disciplinary officers, it would have been difficult not to invoke Article 58-8 (on terrorism), and the camp authorities could have easily responded with death sentences. Under the present circumstances, however, our masters were presented with an irresistible opportunity to camouflage what was happening in the Special Camps as part of the "bitches' war," which was then shaking the Corrective Labor Camps to their foundations, and was of course engineered by the Gulag administration.[4]

4. The "bitches' war" deserves a chapter of its own in this book, but a great deal of additional material would have to be found. Let me refer the reader to Varlam Shalamov's study *Ocherki Prestupnogo Mira (Essays on the Criminal World)*, although this, too, is incomplete.

Briefly, the "bitches' war" flared up somewhere around 1949 (if we discount sporadic minor clashes between thieves and "bitches"). In 1951–1952 it was at its fiercest. The criminal world was subdivided into many different sects: apart from thieves proper and "bitches," there were also the No-Limiters, the Makhrovtsy, the Uporovtsy, the Pirovarovtsy, the Red Riding Hoods, the Fuli Nam, the Crowbar-Belted—and that is not the end of them.

By this time the Gulag administrators had lost faith in infallible theories about the re-education of criminals and had evidently decided to lighten their load by playing on these differences, supporting first one group, then another, and using their knives to destroy others. The butchery went on openly and wholesale.

Then the professional criminals who took to murder developed their own technique: either they killed with someone else's hands, or when they themselves killed they made someone else take the blame. So young casual offenders or ex-soldiers or officers, under threat of murder, took other people's murders on themselves, were sentenced to twenty-five years under Article 59-3 (banditry), and are still inside. Whereas the thieves who led the groups came out clean under the "Voroshilov amnesty" in 1953. (But let us not be too downhearted: they have been back inside a time or two since then.)

When our newspapers revived the fashion for sentimental stories about the remolding of criminals, reports—muddled and mendacious, of course—about the butchery in the camps also broke through into the columns of the press, with the "bitches' war," the "chopping" in the Special Camps, and any unexplained bloodletting deliberately mixed up (to confuse history).

This was their way of whitewashing themselves. But they also deprived themselves of the right to shoot the camp murderers, which was the only effective countermeasure. And they could not oppose the growing movement.

The orders were of no avail. The prisoner masses did not start *condemning* and *struggling* on behalf of their masters. The next measure was to put the whole camp on a punitive routine. This meant that in nonworking hours on weekdays, and all day on Sundays, we were under lock and key, and had to use the latrine bucket, and were even fed in our huts. They started carrying the broth and mush around in big tubs, and the mess hall was deserted.

This was a hard regime, but it did not last long. We were lazy at work, and the Mining Trust set up a howl. More important, the warders' work load was quadrupled—they were incessantly rushing from one end of the camp to the other, letting orderlies in with buckets, letting them out again, bringing the food around, escorting groups of prisoners to and from the Medical Section.

The object of the camp administration was to make things so hard for us that we would betray the murderers out of exasperation. But we braced ourselves to suffer, to hang on a bit: it was worth it! Their other object was to keep the huts closed so that murderers could not come from outside, and so would be easier

---

The camp theme interests the whole Soviet people, and such articles are read avidly, but they are no aid to understanding (which is why they are written). Thus the journalist Galich published in *Izvestiya* in July, 1959, a rather suspect "documentary" tale about a certain Kosykh, who is supposed to have touched the hearts of the Supreme Soviet with an eighty-page typewritten letter from a camp. (1. Where did he get his typewriter? Did it belong to the security officer? 2. Who would ever read eighty whole pages—those people yawn their heads off after one.) This Kosykh was in for twenty-five years—a second sentence for something he did in the camps. But what? On this point Galich—and it is our journalists' distinctive characteristic—immediately loses the capacity for clear and articulate speech. It is impossible to understand whether Kosykh had murdered a "bitch," or a stoolie (which would make it "political"). But that is characteristic—in historical retrospect everything is consigned to a single heap and called gangsterism. This is the best a national newspaper can do by way of scientific explanation: "Beria's stooges were then active in the camps." (Then, and not earlier? Then, but not now?) "Rigorous application of the law was undermined by the illegal actions of persons who were supposed to enforce it" (How? Did they go against generally binding instructions?). *"They did all they could to foment hostility* [My italics. This much is true—A.S.] between various groups of prisoners." (The use of informers can also be covered by this formula.) "There was savage, ruthless, artificially fomented enmity."

It of course proved impossible to put a stop to the killings in the camps by means of twenty-five-year sentences—many of the murderers had twenty-five years already—so a decree of 1961 made murder (including, of course, the murder of informers) in the camps punishable by shooting. This Khrushchevian decree was all the Stalinist Special Camps had lacked.

to find. But another murder took place, and still no one was caught—just as before no one had ever seen anything or knew anything. Then somebody's head was smashed in at work—locked huts are no safeguard against that.

They revoked the punitive regime. Instead they had the bright idea of building the "Great Wall of China." This was a wall two bricks thick and four meters high, to cut across the width of the camp. They were preparing to divide the camp into two parts, but left an opening for the time being. (All Special Camps were to do the same. Barriers to break up large camp areas were going up in many other places.) Since the Mining Trust could not pay for this work and since it had no relevance to the settlement, the whole burden—making the adobe bricks, shifting them around to dry, carrying them to the wall, laying them—fell upon us again, upon our Sundays and our light summer evenings after we returned from work. We greatly resented that wall—we knew that the bosses had some dirty trick in store—but we had no choice but to build it. Only a little of us was as yet free—our heads and our mouths—but we were still stuck up to our shoulders in the quagmire of slavery.

All these measures—the threatening orders, the punitive regime, the wall—were crude, and in the best tradition of prison thinking. But what was this? Without warning or explanation, they called teams one after another to the photography room, and photographed them, but politely, without putting number plates on dog collars around their necks, without making them turn their heads to a particular angle—it was just sit comfortably, look just as you please. From a "careless" remark dropped by the head of the Culture and Education Section, the workers learned that they were being "photographed for documents."

For what documents? What documents can prisoners have? . . . A ripple of excitement went around among the credulous: Perhaps they were preparing passes so that prisoners could move without convoy guards? Or perhaps . . . ? Or perhaps . . . ?

Then one warder returned from leave and loudly told another (in the presence of prisoners) that on his way he had seen trainloads of released prisoners going home, with slogans and green branches.

Lord, how our hearts beat! It was high time, of course! They should have started like this after the war! Had it really begun at last?

We heard that someone had received a letter from his family saying that his neighbors had already been released and were at home!

Suddenly one of the photographed brigades was summoned before a board. Go in one by one. At a table with a red cloth on it, under a portrait of Stalin, sat our senior camp personnel, but not alone: there were also two strangers, one a Kazakh, the other a Russian, who had never been in our camp. They were business-like but jolly as they filled in their form: surname, first names, year of birth, place of birth—and then, instead of the usual "article under which sentenced, length of sentence, end of sentence," they asked in detail about the man's family status, his wife, his parents, if he had children, how old they were, where they all lived, to-gether or separately. And all this was taken down! (One or another of the board would tell the clerk to "get that down, too.")

Strange, painfully pleasurable questions! They make the hardest among us feel a glow and want to weep! For years and years he has heard only that abrupt yapping: "Article? Sentence? Court?" —and suddenly he sees sitting there kindly, serious, humane offi-cers, questioning him unhurriedly and sympathetically, yes, sym-pathetically, about things he has buried so deep that he is afraid to touch them himself, things about which he might occasionally say a word or two to his neighbor on the bed platform, or then again might not. . . . And these officers (if you remember it at all, you forgive the first lieutenant there who took a photograph of your family away from you and tore it up last October anniver-sary)—these officers, when they hear that your wife has remarried, that your father is failing fast and has lost hope of seeing his son again—tut sadly, exchange glances, shake their heads.

No, they're not so bad, after all; they're human, too; it's just the lousy service they're in. And when they've written it all down, the last question they ask each prisoner is this:

"Right; now where would you like to *live?* . . . Where your parents are, or where you lived before?"

The zek's eyes pop out. "How do you mean? I'm in No. 7 Barracks. . . ."

"Look, we know that." The officers laugh. "We're asking where you'd *like* to live. Suppose you were let out—where should your documents be made out for?"

The whole world spins before the prisoner's eyes, the sunlight splinters into an iridescent haze. With his mind, he understands that this is a dream, a fairy tale, that it cannot be true, that his sentence is twenty-five years or ten, that nothing has changed, that he is plastered all over with clay and will be back on the job tomorrow—but there sit several officers, two majors among them, calmly, compassionately insisting:

"Where is it to be, then? Name it."

With his heart hammering, and warm waves of gratitude washing over him, like a blushing boy mentioning the name of his girl, he gives away his cherished secret—where he would like to live out peacefully the remainder of his days if he were not a doomed convict with four number patches.

And they . . . write it down! And ask for the next man to be called in. While the first dashes half-crazed into the corridor and tells the other lads what has happened.

The members of the team go in one by one and answer questions for the friendly officers. And there is only one in half a hundred who says with a grin:

"Everything's just fine here in Siberia, only the climate's too hot. Couldn't I go to the Arctic Circle?"

Or: "Write this down: 'In a camp I was born, in a camp I'll die. I know no better place.' "

They had such talks with two or three teams (there were two hundred of them in the camp). The camp was in a state of excitement for some days, here was something to argue about—though half of us didn't really believe it. Those times had passed! But the board never convened again. The photography had cost them little —the cameras clicked on empty cartridges. But sitting in a huddle for heart-to-heart talks with those scoundrels overtaxed their patience. And so nothing came of their shameless trick.

(Let's admit it—this was a great victory! In 1949 camps with a ferocious regime were set up—and intended, of course, to last forever. Yet by 1951 their masters were reduced to this maudlin playacting. What further admission of our success could we ask for? Why did they never have to put on such an act in the Corrective Labor Camps?)

Again and again the knives flashed.

So our masters decided to *make a snatch*. Without stoolies they didn't know exactly whom they wanted, but still they had ideas and suspicions of their own (and perhaps denunciations were somehow arranged on the sly).

Two warders came into a hut after work, and casually told a man, "Get ready and come with us."

The prisoner looked around at the other lads and said, "I'm not going."

In fact, this simple everyday situation—*a snatch, an arrest*—which we had never resisted, which we were used to accepting fatalistically, held another possibility: that of saying, "I'm not going!" Our liberated heads understood that now.

The warders pounced on him. "What do you mean, not going?"

"I'm just not going," the zek answered, firmly. "I'm all right where I am."

There were shouts from all around:

"Where's he supposed to go? . . . What's he got to go for? . . . We won't let you take him! . . . We won't let you! . . . Go away!"

And the wolves understood that we were not the sheep we used to be. That if they wanted to grab one of us now they would have to use trickery, or do it at the guardhouse, or send a whole detail to take one prisoner. With a crowd around, they would never take him.

Purged of human filth, delivered from spies and eavesdroppers, we looked about and saw, wide-eyed, that . . . we were thousands! that we were . . . *politicals!* that we could *resist!*

We had chosen well; the chain would snap if we tugged at this link—the stoolies, the talebearers and traitors! Our own kind had made our lives impossible. As on some ancient sacrificial altar, their blood had been shed that we might be freed from the curse that hung over us.

The revolution was gathering strength. The wind that seemed to have subsided had sprung up again in a hurricane to fill our eager lungs.

Chapter 11

■

Tearing at the Chains

The middle ground had now collapsed, the ditch which ran between us and our custodians was now a deep moat, and we stood on opposite slopes, taking the measure of the situation.

"Stood" is of course a manner of speaking. *We* went to work daily with our new foremen (some of them secretly elected and coaxed into serving the common cause, others not new to the job but now so sympathetic, so friendly, so solicitous as to be unrecognizable), we were never late for work line-up, we never let each other down, there were no shirkers, we chalked up a good day's work—you might think that our masters could be pleased with us. And that we could be pleased with them: they had quite forgotten how to yell and to threaten, they no longer hauled us into the Disciplinary Barracks for petty reasons, they appeared not to notice that we had stopped doffing our caps to them. Major Maksimenko did not get up for work line-up in the morning, but he did like to greet the columns at the guardhouse of an evening, and to crack a joke or two while they were marking time there. He beamed upon us with fat complacency, like a Ukrainian rancher somewhere in the Tavrida surveying his countless flocks as they come home from the steppe. They even started showing us films occasionally on Sundays. There was just one thing. They went on plaguing us with the "Great Wall of China."

All the same, we and they were thinking hard about the next stage. Things could not remain as they were: we could not be satisfied with what had happened, nor could they. Someone had to strike a blow.

But what should be our aim? We could now say out loud,

without looking nervously around, whatever we liked—all those things which had seethed inside us (and freedom of speech, even if it was only there in the camp, even though it had come so late, was a delight!). But could we hope to spread that freedom beyond the camp, or carry it out with us? Of course not. What further *political* demands could we put forward? We just couldn't think of any! Even if it had not been pointless and hopeless, we couldn't think of any! We, from where we were, could not demand that the country should change completely, nor that it should give up the camps: they would have rained bombs on us.

It would have been natural for us to demand that our *cases* be reviewed, and that unjust sentences, imposed without reason, be quashed. But even that looked hopeless. In the foul fog of terror that hung thickly over the land, the cases brought against most of us, and the sentences passed, seemed to our judges fully justified —and they had almost made us believe it ourselves. Besides, judicial review was a phantom process, which the crowd could get no grip on, and there could be no easier way to cheat us: they could make promises, spin out the proceedings, keep coming back to ask more questions—it could drag on for years. Suppose somebody was suddenly declared free and removed: how could we be sure that he was not on his way to be shot, or to another prison, or to be sentenced afresh?

Hadn't the farce of the "Board" already shown us how easy it is to create illusions? The Board was for packing us off home, even without judicial review. . . .

Where we were all of one mind, and had no doubts at all, was that the most humiliating practices must be abolished: the huts must be left unlocked overnight, and the latrine buckets removed; our number patches must be taken off; our labor must not be completely unpaid; we must be allowed to write twelve letters a year. (But all this and more—indeed, the right to twenty-four letters a year—had been ours in the Corrective Labor Camps— and had it made life there livable?)

As to whether we should fight for an eight-hour working day —there was no unanimity among us. . . . We were so unused to freedom that we seemed to have lost all appetite for it.

Ways and means were also discussed. How should we present our demands? What action should we take? Clearly, we could do nothing with bare hands against modern arms, and therefore the course for us to take was not armed rebellion, but a strike. On strike we could, for instance, tear off the number patches ourselves.

But the blood in our veins was still slavish, still servile. . . . For all of us at once to remove the odious numbers from our persons seemed a step as daring, as audacious, as irrevocable as, say, taking to the streets with machine guns. And the word "strike" sounded so terrifying to our ears that we sought firmer ground: by refusing food when we refused to work it seemed to us that our moral right to strike would be reinforced. We felt that we had some sort of right to go on a hunger strike—but to strike in the ordinary sense? Generation after generation in our country had grown up believing that the horrifyingly dangerous and, of course, counter-revolutionary word "strike" belonged with "Entente," "Denikin," "kulak sabotage,"* "Hitler."

So that by voluntarily undertaking an unnecessary hunger strike, we voluntarily agreed to undermine the physical strength which we needed for the struggle. (Fortunately, no other camp seems to have repeated Ekibastuz's mistake.)

We went over and over the details of the proposed work stoppage–hunger strike. The general Disciplinary Code for Camps, recently made applicable to us, told us that they would reply by locking us in our huts. How, then, should we keep in touch with each other, and pass decisions about the further conduct of the strike from hut to hut? Someone had to devise signals, and get the huts to agree from which windows they would be made, and at which windows they would be picked up.

It was talked over in various places, in one group and another; it seemed inevitable and desirable, yet, because it was so novel, somehow impossible. We could not imagine ourselves suddenly assembling, finishing our discussions, resolving, and . . .

But our custodians did not have to organize secretly, they had a clear chain of command, they were more accustomed to action, they were less likely to lose by acting than by failing to act—and they got their blow in first.

After that, events took on a momentum of their own.

At peace and at ease on our familiar bunks, in our familiar sections, we greeted the new year, 1952. Then on Sunday, January 6, the Orthodox Christmas Eve, when the Ukrainians were getting ready to observe the holiday in style—they would make kutya,* fast till the first star appeared, and then sing carols—the doors were locked after morning inspection and not opened again.

No one had expected this! The preparations had been secret and sly! Through the windows we saw them herding a hundred or more prisoners from the next hut through the snow to the guardhouse, with all their belongings.

Were they being moved to another camp?

Then it was our turn. Warders came. And officers with cards. They called out names from the cards. Outside with all your things—including mattresses, just as they are; don't empty them!

So that was it! A regrouping! Guards were posted by the break in the Chinese Wall. It would be bricked up next day. We were taken past the guardhouse and herded, hundreds of us, with sacks and mattresses, like refugees from a burning village, around the boundary fence, past another guardhouse, into the other camp area. Passing those who were being driven in the opposite direction.

All minds were busily trying to work out who had been moved, who had been left behind, what this reshuffle meant. What our masters had in mind became clear soon enough. In one half of the camp (Camp Division No. 2) only the Ukrainian nationalists, some 2,000 of them, were left. In the half to which we had been driven, and which was to be Camp Division No. 1, there were some 3,500 men belonging to all the other ethnic groups—Russians, Estonians, Lithuanians, Latvians, Tatars, Caucasians, Georgians, Armenians, Jews, Poles, Moldavians, Germans, and a variegated sprinkling of other nationalities picked up from the expanses of Europe and Asia. In a word—our country, "one and indivisible." (Curious, this. The thinking of the MVD, which should have been enlightened by a supranational doctrine, called socialism, still followed the same old track: that of dividing nation from nation.)

The old teams were broken up, new ones were mustered; they would go to new work sites, live in new huts—in short, a complete reshuffle! There was enough to think about here for a week, not just one Sunday. Many links were snapped, people were thrown together in different combinations, and the strike, which had seemed imminent, was broken in advance. . . . Oh, they were clever!

The whole hospital, the mess hall, and the club remained in the Ukrainian Camp Division. We were left instead with the camp jail, while the Ukrainians, the Banderists, the most dangerous rebels, had been moved farther away from it. What did this mean?

We soon learned what. Reliable rumors went around the camp (from the working prisoners who took the gruel to the BUR) that the stoolies in the "safe deposit" had grown cheeky. Suspects, picked up here and there, two or three at a time, had been put in with the stoolies, who were torturing them in their common cell, choking them, beating them, trying to make them "sing" and to name names. *"Who's doing the slashing?"* This made the whole scheme as clear as daylight. They were using torture! Not the dog pack themselves—they probably had no authorization for it, and might run into trouble, so they had entrusted the stoolies with the job: find your murderers yourselves! The stoolies were all eagerness—no shot in the arm needed! And this was one way for those parasites to earn their keep. That was why the Banderists had been moved away from the BUR—so that they could not attack it. We were less of a worry: docile people, of different races, we would not make common cause. The rebels were . . . in the other place. And the wall was four meters high.

So many deep historians have written so many clever books—and still they have not learned how to predict those mysterious conflagrations of the human spirit, to detect the mysterious springs of a social explosion, nor even to explain them in retrospect.

Sometimes you can stuff bundle after bundle of burning tow under the logs, and they will not take. Yet up above, a solitary little spark flies out of the chimney and the whole village is reduced to ashes.

Our three thousand had no plans made, were quite unprepared, but one evening on their return from work the prisoners in a hut next to the BUR began dismantling their bunks, seized the long bars and crosspieces, ran through the gloom (there was a darkish place to one side of the BUR) to batter down the stout fence around the camp jail. They had neither axes nor crowbars—because there never are any inside the camp area—unless perhaps they had begged a couple from the maintenance yard.

There was a hammering noise—they worked like a team of good carpenters, levering the planks away as soon as they gave—and the grating protests of 12-centimeter nails could be heard all over the camp. It was hardly the time for carpenters to be working, but at least it was a workmanlike noise, and at first neither the men on the towers nor the warders, nor the other prisoners, thought anything of it. Life was following its usual evening routine: some teams were going to supper, others straggling back from supper, some making for the Medical Section, others for the stores, others for the parcels office.

All the same, the warders were worried, hustled over to the BUR, to the half-dark wall where all the activity was—and raced like scalded cats to the staff barracks. Somebody rushed after a warder with a stick. Then, to provide full musical accompaniment, somebody started breaking windows in the staff barracks with stones or a stick. The staff's windowpanes shattered with a merry menacing crash.

What the lads had in mind was not to raise a rebellion, nor even to capture the BUR (no easy matter: Plate No. 5, taken many years later, shows the door of the Ekibastuz BUR off its hinges), but merely to pour petrol into the stoolies' cell, and toss in something burning—meaning: Watch your step, we'll show you yet! A dozen men did force their way through the gap knocked in the BUR fence. They started tearing around looking for the cell—they had made a guess at the window, but were not sure—then they had to dislodge the muzzle, give someone a leg up, pass the petrol pail —but machine-gun fire from the towers suddenly rattled across the camp, and they never did start their blaze.

The warders and Chief Disciplinary Officer Machekhovsky had fled from the camp and informed Division. (Machekhovsky, too, had been pursued by a prisoner with a knife, had run by way of the shed in the maintenance yard to a corner tower, shouting: "You in the tower, don't fire! I'm a friend," and scrambled through the outer fence.)[1] Division (where can we now inquire the names of the commanders?) gave telephonic instructions for the corner towers to open fire from machine guns—on three thousand unarmed people who knew nothing of what had happened. (Our team, for instance, was in the mess hall, and we were completely mystified when we heard all the shooting outside.)

It was one of fate's little jokes that this took place on January 22 (New Style), January 9 (Old Style), the anniversary of Bloody Sunday, which until that year was marked in the calendar as a day of solemn mourning. For us it proved to be Bloody Tuesday, and the butchers had much more elbow room than in Petersburg: this was not a city square, but the steppe,

---

1. He was hacked to pieces just the same—not, however, by us but by the thieves who replaced us in Ekibastuz in 1954. He was harsh, but courageous; there's no denying that.

with no witnesses, no journalists or foreigners around.[2]

Firing at random in the darkness, the machine-gunners blasted away at the camp area. True, the shooting did not last long, and most of the bullets probably passed overhead, but quite a few of them were lower—and how many does a man need? They pierced the flimsy walls of the huts, and, as always happens, wounded not those who had stormed the BUR, but others, who had no part in it. Nonetheless, they now had to *conceal their wounds,* stay away from the Medical Section, wait like dogs for time to heal them, otherwise they might be identified as participants in the mutiny—somebody, after all, must be plucked out of the faceless mass! In hut No. 9 a harmless old man, nearing the end of a ten-year sentence, was killed in his bed. He was due to be released in a month's time. His grown-up sons were serving in the same army as those who blazed away at us from the towers.

The besiegers left the prison yard and quickly dispersed to their barracks (where they had to put their bunks together again so as to cover their traces). Many others took the shooting as a warning to stay inside their huts. Yet others, on the contrary, poured out excitedly and scurried about the camp, trying to understand what it was all about.

By then there were no warders left in the camp area. The staff barracks was empty of officers, and terrible jagged holes yawned in its windows. The towers were silent. The curious, and the seekers after truth, roamed the camp.

Suddenly the gates of our Camp Division were flung wide and a platoon of convoy troops marched in with Tommy guns at the ready, firing short bursts at random. They fanned out in all directions, and behind them came the enraged warders, with lengths of iron pipe, clubs, or anything else they had been able to lay their hands on.

They advanced in waves on every hut, combing the whole camp area. Then the Tommy-gunners were silent, and halted while the warders ran forward to flush out prisoners in hiding, whether wounded or unhurt, and beat them unmercifully.

All this became clear later, but at the time we could only hear

2. But it is interesting to note that it was about this time that Soviet calendars stopped marking Bloody Sunday—as though it was after all a fairly ordinary occurrence, and not worth commemorating.

heavy firing in the camp area, and could see and understand nothing in the half-dark.

A lethal crush developed at the entrance to our hut: the prisoners were so anxious to shove their way in that no one could enter. (Not that the thin boards of the hut walls gave any protection against bullets, but once inside, a man ceased to be a mutineer.) I was one of those by the steps. I remember very well my state of mind: a nauseated indifference to my fate; a momentary indifference whether I survived or not. Why have you fastened your hooks on us, curse you! Why must we go on paying you till the day we die for the crime of being born into this unhappy world? Why must we sit forever in your jails? The prison sickness which is at once nausea and peace of mind flooded my being. Even my constant fear for the as-yet-unrecorded poem and the play I carried within me was in abeyance. In full view of the death which was wheeling toward us in military greatcoats, I made no effort at all to push through the door. This was the true convict mentality; this was what they had brought us to.

The doorway gradually cleared and I was among the last to go through. Shots rang out at this point, amplified by the hollow building. The three bullets they fired after us lodged in a row in the doorjamb. A fourth ricocheted upward and left a little round hole haloed with hairline cracks in the glass above the door.

Our pursuers did not break into the huts. They locked us in. They hunted down and beat those who had not been quick enough to run inside. A couple of dozen prisoners were wounded or badly beaten: some lay low and hid their wounds, others were passed to the Medical Section for a start, with jail and interrogation as participants in a mutiny to follow.

But all this became known only later. The doors were locked overnight, and on the following morning the inmates of different huts were not allowed to meet in the mess hall and piece the story together. In some huts, where no one was seen to be hurt and nothing was known about the killings, the deluded prisoners turned out to work. Our hut was one of them.

Out we went, but no one was led through the camp gates after us: the midway was empty, there was no work line-up! We had been tricked!

We felt wretched in the engineering shops that day. The lads went from bench to bench, and sat down to discuss what had happened the day before, and how long we would go on working like donkeys and tamely putting up with it all. Camp veterans, who would never straighten their backs again, were skeptical. What else could we do? they asked. Did we suppose that anyone had ever survived unbroken? (This was the philosophy of the 1937 "draft.")

When we returned from work in the dark, the camp area was again deserted. Our scouts ran to the windows of other huts. They found that No. 9, in which there were two dead men and three wounded, and the huts next to it had not gone to work. The bosses had told them about us, hoping that they, too, would turn out tomorrow. But the way things were, we should obviously not be going in the morning ourselves.

Notes were tossed over the wall telling the Ukrainians what we had decided and asking for their support.

The work stoppage–hunger strike had not been carefully prepared, it was not even a coherent concept, and it began impulsively, with no directing center, no signal system.

Those prisoners in other camps who took over the food stores and then stayed away from work of course behaved more sensibly. But our action, if not very clever, was impressive: three thousand men simultaneously swore off both food and work.

Next morning not a single team sent its man to the bread-cutting room. Not a single team went to the mess hall, where broth and mush awaited them. The warders just could not understand it: twice, three times, four times they came into the huts to summon us with brisk commands, then to drive us out with threats, then to ask us nicely—no farther than to the mess hall, to collect our bread, with never a word for the present about work line-up.

But nobody went. They all lay on their bunks, fully dressed, wearing their shoes, and silent. Only we, the foremen (I had become a foreman in that hot year), felt called upon to answer, since the warders kept addressing themselves to us. We lay on our bunks like the rest and muttered from our pillows: "It's no good, boss. . . ."

This unanimous quiet defiance of a power which never forgave, this obstinate, painfully protracted insubordination, was somehow more frightening than running and yelling as the bullets fly.

In the end they stopped coaxing us and locked up the huts.

In the days that followed, no one left the huts except the orderlies—to carry the latrine buckets out and bring drinking water and coal in. Only bed cases in the Medical Section were by general agreement allowed not to fast. Only doctors and medical orderlies were allowed to work. The kitchens cooked one meal and poured it away, cooked another, poured that away, and cooked no more. The trusties who worked there seem to have appeared before the camp authorities on the first day, explained that they simply couldn't carry on, and left the kitchens.

The bosses could no longer see us, no longer peer into our souls. A gulf had opened between the overseers and the slaves!

None of those who took part will ever forget those three days in our lives. We could not see our comrades in other huts, nor the corpses lying there unburied. Nonetheless, the bonds which united us, at opposite ends of the deserted camp, were of steel.

This was a hunger strike called not by well-fed people with reserves of subcutaneous fat, but by gaunt, emaciated men, who had felt the whip of hunger daily for years on end, who had achieved with difficulty some sort of physical equilibrium, and who suffered acute distress if they were deprived of a single 100-gram ration. Even the goners starved with the rest, although a three-day fast might tip them into irreversible and fatal decline. The food which we had refused, and which we had always thought so beggarly, was a mirage of plenty in the feverish dreams of famished men.

This was a hunger strike called by men schooled for decades in the law of the jungle: "You die first and I'll die later." Now they were reborn, they struggled out of their stinking swamp, they consented to die today, all of them together, rather than to go on living in the same way tomorrow.

In the huts roommates began to treat each other with a sort of ceremonious affection. Whatever scraps of food anyone—this meant mainly those who received parcels—had left were pooled, placed on a piece of rag spread out like a tablecloth, and then, by joint decision of the whole room, some eatables were shared out and others put aside for the next day. (Recipients of parcels might also have quite a bit of food in the personal provisions store, but for one thing no one could cross the camp to fetch it, and for another, not everyone would have been happy to bring his left-overs back with him: he might be counting on them to build him up when the strike was over. For this reason the strike, like everything that happened in prison, was an unequal ordeal, and the truly brave were those who had nothing in reserve, no hope of recruiting their strength after the strike.) If there was any meal, they boiled it at the mouth of the stove and distributed the gruel by the spoonful. To make the fire hotter they broke planks off the bunks. The couch provided by the state is gone, but who cares when his life may not last the night!

. What the bosses would do no one could predict. We thought that perhaps they would start firing on the huts again from the towers. The last thing we expected was any concession. We had never in our lives wrested anything from them, and our strike had the bitter tang of hopelessness.

But there was a sort of satisfaction in this feeling of hopelessness. We had taken a futile, a desperate step, it could only end badly—and that was good. Our bellies were empty, our hearts were in our boots—but some higher need was being satisfied. During those long hungry days, evenings, nights, three thousand men brooded over their three thousand sentences, their families, their lack of families, all that had befallen and would yet befall them, and although the hearts in thousands of breasts could not beat together—and there were those who felt only regret, only despair—yet most of them kept time: Things are as they should be! We'll keep it up to spite you! Things are bad! So much the better!

This, too, is a phenomenon which has never been adequately studied: we do not know the law that governs sudden surges of mass emotion, in defiance of all reason. I felt this soaring emotion myself. I had only one more year of my sentence to serve. I might have been expected to feel nothing but dismay and vexation that I was dirtying my hands on a broil from which I should hardly escape without a new sentence. And yet I had no regrets. Damn and blast the lot of you, I'll serve my time all over again if you like!

Next day we saw from our windows a group of officers making their way from hut to hut. A detail of warders opened the door, went along the corridors, looked into the rooms, and called us (not in the old way, as though we were cattle, but gently): "Foremen! You're wanted at the entrance!"

A debate began among us. It was the teams, not their foremen, who had to decide. Men went from room to room to talk it over. Our position was ambiguous. Stoolies had been weeded out from our ranks, but we suspected that there were others, and there were certainly some—like the slippery, bold-faced foreman-mechanic,

Mikhail Generalov. And anyway, knowledge of human nature told us that many of those on strike and starving for freedom's sake today would *spill the beans* tomorrow for the sake of a quiet life in chains. For this reason, those who were steering the strike (and there were leaders, of course) did not show themselves, but remained underground. They did not openly assume power, and the foremen had openly resigned their authority. So that the strikers seemed to be drifting, without a helmsman.

A decision was reached at last in some invisible quarter. We foremen, six or seven of us, went out to the entrance, where the officers were patiently awaiting us. (It was the entranceway of that very same hut, No. 2, until recently a Disciplinary Barracks, from which the "metro" tunnel had run, and the escape hatch itself was a few meters away from the place where we now met.) We leaned back against the walls, lowered our eyes, and stood like men of stone. We lowered our eyes because not one of us could now look at our bosses sycophantically, and rebellious looks would have been foolish. We stood like hardened hooligans called before a teachers' meeting—hunched, hands in pockets, heads lowered and averted—incorrigible, impenetrable, hopeless.

From both corridors, however, a crowd of zeks pressed into the entranceway, and hiding behind those in front, the back rows could speak freely, call out our demands and our answers.

Officially, the officers with blue-edged epaulets (some we knew, others we had never seen before) saw and addressed only the foremen. Their manner was restrained. They did not try to intimidate us, but their tone was still intended to remind us that we were inferior. It would, so they said, be in our own interest to end the strike and the hunger strike. If we did, we would receive not only today's rations but—something unheard of in Gulag!—yesterday's, too. (They were so used to the idea that hungry men can always be bought!) Nothing was said either about punishment or about our demands—they might not have existed.

The warders stood at the sides, keeping their right hands in their pockets.

There were shouts from the corridor:

"Whoever's to blame for the shooting must be brought to justice!"

"Take the locks off the doors!"

"Off with the numbers!"

In other huts they also demanded a review of Special Board cases in the open courts.

While we foremen stood like schoolboy hooligans waiting for the headmaster to finish nagging.

The bosses left, and the hut was locked up again.

Although hunger had begun to get many of us down—our heads were heavy and our thoughts lacked clarity—in our hut not a single voice was raised in favor of surrender. Any regrets remained unspoken.

We tried to guess how high the news of our rebellion would go. They knew already, of course, in the Ministry of Internal Affairs, or would learn today—but did *Whiskers?* That butcher wouldn't stop at shooting the lot of us, all five thousand.

Toward evening we heard the drone of a plane somewhere near, although it was cloudy and not good flying weather. We surmised that someone even higher up had flown in.

A seasoned son of Gulag, Nikolai Khlebunov, who was friendly with some of us, had landed a job somewhere in the kitchens after nineteen years in prison, and as he passed through the camp that day he was quick enough and brave enough to slip us a half-pood sack of millet flour through the window. It was shared out between the seven teams, and cooked at night so that the warders couldn't catch us at it.

Khlebunov passed on some very bad news: Camp Division No. 2, where the Ukrainians were, beyond the Chinese Wall, had not supported us. That day and the day before, the Ukrainians had turned out to work as though things were quite normal. There could be no doubt that they had received our notes; they could hear how quiet we had been for two days; they could see from the tower crane on the building site that our camp area had been deserted since the shooting; they must have missed meeting our columns outside the camp. Nevertheless, they had not supported us! . . . (We learned afterward that the young men who were their leaders, and who still had no experience of practical politics, had argued that the Ukrainians had their own destiny, distinct from that of the Muscovites. They who had begun with such spirit had now fallen back and abandoned us.) So that there were not five thousand of us, but only three.

For the second night, the third morning, and the third day hunger clawed at our guts.

But when on the third morning the Chekists, in still greater force, again summoned the foremen to the entrance, and once again we stood there—sullen, unreachable, hangdog—our general resolve was not to give way! We were carried along by inertial force.

The bosses only gave us new strength. The newly arrived brass hat had this to say:

"The administration of the Peschany camp *requests the prisoners to take their food*. The administration will receive any complaints. It will examine them and eliminate the causes of *conflict* between the administration and the prisoners."

Had our ears deceived us? They were *requesting us to take food!* And not so much as a word about work. We had stormed the camp jail, broken windows and lamps, chased warders with knives, and it now turned out that far from being a mutiny, this was a *conflict between* (!) . . . between equals . . . between the administration and the prisoners!

It had taken only two days and two nights of united action—and look how our serfmasters had changed their tone! Never in our lives, not only as prisoners, but as free men, as members of trade unions, had we heard our bosses speak with such unction!

Nonetheless, we started silently dispersing—no one could take a decision *there*. Nor could anyone there promise a decision. The foremen went away without once raising their eyes or looking around, even when the head of the Separate Camp Site addressed us one after another by name.

That was our answer.

The hut was locked again.

From outside it looked to the bosses as dumb and unyielding as ever. But inside, the sections were the scene of stormy debate. The temptation was too great! Soft speech had affected the undemanding zeks more than any threat would. Voices were heard urging surrender. What more, indeed, could we hope to achieve! . . .

We were tired! We were hungry! The mysterious force which had fused our emotions and borne us aloft was losing height and with tremulous wings bringing us down to earth again.

Yet mouths clamped tight for decades, mouths which had been silent for a lifetime, and should have stayed silent for what was left of it, were now opened. Among those listening to them, of course, were the surviving stoolies. These exhortations, in voices suddenly recovered for a few minutes, voices with a new ring to them (in our room that of Dmitri Panin), would have to be paid for by a fresh term of imprisonment, a noose around the throat in which the pulse of freedom had fluttered. It was a price worth paying, for the vocal cords were for the first time put to the use for which they were created.

Give way now? That would mean accepting someone's word of honor. Whose word exactly? That of our jailers, the camp dog pack. In all the time that prisons had existed, in all the time the camps had been there—had they ever once kept their word?

The sediment of ancient sufferings and wrongs and insults was stirred up anew. For the first time ever we had taken the right road —were we to give in so soon? For the first time we had felt what it was like to be human—only to give in so quickly? A keen, a bracing breeze of mischief blew around us. We would go on! We would go on! They'd sing a different tune before we finished! They would give way! (But when would we ever be able to believe anything they said? This was as unclear as ever. That is the fate of the oppressed: they are forced to *believe* and to yield. . . .)

Once more the emotions of two hundred men were fused in a single passion; the wings of the eagle beat the air—he sailed aloft!

We lay down to conserve our strength, trying to move as little as possible and not to talk unnecessarily. Our thoughts were quite enough to occupy us.

The last crumbs in the hut had been finished long ago. No one had anything to cook or to share. In the general silence and stillness the only sound was the voices of young observers glued to the windows: they told us about all the comings and goings outside in the camp. We admired these twenty-year-olds, their enthusiasm undimmed by hunger, their determination to die on the threshold of life, with everything still before them, rather than surrender. We were envious of them, because the truth had entered our heads so late, and our spines were already setting in a servile arc.

I can, I think, now mention by name Janek Baranovsky, Volodya Trofimov, and Bogdan, the metalworker.

Suddenly, in the late afternoon of the third day, when the western sky was clearing and the setting sun could be seen, our observers shouted in anger and dismay:

"Hut nine! . . . Nine has surrendered! . . . Nine's going to the mess hall!"

We all jumped up. Prisoners from the other side of the corridor ran into our room. Through the bars, from the upper and lower bed platforms, some of us on all fours, some looking over other people's shoulders, we watched, transfixed, that sad procession.

Two hundred and fifty pathetic little figures, darker than ever against the sunset, cowed and crestfallen, were trailing slantwise across the camp. On they went, each of them glimpsed briefly in the rays of the setting sun, a dawdling, endless chain, as though those behind regretted that the foremost had set out, and were loath to follow. Some, feebler than the rest, were led by the arm or the hand, and so uncertain were their steps that they looked like blind men with their guides. Many, too, held mess tins or mugs in their hands, and this mean prisonware, carried in expectation of a supper too copious to gulp down onto constricted stomachs, these tins and cups held out like begging bowls, were more degrading and slavish and pitiable than anything else about them.

I felt myself weeping. I glanced at my companions as I wiped away my tears, and saw theirs.

Hut No. 9 had spoken, and decided for us all. It was there that the dead had been lying around for four days, since Tuesday evening.

They went into the mess hall, and it was as though they had decided to forgive the murderers in return for their bread ration and some mush.

No. 9 was a hungry hut. The teams in it were all general laborers, and very few prisoners received parcels. There were many goners among them. Perhaps they had surrendered for fear that there would be other corpses?

We went away from the windows without a word.

It was then that I learned the meaning of Polish pride, and understood their recklessly brave rebellions. The Polish engineer Jerzy Wegierski, whom I have mentioned before, was now in our team. He was serving his ninth and last year. Even when he was a work assigner no one had ever heard him raise his voice. He was always quiet, polite, and gentle.

But now—his face was distorted with rage, scorn, and suffering, as he tore his eyes away from that procession of beggars, and cried in an angry, steely voice:

"Foreman! Don't wake me for supper! I shan't be going!"

He clambered up onto the top bunk, turned his face to the wall —and didn't get up again. That night we went to eat—but he wouldn't get up! He never received parcels, he was quite alone, he was always short of food—but he wouldn't get up. In his mind's eye the steam from a bowl of mush could not veil the ideal of freedom.

If we had all been so proud and so strong, what tyrant could have held out against us?

■

The following day, January 27, was a Sunday. They didn't drive us out to work to make up for lost time (although the bosses, of course, were itching to get back on schedule) but simply fed us, issued arrears of rations, and let us wander about the camp. We all went from hut to hut, telling each other how we had felt in the past few days, and we were all in holiday mood, as though we had won instead of losing. Besides, our kind masters promised yet again that all legitimate requests (but who knew, who was to define what was legitimate?) would be satisfied.

There was, however, one untoward little event: a certain Volodka Ponomarev, a "bitch" who had been with us throughout the strike, heard many rash speeches, and looked into many eyes, *ran away to the guardhouse,* which meant that he had run to betray us outside the camp area, where he could avoid the knife.

For me the whole essence of the criminal world crystallized in Ponomarev's flight. Their alleged nobility is just a matter of caste obligations. But when they find themselves in the whirlpool of revolution they inevitably behave treacherously. They can understand no principles, only brute force.

It was an easy guess that our bosses were getting ready to arrest the ringleaders. But they announced that, on the contrary, commissions of inquiry had arrived from Karaganda, Alma-Ata, and Moscow to look into things. A table was placed on ground stiff with hoarfrost in the middle of the camp, where we lined up for work assignment, and some high-ranking officers sat at it in sheepskin coats and felt boots and invited us to come forward with our complaints. Many prisoners went and talked to them. Notes were taken.

After work on Tuesday they assembled the foremen "to present complaints." In reality this conference was another low trick, a form of interrogation: they knew that the prisoners were boiling with resentment and let them vent it so that they could be sure of arresting the right people.

This was my last day as a foreman: my neglected tumor was growing rapidly, and for a long time now I had been putting off my operation until, in camp terms, it was "convenient." In January and particularly during the fatal days of the hunger strike, the tumor decided for me that it was now "convenient," and it seemed to get bigger by the hour. The moment the huts were opened I showed myself to the doctors and they set a date for the operation. But I dragged myself to this last conference.

It was convened in the spacious anteroom to the bathhouse. They placed the presidium's table in front of the barbers' chairs, and seated at it were one MVD colonel, several lieutenant colonels, and some smaller fry, with our camp commanders inconspicuous in the second row, behind them. There, too, behind the backs of the presidium, sat the note-takers—making hasty notes throughout the meeting, while the front row helped them by repeating the names of speakers.

One man stood out from the rest, a certain lieutenant colonel from the Special Section or from the Organs—a quick, clever, nimble-witted villain, with a tall brow and a long face: this quick-wittedness, these narrow features, somehow made it difficult to believe that he belonged to that pack of obtuse police officials.

The foremen were reluctant to come forward, and practically had to be dragged to their feet from the close-packed benches. As soon as they started saying something of their own, they were interrupted and invited to explain why *people* were being murdered, and what were the aims of the strikers? And if a hapless foreman tried to give some sort of answer to the questions—reason for murders, nature of demands—the whole pack at once flung itself upon him: And how do you know that? So you're connected with these gangsters? Right, let's have their names!!

This was their idea of a fair and honorable inquiry into the "legitimacy" of our demands. . . .

The lofty-browed villain of a lieutenant colonel was especially quick to interrupt the speakers: he had a nimble tongue, and the advantage of impunity. With his caustic interjections he thwarted each of our attempts to present our case. From the tone which the proceedings were beginning to take you would have thought that only we faced any charges and needed to defend ourselves.

An urge to put a stop to this swelled inside me. I took the floor, and gave my name (which was repeated for the note-taker). I rose from the bench pretty certain that there was no one in that gathering who could trot out a rounded sentence more easily than I. The only difficulty was that I had no idea what to tell them. All that is written in these pages, all that we had gone through, all that we had brooded over in all those years and all those days on hunger strike—I might as well try telling it to orang-utans as to them. They were still in some formal sense Russians, still more or less capable of understanding fairly simple Russian phrases, such as "Permission to enter!" "Permission to speak, sir!" But as they sat there all in a row at the long table, exhibiting their sleek, white, complacent, uniformly blank physiognomies, it was plain that they had long ago degenerated into a distinct biological type, that verbal communication between us had broken down beyond repair, and that we could exchange only . . . bullets.

Only the long-headed one had not yet turned into an orangutan; his hearing and understanding were excellent. The moment I spoke he tried to interrupt me. With the whole audience paying close attention, a duel of lightning-swift repartee began.

"Where do you work?"

(What difference could that make? I wonder.)

"In the engineering shops!" I rapped out over my shoulder, and hurried on with more important things.

He came straight back at me.

"Where they make the knives?"

"No," I said, parrying his thrust. "Where they repair self-propelled excavators!" (I don't know myself how my mind worked so quickly and clearly.)

Hurry, hurry, make them be quiet and listen—that's the main thing.

The brute crouched behind the table and suddenly pounced to sink his fangs in me:

"You are here because the *bandits* delegated you?"

"No, because you invited me!" I snapped back triumphantly, and went on talking and talking.

He sprang at me once or twice more, was beaten off, and sat completely silent. I had won.

Won—but to what purpose? Just one more year! One more year to go, and the thought crushed me. I could not get out the words they deserved to hear. I could have delivered there and then an immortal speech, and been shot next day. I would have delivered it just the same—if they had been broadcasting it throughout the world! But no, the audience was too small.

So I did not tell them that our camps followed the Fascist model, and were a symptom of the regime's degeneration. I limited myself to waving a kerosene-soaked rag under their eagerly sniffing noses. I had learned that the commander of convoy troops was sitting there, and so I deplored the unworthy conduct of the camp guards, who had ceased to resemble *Soviet fighting men,* who joined in pilfering from work sites, *and* they were boors and bullies, *and* they were murderers into the bargain. Then I portrayed the warders in the camps as a gang of greedy rogues who forced zeks to steal building materials for them. (This was quite true, except that it started with the officers sitting and listening to me.) And what a countereducational effect all this had, I said, on prisoners desirous of amendment!

I didn't like my speech myself. The only good thing about it was that we were now setting the pace.

In the interval of silence which I had won, one of the foremen, T., rose and spoke slowly, almost inarticulately, whether because that was natural to him or because he was extremely agitated.

"I used to agree . . . when other prisoners said . . . we live . . . like dogs. . . ."

The brute in the presidium bristled. T. kneaded the cap in his hands, an ugly crop-headed convict, his coarsened features contorted by his struggle to find the right words.

"But now I see that I was wrong. . . ."

The brute's face cleared.

"We live—much worse than dogs," T. rapped out with sudden emphasis, and all the foremen sat bolt upright. "A dog has only one number on his collar; we have four. Dogs are fed on meat; we're fed on fishbones. A dog doesn't get put in the cooler! A dog doesn't get shot at from watchtowers! Dogs don't get *twenty-fivers* pinned on them!"

They could interrupt whenever they liked now—he had said all that mattered.

Chernogorov rose, introducing himself as a Hero of the Soviet Union, then another foreman, and both of them spoke boldly and passionately. Their names were echoed in the presidium with heavy significance.

Maybe this can only lead to our destruction, lads. . . . And maybe it is only by banging our heads against it that we can bring this accursed wall down.

The meeting ended in a draw.

All was quiet for a few days. The commission was seen no more, and life was so peaceful in our Camp Division that there might never have been any trouble.

An escort took me off to a hospital on the Ukrainian side. I was the first to be taken there since the hunger strike, the first swallow. Yanchenko, the surgeon who was to operate on me, had called me in for examination, but his questions and my answers were not about my tumor. He was not interested in my tumor, and I was glad to have such a reliable doctor. There was no end to his questions. His face was dark with the pain we all shared.

The same experience, in different lives, can be seen in very different perspectives. This tumor, which was to all appearances malignant—what a blow it would have been if I were a free man; how I should have suffered, how my loved ones would have wept. But in this place, where heads were so casually severed from trunks, the same tumor was just an excuse to stay in bed, and I didn't give it much thought.

I was lying in the hospital among those wounded and maimed on that bloody night. There were men beaten by the warders to a bloody pulp: they had *nothing left to lie on*—their flesh was in ribbons. One burly warder had been particularly brutal with his length of iron piping. (Memory! memory! I cannot now recall his name.) One man had already died of his wounds.

News came in thick and fast. The punitive operation had begun in the "Russian" Camp Division. Forty men had been arrested. For fear of a fresh mutiny, they did it this way: Until the very last day the bosses showed nothing but kindness, and you could only suppose that they were trying to decide which of their own number were to blame. But on the appointed day, as the work teams were passing through the gates, they noticed that the escort party which took charge of them was twice or three times its normal strength. The plan was to seize the victims where prisoners could not help one another, nor could the walls of huts or buildings under construction help them. The escort marched the columns out of the camp, and took them by different ways into the steppe, but before they had brought any of them to their destination, the officers in command gave their orders. "Halt! Weapons at the ready! Chamber cartridges! Prisoners, sit down! I shall count to three and fire if you aren't sitting down! Everybody sit!"

Once again, as at Epiphany the year before, tricked and help-less, the slaves were pinned to the snow. Then, too, an officer had unfolded a piece of paper and read out the names and numbers of those who had to rise, leave the unresisting herd, and pass through the cordon. Next, this handful of mutineers was marched back under separate escort, or else a Black Maria rolled up to collect them. The herd, now purged of fermenting agents, was then brought to its feet and driven to work.

Our educators had shown us whether we could ever believe anything they said.

They also *plucked out* candidates for jail in the camp grounds while they were deserted for the day. And arrests easily flitted over that four-meter wall which the strike had been unable to surmount and pecked at the Ukrainian Camp Division. The very day before my operation was due, Yanchenko the surgeon was arrested and taken off to jail.

Prisoners continued to be arrested or posted to other camps—it was always difficult to know which—without the precautions observed at the beginning. Small groups of twenty or thirty were sent off somewhere. Then suddenly on February 19 they began assembling an enormous transport, some seven hundred strong. They were under special discipline: as they left the camp they were handcuffed. Fate had exacted retribution! The Ukrainians, who had taken such good care not to help the Muscovites, were thicker on the ground in this transport than we were.

True, on the point of leaving, they saluted our shattered strike. A new wood-processing plant, itself oddly enough built entirely of wood (in Kazakhstan, where there is no timber and lots of stone!), for reasons which remained officially unexplained (but I know for certain that there was arson), burst into flame at several points simultaneously, and within two hours three million rubles had gone up in smoke. For those on their way to be shot it was like a Viking's funeral—the old Scandinavian custom of burning the hero's boat together with his body.

I was lying in the recovery room. I was alone in the ward: the camp was in such a turmoil that no one could be admitted, the hospital had come to a standstill. After my room, which was at the butt end of the hut, came the morgue, where Dr. Kornfeld's body had been lying for I don't know how many days because no one had time to bury him. (Morning and evening, a warder nearing the end of his round would stop outside and to simplify the count embrace my room and the morgue in one inclusive gesture: "Two more *here.*" And tick us off on his clipboard.)

Pavel Boronyuk, who had also been called to join the large transport, broke through all the cordons and came to embrace me before he left. Not just our camp but the whole of creation seemed to us to be shaken, reeling before the storm. We were storm-tossed and we could not realize that outside the camp all was as calm and stagnant as ever. We felt as though we were riding great waves, on something that might sink under our feet, and that if we ever saw each other again, it would be in quite a different country!

But just in case—farewell, my friend! Farewell, all my friends!

■

That year of tedium and stupidity—my last year in Ekibastuz, and the last Stalinist year on the Archipelago—dragged on. After they had been kept in jail for a while, and no evidence against them found, a few—but only a few—were sent back into the camp. While many, very many, whom we had come to know and love over the years were taken away: some for further investigation and trial, others to the isolator* because of some indelible black mark in their dossiers (although they might long since have been more like angels than prisoners); others again to the Dzhezkazgan mines; and there was even a transport of the "mentally defective" —Kishkin the joker was squeezed in with this lot, and the doctors also fixed up Volodya Gershuni.

To replace those who had left, the stoolies crawled one by one out of the "safe deposit": timidly and apprehensively at first, then more and more brazenly. One who returned to the body of the camp was the venal "bitch" Volodka Ponomarev, once a mere lathe operator, but now in charge of the parcels room. The distribution of the precious crumbs collected by destitute families was a task which the old Chekist Maksimenko naturally entrusted to a notorious thief!

The security officers again started summoning anyone and everyone to their offices as often as they pleased. It was an airless spring. Anyone whose horns or ears stuck out too much quickly learned to keep his head down. I did not go back to my foreman's job (there were now plenty of foremen again), but became a smelter's mate in the foundry. We had to work hard that year, for reasons which I shall explain. The one and only concession which the administration made when all our hopes and demands lay in ruins was to make us self-financing: under this system what we produced did not simply vanish into the maw of Gulag, but was priced, and 45 percent of its value was counted as our *earnings* (the rest went to the state). Of these "earnings" the camp appropriated 70 percent for the maintenance of guards, dogs, barbed wire, the camp jail, security officers, the officers responsible for discipline, for censorship, for education—in a word, all the things without which our lives would be unlivable—but the remaining 30 percent (13 ½ percent of the whole product) was credited to the personal account of the prisoner, and part of this money, though not all of it (provided you had not misbehaved, not been late, not been rude, not been a disappointment to your bosses), you could on application once a month convert into a new camp currency —vouchers—and these vouchers you could spend. The system was so contrived that the more sweat you lost, the more blood you gave, the closer you came to that 30 percent, but if you didn't feel like breaking your back, all your labor went to the camp and you got zero.

And the majority—ah, what a part the *majority* plays in our history, especially when it is carefully prepared by *weeding*— gratefully gulped down this sop from its bosses and risked working itself to death to buy condensed milk, margarine, and nasty sweets at the food counter, or get itself a second supper in the "commercial" dining room. And since work sheets were made up for the team as a whole, not for individual members, all those who didn't want to sacrifice their health for margarine still had to do it, so that their comrades could earn more.

They also started bringing films to the camp much more frequently than before. As is always the case in the camps, or in villages, or in remote workers' settlements, no one had enough respect for the spectators to announce the titles in advance—a pig, after all, is not informed in advance what is going to be poured into his trough. Nonetheless, the prisoners—could they be the same prisoners who had kept up the hunger strike so heroically that winter?!—now flocked in, grabbed seats an hour before the windows were draped, without worrying one little bit whether the film was worth it.

Bread and circuses! . . . Such a cliché that it's embarrassing to repeat it.

No one could blame people for wanting to eat their fill after so many years of hunger. But while we were there filling our bellies, comrades of ours—some who had taught us to fight, some who had shouted "No surrender!" to their hutmates in those January days, and some who had not been involved at all—were at that moment on trial somewhere, comrades of ours were being shot, or carried off to begin new sentences in isolation camps, or broken by interrogation after interrogation, bundled into cells where condemned men had scratched a forest of crosses on the walls, and the snake of a major looked in to smile a promise: "Ah, Panin! I remember you—oh, yes, I remember you! The wheels are turning, don't worry! We'll soon process you!"

A fine word that—process! You can process a man for the next world, process a man into the cooler for twenty-four hours, and a chit for a pair of secondhand trousers may also be processed. But the door slams shut, the snake has gone, smiling enigmatically, leaving you to guess, to spend a month without sleep, to beat your head against the stones wondering how exactly they intend to process you. . . .

Talking about it is easy enough.

Suddenly in Ekibastuz they got together another party of twenty men for transportation. Rather a strange party. They were gathered unhurriedly, they were not treated harshly, they were not isolated—it was almost as though they were being assembled for release. Not one of them, however, was anywhere near the end of his sentence. Nor were there any of those hard-boiled zeks among them whom the bosses try to break with spells in the cooler and special punishments; no, they were all *good* prisoners, in good standing with their superiors: there, once again, was the slippery and self-assured foreman of the vehicle repair shop, Mikhail Mikhailovich Generalov; the crafty simpleton Belousov, a foreman machinist; the engineer and technician Gultyaev; the Moscow designer Leonid Raikov, a grave and steady man with the face of a statesman; the very amiable, universally friendly Zhenka Milyukov, a lathe operator with a pert pancake face; and another lathe operator, the Georgian Kokki Kocherava, a great lover of truth, hot in defense of justice when the crowd was looking.

Where were they going? From the party's composition, obviously not to a maximum punishment prison. "Must be a nice place you're going to," they were told. "They'll be taking the guards off you." But not one of them showed a glimmer of happiness, not for a single moment. They wagged their heads miserably, reluctantly gathered their belongings, in two minds as to whether to take them or leave them. They looked like beaten curs. Could they really have grown so fond of turbulent Ekibastuz? They even said good-bye with lips that seemed numb, and unconvincing intonations.

They were taken away.

We were not given time to forget them. Three weeks later the word went around: they've been brought back! Back here? Yes. All of them? Yes . . . only they're sitting in the staff barracks and won't go to their own huts.

This put the finishing touch to the strike of three thousand at Ekibastuz—the strike of the traitors! . . . So much for their reluctance to go! In the interrogators' offices, when they were snitching on our friends and signing their perfidious statements, they had hoped that it would all be kept under the seal of the confessional. It had of course been that way for decades: a political denunciation was regarded as an unchallengeable document, and the informer's identity was never revealed. But something about our strike—the need, perhaps, to vindicate themselves in the eyes of their superiors?—had compelled our bosses to mount a full-dress trial somewhere in Karaganda. *These* creatures were taken off one day, and when they looked into each other's anxious eyes each of them realized that he and all the others were on their way to testify in court. That wouldn't have bothered them, but they knew Gulag's postwar rule: a prisoner called away for some temporary purpose must be returned to his former camp. They were, however, promised that, by way of exception, they would be left in Karaganda! An order was in fact drafted, but incorrectly, not in due form, and Karaganda refused to have them.

They were three weeks on the road. Their guards herded them from Stolypin cars into transit camps and from transit camps into Stolypin cars, yelled at them to "sit on the ground," searched them, took away their belongings, rushed them into the bathhouse, fed them on herrings and gave them no water—they received the full treatment used to wear down ordinary uncooperative prisoners. Then they were taken under guard into the courtroom, where they faced yet again those whom they had denounced, this time to drive the final nails into their coffins, hang the locks on the doors of their solitary cells, wind back their sentences so that they would have long years to run—after which they were brought home via all those transit prisons, and flung, without their masks, into their old camp.

They were no longer needed. Informers are like ferrymen. . . .

But was not the camp now pacified? Had not nearly a thousand men been moved out? Could anybody now prevent them from going along to the godfather's office? . . . Nevertheless, they wouldn't leave the staff building! They were on strike—they refused to enter the camp grounds! Only Kocherava made up his mind to brazen it out in his old role of lover of truth. He went to his team and said:

"We don't know why they took us! They took us all over the place, and then brought us back. . . ."

But his daring lasted just one night and one dawn. Next day he fled to the staff room and his friends.

So that what had happened had not gone for nothing, and our comrades had not fallen in vain. The atmosphere in the camp would never be as oppressive as before. Meanness was back on its throne, but very precariously. Politics were freely discussed in the huts. No work assigner or foreman would dare kick a zek or take a swing at him. Because everybody knew now how easy it is to make knives and how easily they sink home between the ribs.

Our little island had experienced an earthquake—and ceased to belong to the Archipelago.

This was how Ekibastuz felt. It is doubtful whether Karaganda felt the same. And certain that Moscow did not. The Special Camp system was beginning to collapse in one place after another, but our Father and Teacher had no inkling of it—it was not, of course, reported to him (and in any case, incapable as he was of giving up anything, he would only have relinquished *katorga* on the day his chair burst into flames beneath him). On the contrary, he planned a new great wave of arrests for 1953, perhaps in connection with a new war, and in 1952 expanded the Special Camp network accordingly. Thus it was decreed that the Ekibastuz camp should be converted from a division of Steplag, or at times Peschanlag, into the headquarters of a big new Special Camp complex in the Irtysh basin (provisionally called Dallag). So that over and above the numerous slavedrivers already there, a whole new administration of parasites arrived in Ekibastuz, and these as well we had to support by our labor.

New prisoners, too, were expected any day.

■

Meanwhile the germ of freedom was spreading. Where, though, could it go from the Archipelago? Just as the Dubovka prisoners had once brought it to us, so our comrades now carried it farther. That spring you could see this inscription written, scratched, or chiseled on every lavatory wall in Kazakhstan:

"Hail to the fighters of Ekibastuz!"

The first culling of the "center mutineers," about forty men, and the 250 most "hardened cases" among the big February transport, were taken all the way to Kengir. (The settlement was called Kengir, and the station Dzhezkazgan. This was the Third Steplag Camp Division, where the Steplag Administration and the big-bellied Colonel Chechev in person were to be found.) The other Ekibastuz prisoners to be punished were shared between the First and Second Divisions of Steplag (Rudnik).

To warn them off, the eight thousand zeks of Kengir were informed that the new arrivals were *bandits.* They were marched all the way from the station to Kengir jail's new building in handcuffs. In this way, like a legend in chains, our movement entered still servile Kengir, to awaken it, too. Here, as in Ekibastuz a year back, the bully and the informer still reigned supreme.

When he had kept our quarter thousand in jail till April, the commander of the Kengir Camp Division, Lieutenant Colonel Fedotov, decided that they had been sufficiently intimidated, and gave instructions for them to be taken out to work. The center had supplied 125 pairs of brand-new nickel-plated handcuffs, latest Fascist design—just enough, if you handcuffed them two together, for 250 prisoners (which was probably how they had determined Kengir's allocation).

With one hand free, life is not so bad! Quite a few of the lads in the column had experience of camp jails, and there were also old escapers among them (Tenno, too, was included in the transport), who knew all the peculiarities of handcuffs and explained to neighbors in the column that with one hand free there was nothing to getting these cuffs off—with a pin, or even without one.

When they got near the working area, the warders began removing handcuffs at several places in the column simultaneously so as to start the day's work without delay. Whereupon those who knew how hurriedly took off their own handcuffs and those of other prisoners and hid them under their coats: "Another warder took ours off!" It never occurred to the warders to count the handcuffs before they let the column pass, and prisoners were never searched on entering their place of work.

So that on the very first morning, out of 125 pairs of handcuffs, our lads carried off 23! There, in the work zone, they started by smashing the cuffs with stones and hammers, but soon they had a brighter idea: wrapping them in greased paper, so that they would last better, and bricking them up in the walls and foundations of the buildings on which they were working that day (residential block No. 20, opposite the Kengir Palace of Culture), together with ideologically uninhibited covering notes: "Descendants! These houses were built by Soviet slaves! Here you see the sort of handcuffs they wore!"

The warders abused and cursed the *bandits,* and produced some rusty old cuffs for the return journey. They were very much on their guard now, but the lads still pinched another six pairs on the way in to camp. On each of the two following working days they stole a few more. And every pair cost 93 rubles.

So the bosses of Kengir declined to march the lads about in handcuffs.

A man must fight for his rights!

At about the beginning of May they gradually started transferring the Ekibastuz group from the jail to the main camp.

The time had come for them to teach the locals a little sense. As a beginning, they mounted a small demonstration: a trusty, barging in at the head of a queue, as was his *right,* was strangled, not quite fatally. This was enough to start people talking. Things are going to change around here! The new lot aren't like us. (It would be untrue to say that in the nest of camps around Dzhezkazgan stoolies had never been touched, but this had not become a *trend.* In 1951, in the jail at Rudnik, prisoners once snatched a warder's keys, unlocked the cell they wanted, and knifed Kozlauskas.)

Underground *centers* were now set up in Kengir, one Ukrainian, one "All-Russian." Knives and masks were made, ready for the *chopping*—and the whole story began all over again.

Voinilovich "hanged himself" from the bars of his cell. Others killed were the foreman Belokopyt and the loyalist stoolie Lifshitz (a member of the Revolutionary Military Committee with the forces facing Dutov during the Civil War). (Lifshitz had lived happily in the Rudnik Camp Division, where he was librarian in the Culture and Education Section, but his fame had preceded him, and he was knifed the day after his arrival at Kengir.) A

Hungarian maintenance orderly was hacked to death with axes near the bathhouse. The first to flee and blaze a trail to the "safe deposit" was Sauer, a former minister in Soviet Estonia.

But by now the camp bosses, too, knew what to do. For a long time past there had been walls between the four Camp Divisions at Kengir. The idea now was to surround each hut with a wall of its own—and eight thousand men started working on it in their spare time. They also partitioned every hut into four sections, with no communication between them. Each miniature camp area and each section was regularly locked. Ideally, of course, they would have liked to divide the whole world into one-man compartments.

The sergeant major in charge of the Kengir jail was a professional boxer. He used prisoners as punching bags. In his jail they had also invented a technique of beating prisoners with mallets through a layer of plywood, so as to leave no marks. (These *practical* MVD personnel knew that re-education was impossible without beatings and murders; and any *practical* public prosecutor would agree. But there was always the danger that some *theorist* might descend on them! It was this rather improbable visiting theorist who made the interposed plywood necessary.) One Western Ukrainian, tortured beyond endurance and afraid that he might betray his friends, hanged himself. Others behaved worse. And both *centers* were put out of business.

What is more, there were among the "fighters" some greedy rascals interested not in the success of the movement but in feathering their nests. They wanted extra food brought to them from the kitchen, and a share in other prisoners' food parcels.[3] This helped the authorities to discredit the movement and put a stop to it.

Or so they thought. But the stoolies, too, sang smaller after this first rehearsal. At least the atmosphere in Kengir was cleaner.

The seed had been sown. But the crop would be late—and a surprise.

3. Among those who take the path of violence this is probably inevitable. I do not see Kamo's raiders leaving themselves with empty pockets when they paid the proceeds of their bank robberies into Party funds. And can we imagine Koba, who directed their operations, leaving himself without money for wine? During the Civil War, when consumption of wine and spirits was prohibited throughout Soviet Russia, he kept a wine cellar in the Kremlin, more or less openly.

We are forever being told that individuals do not mold history, especially when they resist the course of progress, but for a quarter of a century one such individual twisted our tails as if we were sheep, and we did not even dare to squeal. Now they say that nobody understood—the rear didn't understand, the vanguard didn't understand, only the oldest of the Old Guard understood, and they chose to poison themselves in corners, shoot themselves in the privacy of their homes, or end their days as meek pensioners, rather than cry out to us from a public platform.

So the lot of the liberator fell upon us little ones. In Ekibastuz, by putting five thousand pairs of shoulders under those prison vaults, and heaving, we had at least caused a crack. Only a little one, perhaps unnoticeable at a distance, and perhaps we had overstrained ourselves—but cracks make caves collapse.

There were other disturbances besides ours, besides those in the Special Camps, but the whole bloody past has been so carefully cleaned up and painted and polished that it is impossible for me now to establish even a bare list of disorders in the camps. I did learn by chance that in 1951, in the Vakhrushevo Corrective Labor Camp on Sakhalin, five hundred men were on hunger strike for five days, with excitement running high and selective arrests, after three runaways had been savagely bayoneted outside the guardhouse. We know of a serious disturbance in Ozerlag, on September 8, 1952, after a man had been killed in the ranks at the guardhouse.

Evidently, the Stalinist camp system, particularly in the Special Camps, was nearing a crisis at the beginning of the fifties. Even in the Almighty One's lifetime the natives were beginning to tear at their chains.

There is no knowing how things would have gone if he had lived. As it was—for reasons which had nothing to do with the laws of economics or society—the sluggish and impure blood suddenly stopped flowing in the senile veins of that undersized and pockmarked *individual*.

According to the Vanguard Doctrine, no change should have resulted from this; nor did the bluecaps fear any change, though they wept outside the camp gates on March 5;* nor did the men in black jerkins dare to hope for change, though they strummed on their balalaikas (they were not let out of the camp grounds that day) when they discovered that funeral marches were being broadcast, and that black-bordered flags had been hung out—yet some obscure convulsion, some slippage was started underground.

True, the amnesty at the end of March, 1953, known to the camps as the "Voroshilov amnesty," was utterly faithful to the spiritual legacy of the deceased—in its tenderness for thieves and its viciousness toward politicals. To curry favor with the underworld, the authors of the amnesty released the thieves upon the land like a plague of rats, leaving ordinary citizens to suffer, to bar their windows and make jails of their homes, and leaving the militia to hunt down all over again all those it had ever caught. Whereas 58's were released in the normal ratio: of the three thousand men at No. 2 Camp Division in Kengir, the number set free was . . . three.

An amnesty like this could convince those in *katorga* of one thing only: that Stalin's death had changed nothing. No mercy had ever been shown them, and it would not be shown now. If they wanted some sort of life on this earth they must fight for it!

Disturbances in the camps continued in various places in 1953 —minor brawls like that in Karlag, Camp Division No. 12, and a major rebellion at Gorlag (Norilsk), about which a separate chapter would now follow if we had any material at all. But there is none.

However, the tyrant did not die in vain. Something hidden from view slipped and shifted—and suddenly, with a tinny clatter like an empty bucket falling, yet another *individual* came tumbling headlong from the very top of the ladder into the muckiest of bogs.

And now everyone—the vanguard, the rear, even the most wretched natives of Gulag—realized that a new age had arrived.

To us on the Archipelago, Beria's fall was like a thunderclap: he was the Supreme Patron, the Viceroy of the Archipelago! MVD officers were perplexed, embarrassed, dismayed; when the news was announced over the radio, they would have liked to stuff this horror back into the loudspeaker, but had instead to lay hands on the portraits of their dear, kind Protector, take them down from the walls at Steplag's headquarters. "It's all over now," Colonel Chechev said with quivering lips. (But he was mistaken. He thought that they would all be put on trial the very next day.)[4] The officers

4. As Klyuchevsky notes, the very day after the emancipation of the gentry (Decree on Rights of the Gentry, February 18, 1762), the peasants were also freed (February 19, 1861) —but after an interval of ninety-nine years.

and warders suddenly showed an uncertainty, a bewilderment even, of which the prisoners were keenly aware. The disciplinary officer of Camp Division No. 3 at Kengir, from whom no prisoner had ever received a kind look, suddenly came up to a team from the Disciplinary Barracks while they were working, sat down, and started offering them cigarettes. (He wanted to see what sparks were flying in that turbid atmosphere, and what danger could be expected from them.) "What do you say to that, then?" they asked him mockingly. "Was your top boss really an enemy of the people?" "Yes, as it turns out," said the disciplinary officer dolefully. "He was Stalin's right hand, though," said the maliciously grinning prisoners. "So that means even Stalin slipped up, doesn't it?" "Ye-e-es," said the amiable chatterbox. "Well, lads, it looks as though they'll be letting you out, if you're patient. . . ."

Beria had fallen, and he had bequeathed the "blot" on his name to his faithful Organs. Until then, no prisoner and no *free man,* if he valued his life, had dared even to think of doubting the crystalline purity of each and every MVD officer, but now it was enough to call one of these reptiles a "Beria-ite," and he was defenseless!

In Rechlag (Vorkuta) in June, 1953, the great excitement caused by Beria's removal coincided with the arrival of the mutineers transported from Karaganda and Taishet (most were Western Ukrainians). Vorkuta was still servile and downtrodden and the newly arrived zeks astounded the locals with their intransigence and their audacity.

And the process that had taken us several long months was completed here in one month's time. On July 22, the cement works, building project TETS-2, and pits No. 7, No. 29, and No. 6 struck. The prisoners at these work sites could see each other stopping work, see the wheels in the pit frames coming to a standstill. This time there was no repetition of our mistake at Ekibastuz —no hunger strike. The warders to a man immediately fled from the camp grounds, but every day, to yells of "Hand over the rations, boss man!" they trundled provisions up to the fence and shoved them through the gates. (I suppose the fall of Beria had made them so conscientious—but for that, they would have starved the prisoners out.) Strike committees were set up in the Camp Divisions affected, "revolutionary order" was established, the mess hall staff immediately stopped stealing, and although rations were not increased, the food improved noticeably. At pit No. 7 they hung out a red flag, and at No. 29, on the side facing a nearby railway line, they put up . . . portraits of the Politburo. What could they display? . . . and what could they demand? They demanded that number patches, window bars, and locks be taken off, but touched none of these things themselves. They demanded the right to correspond, the right to receive visits, and a review of their cases.

On the first day only, attempts were made to talk the strikers out of it. Then nobody came near for a week, but machine guns were set up on the watchtowers and the Camp Divisions on strike were cordoned off. No doubt the brass was scurrying back and forth between Vorkuta and Moscow—it was hard to know what to do in the new circumstances. At the end of that week General Maslennikov; the head of Rechlag, General Derevyanko; and the Prosecutor General, Rudenko, started going around the camp with a large suite of officers (as many as forty). Everyone was assembled on the camp parade ground to meet this glittering train. The prisoners sat on the ground while the generals showered abuse on them for "sabotage" and "disgraceful behavior." At the same time, they conceded that "some of the demands are well founded" ("You can take off your number patches"; "Orders have been given" about the window bars). But the prisoners must return to work at once: "The country needs coal!" At pit No. 7 somebody shouted from the back: "And what we need is freedom, you dirty . . . !" and prisoners began to rise from the ground and disperse, leaving the generals with no audience.[5]

At this point they tore off number patches and began levering out window bars. However, a schism immediately developed, and their spirits fell. Perhaps we've gone far enough? We shan't get any more out of them. Part of the night shift reported, and the whole day shift. The pit wheels started turning again, and the various sites watched each other resume work.

Pit 29, however, was behind a hill and could not see the others. It was told that all the rest had started work—but did not believe it and did not go back. It would obviously have been no trouble

---

5. According to other accounts, they actually put up this slogan: "Freedom for us, coal for our country!" "Freedom for us" is in itself seditious, of course, so they hastily added "coal for our country" by way of apology.

to take delegates from pit 29 over to the other pits. But to make such a fuss over prisoners would have been demeaning, and anyway the generals were thirsting for blood: without blood there's no victory; without blood these dumb brutes would never learn.

On August 11, eleven truckloads of soldiers drove up to pit 29. The prisoners were called out onto the parade ground, toward the gate. On the other side of the gate was a serried mass of soldiers. "Report for work, or we shall take harsh measures!"

Never mind what measures. Just look at the Tommy guns. There was silence. Then the movement of human molecules in the crowd. Why risk your neck? Especially if you have a short sentence . . . Those with a year or two to go pushed their way forward. But there were others, who forced a path through the ranks—to stand in the front row, link arms, and form a barrier against the strikebreakers. The crowd was undecided. An officer tried to break the cordon, and was struck with an iron bar. General Derevyanko withdrew to one side and gave the order "Fire!"—on the crowd.

There were three volleys—with machine-gun fire in between. Sixty-six men were killed. (Who were the victims? The front rows —that is to say the most fearless, and also those who had weakened first. This is a law with a wide application—you will even find it expressed in proverbs.) The rest ran away. Guards with clubs and iron bars rushed after the zeks, beating them and driving them out of camp.

Arrests continued for three days (August 1–3) in all the Camp Divisions which had been on strike. But what could be done with those arrested? The Organs had lost their cutting edge since the death of their breadwinner. They could not rise to a formal investigation. More special trains, more transfers hither and thither, to spread the epidemic more widely. The Archipelago was becoming uncomfortably small.

For those who were left behind, there was a special punitive regime.

A number of thin wooden patches appeared on the roofs of huts at pit No. 29, covering the bullet holes made by soldiers firing over the heads of the crowd. Unknown soldiers who refused to become murderers.

But there were plenty of others who hit the target.

Near the slag heap at pit 29, somebody in Khrushchev's day

raised a cross—with a tall stem like a telegraph pole—on the communal grave. Then it was knocked down. And someone put it up again.

I do not know whether it is still standing.

Chapter 12

■

The Forty Days of Kengir

For the Special Camps there was another side to Beria's fall: by raising their hopes it confused, distracted, and disarmed the *katorzhane*. Hopes of speedy change burgeoned—and the prisoners lost their interest in hunting stoolies, and sitting in the hole for them, in strikes and rebellions. Their anger cooled. Things seemed to be improving anyway, and all they had to do was wait.

There was another aspect, too. The epaulets with blue borders (but without air force wings), hitherto the most respected, the least questionable in the armed forces at large, had suddenly become a stigma, not just in the eyes of prisoners or prisoners' relatives (who gives a damn for them?), but even perhaps in the eyes of the government.

In that fateful year, 1953, MVD officers lost their second wage ("for their stars"), which meant that henceforward they received only one salary, plus increments for length of service, polar allowances, and of course bonuses. This was a great blow to their pockets, but a still greater one to their expectations: did it mean that they were less *needed?*

The fall of Beria made it urgent for the security ministry to prove its devotion and its usefulness in some signal way. But how?

The mutinies which the security men had hitherto considered a menace now shone like a beacon of salvation. Let's have more disturbances and disorders, so that *measures will have to be taken.* Then staffs, and salaries, will not be reduced.

In less than a year the guards at Kengir opened fire several times on innocent men. There was one incident after another: and it cannot have been unintentional.[1]

They shot Lida, the young girl from the mortar-mixing gang who hung her stockings out to dry near the boundary fence.

They winged the old Chinaman—nobody in Kengir remembered his name, and he spoke hardly any Russian, but everybody knew the waddling figure with a pipe between his teeth and the face of an elderly goblin. A guard called him to a watchtower, tossed a packet of makhorka near the boundary fence, and when the Chinaman reached for it, shot and wounded him.

There was a similar incident in which the guard threw some cartridges down from the tower, ordered a prisoner to pick them up, and shot him.

Then there was the famous case of the column returning to camp from the ore-dressing plant and being fired on with dumdum bullets, which wounded sixteen men. (Another couple of dozen concealed light wounds to keep their names out of reports and avoid the risk of punishment.)

This the zeks did not take quietly—it was the Ekibastuz story over again. Kengir Camp Division No. 3 did not turn out for work three days running (but did take food), demanding punishment of the culprits.

A commission arrived and persuaded them that the culprits would be prosecuted (as though the zeks would be invited to the trial to check! . . .). They went back to work.

But in February, 1954, another prisoner was shot at the woodworking plant—"the Evangelist," as all Kengir remembered him (Aleksandr Sisoyev, I think his name was). This man had served nine years and nine months of his *tenner*. His job was fluxing arc-welding rods and he did this work in a little shed which stood near the boundary fence. He went out to relieve himself near the shed—and while he was at it was shot from a watchtower. Guards quickly ran over from the guardhouse and started dragging the dead man into the boundary zone, to make it look as though he had trespassed on it. This was too much for the zeks, who grabbed picks and shovels and drove the murderers away from the murdered man. (All this

1. The camp authorities, of course, acted similarly to speed up events in other places, for instance in Norilsk.

time near the woodworking plant stood a saddled horse belonging to Security Officer Belyaev—known as "the Wart" because he had one on his left cheek. Captain Belyaev was an enterprising sadist, and engineering a murder like this was just his style.)

The woodworking plant was in an uproar. The prisoners said that they would carry the dead man into camp on their shoulders. The camp officers would not permit it. "Why did you kill him?" shouted the prisoners. The bosses had their explanation ready: the dead man himself was to blame—he had started it by throwing stones at the tower. (Can they have had time to read his identity card; did they know that he had three months more to go and was an Evangelical Christian? . . .)

The march back was grim, and there were reminders that the bosses meant business. Machine-gunners lay here and there in the snow, ready to shoot (only too ready, as the men of Kengir had learned). Machine-gunners were also posted on the roofs of the escort troops' quarters.

This was at the same Camp Division, No. 3, which had already seen sixteen men wounded at once. Although only one man was killed on this occasion, they felt more painfully than ever that they were defenseless, doomed. Nearly a year had gone by since Stalin's death, but his dogs had not changed. In fact, nothing at all had changed.

In the evening after supper, what they did was this. The light would suddenly go out in a section, and someone invisible said from the doorway: "Brothers! How long shall we go on building and taking our wages in bullets? Nobody goes to work tomorrow!" The same thing happened in section after section, hut after hut.

A note was thrown over the wall to the Second Camp Division; they had some experience by now, and had thought about it often enough, so that they were able to call a strike there, too. In the Second Camp Division, which was multinational, the majority had *tenners* and many were coming to the end of their time—but they joined in just the same.

In the morning the men's Camp Divisions, 2 and 3, did not report for work.

This bad habit—striking without refusing the state's bread and slops—was becoming more and more popular with prisoners, and less and less popular with their bosses. They had an idea: warders and escort troops went unarmed into the striking Camp Divisions, where two of them at a time took hold of a single zek and tried shoving and jostling him out of the hut. (Far too humane a method: only thieves deserve to be nannied like this, not enemies of the people. But since Beria's execution, no general or colonel dared take the lead and order machine-gunners to fire into a camp.) This was wasted labor: the prisoners just went off to the latrines, or sloped about the camp ground—anything rather than report for work.

They held out like this for two days. The simple idea of punishing the guard who had killed "the Evangelist" did not seem at all simple, or just, to the bosses. Instead, a colonel from Karaganda, with a large retinue, went around the camp on the second night of the strike, confident that he was in no danger, roughly waking everybody up. "How long do you intend to carry on slacking?"[2] Then, knowing nobody there, he pointed at random: "You there —outside! . . . And you . . . And you . . . Outside!" And these chance people the valiant and forceful manager of men consigned to jail, imagining that this was the most sensible way to deal with slackers. The Latvian Will Rosenberg, when he saw this senseless high-handedness, said to the colonel: "I'll go, too!" "Go on, then," the colonel readily agreed. He probably did not even realize that this was a protest, or that there were any grounds for protest.

That same night it was announced that the liberal feeding policy was at an end and that those who did not go out to work would be put on short rations. Camp Division No. 2 went to work in the morning. No. 3 didn't turn out for the third morning running. The jostling and shoving tactics were now applied to them, but with heavier forces: all the officers serving in Kengir, those who had come in to help, and those who were with investigating commissions, were mobilized. The officers picked a hut and entered in strength, dazzling the prisoners with the coming and going of white fur hats and the brilliance of their epaulets, made their way, stooping, among the bunks, and with no sign of distaste sat down in their clean breeches on dirty pillows stuffed with shavings. "Come on, move up a bit—can't you see I'm a lieutenant colonel!" The lieutenant colonel kept this up, shifting his seat with arms akimbo, until he shoved the owner of the mattress out into the

2. "Slacking" was a word much used in official language after the Berlin disturbances of June, 1953. If ordinary people somewhere in Belgium fight for a raise, it's called "the righteous anger of the people," but if simple people in our country struggle for black bread they are "slackers."

passageway, where warders grabbed him by his sleeves and hustled him along to the work-assignment area or, if he was still too stubborn, into jail. (The limited capacity of the two Kengir jails was a great nuisance to the staff: they held about five hundred men.)

The strike was mastered, regardless of cost to the dignity and privileges of officers. This sacrifice was forced on them by the ambiguities of the time. They had no idea what was required of them, and mistakes could be dangerous! If they showed excessive zeal and shot down a crowd, they might end up as henchmen of Beria. But if they weren't zealous enough, and didn't energetically push the strikers out to work—exactly the same thing could happen.[3] Moreover, by their massive personal participation in putting down the strike, the MVD officers demonstrated as never before the importance of their epaulets to the defense of holy order, the impossibility of reducing staffs, and their individual bravery.

All previously proved methods were also employed. In March and April, several contingents of prisoners were transferred to other camps. (The plague crept further!) Some seventy men (Tenno among them) were sent to maximum security prisons, with the classic formula: "All *measures of correction* exhausted, corrupts other prisoners, not suitable for labor camp." Lists of those dispatched to maximum security jails were posted in the camp to deter others. And to make the self-financing system—Gulag's New Economic Policy, as it were—a more satisfactory substitute for freedom and justice, a wide selection of foodstuffs was delivered to the previously ill-stocked sale points. They even—incredibly!—started giving prisoners advances so that they could buy these provisions. (Gulag giving the natives credit! Who had ever heard of such a thing!)

So, for the second time in Kengir, a ripening abscess was lanced before it could burst.

But then the bosses went too far. They reached for the biggest stick they could use on the 58's—for the thieves! (Why, indeed, should they dirty their hands and sully their epaulets when they had the "class allies"?)

---

3. Colonel Chechev, for one, was defeated by this conundrum. He retired after the February events and we lose track of him—to discover him later living on his pension in Karaganda. We do not know how soon the camp commander, Colonel Yevsigneyev, left Ozerlag. "An excellent manager . . . a modest comrade," he became deputy head of the Bratsk hydroelectric station. (No hint of this in Yevtushenko's poem.)*

The bosses now renounced the whole principle of the Special Camps, acknowledged that if they segregated political prisoners they had no means of making themselves *understood,* and just before the May Day celebrations brought in and distributed throughout the mutinous No. 3 Camp Division 650 men, most of them thieves, some of them petty offenders (including many minors). "A *healthy batch* is joining us!" the bosses spitefully warned the 58's. "Now you won't dare breathe." And they called on the new arrivals to "put our house in order!"

The bosses understood well enough how the restorers of order would begin: by stealing, by preying on others, and so setting every man against his fellows. And the bosses smiled the friendly smile which they reserve for such people when the thieves heard that there was a women's camp nearby and asked in their impudent beggar's whine for a "look at the women, boss man!"

But here again we see how unpredictable is the course of human emotions and of social movements! Injecting in Kengir No. 3 a mammoth dose of tested ptomaine, the bosses obtained not a pacified camp but the biggest mutiny in the history of the Gulag Archipelago.

■

Though they seem to be so scattered and so carefully sealed off, the islands of Gulag are linked by the transit prisons, so that they breathe the same air and the same vital fluids flow in their veins. Thus the massacre of stoolies, the hunger strikes, the strikes, the disturbances in the Special Camps, had not remained unknown to the thieves. By 1954, so we are told, it was noticeable in transit prisons that *the thieves came to respect the politicals.*

If this is so—what prevented us from gaining their respect earlier? All through the twenties, thirties, and forties, we blinkered philistines, preoccupied as we were with our own importance to the world, with the contents of our duffel bags, with the shoes or trousers we had been allowed to retain, had conducted ourselves in the eyes of the thieves like characters on the comic stage: when they plundered our neighbors, intellectuals of world importance like ourselves, we shyly looked the other way and huddled together in our corners; and when the submen crossed the room to give us the treatment, we expected, of course, no help from neighbors, but obligingly surrendered all we had to these ugly customers in case they bit our heads off. Yes, our minds were busy elsewhere, and our hearts were trained for other things! We had never expected to meet an enemy so vile and so cruel! We who were racked by the twists and turns of Russian history, were ready to die only in public, beautifully, with the whole world looking on, and only for the final salvation of all mankind. It might have been better if we had been far less clever. Perhaps when we first stepped into the cell of a transit prison we should have been prepared, every man of us in the place, to take a knife between the ribs and slump in a wet corner on the slime around the latrine bucket, in a sordid brawl with those ratmen whom the boys in blue had thrown in to gnaw our flesh. If we had, perhaps we should have suffered far fewer losses, found our courage sooner, and, who knows, shoulder to shoulder with these very same thieves smashed Stalin's camps to smithereens? What reason, indeed, had the thieves to *respect* us? . . .

Well, then, when they arrived in Kengir the thieves had already heard a thing or two; they came expecting to find a fighting spirit among the politicals. And before they could get their bearings, and exchange doggy compliments with the camp authorities, their atamans were visited by some calm, broad-shouldered lads who sat down to *talk about life* and told them this: "We are *representatives.* You've heard all about the *chopping* in the Special Camps, or if you haven't we'll tell you. We can make knives as good as yours now. There are six hundred of you, two thousand six hundred of us. Think it over, and take your choice. If you try squeezing us we'll cut the lot of you up."

Now, this was a wise step, if ever there was one, and long overdue—rounding on the thieves with everything they had. Seeing them as the *main enemy!*

Of course, nothing would have suited the boys in blue better than a free-for-all. But the thieves looked at the odds and saw that it wouldn't pay to take on the newly emboldened 58's one against four. Their protectors, after all, were beyond the camp limits, and a fat lot of use anyway! What thief had ever respected them? Whereas the alliance which our lads offered was a novel and jolly adventure, which might also, they thought, clear a way over the fence into the women's camp.

Their answer, then, was: "No, we're wiser than we used to be. We're with you fellows!"

The conference has not been recorded for history and the names of its participants are not preserved in protocols. This is a pity. They were clever lads.

In their first huts while they were still in the quarantine period, the *healthy contingent* held a housewarming party—making bonfires of their bunks and lockers on the cement floor, and letting smoke pour through the windows. They expressed their disapproval of locks on hut doors by stuffing the keyholes with wood chips.

For two weeks the thieves behaved as though they were at a health resort: they reported for work, but all they did was sun themselves. The bosses, of course, would not dream of putting them on short rations, but for lack of funds could not pay wages to those of whom they had such bright hopes. Soon, however, vouchers turned up in the possession of the thieves, and they went to the stall to buy food. The bosses were heartened by this sign that the *healthy element* had begun thieving after all. But they were ill-informed, they were mistaken: a collection in aid of the thieves had been taken up among the politicals (also, no doubt, part of the compact—otherwise the thieves would not have been interested) and this was how they had come by their vouchers! An event too far out of the ordinary for the bosses to guess at it!

No doubt the novelty and unfamiliarity of the game made it great fun for the criminals, especially the juveniles: treating "Fascists" politely, not entering their sections without permission, not sitting down on their bunks without invitation.

Paris in the last century took some of its criminals (it seems to have had plenty of them) into the militia, and called them the *mobiles*. A very apt description! They are such a mobile breed that they cannot rest quietly inside the shell of an ordinary humdrum existence, but inevitably break it. They had made it a rule not to steal, and it was unethical to slog away for the government, but they had to do something! The young cubs amused themselves by snatching off warders' caps, prancing over the hut roofs and over the high wall from Camp Division No. 3 to No. 2 during evening roll call, confusing the count, whistling, hooting, scaring the *towers* at night. They would have gone further and climbed into the women's camp if the service yard and its sentries had not been in the way.

When disciplinary officers, or education officers, or security men, looked in for a friendly chat with thieves in their hut, the juveniles hurt their feelings badly by pulling notebooks and purses out of their pockets, or suddenly leaning out from a top bunk and switching godfather's cap peak-backward. Gulag had never encountered such conduct—but then the whole situation was unprecedented! They had always in the past regarded their foster fathers in Gulag as fools, and the more earnestly those turkey cocks believed that they were successfully *reforging* the thieves, the more the thieves despised them. They were ready to burst with scornful laughter as they stepped onto a platform or before a microphone to talk about beginning a new life behind a wheelbarrow. But so far there had been no point in quarreling with the bosses. Now, however, the compact with the politicals turned the newly released forces of the thieves against the bosses alone.

Thus the Gulag authorities, because they had only the mean intelligence of bureaucrats, and lacked the higher intellectual powers of human beings, had themselves prepared the Kengir explosion: to begin with, by the senseless shootings, and then by pouring the thieves into the camp like petrol fumes into an overheated atmosphere.

Events followed their inevitable course. It was *impossible* for the politicals not to offer the thieves a choice between war and alliance. It was *impossible* for the thieves to refuse an alliance. And it was *impossible* for the alliance, once concluded, to remain inactive—if it had, it would have fallen apart and civil war would have broken out.

They had to start something, no matter what! And since those who *start something* are strung up if they are 58's, with nooses around their necks, whereas if they are thieves they are only mildly rebuked in their political discussion period, the thieves made the obvious suggestion: we'll start, and you join in!

It should be noted that the whole Kengir camp complex formed a single rectangle, with one common outer fence, and was subdivided across its width into separate camp areas. First came Camp Division No. 1 (the women's camp), then the service yard (we have already talked about the industrial importance of its workshops), then No. 2, then No. 3, and then the prison area, with its two jailhouses, an old and a new building in which not only inmates of the camp but free inhabitants of the settlement were locked up from time to time.

The obvious first objective was to capture the service yard, in which all the camp's food stores were also situated. They began the operation in the afternoon of a nonworking day (Sunday, May 16, 1954). First the *mobiles* climbed onto the roofs of their huts and perched at intervals along the wall between Camp Divisions 2 and 3. Then at the command of their leaders, who stayed up aloft, they jumped down into Camp Division No. 2 with sticks in their hands, formed up in a column, and marched in line along the central road. This ran along the axis of No. 2 right up to the inner gates of the service yard, which brought them to a halt.

All these quite undisguised operations took a certain time, during which the warders managed to get themselves organized and obtain instructions. And here is something extremely interesting! The warders started running around to the huts of the 58's, appealing to these men whom they had treated like dirt for thirty-five years: "Look out, lads! The thieves are on their way to break into the women's camp. They are going to rape your wives and daughters! Come and help us. Let's stop them at it!" But a treaty is a treaty, and those who, not knowing about it, seemed eager to follow the bosses were stopped. Normally the 58's would have risen to this bait, but this time the warders found no helpers among them.

Just how the warders would have defended the women's camp against their favorites, no one knows—but first they had to think of defending the storerooms around the service yard. The gates of the service yard were flung open and a platoon of unarmed soldiers came out to meet the attackers, with Belyaev the Wart leading them from behind—perhaps devotion to duty had kept him inside the camp on a Sunday out of zeal, or perhaps he was officer of the day. The soldiers started pushing the *mobiles* away, and broke their lines. Without resorting to their clubs, the thieves began retreating to their own Camp Division No. 3, scaling the wall once again, from which the rear guard covered their retreat by throwing stones and mud bricks at the soldiers.

No thief, of course, was arrested as a result of this. The authorities still saw it as nothing but high-spirited mischief, and let the camp Sunday quietly run its course. Dinner was handed around without incident, and in the evening as soon as it was dark they started showing a film, *Rimsky-Korsakov*, using a space near the mess hall as an open-air cinema.

But before the gallant composer could withdraw from the conservatory in protest against oppression and persecution, the tinkle of broken glass could be heard from the lamps around the boundary zone: the *mobiles* were shooting at them from sling-shots to put out the lights over the camp area. They swarmed all over Camp Division No. 2 in the darkness, and their shrill bandit whistles rent the air. They broke the service yard gate down with a beam, and from there made a breach in the wall with a section of railway line and were through to the women's camp. (There were also some of the younger 58's with them.)

In the light of the flares fired from the towers, our friend Captain Belyaev, the security officer, broke into the service yard from outside the camp, through the guardhouse, with a platoon of Tommy-gunners, and for the first time in the history of Gulag opened fire on the "class allies"! Some were killed and dozens wounded. Behind them came red tabs to bayonet the wounded. Behind them again, observing the usual division of punitive labor, already adopted in Ekibastuz, in Norilsk, and in Vorkuta, ran warders with iron crowbars, and with these they battered the wounded to death. (That night the lights went up in the operating room of the hospital in Camp Division No. 2, and Fuster, the surgeon, a Spanish prisoner, went to work.)

The service yard was now firmly held by the punitive forces, and machine-gunners were posted there. But the Second Camp Division (the mobiles had played their overture, and the politicals now came onto the stage) erected a barricade facing the service yard gate. The Second and Third Camp Divisions had been joined together by a hole in the wall, and there were no longer any warders, any MVD authority, in them.

But what of those who had succeeded in breaking through to the women's camp, and were now cut off there? Events outsoared the casual contempt which the thieves feel for *females.* When shots rang out in the service yard, those who had broken into the women's camp ceased to be greedy predators and became comrades in misfortune. The women hid them. Unarmed soldiers came in to catch them, then others with guns. The women got in the way of the searchers, and resisted attempts to move them. The soldiers punched the women and struck them with their gun butts, dragged some of them off to jail (thanks to someone's foresight, there was a jailhouse in the women's camp area), and shot at some of the men.

Finding its punitive force under strength, the command brought into the women's camp some "black tabs"—soldiers from a construction battalion stationed in Kengir. But they would have nothing to do with this *"unsoldierly work"* and had to be taken away.

At the same time, here in the women's camp was the best political excuse which the executioners could offer their superiors in self-defense! They were not at all stupid! Whether they had read something of the sort or thought of it for themselves, on Monday they let photographers into the women's camp, together with two or three of their own apes, disguised as prisoners. The impostors started pulling women about, while the photographers took pictures. Obviously it was to save defenseless women from such bullying that Captain Belyaev had been compelled to open fire!

In the morning hours on Monday, there was growing tension on both sides of the barricade and the broken gates to the service yard. The yard had not been cleared of bodies. Machine-gunners lay at their guns, which were trained on the gate. In the liberated men's camps they were breaking bunks to arm themselves, and making shields out of boards and mattresses. Prisoners shouted across the barricade at the butchers, who shouted back. Something had to give; the position was far too precarious. The zeks on the barricade were thinking of taking the offensive themselves. Some emaciated men took their shirts off, got up on the barricade, pointed to their bony chests and ribs, and shouted to the machine-gunners: "Come on, then, shoot! Strike down your fathers! Finish us off!"

Suddenly a soldier ran into the yard with a note for the officer. The officer gave orders for the bodies to be taken up and the red tabs left the yard with them.

For five minutes the barricade was silent and mistrustful. Then the first zeks peeped cautiously into the yard. It was empty, except for the black prison caps lying around, dead men's caps with stitched-on number patches.

(They discovered later that the order to clear the yard had been given by the Minister of Internal Affairs of the Kazakh Republic, who had just flown in from Alma-Ata. The bodies carried away were driven out into the steppe and buried, to rule out postmortem examination if someone later called for it.)

Shouts of "Hurrah" went through the ranks, and they poured into the yard and on into the women's camp. They enlarged the breach. Then they freed the women in the jailhouse—and the whole camp was united! The whole of the main camp area was free —only No. 4 Division (the camp jail) was left.

There were *four* red-tabbed sentries on every tower—no lack of ears to cram with insults! Prisoners stood facing the towers and shouted at them (the women, naturally, louder than anyone): "You're worse than Fascists! . . . Bloodsuckers! . . . Murderers!"

A priest or two were of course easily found in the camp—and there in the morgue a requiem was sung for those who had been killed or died later from their wounds.

How can we say what feelings wrung the hearts of those eight thousand men, who for so long and until yesterday had been slaves with no sense of fellowship, and now had united and freed themselves, not fully perhaps, but at least within the rectangle of those walls, and under the gaze of those quadrupled guards? Even the bedridden fast in locked huts at Ekibastuz was felt as a moment of contact with freedom! This now was the February Revolution! So long suppressed, the brotherhood of man had broken through at last! We loved the thieves! And the thieves loved us! (There was no getting away from it: they had sealed the friendship in blood! They had departed from their *code!*) Still more, of course, we loved the women, and now we were living as human beings should, there were women at our side once more, and they were our sisters and shared our fate.

Proclamations appeared in the mess hall: "Arm yourselves as best you can, and attack the soldiers first!" The most passionate among them hastily scrawled their slogans on scraps of newspaper (there was no other paper) in black or colored letters: "Bash the Chekists, boys!" "Death to the stoolies, the Cheka's stooges!" Here, there, everywhere you turned there were meetings and orators. Everybody had suggestions of his own. Come on, think— you're permitted to think now: Who gets your vote? What demands shall we put forward? What is it we want? Put Belyaev on trial—that's understood! Put the murderers on trial!—goes without saying. What else? . . . No locking huts; take the numbers off! But beyond that? . . . Beyond that came the most frightening thing —the real reason why they had *started it all,* what they really wanted. We want freedom, of course, just freedom—but who can give it to us? The judges who condemned us in Moscow. As long as our complaints are against Steplag or Karaganda, they will go on talking to us. But if we start complaining against Moscow . . . we'll all be buried in this steppe.

Well, then—what do we want? To break holes in the walls? To run off into the wilderness? . . .

Those hours of freedom! Immense chains had fallen from our arms and shoulders! No; whatever happened, there could be no regrets! That one day made it all worthwhile!

Late on Monday, a delegation from command HQ arrived in the seething camp. The delegation was quite well disposed, they did not glare savagely at the prisoners, they had no Tommy-gunners with them, no one would ever take them for henchmen of the bloody Beria. Our side learned that generals had flown in from Moscow—Bochkov, from Gulag HQ, and the Deputy Procurator General, Vavilov. (They had served in Beria's time, but why reopen old wounds?) They found the prisoners' demands *fully justified!* (We simply gasped: justified? We aren't rebels, then? No, no, they're *quite* justified!) "Those responsible for the shooting will be made to answer for it!" "But why did they beat up women?" "Beat up women?" The delegation was shocked. "That can't be true." Anya Mikhalevich brought in a succession of battered women for them to see. The commission was deeply moved: "We'll look into it, never fear!" "Beasts!" Lyuba Bershadskaya shouts at the general. There were other shouts: "No locks on huts!" "We won't lock them any more." "Take the numbers off!" "Certainly we'll take them off," comes the assurance from a general whom the prisoners had never laid eyes on (and would never see again). "The holes in the wall between camp areas must remain!" They were getting bolder. "We must be allowed to mix with each other." "All right, mix as much as you like," the general agreed. "Let the holes remain." Right, brothers, what else do we want? We've won, we've won! We raised hell for just one day, enjoyed ourselves, let off steam—and we won! Although some among us shake their heads and say, "It's a trick, it's all a trick," we believe it! We believe our bosses; they're not so bad, on the whole! We believe because that's our easiest way out of the situation. . . .

All that the downtrodden can do is go on hoping. After every disappointment they must find fresh reason for hope.

So on Tuesday, May 18, all the Kengir Camp Divisions went out to work, reconciling themselves to thoughts of their dead.

That morning the whole affair could still have ended quietly. But the exalted generals assembled in Kengir would have considered such an outcome a defeat for themselves. They could not seriously admit that prisoners were in the right! They could not seriously punish soldiers of the MVD! Their mean understanding could draw only one moral: the walls between Camp Divisions were not strong enough! They must ring them with hoops of fire!

And that day the zealous commanders harnessed for work people who had lost the habit years or decades ago. Officers and warders donned aprons; those who knew how to handle them took up trowels; soldiers released from the towers wheeled barrows and carried hods; discharged soldiers who had stayed around the camps hauled and handed up mud bricks. And by evening the breaches were bricked up, the broken lamps were replaced, prohibited zones had been marked along inside walls and sentries posted at the ends with orders to fire!

When the columns of prisoners returned to camp in the evening after giving a day's work to the state, they were hurried in to supper before they knew what was happening, so that they could be locked up quickly. On orders from the general, the jailers had to play for time that first evening—that evening of blatant dishonesty after yesterday's promises; later on the prisoners would get used to it and slip back into the rut.

But before nightfall the long-drawn whistles heard on Sunday shrilled through the camp again—the Second and Third Camp Divisions were calling to each other like hooligans on a spree. (These whistles were another useful contribution from the thieves to the common cause.) The warders took fright, and fled from the camp grounds without finishing their duties. Only one officer slipped up. Medvedev, a first lieutenant in the quartermaster service, stayed behind to finish his business and was held prisoner till morning.

The camp was in the hands of the zeks, but they were divided. The towers opened fire with machine guns on anyone who approached the inside walls. They killed several and wounded several. Once again zeks broke all the lamps with slingshots, but the towers lit up the camp with flares. This was where the Second Camp Division found a use for the quartermaster: they tied him, with one of his epaulets torn off, to a table, and pushed it up to the strip near the boundary fence, with him yelling to his people: "Don't shoot, it's me! This is me, don't shoot!"

They battered at the barbed wire, and the new fence posts, with long tables, but it was impossible, under fire, either to break through the barrier or to climb over it—so they had to burrow under. As always, there were no shovels, except those for use in case of fire, inside the camp. Kitchen knives and mess tins were put into service.

That night—May 18–19—they burrowed under all the walls and again united all the divisions and the service yard. The towers had stopped shooting now, and there were plenty of tools in the service yard. The whole daytime work of the epauleted masons had gone to waste. Under cover of night they broke down the boundary fences, knocked holes in the walls, and widened the passages, so that they would not become traps (in the next few days they made them twenty meters wide).

That same night they broke through the wall around the Fourth Camp Division—the prison area—too. The warders guarding the jails fled, some to the guardhouse, some to the towers, where ladders were let down for them. The prisoners wrecked the interrogation offices. Among those released from the jail were those who on the morrow would take command of the rising: former Red Army Colonel Kapiton Kuznetsov (a graduate of the Frunze Academy, no longer young; after the war he had commanded a regiment in Germany, and one of his men had run away to the Western part—this was why he had been imprisoned; he was in the camp jail for "slanderous accounts of camp life" in letters sent out through free workers); and former First Lieutenant Gleb Sluchenkov (he had been a prisoner of war, and some said a Vlasovite).

In the "new" jailhouse were some inhabitants of the Kengir settlement, minor offenders. At first they thought that nationwide revolution had broken out, and rejoiced in their unexpected freedom. But they quickly discovered that the revolution was too localized, and the minor offenders loyally returned to their stone sack and dutifully lived there without guards throughout the rebellion—though they did go to eat in the mutinous zeks' mess hall.

Mutinous zeks! Who three times already had tried to reject this mutiny and this freedom. They did not know how to treat such gifts, and feared rather than desired them. But a force as relentless as the surf breaking on the shore had carried them helplessly into this rebellion.

What else could they do? Put their faith in promises? They would be cheated again, as the slavemasters had shown so clearly the day before, and so often in the past. Should they kneel in submission? They had spent years on their knees and earned no clemency. Should they give themselves up and take their punishment today? Today, or after a month of freedom, punishment would be equally cruel at the hands of those whose courts functioned like clockwork: if *quarters* were given out, it would be all around, with no one left out!

The runaway escapes to enjoy just one day of freedom! In just the same way, these eight thousand men had not so much raised a rebellion as *escaped to freedom,* though not for long! Eight thousand men, from being slaves, had suddenly become free, and now was their chance to . . . live! Faces usually grim softened into kind smiles.[4] Women looked at men, and men took them by the hand. Some who had corresponded by ingenious secret ways, without even seeing each other, met at last! Lithuanian girls whose weddings had been solemnized by priests on the other side of the wall now saw their lawful wedded husbands for the first time—the Lord had sent down to earth the marriages made in heaven! For the first time in their lives, no one tried to prevent the sectarians and believers from meeting for prayer. Foreigners, scattered about the Camp Divisions, now found each other and talked about this strange Asiatic revolution in their own languages. The camp's food supply was in the hands of the prisoners. No one drove them out to work line-up and an eleven-hour working day.

The morning of May 19 dawned over a feverishly sleepless camp which had torn off its number patches. Posts with broken lamps sprawled against the wire fences. Even without their help the zeks moved freely from zone to zone by the trenches dug under the wires. Many of them took their street clothes from the storerooms and put them on. Some of the lads crammed fur hats on their heads; shortly there would be embroidered shirts, and on the Central Asians bright-colored robes and turbans. The gray-black camp would be a blaze of color.

Orderlies went around the huts summoning us to the big mess hall to elect a commission for negotiations with the authorities and for self-government, as it modestly and timidly described itself.

For all they knew, they were electing it just for a few hours, but it was destined to become the government of Kengir camp for forty days.

■

4. A hostile witness, Makeyev, noted this.

Had all this taken place two years earlier, if only for fear that *He* would find out, the bosses of Steplag would not have delayed a moment before giving the classic order "Don't spare the bullets!" and shooting the whole crowd penned within these walls. If it had proved necessary to knock off four thousand—or all eight thousand—they would not have felt the slightest tremor, because they were shockproof.

But the complexity of the situation in 1954 made them vacillate. Even Vavilov, even Bochkov, sensed that a new breeze was stirring in Moscow. Quite a few prisoners had been shot in Kengir already, and they were still wondering how to make it look legal. So a delay was created, which meant time for the rebels to begin their new life of independence.

In its very first hours the political line of the revolt had to be determined: to be or not to be? Should it follow in the wake of the simple-hearted messages scrawled over the columns of the robot press: "Bash the Chekists, boys"?

As soon as he left the jail and began to take charge—whether through force of circumstances, or because of his military skill, or on the advice of friends, or moved by some inner urge—Kapiton Ivanovich immediately adopted the line of those orthodox Soviet citizens who were not very numerous in Kengir and were usually pushed into the background. "Cut out all this scribbling" (of leaflets), "nip the evil of counter-revolution in the bud, frustrate those who want to *take advantage* of events in our camp!" These phrases I quote from notes kept by another member of the commission, A. F. Makeyev, of an intimate discussion in Pyotr Akoyev's storeroom. The orthodox citizens nodded approval of Kuznetsov: "Yes, if we don't stop those leaflets, we shall all get extra time."

In the first hours, during the night, as he went around all the huts haranguing himself hoarse, again at the meeting in the mess hall next morning, and on many subsequent occasions, whenever he encountered extremist sentiments and the bitter rage of men whose lives were trodden so deep into the mire that they felt they had nothing more to lose, Kuznetsov endlessly, tirelessly repeated the same words:

"Anti-Sovietism will be the death of us. If we display anti-Soviet slogans we shall be crushed immediately. They're only waiting for an excuse to crush us. If we put out leaflets like that, they will be fully justified in shooting. Our salvation lies in loyalty. We must talk to Moscow's representatives *in a manner befitting Soviet citizens!*"

Then, in a louder voice: "We shall not permit such behavior on the part of a few provocateurs!" (However, while he was making these speeches, people on the bunks were loudly kissing. They didn't take in much of what he said.)

When a train carries you in the wrong direction and you decide to jump off, you have to jump *with* the motion of the train, and not *against* it. The inertial force of history is just as hard to resist. By no means everyone wanted it that way, but the reasonableness of Kuznetsov's line was immediately perceived, and it prevailed. Very soon slogans were hung up all over the camp, in big letters easily legible from the towers and the guardhouse:

"Long live the Soviet Constitution!"

"Long live the Presidium of the Central Committee!"

"Long live the Soviet regime!"

"The Central Committee must send one of its members and review our cases!"

"Down with the murdering Beria-ites!"

"Wives of Steplag officers! Aren't you ashamed to be the wives of murderers?"

Although it was clear as could be to the majority in Kengir that all the millions of acts of rough justice, near and distant, had taken place under the watery sun of that very constitution, and had been sanctioned by a Politburo consisting of the *very same members,* all they felt able to do was write "long live" *that* constitution and *that* Politburo. As they reread their slogans, the rebel prisoners now felt themselves on firm legal ground and began to be less anxious: their movement was not hopeless.

Over the mess hall, where the elections had just taken place, a flag was raised which the whole settlement could see. It hung there long afterward: white, with a black border, and the red hospitalers' cross in the middle. In the international maritime code this flag means:

"Ship in distress! Women and children on board."

Twelve men were elected to the Commission, with Kuznetsov at their head. The members of the Commission assumed individual responsibilities, and created the following departments:

Agitation and propaganda (under the Lithuanian Knopkus, who had been sent for punishment from Norilsk after the rising there).

Services and maintenance.

Food.

Internal security (Gleb Sluchenkov).

Military.

Technical (perhaps the most remarkable branch of this camp government).

Former Major Mikheyev was made responsible for contacts with the authorities. The Commission also had among its members one of the atamans of the thieves, and he, too, was in charge of something. There were also some women (apparently, Shakhnovskaya, an economist and Party member already gray; Suprun, an elderly teacher from the Carpathians; and Lyuba Bershadskaya).

But did the real moving spirits behind the revolt join this Commission? It seems that they did not. The *centers,* especially the Ukrainian Center (not more than a quarter of those in the whole camp were Russian), evidently kept themselves to themselves. Mikhail Keller, a Ukrainian partisan, who from 1941 had fought alternately against the Germans and the Soviet side, and had publicly axed a stoolie in Kengir, appeared at meetings of the Commission as an observer from the *other* staff.

The Commission worked openly in the offices of the Women's Camp Division, but the Military Department moved its command post (field staff) out to the bathhouse in Camp Division No. 2. The departments set to work. The first few days were particularly hectic: there was so much planning and arranging to be done.

First of all they had to fortify their position. (Mikheyev, in expectation of the inevitable military action to crush them, was against creating defenses of any kind, but Sluchenkov and Knopkus insisted.) Great piles of mud bricks had risen, where the breaches in the inner walls were widened and cleared. They used these bricks to make barricades facing all the guard points—exits from within, entrances from without—which remained in the hands of the jailers, and any one of which might open at any minute to admit the punitive force. There were plenty of coils of barbed wire in the service yard. With this they made entanglements and scattered them about the threatened approaches. Nor did they omit to put out little boards saying: "Danger! Minefield!"

This was one of the first of the Technical Department's bright ideas. The department's work was surrounded by great secrecy. In the occupied service yard the Technical Department set aside secret premises, with a skull and crossbones drawn on the door, and the legend "High Tension—100,000 volts." Only the handful of men who worked there were allowed in. So that even the prisoners did not get to know what the Technical Department's activities were. A rumor was very soon put about that it was making secret weapons of a chemical nature. Since both zeks and bosses knew very well what clever engineers there were in the camp, it was easy for a superstitious conviction to get around that they could do anything, even invent a weapon which no one had yet thought of in Moscow. As for making a few miserable mines, using the reagents which were there in the service yard—what was to stop them? So the boards saying "Minefield" were taken seriously.

Another weapon was devised: boxes of ground glass at the entrance to every hut (to throw in the eyes of the Tommy-gunners).

The teams were kept just as they had been, but were now called platoons, while the huts were called detachments, and detachment commanders were named, subordinate to the Military Department. Mikhail Keller was put in charge of all guard duties. Vulnerable places were occupied, according to a precise roster, by pickets, which were reinforced for the night hours. A man will not run away and in general will show more courage in the presence of a woman: with this masculine psychological trait in mind, the rebels organized mixed pickets. Besides the many loud-mouthed women in Kengir, there proved to be many brave ones, especially among the Ukrainian girls, who were the majority in the women's camp.

Without waiting this time for master's kind permission, they began taking the window bars down themselves. For the first two days, before the bosses thought of cutting the power supply to the camp, the lathes were still working in the shops and they made a large number of *pikes* from these window bars by grinding down and sharpening their ends. The smiths and the lathe operators worked without a break in those first days, making weapons: knives, halberds, and sabers (which were particularly popular with the thieves; they decorated the hilts with jingles and colored leather). Others were seen with bludgeons in their hands.

The pickets shouldered pikes as they went to take up their posts for the night. The women's platoon, on their way at night to rooms provided in the men's camp, so that they could rush out to meet the attackers if the alarm was raised (it was naïvely assumed that the butchers would be ashamed to hurt women), also bristled with pikes.

All this would have been impossible, would have been ruined by mockers or lechers, if the stern and cleansing wind of rebellion had not been blowing through the camp. In our age these pikes, these sabers, were children's toys, but for these people, prison—prison behind them, and prison before them—was no game. The pikes were playthings, but this was what fate had sent—their first chance to defend their freedom. In the puritanical air of that revolutionary springtime, when the presence of women on the barricades itself became a weapon, men and women behaved with proper dignity, and with dignity carried their pikes, points skyward.

If anyone during those days entertained hopes of vile orgies, it was the blue-epauleted bosses, on the other side of the fence. Their calculation was that left alone for a week, the prisoners would drown in their debauchery. That was the picture they painted for the inhabitants of the settlement—prisoners mutinying for the sake of sexual indulgence. (Obviously, there was no other lack they could feel in their comfortable existence.)[5]

The main hope of the authorities was that the thieves would start raping women, that the politicals would intervene, and that there would be a massacre. But once again the MVD psychologists were wrong! And we ourselves may well be surprised. All witnesses agree that the thieves behaved *like decent people,* but in our sense of the term, not in their own traditional sense. On their side, the politicals and the women themselves emphasized by their behavior that they regarded the thieves as friends, and trusted them. What lay below these attitudes need not concern us here. Perhaps the thieves kept remembering their comrades bloodily murdered that first Sunday.

5. After the mutiny the bosses had the effrontery to carry out a general medical examination of all the women. When they discovered that many were still virgins they were flabbergasted. Eh? What were you thinking of? All those days together! . . . In their judgment of events they could not rise above their own moral plane.

If we can speak of the strength of the Kengir revolt at all, its strength was in this unity.

Nor did the thieves touch the food stores, which, for those who know them, is no less surprising. Although there was enough food for many months in the storerooms, the Commission, after due consideration, decided to leave the allowances of bread and other foodstuffs as before. The honest citizen's fear of eating more than his share of public victuals, and having to answer for this waste! As though the state owed the prisoners nothing for all those hungry years! On the other hand (as Mikheyev recalls), when there was a shortage outside the camp, the supply section of the camp administration asked the prisoners to release certain foodstuffs. There was some fruit, intended for those on higher norms (free workers!), and the zeks released it.

The camp bookkeepers allocated foodstuffs on the old norms, the kitchen took them and cooked them, but in the new revolutionary atmosphere did not pilfer, nor did emissaries from the thieves appear with instructions to *bring the stuff for the people*. Nothing extra was ladled out for the trusties. And it was suddenly found that though the norms were the same, there was noticeably more to eat!

If the thieves sold things (things previously stolen in other places), they did not, as had been their custom, immediately come along to take them back again. "Things are different now," they said.

The stalls belonging to the local Workers' Supply Department went on trading inside the camp. The staff guaranteed the safety of the cashier (a free woman). She was allowed into the camp and went around the stalls with two girls, collecting the takings (in vouchers) from the salespeople. (But the vouchers, of course, soon ran out, and the bosses did not let new stocks through into the camp.)

The supply of three items needed in the camp remained in the hands of the bosses: electricity, water, and medicines. They did not, of course, control the air supply. As for medicines, they gave the camp not a single powder nor a single drop of iodine in forty days. The electricity they cut off after two or three days. The water supply they left alone.

The Technical Department began a fight for light. Their first idea was to fix hooks to fine wires and sling them forcibly over the outside cable, which ran just beyond the camp wall—and in this way they stole current for a few days, until the tentacles were discovered and cut. This had given the Technical Department time to try out a windmill, abandon it, and begin installing in the service yard (in a spot concealed from prying eyes on the towers, or low-flying observation planes) a hydroelectric station, worked by . . . a water tap. A motor which happened to be in the yard was converted into a generator, and they started supplying the camp telephone network, the lighting in staff headquarters, and . . . the radio transmitter! (The huts were lit by wood splints.) This unique hydroelectric station went on working till the last day of the revolt.

In the very early days of the mutiny, the generals would come into the camp as though they owned it. True, Kuznetsov was not at a loss. When the first parleys took place he ordered his men to bring the bodies from the morgue and loudly ordered: "Caps off!" The zeks bared their heads, and the generals, too, had to take off their peaked caps in the presence of their victims. But the initiative remained with Gulag's General Bochkov. After approving the election of a commission ("You can't talk to everybody at once"), he demanded that the negotiators tell him first how their *cases* had been investigated (and Kuznetsov began lengthily and perhaps eagerly presenting his story); and further insisted that prisoners should stand up to speak. When somebody said, "The prisoners demand . . ." Bochkov touchily retorted that "Prisoners cannot demand, they can only request!" And "The prisoners request" became the established formula.

Bochkov replied to the prisoners' *requests* with a lecture on the building of socialism, the unprecedentedly rapid progress of the economy, and the successes of the Chinese revolution. That complacent irrelevancy, that driving of screws into the brain which always leaves us weak and numb. He had come into the camp to look for evidence that the use of firearms had been justified. (They would shortly declare that in fact there had been no shooting in the camp: this was just a lie told by the gangsters; nor had there been any beatings.) He was simply amazed that they should dare ask him to infringe the "instruction concerning the separate housing of men and women prisoners." (They talk this way about their instructions, as though they were laws for and from all time.)

Shortly, other, more important generals arrived in Douglas aircraft: Dolgikh (at that time allegedly head of Gulag) and Yegorov (Deputy Minister of Internal Affairs for the U.S.S.R.).

A meeting was called in the mess hall, and some two thousand prisoners assembled. Kuznetsov gave the orders: "Silence, please! All stand! Atten-tion!" and respectfully invited the generals to sit on the dais, while he, as befitted his subordinate rank, stood to one side! (Sluchenkov behaved differently. When one of the generals casually spoke of *enemies* present, Sluchenkov answered in a clear voice: "How many of your sort turned out to be enemies? Yagoda was a public enemy, Yezhov was a public enemy, Abakumov was a public enemy, Beria was a public enemy. How do we know that Kruglov is any better?")

Makeyev, to judge from his notes, drew up a draft agreement in which the authorities promised not to transfer people to other camps, or otherwise victimize them, and to begin a thorough investigation, while the zeks in return agreed to return to work immediately. However, when he and his supporters started going around the huts and asking prisoners to accept his draft, they were jeered at as "bald-headed Komsomols," "procurement agents," and "lackeys of the Cheka." They got a particularly hostile reception in the women's camp, and found that the separation of the men's and women's divisions was by now the last thing the zeks would agree to. (Makeyev angrily answered his opponents: "Just because you've had your hand on some wench's tits, d'you think the Soviet regime is a thing of the past? The Soviet regime will get its own way, whatever happens!")

The days ran on. They never took their eyes off the camp grounds— soldiers' eyes from the towers, and warders' eyes, too (the warders, knowing the zeks by sight, were supposed to identify them and remember who did what), and even the eyes of airmen (perhaps equipped with cameras)—and the generals were regretfully forced to conclude that there were no massacres, no pogroms, no violence in there, that the camp was not disintegrating of its own accord, and that there was no excuse to send troops in to the rescue.

The camp *stood fast* and the negotiations changed their character. Golden-epauleted personages, in various combinations, continued coming into the camp to argue and persuade. They were all allowed in, but they had to pick up white flags, and between the outer gate of the service yard (now the main entrance) and the barricade, they had to undergo a body search, with some Ukrainian girl slapping the generals' pockets in case there was a pistol or a hand grenade in them. In return, the rebel staff *guaranteed* their personal safety! . . .

They showed the generals around, wherever it was allowed (not, of course, around the *secret* sector of the service yard), let them talk to prisoners, and called big meetings in the Camp Divisions for their benefit. Their epaulets flashing, the bosses took their seats in the presidium as of old, as though nothing were amiss.

The prisoners put up speakers. But speaking was so difficult! Not only because each of them, as he spoke, was writing his future sentence, but also because in their experience of life and their ideas of truth, the grays and the blues had grown too far apart, and there was hardly any way of penetrating, of letting some light into, those plump and prosperous carcasses, those glossy, melon-shaped heads. They seem to have been greatly angered by an old Leningrad worker, a Communist who had taken part in the Revolution. He asked them what chance Communism had when officers got fat on the output of camp workshops, when they had lead stolen from the separating plant and made into shot for their poaching trips; when prisoners had to dig their kitchen gardens for them; when carpets were laid in the bathhouse for the camp commander's visits, and an orchestra accompanied his ablutions.

To cut out some of this disorganized shouting, the discussions sometimes took the form of direct negotiations on the loftiest diplomatic model. Sometime in June a long mess table was placed in the women's camp, and the golden epaulets seated themselves on a bench to one side of it, while the Tommy-gunners allowed in with them as a bodyguard stood at their backs. Across the table sat the members of the Commission, and they, too, had a bodyguard—which stood there, looking very serious, armed with sabers, pikes, and slingshots. In the background crowds of prisoners gathered to listen to the powwow and shout comments. (Refreshments for the guests were not forgotten! Fresh cucumbers were brought from the hothouses in the service yard, and kvass from the kitchen. The golden epaulets crunched cucumbers unselfconsciously. . . .)

The rebels had agreed on their demands (or requests) in the first two days, and now repeated them over and over again:

- Punish the Evangelist's murderer.
- Punish all those responsible for the murders on Sunday night in the service yard.
- Punish those who beat up the women.
- Bring back those comrades who had been illegally sent to closed prisons for striking.
- No more number patches, window bars, or locks on hut doors.
- Inner walls between Camp Divisions not to be rebuilt.
- An eight-hour day, as for free workers.
- An increase in payment for work (here there was no question of equality with free workers).
- Unrestricted correspondence with relatives, periodic visits.
- Review of cases.

Although there was nothing unconstitutional in any of these demands, nothing that threatened the foundations of the state (indeed, many of them were requests for a return to the old position), it was impossible for the bosses to accept even the least of them, because these bald skulls under service caps and supported by close-clipped fat necks had forgotten how to admit a mistake or a fault. Truth was unrecognizable and repulsive to them if it manifested itself not in secret instructions from higher authority but on the lips of common people.

Still, the obduracy of the eight thousand under siege was a blot on the reputation of the generals, it might ruin their careers, and so they made promises. They promised that nearly all the demands would be satisfied—only, they said (to make it more convincing), they could hardly leave the women's camp open, that was against the rules (forgetting that in the Corrective Labor Camps it had been that way for twenty years), but they could consider arranging, should they say, *meeting days.* To the demand that the Commission of Inquiry (into the circumstances of the shooting) should start its work inside the camp, the generals unexpectedly agreed. (But Sluchenkov guessed their purpose, and refused to hear of it: while making their statements, the stoolies would expose everything that was happening in the camp.) Review of cases? Well, of course, cases would be re-examined, but prisoners would *have to be patient.* There was one thing that couldn't wait at all—the prisoners must get back to work! to work! to work!

But the zeks knew that trick by now: dividing them up into columns, forcing them to the ground at gunpoint, arresting the ringleaders.

No, they answered across the table, and from the platform. No! shouted voices from the crowd. The administration of Steplag have behaved like provocateurs! We do not trust the Steplag authorities! We don't trust the MVD!

"Don't trust *even* the MVD?" The vice-minister was thrown into a sweat by this treasonable talk. "And who can have inspired in you such hatred for the MVD?"

A riddle, if ever there was one.

"Send us a member of the Central Committee Presidium! A member of the Presidium! Then we'll believe you!" shouted the zeks.

"Be careful," the generals threatened. "You'll make it worse for yourselves!"

At this Kuznetsov got up. He spoke calmly and precisely, and he held himself proudly.

"If you enter the camp with weapons," he warned them, "don't forget that half of those here had a hand in the capture of Berlin. They can cope even with your weapons!"

Kapiton Kuznetsov! Some future historian of the Kengir mutiny must help us to understand the man better. What were his thoughts, how did he feel about his imprisonment? What stage did he imagine his appeal to have reached? How long was it since he had asked for a review, if the *order of release* (with rehabilitation, I believe) arrived from Moscow during the rebellion? His pride in keeping the mutinous camp in such good order—was it only the professional pride of a military man? Had he put himself at the head of the movement because it captured his imagination? (I reject that explanation.) Or, knowing his powers of leadership, had he taken over to restrain the movement, tame the flood, and channel it, to lay his chastened comrades at their masters' feet? (That is my view.) In meetings and discussions, and through people of lesser importance, he had opportunities to tell those in charge of the punitive operation anything he liked, and to hear things from them. On one occasion in June, for instance, the artful dodger Markosyan was sent out of camp on an errand for the Commission. Did Kuznetsov exploit such opportunities? Perhaps not. His position may have been a proud and independent one.

Two bodyguards—two enormous Ukrainian lads—accompanied Kuznetsov the whole time with knives in their belts.

To defend him? To settle scores?

(Makeyev claims that during the rebellion Kuznetsov also had a temporary wife—another Ukrainian nationalist.)

Gleb Sluchenkov was about thirty. Which means that he must have been captured by the Germans when he was nineteen or so. Like Kuznetsov, he now went around in his old uniform, which had been kept in storage, acting or overacting the old soldier. He had a slight limp, but the speed of his movements made it unimportant.

In the negotiations he was precise to the point of curtness. The authorities had the idea of calling all "former juvenile offenders" out of the camp (youngsters jailed before they were eighteen, some of them now twenty or twenty-one) and releasing them. This may not have been a trick—at about this time they were releasing such prisoners or reducing their sentences all over the country. Sluchenkov's answer was:

"Have you asked the minors whether they *want* to move from one camp to another, and leave their comrades in the lurch?" (In the Commission, too, he insisted that "These kids carry out our guard duties—we can't hand them over!" This, indeed, was the unspoken reason why the generals wanted to release these youngsters while Kengir was in revolt: and for all we know, they would have shoved them in cells outside the camp.) The law-abiding Makeyev nonetheless began rounding up former juveniles for the "discharge tribunal," and he testifies that out of 409 persons with a claim to be released, he only succeeded in collecting thirteen who were willing to leave. If we bear in mind Makeyev's sympathy with the authorities, and his hostility to the rising, his testimony is amazing: four hundred youngsters in the bloom of youth, and in their great majority not politically minded, renounced not merely freedom but salvation! And stayed with the doomed revolt . . .

Sluchenkov's reply to threats that troops would be used to put down the revolt was "Send them in! Send as many Tommy-gunners as you like! We'll throw ground glass in their eyes and take their guns from them! We'll trounce your Kengir troops. Your bowlegged officers we'll chase all the way to Karaganda—we'll ride into Karaganda on your backs! And once there, we're among friends!"

There is other evidence about him which seems reliable. "Anybody who runs away will get this in the chest!"—flourishing a hunting knife in the air. "Anybody who doesn't turn out to defend the camp will get the knife," he announced in one hut. The inevitable logic of any military authority and any war situation.

The newborn camp government, like all governments through the ages, was incapable of existing without a security service, and Sluchenkov headed this (occupying the security officer's room in the women's camp). Since there could be no victory over the outside forces, Sluchenkov realized that this post meant certain execution. In the course of the revolt he told people in the camp that the bosses had secretly urged him to provoke a racial bloodbath (the golden epaulets were banking heavily on this) and so provide a plausible excuse for troops to enter the camp. In return the bosses promised Sluchenkov his life. He rejected their proposal. (What approaches were made to others? They haven't told us.) Moreover, when a rumor went around the camp that a Jewish pogrom was imminent, Sluchenkov gave warning that the rumormongers would be publicly flogged. The rumor died down.

A clash between Sluchenkov and the loyalists seemed inevitable. And it happened. It should be said that all these years, in all the Special Camps, orthodox Soviet citizens, without even consulting each other, unanimously condemned the massacre of the stoolies, or any attempt by prisoners to fight for their rights. We need not put this down to sordid motives (though quite a few of the orthodox were compromised by their work for the godfather) since we can fully explain it by their theoretical views. They accepted all forms of repression and extermination, even wholesale, provided they came *from above*—as a manifestation of the dictatorship of the proletariat. Even impulsive and uncoordinated actions of the same kind but *from below* were regarded as *banditry*, and what is more, in its "Banderist" form (among the loyalists you would never get one to admit the right of the Ukraine to secede, because to do so was bourgeois nationalism). The refusal of the *katorzhane* to be slave laborers, their indignation about window bars and shootings, depressed and frightened the docile camp Communists.

In Kengir as elsewhere, there was a nest of loyalists. (Genkin, Apfelzweig, Talalayevsky, and evidently Akoyev. We have no more names.) There was also a malingerer who spent years in the hospital pretending that "his leg kept going around." Intellectual methods of struggle such as this were deemed permissible. In the

Commission itself Makeyev was an obvious example. All of them were reproachful from the beginning: "You shouldn't have started it"; when the passages under the walls were blocked off, they said that they shouldn't have tunneled; it was all a stunt thought up by the Banderist scum, and the thing to do now was to back down quickly. (Anyway, the sixteen killed were not from their camp, and it was simply silly to shed tears over the Evangelist.) All their bile and bigotry is blurted out in Makeyev's notes. Everything and everybody in sight is bad, and there are dangers on every hand; it's either a new sentence from the bosses, or a knife in the back from the Banderists. "They want to frighten us all with their bits of iron and drive us to our deaths." Makeyev angrily calls the Kengir revolt a "bloody game," a "false trump," "amateur dramatics" on the part of the Banderists, and more often than any of these, "the wedding party." The hopes and aims of the leaders, as he sees them, were fornication, evasion of work, and putting off the day of reckoning. (That there was a reckoning to be paid, he tacitly assumes to be just.)

This very accurately expresses the attitude of the loyalists in the fifties to the freedom movement in the camps. Whereas Makeyev was very cautious, and was indeed among the leaders of the revolt, Talalayevsky poured out such complaints quite openly, and Sluchenkov's internal security service locked him in a cell in the Kengir jailhouse for agitation hostile to the rebels.

Yes, this really happened. The rebels who had liberated the jail now set up one of their own. The old, old ironical story. True, only four men were put inside for various reasons (usually for dealings with the bosses), and none of them was shot (instead, they were presented with the best of alibis for the authorities).

This incident apart, the jailhouse—a particularly gloomy old place, built in the thirties—was put on display to a wide public: it had windowless solitary-confinement cells, with nothing but a tiny skylight; legless beds, mere wooden boards on the cement floor, where it was still colder and damper than elsewhere in the cold cell; and beside each bed, which means down on the floor, a rough earthenware bowl like a dog dish.

To this place the Agitation Department organized sightseeing trips for their fellow prisoners who had never been inside and perhaps never would be. They also took there visiting generals (who were not greatly impressed). They even asked that sightseers from among the free inhabitants of the settlement be sent along:

with the prisoners absent, they could in any case do nothing at the work sites. The generals actually sent such a party—not, of course, ordinary workingmen, but hand-picked personnel who found nothing to excite indignation.

In reply, the authorities offered to arrange a prisoners' outing to Rudnik (Divisions 1 and 2 of Steplag), since according to camp rumors a revolt had broken out there, too. (Incidentally, for their own good reasons both slaves and slavemasters shunned the word "revolt," or still worse, "rising," replacing them with the bashful euphemism "horseplay.") The delegates went, and saw for themselves that all was as it had been, that prisoners were going to work.

Their hopes largely depended on strikes like their own spreading. Now the delegates returned with cause for despondency.

(The authorities, in fact, had taken them there only just in time. Rudnik was of course worked up. Prisoners had heard from free workers all sorts of facts and fantasies about the Kengir revolt. In the same month [June] it so happened that many appeals for judicial review were turned down simultaneously. Then some half-crazy lad was wounded in the prohibited area. And there, too, a strike started, the gates between Camp Divisions were knocked down, and prisoners poured out onto the central road. Machine guns appeared on the towers. Somebody hung up a placard with anti-Soviet slogans on it, and the rallying cry "Freedom or death!" But this was taken down and replaced by one with *legitimate* demands and an undertaking to make up for losses caused by the stoppage once the demands were satisfied. Lorries came to fetch flour from the storerooms; the prisoners wouldn't let them have it. The strike lasted for something like a week, but we have no precise information about it: this is all at third hand, and probably exaggerated.)

There were weeks when the whole war became a war of propaganda. The outside radio was never silent: through several loudspeakers set up at intervals around the camp it interlarded appeals to the prisoners with information and misinformation, and with a couple of trite and boring records that frayed everybody's nerves.

> Through the meadow goes a maiden,
> She whose braided hair I love.

(Still, to be thought worthy even of that not very high honor—having records played to them—they had to rebel. Even rubbish like that wasn't played for men on their knees.) These records also served, in the spirit of the times, as a *jamming device*—drowning the broadcasts from the camps intended for the escort troops.

On the outside radio they sometimes tried to blacken the whole movement, asserting that it had been started with the sole aim of rape and plunder. (In the camp itself the zeks just laughed, but the free inhabitants of the settlement also listened, willy-nilly, to the loudspeakers. Of course, the slavemasters could not rise to any other explanation—an admission that this rabble was capable of seeking justice was far beyond the reach of their minds.) At other times they tried telling filthy stories about members of the Commission. (They even said that one of the "old ones," when he was being transported by barge to the Kolyma, had made a hole below the waterline and sunk the boat with three hundred zeks in it. The emphasis was on the fact that it was poor zeks he had drowned, practically all of them 58's, too, and not the escort troops; how he had survived himself was not clear.) Or else they would taunt Kuznetsov, telling him that his discharge had arrived, but was now canceled. Then the appeals would begin again. Work! Work! Why should the Motherland keep you for nothing? By not going to work you are doing enormous damage to the state! (This was supposed to pierce the hearts of men doomed to eternal *katorga!*) Whole trainloads of coal are standing in the siding, there's nobody to unload it! (Let them stand there—the zeks laughed—you'll give way all the sooner! Yet it didn't occur even to them that the golden epaulets could unload it themselves if it troubled them so much.)

The Technical Department, however, gave as good as it got. Two portable film projectors were found in the service yard. Their amplifiers were used for loudspeakers, less powerful, of course, than those of the other side. They were fed from the secret hydroelectric station! (The fact that the camp had electricity and radio greatly surprised and troubled the bosses. They were afraid that the rebels might rig up a transmitter and start broadcasting news about their rising to foreign countries. Rumors to this effect were also put around inside the camp.)

The camp soon had its own announcers (Slava Yarimovskaya is one we know of). Programs included the latest news, and news features (there was also a daily wall newspaper, with cartoons). "Crocodile Tears" was the name of a program ridiculing the anxiety of the MVD men about the fate of women whom they themselves had previously beaten up. Then there were programs for the escort troops. Apart from this, prisoners would approach the towers at night and shout to the soldiers through megaphones.

But there was not enough power to put on programs for the only potential sympathizers to be found in Kengir—the free inhabitants of the settlement, many of them exiles. It was they whom the settlement authorities were trying to fool, not by radio but, in some place which the prisoners could not reach, with rumors that bloodthirsty gangsters and insatiable prostitutes (this version went down well with the women)[6] were ruling the roost inside the camp; that over there innocent people were being tortured and burned alive in furnaces. (In that case, it was hard to see why the authorities did not intervene! . . .)

How could the prisoners call out through the walls, to the workers one, or two, or three kilometers away: "Brothers! We want only justice! They were murdering us for no crime of ours, they were treating us worse than dogs! Here are our demands"?

The thoughts of the Technical Department, since they had no chance to outstrip modern science, moved backward instead to the science of past ages. Using cigarette paper (there was everything you could think of in the service yard; we have talked about that already:[7] for many years it provided the Dzhezkazgan officers with their own Moscow tailor's shop, and a workshop for every imaginable article of consumer goods), they pasted together an enormous air balloon, following the example of the Montgolfier brothers. A bundle of leaflets was attached to the balloon, and slung underneath it was a brazier containing glowing coals, which sent a current of warm air into the dome of the balloon through an opening in its base. To the huge delight of the assembled crowd (if prisoners ever do feel happy they are like children), the marvelous aeronautical structure rose and was airborne. But alas! The speed of the wind was greater than the speed of its ascent, and as it was flying over the boundary fence the brazier caught on the barbed wire. The balloon, denied its current of warm air, fell and burned to ashes, together with the leaflets.

After this failure they started inflating balloons with smoke.

6. When it was all over and the women's column was marched through the settlement on the way to work, married women, Russian women, gathered along the roadside and shouted at them: "Prostitutes! Dirty whores! Couldn't do without it, could you . . . !" and other, still stronger remarks. The same thing happened next day, but the women prisoners had left the camp prepared, and replied to these insulting creatures with a bombardment of stones. The escort troops just laughed.

7. Part III, Chapter 22.

With a following wind they flew quite well, exhibiting inscriptions in large letters to the settlement:

"Save the women and old men from being beaten!"

"We demand to see a member of the Presidium."

The guards started shooting at these balloons.

Then some Chechen prisoners came to the Technical Department and offered to make kites. (They are experts.) They succeeded in sticking some kites together and paying out the string until they were over the settlement. There was a percussive device on the frame of each kite. When the kite was in a convenient position, the device scattered a bundle of leaflets, also attached to the kite. The kite fliers sat on the roof of a hut waiting to see what would happen next. If the leaflets fell close to the camp, warders ran to collect them; if they fell farther away, motorcyclists and horsemen dashed after them. Whatever happened, they tried to prevent the free citizens from reading an independent version of the truth. (The leaflets ended by requesting any citizen of Kengir who found one to deliver it to the Central Committee.)

The kites were also shot at, but holing was less damaging to them than to the balloons. The enemy soon discovered that sending up counter-kites to tangle strings with them was cheaper than keeping a crowd of warders on the run.

A war of kites in the second half of the twentieth century! And all to silence a word of truth.

(Perhaps it will help the reader to place the events at Kengir chronologically if we recall what was happening outside during the days of the mutiny. The Geneva Conference on Indochina was in session. The Stalin Peace Prize was conferred on Pierre Cot. Another progressive French man, the writer Sartre, arrived in Moscow to join in the life of our progressive society. The third centenary of the reunification of Russia and the Ukraine was loudly and lavishly celebrated.[8] On May 31 there was a solemn parade on Red Square. The Ukrainian S.S.R. and the Russian S.F.S.R. were awarded the Order of Lenin. On June 6 a monument to Yuri Dolgoruky was unveiled in Moscow. A Trade Union Congress opened on June 8 [but nothing was said there about Kengir]. On the tenth a new state loan was launched. The twentieth was Air Force Day, and there was a splendid parade at Tushino. These months of 1954 were also marked by a powerful

8. The Ukrainians at Kengir declared it a day of mourning.

offensive on the literary *front,* as people call it: Surkov, Kochetov, and Yermilov came out with very tough admonitory articles. Kochetov even asked: *"What sort of times are we living in?"* And nobody answered: *"A time of prison camp risings!"* Many *incorrect* plays and books were abused during this period. And in Guatemala the imperialistic United States met with the rebuff it deserved.)

There were Chechen exiles in the settlement, but it is unlikely that they made the other kites. You cannot accuse the Chechens of ever having served oppression. They understood perfectly the meaning of the Kengir revolt, and on one occasion brought a bakery van up to the gates. Needless to say, the soldiers drove them away.

(There is more than one side to the Chechens. People among whom they live—I speak from my experience in Kazakhstan—find them hard to get along with; they are rough and arrogant, and they do not conceal their dislike of Russians. But the men of Kengir only had to display independence and courage—and they immediately won the good will of the Chechens! When we feel that we are not sufficiently *respected,* we should ask ourselves whether we are living as we should.)

In the meantime the Technical Department was getting its notorious "secret" weapon ready. Let me describe it. Aluminum corner brackets for cattle troughs, produced in the workshops and awaiting dispatch, were packed with a mixture of sulfur scraped from matches and a little calcium carbide (every box of matches had been carried off to the room with the 100,000-volt door). When the sulfur was lit and the brackets thrown, they hissed and burst into little pieces.

But neither these star-crossed geniuses nor the field staff in the bathhouse were to choose the hour, place, and form of the decisive battle. Some two weeks after the beginning of the revolt, on one of those dark nights without a glimmer of light anywhere, thuds were heard at several places around the camp wall. This time it was not escaping prisoners or rebels battering it down; the wall was being demolished by the convoy troops themselves! There was commotion in the camp, as prisoners charged around with pikes and sabers, unable to make out what was happening and expecting an attack. But the troops did not take the offensive.

In the morning it turned out that the enemy without had made about a dozen breaches in the wall in addition to those already there and the barricaded gateway. (Machine-gun posts had been set up on the other side of the gaps, to prevent the zeks from pouring through them.)[9] This was of course the preliminary for an assault through the breaches, and the camp was a seething anthill as it prepared to defend itself. The rebel staff decided to pull down the inner walls and the mud-brick outhouses and to erect a second circular wall of their own, specially reinforced with stacks of brick where it faced the gaps, to give protection against machine-gun bullets.

How things had changed! The troops were demolishing the boundary wall, the prisoners were rebuilding it, and the thieves were helping with a clear conscience, not feeling that they were contravening their *code*.

Additional defense posts now had to be established opposite the gaps, and every platoon assigned to a gap, which it must run to defend should the alarm be raised at night. Bangs on the buffer of a railroad car, and the usual whistles, were the agreed-upon alarm signals.

The zeks quite seriously prepared to advance against machine guns with pikes. Those who shied at the idea to begin with soon got used to it.

There was one attack in the daytime. Tommy-gunners were moved up to one of the gaps, opposite the balcony of the Steplag Administration Building, which was packed with important personages sheltering under broad army epaulets or the narrow ones of the Public Prosecutor's department, and holding cameras or even movie cameras. The soldiers were in no hurry. They merely advanced just far enough into the breach for the alarm to be given, whereupon the rebel platoons responsible for the defense of the breach rushed out to man the barricade—brandishing their pikes and holding stones and mud bricks—and then, from the balcony, movie cameras whirred and pocket cameras clicked (taking care to keep the Tommy-gunners out of the picture). Disciplinary officers, prosecutors, Party officials, and all the rest of them—Party members to a man, of course—laughed at the bizarre spectacle of the impassioned savages with pikes. Well-fed and shameless, these

9. The precedent is said to have been set by Norilsk; there, too, they made breaches in the wall, to lure out the fainthearted, who would be used to stir up the thieves, and so provide an excuse to introduce troops and restore order.

grand personages mocked their starved and cheated fellow citizens from the balcony, and found it *all very funny.*[10]

Then warders, too, stole up to the gaps and tried to slip nooses with hooks over the prisoners, as though they were hunting wild animals or the abominable snowman, hoping to drag out a *talker.*

But what they mainly counted on now were deserters, rebels with cold feet. The radio blared away. Come to your senses! Leave the camp through the gaps in the wall! At those points we shall not fire! Those who come over will not be tried for mutiny!

The Commission's response, over the camp radio, was this: Anybody who wants to run away can go right ahead, through the main gate if he likes; we are holding no one back!

One who did so was . . . a member of the Commission itself, former Major Makeyev, who walked up to the main guardhouse as though he had business there. (*As though,* not because they would have detained him, nor because they had the means of shooting him in the back—but because it is almost impossible to play the traitor with your comrades looking on and howling their contempt!)[11] For three weeks he had kept up a pretense; now at last he could give free rein to his defeatism, and his anger with the rebels for wanting the freedom which he, Makeyev, did not want. Now, working off his debts to the bosses, he broadcast an appeal to surrender, and reviled those who favored holding out longer. Here are some sentences from his own written account of this broadcast: "Somebody has decided that freedom can be won with the help of sabers and pikes. They want to expose to bullets people who won't take their bits of iron. . . . We have been promised a review of our cases. The generals are patiently negotiating with us, but Sluchenkov regards this as a sign of weakness on their part. The Commission is a screen for gangster debauchery. . . . Conduct the negotiations in a manner worthy of political prisoners, and do not (!!) prepare for senseless resistance."

The holes in the wall gaped for weeks; the wall had not remained whole so long while the revolt was on. And in all those weeks only about a dozen men fled from the camp.

___

10. These photographs must still exist somewhere, gummed to reports on the punitive operation. Perhaps somebody will not be swift enough to destroy them before posterity sees them.

11. Even ten years later he was so ashamed that in his memoirs, which were probably designed to serve as an apologia, he writes that he chanced to put his head out of the gate, where the other side pounced on him and tied his hands. . . .

Why? Surely the rest did not believe in victory. Were they not appalled by the thought of the punishment ahead? They were. Did they not want to save themselves for their families' sake? They did! They were torn, and thousands of them perhaps had secretly considered this possibility. The invitation to the former juveniles had a firm legal base. But the social temperature on this plot of land had risen so high that if souls were not transmuted, they were purged of dross, and the sordid laws saying that "we only live once," that being determines consciousness, and that every man's a coward when his neck is at stake, ceased to apply for that short time in that circumscribed place. The laws of survival and of reason told people that they must all surrender together or flee individually, but they did not surrender and they did not flee! They rose to that spiritual plane from which executioners are told: "The devil take you for his own! Torture us! Savage us!"

And the operation, so beautifully planned, to make the prisoners scatter like rats through the gaps in the wall till only the most stubborn were left, who would then be crushed—this operation collapsed because its inventors had the mentality of rats themselves.

In the rebel wall newspaper, next to a drawing of a woman showing a child a pair of handcuffs in a glass case—"like the ones they kept your father in"—appeared a cartoon of the "Last Renegade" (a black cat running through one of the holes in the wall).

Cartoonists can always laugh, but the people in the camp had little to laugh about. The second, third, fourth, fifth week went by. . . . Something which, according to the laws of Gulag, could not last an hour had lasted for an incredibly, indeed an agonizingly long time—half of May and almost the whole of June. At first people were intoxicated with the joy of victory, with freedom, meetings, and schemes, then they believed the rumors that Rudnik had risen; perhaps Churbay-Nura, Spassk—all Steplag would follow. In no time at all Karaganda would rise! The whole Archipelago would erupt and fall in ash over the face of the land! But Rudnik put its hands behind its back, lowered its head, and reported as before for its eleven-hour shifts, contracting silicosis, with never a thought for Kengir, or even for itself.

No one supported the island of Kengir. It was impossible by now to take off into the wilderness: the garrison was being steadily reinforced; troops were under canvas out on the steppe. The whole camp had been encircled with a double barbed-wire fence outside the walls. There was only one rosy spot on the horizon: the lord and master (they were expecting Malenkov) was coming to dispense justice. He would come, kind man, and exclaim, and throw up his hands: "However could they live in such conditions? and why did you treat them like this? Put the murderers on trial! Shoot Chechev and Belyaev! Sack the rest...." But it was too tiny a spot, and too rosy.

They could not hope for pardon. All they could do was live out their last few days of freedom, and submit to Steplag's vengeance.

There are always hearts which cannot stand the strain. Some were already morally crushed, and were in an agony of suspense for the crushing proper to begin. Some quietly calculated that they were not really involved, and need not be if they went on being careful. Some were newly married (what is more, with a proper religious ceremony—a Western Ukrainian girl, for instance, will not marry without one, and thanks to Gulag's thoughtfulness, there were priests of all religions there). For these newlyweds the bitter and the sweet succeeded each other with a rapidity which ordinary people never experience in their slow lives. They observed each day as their last, and retribution delayed was a gift from heaven each morning.

The believers ... prayed, and leaving the outcome of the Kengir revolt in God's hands, were as always the calmest of people. Services for all religions were held in the mess hall according to a fixed timetable. The Jehovah's Witnesses felt free to observe their rules strictly and refused to build fortifications or stand guard. They sat for hours on end with their heads together, saying nothing. (They were made to wash the dishes.) A prophet, genuine or sham, went around the camp putting crosses on bunks and foretelling the end of the world. Conveniently for him, a severe cold spell set in, of the sort that a shift in the wind sometimes brings to Kazakhstan even in summer. The old women he had gathered together sat, not very warmly dressed, on the cold ground shivering and stretching out their hands to heaven. Where else could they turn?

Some knew that they were fatally compromised and that the few days before the troops arrived were all that was left of life. The theme of all their thoughts and actions must be how to hold out longer. These people were not the unhappiest. (The unhappiest were those who were not involved and who prayed for the end.)

But when all these people gathered at meetings to decide whether to surrender or to hold on, they found themselves again in that heated climate where their personal opinions dissolved, and ceased to exist even for themselves. Or else they feared ridicule even more than the death that awaited them.

"Comrades," the majestic Kuznetsov said confidently, as though he knew many secrets, and all to the advantage of the prisoners, "we have *defensive firepower*, and the enemy will suffer fifty percent of our own losses."

He also said: "Even our destruction will not be in vain."

(In this he was absolutely right. The social temperature had its effect on him, too.)

And when they voted for or against holding out, the majority were *for*.

Then Sluchenkov gave an ominous warning. "Just remember, if anyone remains in our ranks now and wants to surrender later, we shall settle accounts with him five minutes before he gets there!"

One day the outside radio broadcast an "order of the day to Gulag": for refusal to work, for sabotage, for this, that, and the other, the Kengir Camp Division of Steplag was to be disbanded and all prisoners sent to Magadan. (Clearly, the planet was getting too small for Gulag. And those who had been sent to Magadan previously—what were they there for?) One last chance to go back to work . . .

Once more their last chance ran out, and things were as before.

All was as it had been, and the dreamlike existence of these eight thousand men, suspended in midair, was rendered all the more startlingly improbable and strange by the regularity of the camp routine; fresh linen from the laundry; haircuts; clothes and shoes repaired. There were even conciliation courts for disputes. Even . . . even a release procedure!

Yes. The outside radio sometimes summoned prisoners due for release: these were either foreigners from some country which had earned the right to gather in its citizens, or else people whose sentence was (or was said to be?) nearing its end. Perhaps this was the administration's way of picking up "tongues" without the use of warders' ropes and hooks? The Commission sat on it, but had no means of verification, and let them all go.

Why did it drag on so long? What can the bosses have been waiting for? For the food to run out? They knew it would last a long time. Were they considering opinion in the settlement? They had no need to. Were they carefully working out their plan of repression? They could have been quicker about it. (True, it was learned later that they had sent for a "special purposes"—meaning punitive—regiment from somewhere around Karaganda. It's a job not everyone can do.) Were they having to seek approval for the operation *up top?* How high up? There is no knowing on what date and at what level the decision was taken.

On several occasions the main gate of the service yard suddenly opened—perhaps to test the readiness of the defenders? The duty picket sounded the alarm, and the platoons poured out to meet the enemy. But no one entered the camp grounds.

The only field intelligence service the defenders had were the observers on the hut roofs. Their anticipations were based entirely on what the fence permitted them to see from the rooftops.

In the middle of June several tractors appeared in the settlement. They were working, shifting something perhaps, around the boundary fence. They began working even at night. These nocturnal tractor operations were baffling. Just in case, the prisoners started digging ditches opposite the gaps, as an additional defense. (They were all photographed or sketched from an observation plane.)

The unfriendly roar made the night seem blacker.

Then suddenly the skeptics were put to shame! And the defeatists! And all who had said that there would be no mercy, and that there was no point in begging. The orthodox alone could feel triumphant. On June 22 the outside radio announced that the prisoners' demands had been accepted! A member of the Presidium of the Central Committee was on his way!

The rosy spot turned into a rosy sun, a rosy sky! It is, then, possible to get through to them! There *is*, then, justice in our country! They will give a little, and we will give a little. If it comes to it, we can walk about with number patches, and the bars on the windows needn't bother us, we aren't thinking of climbing out. You say they're tricking us again? Well, they aren't asking us to report for work *beforehand!*

Just as the touch of a stick will draw off the charge from an electroscope so that the agitated gold leaf sinks gratefully to rest, so did the radio announcement reduce the brooding tension of that last week.

Even the loathsome tractors, after working for a while on the evening of June 24, stopped their noise.

Prisoners could sleep peacefully on the fortieth night of the revolt. *He* would probably arrive tomorrow; perhaps he had come already. . . .[12] Those short June nights are too short to have your sleep out, and you are fast asleep at dawn. It was like that summer thirteen years before.*

In the early dawn of Friday, June 25, parachutes carrying flares opened out in the sky, more flares soared from the watchtowers, and the observers on the rooftops were picked off by snipers' bullets before they could let out a squeak! Then cannon fire was heard! Airplanes skimmed the camp, spreading panic. Tanks, the famous T-34's, had taken up position under cover of the tractor noise and now moved on the gaps from all sides. (One of them, however, fell into a ditch.) Some of the tanks dragged concatenations of barbed wire on trestles so that they could divide up the camp grounds immediately. Behind others ran helmeted assault troops with Tommy guns. (Both Tommy-gunners and tank crews had been given vodka first. However *special* the troops may be, it is easier to destroy unarmed and sleeping people with drink inside you.) Operators with walkie-talkies came in with the advancing troops. The generals went up into the towers with the snipers, and from there, in the daylight shed by the flares (and the light from a tower set on fire by the zeks with their incendiary bombs), gave their orders: "Take hut number so-and-so! . . . That's where Kuznetsov is!" They did not hide in observation posts, as they usually do, because no bullets threatened them.[13]

From a distance, from their building sites, free workers watched the operation.

The camp woke up—frightened out of its wits. Some stayed where they were in their huts, lying on the floor as their one chance of survival, and because resistance seemed senseless. Others tried to make them get up and join in the resistance. Yet others ran right into the line of fire, either to fight or to seek a quicker death.

The Third Camp Division fought—the division which had

12. Perhaps he really had come? Perhaps it was he who had given the instructions?
13. They only hid from history. Who were these war lords, so swift and so sure? Why has our country not saluted their glorious victory at Kengir? With some difficulty we can discover the names not of the most important there, but of some by no means unimportant: for instance, Colonel Ryazantsev, head of the Security Operations Section; Syomushkin, head of the Political Department of Steplag. . . . Please help! Add to the list.

started it all. (It consisted mainly of 58's with a large majority of Banderists.) They hurled stones at the Tommy-gunners and warders, and probably sulfur bombs at the tanks. . . . Nobody thought of the powdered glass. One hut counterattacked twice, with shouts of "Hurrah."

The tanks crushed everyone in their way. (Alla Presman, from Kiev, was run over—the tracks passed over her abdomen.) Tanks rode up onto the porches of huts and crushed people there (including two Estonian women, Ingrid Kivi and Makhlapa.)[14] The tanks grazed the sides of huts and crushed those who were clinging to them to escape the caterpillar tracks. Semyon Rak and his girl threw themselves under a tank clasped in each other's arms and ended it that way. Tanks nosed into the thin board walls of the huts and even fired blank shells into them. Faina Epstein remembers the corner of a hut collapsing, as if in a nightmare, and a tank passing obliquely over the wreckage and over living bodies; women tried to jump and fling themselves out of the way: behind the tank came a lorry, and the half-naked women were tossed onto it.

The cannon shots were blank, but the Tommy guns were shooting live rounds, and the bayonets were cold steel. Women tried to shield men with their own bodies—and they, too, were bayoneted! Security Officer Belyaev shot two dozen people with his own hand that morning; when the battle was over he was seen putting knives into the hands of corpses for the photographer to take pictures of dead *gangsters*. Suprun, a member of the Commission, and a grandmother, died from a wound in her lung. Some prisoners hid in the latrines, and were riddled with bullets there.[15]

Kuznetsov was arrested in the bathhouse, his command post, and made to kneel. Sluchenkov was lifted high in the air with his hands tied behind his back and dashed to the ground (a favorite trick with the thieves).

Then the sound of shooting died away. There were shouts of "Come out of your huts; we won't shoot." Nor did they—they merely beat prisoners with their gun butts.

As groups of prisoners were taken, they were marched through

14. In one of the tanks sat Nagibina, the camp doctor, drunk. She was there not to help but to watch—it was interesting.

15. Attention, Bertrand Russell and Jean-Paul Sartre, with your War Crimes Tribunal! Attention, philosophers. Here's material for you! Why not hold a session? They can't hear me. . . .

the gaps onto the steppe and between files of Kengir convoy troops outside. They were searched and made to lie flat on their faces with their arms stretched straight out. As they lay there thus crucified, MVD fliers and warders walked among them to identify and pull out those whom they had spotted earlier from the air or from the watchtowers. (So busy were they with all this that no one had leisure to open *Pravda* that day. It had a special theme—a day in the life of our Motherland: the successes of steelworkers; more and more crops harvested by machine. The historian surveying our country as it was *that day* will have an easy task.)

Curious officers could now inspect the secrets of the service yard —see where the electric power had come from, and what "secret weapons" there were.

The victorious generals descended from the towers and went off to breakfast. Without knowing any of them, I feel confident that their appetite that June morning left nothing to be desired and that they drank deeply. An alcoholic hum would not in the least disturb the ideological harmony in their heads. And what they had for hearts was something installed with a screwdriver.

The number of those killed or wounded was about six hundred, according to the stories, but according to figures given by the Kengir Division's Production Planning Section, which became known some months later, it was more than *seven hundred*.[16] When they had crammed the camp hospital with wounded, they began taking them into town. (The free workers were informed that the troops had fired only blanks, and that prisoners had been killing each other.)

It was tempting to make the survivors dig the graves, but to prevent the story from spreading too far, this was done by troops. They buried three hundred in a corner of the camp, and the rest somewhere out on the steppe.

All day on June 25, the prisoners lay face down on the steppe in the sun (for days on end the heat had been unmerciful), while in the camp there was endless searching and breaking open and shaking out. Later bread and water were brought out onto the steppe. The officers had lists ready. They called the roll, put a tick by those who were still alive, gave them

16. On January 9, 1905, the number killed was about 100. In 1912, in the famous massacre on the Lena goldfields, which shocked all Russia, there were 270 killed and 250 wounded.

their bread ration, and consulting their lists, at once divided the prisoners into groups.

The members of the Commission and other suspects were locked up in the camp jail, which was no longer needed for sightseers. More than a thousand people were selected for dispatch either to closed prisons or to Kolyma (as always, these lists were drawn up partly by guesswork, so that many who had not been involved at all found their way into them).

May this picture of the pacification bring peace to the souls of those on whom the last chapters have grated. Hands off, keep away! No one will have to take refuge in the "safe deposit," and the punitive squads will never face retribution!

On June 26, the prisoners were made to spend the whole day taking down the barricades and bricking in the gaps.

On June 27, they were marched out to work. Those trains in the sidings would wait no longer for working hands!

The tanks which had crushed Kengir traveled under their own power to Rudnik and crawled around for the zeks to see. And draw their conclusions . . .

The trial of the rebel leaders took place in autumn, 1955, in camera, of course, and indeed we know nothing much about it. . . . Kuznetsov, they say, was very sure of himself, and tried to prove that he had behaved impeccably and could have done no better. We do not know what sentences were passed. Sluchenkov, Mikhail Keller, and Knopkus were probably shot. I say probably because they certainly would have been shot earlier—but perhaps 1955 softened their fate?

Back in Kengir all was made ready for a life of honest toil. The bosses did not fail to create teams of *shock workers* from among yesterday's rebels. The "self-financing" system flourished. Food stalls were busy, rubbishy films were shown. Warders and officers again sneaked into the service yard to have things made privately —a fishing reel, a money box—or to get the clasp mended on a lady's handbag. The rebel shoemakers and tailors (Lithuanians and Western Ukrainians) made light, elegant boots for the bosses, and dresses for their wives. As of old, the zeks at the separating plant were ordered to strip lead from the cables and bring it back to the camp to be melted down for shot, so that the comrade officers could go hunting antelopes.

By now disarray had spread throughout the Archipelago and reached Kengir. Bars were not put back at the windows, huts were no longer locked. The "two-thirds" parole system was introduced, and there was even a quite unprecedented re-registration of 58's —the half-dead were released.

The grass on graves is usually very thick and green.

In 1956 the camp area itself was liquidated. Local residents, exiles who had stayed on in Kengir, discovered where *they* were buried—and brought steppe tulips to put on their graves.

Whenever you pass the Dolgoruky monument, remember that it was unveiled during the Kengir revolt—and so has come to be in some sense a memorial to Kengir.

END OF PART V

PART VI

Exile

■

Chapter 1

■

Exile in the First Years of Freedom

Humanity probably invented exile first and prison later. Expulsion from the tribe was of course exile. We were quick to realize how difficult it is for a man to exist, divorced from his own place, his familiar environment. Everything is wrong and awkward, everything is temporary and unreal, even if there are green woods around, not permafrost.

In the Russian Empire as elsewhere they were not slow to discover exile. It was given legal sanction under Tsar Aleksei Mikhailovich by the Code of Laws of 1648. But even earlier, at the end of the sixteenth century, people were exiled without legal sanction: the disgraced people of Kargopol, for instance, then those inhabitants of Uglich who had witnessed the murder of the Tsarevich Dmitri. Our great spaces gave their blessing—Siberia was ours already. By 1645 the number of exiles there had reached one and a half thousand. Peter exiled many hundreds at a time. As we have already said, Elizabeth replaced the death penalty by perpetual exile to Siberia. At this point the term was debased: exile came to mean not only free settlement but also *katorga,* which was a very different matter. Alexander's Regulations on Exiles in 1822 gave legal status to this misuse. For this reason the figures for exiles in the nineteenth century must obviously be taken to include convicts sentenced to hard labor. At the beginning of the nineteenth century, between two thousand and six thousand people were exiled year in and year out. From 1820 they also started exiling vagrants (the "parasites," in our terminology), so that in some years they weeded out as many as ten thousand. In 1863 the desert island of Sakhalin, so conveniently cut off from the mainland, found favor and was equipped as a place of exile, so that the resources of the system were still further enlarged. Altogether in the course of the nineteenth century *half a million* were exiled, and at the end of the century those in exile at any one time numbered 300,000.[1]

By the end of the century the ramifications of the exile system were becoming more and more diverse. Milder forms also appeared: "banishment to a distance of two provinces," and even "banishment abroad"[2] (not then considered such a ruthless punishment as after October). *Administrative* banishment, which usefully supplemented banishment by the courts, also took root. However: the term of banishment was expressed in clear and precise figures, and even exile for life was not in reality a life sentence. Chekhov writes in *Sakhalin* that after ten years in exile (or six years, if his "conduct had been completely satisfactory"— a vague criterion, but according to Chekhov one widely applied), the exile was re-registered as a peasant and could go back to any part of the country he wished except his native place.

A feature of the exile system in the Tsarist nineteenth century which was taken for granted and seemed natural to everyone, but to us now seems surprising, was that it concerned itself only with individuals: whether he was dealt with by the courts or administratively, each man was sentenced separately and not as a member of some group.

The conditions of life in exile, the degree of harshness, changed from one decade to the next, and different generations of exiles have left us a variety of evidence. Transported prisoners traveling from transit prison to transit prison had a hard time of it; but we learn from P. F. Yakubovich and from Lev Tolstoi that politicals were transported in quite tolerable conditions. F. Kon adds that in the presence of politicals escort troops treated *even* common criminals well, and that the criminals therefore thought highly of politicals. For many generations the population of Siberia was

1. All this information is taken from Volume XVI ("Western Siberia") of Semyonov-Tyan-Shansky's multivolume geographical study, *Russia.* Not only the celebrated geographer himself, but his brothers, too, were staunch and selfless liberals in their public life, who did a great deal to spread the light of freedom in our land. During the Revolution their whole family was destroyed. One brother was shot on their comfortable estate on the river Ranov, the house was burned and the great orchard, together with the avenues of limes and poplars, chopped down.

2. P. F. Yakubovich, *V Mire Otverzhennykh (In the World of Outcasts).*

hostile to exiles: they were assigned the poorest plots of land, they were left with the worst and poorest-paid jobs, and peasants would not let their daughters marry such people. Unprovided for, ill clad, branded, and hungry, they formed gangs and lived by stealing—further exacerbating the local inhabitants. None of this, however, applies to the politicals, who became a significant stream from the seventies. F. Kon again writes that the Yakuts received the politicals amicably and hopefully, looking to them as doctors, teachers, and legal advisers in their struggles with authority. The conditions in which political exiles lived were certainly such that many scholars (scholars who only began their studies in exile) sprang from their ranks—experts on particular regions, ethnographers, philologists,[3] scientists, as well as publicists and writers. Chekhov did not see the politicals on Sakhalin and has left us no description of them,[4] but F. Kon, for instance, when he was exiled to Irkutsk, began working on the progressive newspaper *Vostochnoye Obozrenie (Eastern Review),* to which Populists, supporters of the People's Will Party, and Marxists (Krasin) all contributed. This was no ordinary Siberian town but a major provincial capital, to which, according to the Regulations on Exiles, politicals were not supposed to be sent at all—in spite of which they were to be found working in banks and commercial firms, teaching, rubbing shoulders with the local intelligentsia at receptions. And exiles contrived to get into the Omsk newspaper *Stepnoi Krai* articles the likes of which the censorship would never have let through anywhere in Russia proper. The Omsk exiles even supplied the Zlatoust strikers with their newspaper. Through the efforts of the exiles Krasnoyarsk, too, became a radical town. While in Minusinsk, exile activists grouped around the Martyanov Museum enjoyed such respect and such freedom from administrative interference that they not only created without hindrance a network of hideouts and transit points for fugitives (we have previously described how easy it was to escape in those days), but even controlled the proceedings of the "Witte" committee* in the

3. Tan-Bogoraz, V. I. Jochelson, L. Y. Shternberg.

4. In his innocence of legal matters, or perhaps simply in the spirit of his day, Chekhov did not obtain any formal authorization, or any official piece of paper, for his journey to Sakhalin. Nonetheless he was allowed to make a census of the convicts and even given access to prison documents. (Compare that with our experience! Try going to inspect a nest of camps without authorization from the NKVD!) The one restriction was that he could not meet the politicals.

town.[5] And if Chekhov exclaims that the regime for criminals in Sakhalin had been simplified "in the most mean and stupid fashion" until it was in fact "serfdom," this cannot be said of political exile from early times and until the end of the Tsarist period. By the beginning of the twentieth century administrative exile for political offenders in Russia had ceased to be a punishment and become a meaningless, purely formal "antiquated device which has proved its ineffectuality" (Guchkov). In 1906 Stolypin began taking steps to abolish it altogether.

What did exile mean to Radishchev? In the Ust-Ilimsky Ostrov settlement he bought a two-story wooden house (for ten rubles, incidentally), and lived there with his young children and a sister-in-law who took the place of his wife. Nobody even thought of making him *work;* he spent his time in any way he saw fit and he was free to move about anywhere within the Ust-Ilim administrative region. What Pushkin's banishment to Mikhailovskoye meant, the very many people who have been there as sightseers can imagine for themselves. The life in exile of many other writers and public figures was similar—that of Turgenev at Spasskoye-Lutovinovo, that of Aksakov at Varvarino (his own choice). Trubetskoi even had his wife living with him in his cell in Nerchinsk Prison (where a son was born to them), and when after some years he was transferred to Irkutsk as an exile they had a huge house of their own, their own carriage, footmen, French tutors for the children. (Legal thought was still immature and knew nothing of "enemies of the people" or "confiscation of all property.") Herzen, exiled to Novgorod, held a position in the guberniya in which he received reports from the police chief.

This mild treatment of exiles was not confined to socially distinguished or famous people. In the twentieth century, too, it was enjoyed by many revolutionaries and frondeurs—by the Bolsheviks in particular, who were not thought dangerous. Stalin, with four escapes behind him, was exiled for a fifth time . . . all the way to Vologda. Vadim Podbelsky was banished for his outspoken articles against the government . . . from Tambov to Saratov! What cruelty! Needless to say, nobody tried to drive him out to forced labor when he got there.[6]

---

5. Feliks Kon, *Za Pyat'desyat Let.* Tom 2. *Na Poselenii. (Fifty Years.* Vol. 2. *Banishment.)*

6. This revolutionary, after whom the Post Office Street in many a Russian town was

Yet exile even under these conditions, lenient as it seems to us, exile with no danger of starving to death, was sometimes taken hard by the exile himself. Many revolutionaries recall how painful they found the move from prison, where they were assured of bread, warmth, shelter, and leisure for their "universities" and their party wrangles, to a place of exile, where you were all by yourself among strangers, and had to use your own ingenuity to find food and shelter. Where there was no need to worry about these things it was, so we are informed (by F. Kon, for instance), still worse: "the horrors of idleness . . . The most dreadful thing of all is that people are condemned to inactivity." Indeed, some of them even abandoned study for moneymaking, for trade, and some simply despaired and took to drink.

But what made this inactivity possible? The local inhabitants, after all, did not complain of it; they were lucky if they managed to straighten their backs for the evening. So that putting it more precisely, what the exiles suffered from was change of place, the disruption of their accustomed way of life, the tearing up of roots, the severing of links with other living beings.

The journalist Nikolai Nadezhdin needed only two years of exile to lose his taste for libertarianism and change into an honest servant of the throne. The wild rake Menshikov, exiled to Beryozov in 1727, built a church there, discoursed with the local inhabitants on the vanity of this world, grew a long beard, went around in a peasant smock, and died within two years. What, we may wonder, did Radishchev find so intolerable, so soul-destroying in his comfortable place of exile? Yet when, back in Russia, he was threatened with a second term, he was so terrified that he committed suicide. Pushkin, again, wrote to Zhukovsky from the village of Mikhailovskoye, that earthly paradise in which a man might pray to live forever: "Rescue me [i.e., from banishment—A.S.] even if it means the fortress or the monastery at Solovki!" Nor was this mere talk, because he wrote to the governor also, begging that his exile be commuted to imprisonment.

We who have learned what Solovki is like may now find it incomprehensible: what eccentric impulse, what combination of despair and naïveté, could make the persecuted poet give up Mikhailovskoye and ask to go to the isles of Solovki? . . .

---

renamed, had evidently never acquired the habit of work—so much so that he blistered his hands taking part in the first *subbotnik** and . . . died from the infection.

Here we see that the threat of exile—of mere displacement, of being set down with your feet tied—has a somber power of its own, the power which even ancient potentates understood, and which Ovid long ago experienced.

Emptiness. Helplessness. A life that is no life at all. . . .

■

In the inventory of instruments of oppression which the glorious revolution was to sweep away forever, exile was also of course to be found, somewhere around fourth place.

But the revolution had scarcely taken its first steps on legs still infirm, it was still in its infancy, when it realized that exile was indispensable. Exile did not exist in Russia for a year or so, perhaps three years. But very shortly what are now called deportations began—the export of undesirables. Let me quote verbatim the words of a national hero, later a Marshal of the Soviet Union, about the year 1921 in the province of Tambov. "It was decided to organize large-scale deportation of the families of bandits [i.e., "partisans"—A.S.]. Extensive *concentration camps* were organized, in which these families could be confined while waiting." [My emphasis—A.S.][7]

Only the fact that it was so convenient to shoot people on the spot rather than carry them elsewhere, guard them and feed them on the way, resettle them and go on guarding them—only this delayed the introduction of a regular exile system until the end of War Communism.* Soon after that, on October 16, 1922, a permanent Exile Commission was set up under the Commissariat of the Interior to deal with "socially dangerous persons and active members of anti-Soviet parties"—meaning all parties except the Bolsheviks—with a budgetary period of three years.[8] Thus even in the early twenties exile was a familiar and smoothly operating institution.

True, exile as a punishment for criminals was not revived: the Corrective Labor Camps had already been invented, and swallowed the lot. Political exile, on the other hand, became more

7. Tukhachevsky, "Borba s Kontrrevolyutsionnymi Vostaniyami" ("The Struggle Against Counter-Revolutionary Revolts"), in *Voina i Revolyutsiya (War and Revolution)*, 1926, No. 7/8, p. 10.

8. *Sobraniye Uzakonenii RSFSR (Collection of Decrees of the R.S.F.S.R.)*, 1922, No. 65, p. 844.

convenient than ever before: in the absence of opposition newspapers, banishment no longer received publicity, and for those who were near to the exiles or knew them closely, a three-year sentence of banishment passed neither in haste nor in anger looked like a sentimental educational measure after the shootings of War Communism.

However, people did not return to their native places from this almost apologetic prophylactic banishment, or if they did, they were quickly picked up again. Those drawn in began circling around the Archipelago, and the last broken arc invariably ended in a ditch.

Because people are so complacently gullible, the regime's intentions dawned on them slowly: the regime was simply not strong enough yet to eradicate all the unwanted at once. So for the time being they were uprooted not from life itself, but from the memory of their fellows.

What made it all the easier to re-establish the exile system was that the old transportation roads had not become blocked, or fallen in, while the places of exile in Siberia, around Archangel, around Vologda, had not changed in the slightest and were not a bit surprised. (Nor would statesmanlike thought stop there: someone's finger would slide over a map of a sixth part of the earth's land surface, and spacious Kazakhstan, the moment it adhered to the Union of Republics, would fit neatly into the exile system with its great expanses, while even in Siberia itself many suitably God-forsaken places awaited discovery.)

There was, however, in the exile system one residual snag: the parasitical attitude of the exiles, who thought the state had an obligation to feed them. The Tsarist government did not dare to try compelling the exiles to increase the national product. And professional revolutionaries considered it beneath them to work. In Yakutya an exile settler had the right to 15 desyatins of land (65 times as much as a kolkhoznik now). If the revolutionaries did not fling themselves into the task of cultivation, the Yakuts were very attached to the land and "bought them off," leasing the land and paying with produce and the use of horses. Thus the revolutionary arrived with empty hands and immediately became a creditor of the Yakuts.[9] Apart from this, the Tsarist government paid its ex-

9. F. Kon, *op. cit.*

iled political enemy 12 rubles a month subsistence money, and 22 rubles a year clothing money. Lepeshinsky writes[10] that Lenin, too, received (and did not refuse) the 12 rubles a month, while Lepeshinsky himself got 16 rubles, because he was not just an exile but an exiled civil servant. F. Kon now assures us that this money was very much too little: but we know that prices in Siberia were one-half or one-third those in Russia, so that the state's maintenance allowance for exiles was in fact overgenerous. It enabled Lenin, for instance, to spend three whole years comfortably studying the theory of revolution, not worrying at all about the source of his livelihood. Martov writes that he paid his landlord 5 rubles a month for accommodation with full board, and spent the rest on books or put it aside for his escape. The Anarchist A. P. Ulanovsky says that in exile (in the Turukhan region, where he was with Stalin) he had money to spare for the first time in his life. He used to send part of it to a free girl whom he had met somewhere on the way, and could afford to sample cocoa for the first time. Elk meat and sturgeon could be had for nothing out there, and a good solid house cost 12 rubles (one month's subsistence money!). None of the politicals wanted for anything and all administrative exiles received a money allowance. And they were all well clad (they arrived in warm clothes).

True, exiles settled for life (the *bytoviki* of Soviet times) did not receive a cash allowance, but sheepskin coats, all their clothing and boots, came to them gratis from public funds. Chekhov established that on Sakhalin settlers received their basic needs from the state in kind, without payment, the men for two or three years, the women for the whole period of their sentence, and that this included 40 zolotniks (i.e., 200 grams) of meat a day, and three pounds (i.e., "a kilo two hundred") of bakery bread. As much as the Stakhanovites* in our Vorkuta mines got for producing 50 percent above the norm. (True, Chekhov finds that the bread is made from poor flour and underbaked, but it's no better in the camps!) They were issued with one sheepskin coat, a cloth coat, and several pairs of boots. Another of the Tsarist government's practices was to pay the exiles deliberately inflated prices for their wares to encourage them to produce. (Chekhov came to the conclusion that instead of the colony of Sakhalin bringing

10. Lepeshinsky, *Na Perelome (At the Turning Point).*

Russia any profit, Russia fed its colony.)

Needless to say, our Soviet political exile system could not rest on such unsound foundations. In 1928 the Second All-Russian Congress of Administrative Workers recognized that the existing system was unsatisfactory and petitioned for the "organization of places of exile in the form of *colonies* in remote and isolated localities, and also the introduction of a system of unspecified [i.e., indefinite] sentences."[11] From 1929 they started elaborating a system of exile in conjunction with forced labor.[12]

"He that does not work shall not eat"—that is the socialist principle. And the Soviet exile system could only be constructed on the basis of this socialist principle. But socialists were just the people who were used to receiving their food without payment in exile! Because the Soviet government did not dare to break this tradition immediately, its treasury also began paying political exiles—only of course not all of them, not the KR's (Counter-Revolutionaries), but the (socialist) "polits," making some gradations among them: for instance, in Chimkent in 1927, SR's and SD's* were given 6 rubles a month, and Trotskyites 30 (they were still comrades, after all, still Bolsheviks). Only these were no longer Tsarist rubles; the rent for the smallest rooms was 10 rubles a month, and 20 kopecks a day bought only meager fare. As time went by things were harder. By 1933 they were paying the "polits" a grant of 6 rubles 25 a month. But in that year, as I myself very well remember, a kilogram of half-baked "commercial" rye bread (off the ration) cost 3 rubles. There was one thing left for the socialists—not language teaching, not writing theoretical works, but *back-breaking work*. From those who did go to work the GPU immediately took away whatever miserable allowance they were still receiving.

However, even if they were willing to work it was not so easy for exiles to earn a wage. The end of the twenties is of course memorable as a period of high unemployment in our country. Getting a job was the prerogative of persons with unblemished records and trade union cards, and exiles could not compete by exhibiting their education and experience. They also had the local police headquarters hanging menacingly over them—without its permission no employer would ever dare take on an exile. (Even

11. TsGAOR (Central State Archives of the October Revolution), 4042/38/8, pp. 34–35.
12. TsGAOR, 393/84/4, p. 97.

a former exile had little hope of a good job: the brand in his passport was an impediment.) In Kazan in 1934, as P. S——va remembers, a group of educated exiles in desperation hired themselves out as roadmakers. They were rebuked by police headquarters: what was the meaning of this demonstration? But no one offered to help them find other work, and Grigory B. paid the security officer in his own coin. "Haven't you got a *nice little trial* coming up? We could sign on as paid witnesses."

They had to count every crumb twice. This was what the Russian political exile system had come to! No time left for debate or for writing protests against the "Credo." One worry they were free from was how to cope with senseless idleness. . . . Their one concern now was: how to avoid dying of hunger. And how not to let themselves down by becoming stoolies.

In the first Soviet years in a land freed at last from age-old slavery, the pride and independence of political exiles sagged like a punctured balloon. It turned out that the strength which the previous regime had uneasily acknowledged in political prisoners was imaginary. What had created and maintained that strength was only *public opinion* in the country. Once public opinion was supplanted by *organized opinion,* the exiles with their protests and their rights were helplessly subjected to the tyranny of stupid, flustered GPU men and inhuman secret instructions. (Dzerzhinsky lived long enough to set his hand to the first of these instructions.) A single hoarse cry, one strangled word, to remind the free of your existence was now impossible: if an exiled worker sent a letter to his old factory, the worker who read it out (as Vasily Kirillovich Yegoshin did in Leningrad) was immediately exiled himself. Exiles now lost not only monetary allowances, the means of existence, but all rights whatsoever; the GPU could get at them to prolong their banishment, arrest them, or transport them even more easily than when they were supposed to be free, uninhibitedly treat them like rubber dolls rather than people.[13] Nothing was easier than to shake them up by means such as those once used in Chimkent: it was announced that the exile colony was to be wound up within twenty-four hours. In twenty-four hours

13. Those Western socialists who waited till 1967 to feel "ashamed of being socialists side by side with the Soviet Union" could very well have come to that conclusion some forty or forty-five years earlier. At that time Russian Communists were already destroying Russian socialists. But nobody groans when another man's tooth aches.

the exiles had to: hand over any business in which they were engaged, pull down their houses, get rid of their household goods, make their preparations for the journey, and set off by a prescribed route. It was not much easier than being a transported prisoner! Nor was the exile's future much more certain!

But forgetting for a moment the silence of the public and the heavy pressures of the GPU—what were the exiles themselves like? These Party members, so called, without parties? I am not thinking of the Cadets—there were no Cadets left; they had all been done to death—but asking what it meant by 1927 or 1930 to be nominally an SR or a Menshevik. Nowhere in the country were there any active groups corresponding to these names. For a long time now, ever since the Revolution, in ten storm-fraught years their programs had not been re-examined, and even if these parties had suddenly risen from the dead, nobody knew how they would interpret all that had happened and what they would propose. For a long time now the whole press had mentioned them only in the past tense, and surviving members lived with their families, worked at their jobs, and thought no more about their parties. But names once inscribed in the tablets of the GPU are indelible. At a sudden nocturnal signal these scattered rabbits were plucked from their burrows and transported via transit jails to, for instance, Bukhara.

Thus, I. V. Stolyarov arrived there in 1930 to meet aging SR's and SD's gathered in from all corners of the country. Rudely snatched from the lives they were used to, they had nothing better to do than begin arguing and assessing the current political situation and speculating what course history would have taken if only . . . and if only . . .

So their enemies knocked together from them not, of course, a political party, but a target for sinking.

Exiles of the twenties recall that the only live and militant party at that time was the Zionist Socialist Party with its vigorous youth organization, Hashomer, and its legal "Hehalutz" organization, which existed to establish Jewish agrarian communes in the Crimea. In 1926 the whole Central Committee was jailed, and in 1927 indomitably cheerful boys and girls of fifteen, sixteen, and under were taken from the Crimea and exiled. They were sent to Turtkul and other strict places. This really was a party—close-knit, determined, sure that its cause was just. Their aim, however, was not one which all could share, but private and particular: to live as a nation, in a Palestine of their own. The Communist Party, which had voluntarily disowned its fatherland, could not tolerate narrow nationalism in others. . . .[14]

Mutual aid was still practiced among the exiles until the beginning of the thirties. Thus, SR, SD, and Anarchist exiles in Chimkent, where work was easy to find, set up a secret mutual aid fund to help their workless "Northern" party comrades. Communal arrangements for the preparation of food and the care of children, and all the gatherings and exchanges of visits which naturally go with this, were still found in places. They still joined together to celebrate May Day, there in exile (while demonstratively refusing to observe the anniversary of the October Revolution). But as the thirties reached their climax, all this would disappear: everywhere the buzzard's eye of the secret police would fix on exiles meeting in groups. Prisoners would begin to shun one another in case the NKVD suspected them of "organizing" and *picked them up "on a new charge"* (this fate awaited them anyhow). Thus in the pale of exile marked out by the state they would withdraw into a second, voluntary exile—into solitude. (This for the present would be just what Stalin wanted of them.)

The exiles were further weakened by the estrangement of the local population: the local people were persecuted for showing friendship to exiles, offenders were themselves exiled to other places, and youngsters expelled from the Komsomol.

They were still further weakened by the unfriendly relations between parties which developed in the Soviet period and became particularly acute from the mid-twenties, when large numbers of Trotskyites, who acknowledged no one but themselves as politicals, suddenly appeared in exile.

But it was of course not only socialists who were kept in exile in the twenties; most of the exiles, in fact (and this was truer with every year that passed), were not socialists at all. There was an influx of non-party intellectuals—those spiritually independent

14. It might be supposed that this natural and noble aspiration of the Zionists to recreate the land of their ancestors, to confirm the faith of their ancestors, and reassemble there after three thousand years of diaspora would elicit the united support and aid at least of the European peoples. Admittedly a national home in the Crimea was not the Zionist idea in its purest form, and perhaps it was Stalin's joke to invite this Mediterranean people to adopt Birobidian, on the edge of the taiga, as their second Palestine. He was a great master of slow and secret scheming, and perhaps this kind invitation was a rehearsal for the exile which he would have in mind for them in 1953.

people who were making it difficult for the new regime to establish itself firmly. And of *former people* (i.e., members of the pre-Revolutionary establishment) not destroyed in the Civil War. And even of young boys—"for fox-trotting."[15] And spiritualists. And occultists. And clerics—at first with the right to conduct services in exile. And simply believers, simply Christians or *krestyane* (peasants) to use the word as modified by the Russians many centuries ago.

They all came under the eye of the same security police sector, were divided from each other, and grew numb.

Robbed of all strength by the indifference of the country at large, the exiles lost even the will to escape. For exiles in Tsarist times escaping was a merry sport: think of Stalin's five, or Nogin's six escapes. The threat hanging over them was not a bullet, not *katorga*, but merely being resettled in their old place of exile after a diverting journey. But the MVD, as it grew more stupid and heavy-handed, from the mid-twenties imposed collective responsibility on exiles belonging to the same party: if one of you runs away all his party comrades will answer for him! By now the air was so stifling, the pressures so overbearing, that the socialists, so recently proud and indomitable, accepted this responsibility! They *themselves*, in party resolutions, *forbade themselves to escape!*

Where, in any case, could they run? To whom could they run? Was there a *people* to whom they might run?

Specialists in theoretical rationalization deftly put foundations under the decision: *This is not the right time to run away; we must wait.* In fact, it was *not the right time* to fight at all; again they must wait. At the beginning of the thirties, N. Y. Mandelstam noted that socialist exiles in Cherdynsk utterly refused to resist. Even when they sensed that their destruction was inevitable. Their only practical hope was an extension of their term of banishment, so long as they were not arrested, and were allowed to *sign on* again there, on the spot—then at least their modestly comfortable existence would not be ruined. Their only moral aim was to preserve their human dignity in the face of destruction.

It is strange for us, after *katorga*, where we changed from crushed and isolated individuals into a powerful whole, to learn that these socialists, once an articulated whole, tried and proved in action, disintegrated and became helpless units. But in the last

15. Siberia, in 1926. Vitkovsky's testimony.

two decades our life as social beings has steadily expanded, filled out. We are inhaling. Then it was just as steadily contracting, shrinking under pressure: we were exhaling.

So that it is unseemly for our age to judge that other.

There were, moreover, many different degrees of severity in the treatment of exiles, and this, too, disunited and weakened them. The intervals at which identity cards were exchanged varied (for some it was monthly, and the formalities were excruciatingly slow). Unless you didn't care about sinking to the lowest category, you had to observe the rules.

The mildest form of banishment also survived until the beginning of the thirties: this was not exile, but *"free residence . . . minus."* In this case the person penalized (the *minusnik*) was not told to live in one specified place, but allowed to choose any town, with certain exceptions, or "minuses." Once having chosen, he was tied to that place for the same three-year period. The *minusnik* did not have to report regularly to the GPU, but did not have the right to leave the area. In years of widespread unemployment, labor exchanges would not give a *minusnik* work: if he contrived to find himself a job, pressure was put on the management to dismiss him.

The *minus* pinned the dangerous insect down, and it submissively waited its turn to be arrested properly.

Then again, people still believed in a progressive order which could not and would not need to exile citizens! Believed in an amnesty, especially as the splendid tenth anniversary of October drew near! . . .

The amnesty came; the amnesty . . . struck. They started reducing terms of banishment by a quarter (three-year sentences by nine months), and that not for everyone. But since the Great Game of Patience was being laid out, and the three years of exile were followed by three years in a political isolation prison, followed in turn by another three years in exile, this acceleration of the process by three months did nothing to make life more beautiful.

Then it would be time for the next trial. The anarchist Dmitri Venediktov, toward the end of his three-year exile in Tobolsk (1937), was arrested, and the charge against him was categorical and precise: "disseminating rumors about loans" (What can they have meant by rumors, when the loan comes around each year as infallibly as the flowers in May?) "and dissatisfaction with the

Soviet regime" (an exile, of course, should be content with his lot!). And what happened to him next for these obscene crimes? He was sentenced to be shot within seventy-two hours, with no right of appeal! (His surviving daughter, Galina, has already made a brief appearance in the pages of this book.)

We have seen what exile meant in the first years of newly won freedom, and seen, too, the only way of escaping completely.

Exile was a temporary pen to hold sheep marked for slaughter. Exiles in the first Soviet decades were not meant to settle but to await the summons—elsewhere. There were clever people—"former" people, and also some simple peasants—who already realized in the twenties all that lay before them. And when they reached the end of their first three-year term they stayed exactly where they were—in Archangel, for instance—just in case. Sometimes this helped them not to be caught under the nit comb again.

This was what exile had become in our time, instead of the peace and quiet of Shushenskoye, or cocoa drinking in Turukhan.

You will see that we had heavier burdens than Ovid's homesickness to bear.

Chapter 2

■

The Peasant Plague

This chapter will deal with a small matter. Fifteen million souls. Fifteen million lives.

They weren't educated people, of course. They couldn't play the violin. They didn't know who Meyerhold was, or how interesting it is to be a nuclear physicist.

In the First World War we lost in all three million killed. In the Second we lost twenty million (so Khrushchev said; according to Stalin it was only seven million. Was Nikita being too generous? Or couldn't Iosif keep track of his capital?). All those odes! All those obelisks and eternal flames! Those novels and poems! For a quarter of a century all Soviet literature has been drunk on that blood!

But about the silent, treacherous Plague which starved fifteen million of our peasants to death, choosing its victims carefully and destroying the backbone and mainstay of the Russian people—about that Plague there are no books. No bugles bid our hearts beat faster for them. Not even the traditional three stones mark the crossroads where they went in creaking carts to their doom. Our finest humanists, so sensitive to today's injustices, in those years only nodded approvingly: Quite right, too! Just what they deserve!

It was all kept so dark, every stain so carefully scratched out, every whisper so swiftly choked, that whereas I now have to refuse kind offers of material on the camps—"No more, my friends, I have masses of such stories, I don't know where to put them!"—nobody brings me a thing about the deported peasants. Who is the person that could tell us about them? Where is he?

I know, all too well, that what is wanted here is not a chapter, nor even a book by one single man. And I cannot document even one chapter thoroughly.

All the same, I shall make a beginning. Set my chapter down as a marker, like those first stones—simply to mark the place where the new Temple of Christ the Saviour will someday be raised.

Where did it all start? With the dogma that the peasantry is *petit bourgeois?* (And who in the eyes of these people is not petit bourgeois? In their wonderfully clear-cut scheme, apart from factory workers [not the skilled workers, though] and big-shot businessmen, all the rest, the whole people—peasants, office workers, actors, airmen, professors, students, doctors—are nothing but the "petite bourgeoisie.") Or did it start with a criminal scheme in high places to rob some and terrorize the rest?

From the last letters which Korolenko wrote to Gorky in 1921, just before the former died and the latter emigrated, we learn that this villainous assault on the peasantry had begun even then, and was taking almost the same form as in 1930.

But as yet their strength did not equal their impudence, and they backed down.

The plan, however, remained in their heads, and all through the twenties they bullied and prodded and taunted: "Kulak! Kulak! Kulak!" The thought that it was impossible to live in the same world as the kulak was gradually built up in the minds of townspeople.

The devastating peasant Plague began, as far as we can judge, in 1929—the compilation of murder lists, the confiscations, the deportations. But only at the beginning of 1930 (after rehearsals were complete, and necessary adjustments made) was the public allowed to learn what was happening—in the decision of the Central Committee of the Party dated January 5. (The Party is "justified in shifting from a policy of restricting the exploiting tendencies of the kulaks to a policy of liquidating the kulaks as a class." And the admission of kulaks to the kolkhoz was immediately . . . prohibited. Would anyone like to attempt a coherent explanation?)

The dutifully concurring Central Executive Committee of the Soviets and the Council of People's Commissars were not far behind the Central Committee, and on February 1, 1930, they gave legislative form to the will of the Party. Provincial Executive Committees were required to "use all necessary methods in the struggle with the kulaks, up to and including [in reality no other method was used] complete confiscation of the property of kulaks and their removal to points beyond the boundaries of certain regions and provinces."

Only in those last words was the Butcher overcome with shame. He specified *from which* boundaries. But he did not say *to which*. If you were inattentive enough you could take it to mean thirty versts away, in the same neighborhood.

In the Vanguard Doctrine, as far as I know, there was no such person as the *henchman* of kulaks. But as soon as the Party put its hand to the mowing machine, there was obviously no doing without him. We have seen already what the word is worth. A "sack collection" was announced, Young Pioneers went from hut to hut collecting from the peasants on behalf of the indigent state, you wouldn't give up your sack because it was like parting with your lifeblood (there were none in the shops, of course), and there you were, a "henchman of the kulaks." Ripe for deportation.

Names like this rampaged through a Soviet Russia with the bloody exhalations of the Civil War still warm in its nostrils. Words were put into circulation, and although they meant nothing they were easily remembered, they simplified matters, they made thought completely unnecessary. The savage law of the Civil War (Ten for every one! A hundred for every one!) was reinforced —to my mind an un-Russian law: where will you find anything like it in Russian history? For every activist (which usually meant big-mouthed loafer: A. Y. Olenyev is not the only one to recall that thieves and drunkards were in charge of "dekulakization")— for every *activist* killed in self-defense, hundreds of the most industrious, enterprising, and level-headed peasants, those who should keep the Russian nation on an even keel, were eliminated.

Yells of indignation! What's that? What do you say? What about the *bloodsuckers*? Those who squeezed their neighbors dry? "Take your loan—and pay me back with your hide"?

I suppose that bloodsuckers were a small part of the whole number (but were all the bloodsuckers there among them?). And were they bloodsuckers born? we may ask. Bloodsuckers through and through? Or was it just that all wealth—and all power— corrupts human beings? If only the "cleansing" of mankind, or of a social estate, were so simple! But if they had "cleansed" the peasantry of heartless bloodsuckers with their fine-toothed iron comb, cheerfully sacrificing fifteen millions for the purpose—whence all those vicious, fat-bellied rednecks who preside over collectivized villages (and District Party Committees) today? Those pitiless oppressors of lonely old women and all defenseless people? How was the root of this predatory weed missed during dekulakization? Surely, heaven help us, they can't have sprung from the *activists?* . . .

He who grew up robbing banks could not think about the peasantry either as a brother or as a husbandman. He could only whistle like Nightingale the Robber, and millions of toiling peasants were dragged off to the taiga, horny-handed tillers of the soil, the very same who had set up the Soviet power simply to get land, and having obtained it, quickly tightened their grip on it. ("The land belongs to those who work it.")

The word "bloodsuckers" loses all resonance—the tongue that uses it is a clapper in a wooden bell—when we remember what a *clean sweep* they made of some villages in the Kuban, Urupin-skaya for instance: they deported every soul in it, from babes in arms to aged men (and resettled it with demobilized soldiers). Here we see clearly what the "class principle" really meant. (Let us remember that the Kuban gave hardly any support to the Whites in the Civil War, began of its own accord to wreck Denikin's supply lines, and sought agreement with the Reds. Then, suddenly, there they were—"the saboteurs of the Kuban.") The village of Dolinka, renowned throughout the Archipelago as a prosperous agricultural center—where did it come from? *All* its inhabitants were Germans, "dekulakized" and deported in 1929. Who had been exploiting whom is a mystery.

The principle underlying dekulakization can also be clearly seen in the fate of the children. Take Shurka Dmitriyev, from the village of Masleno (Selishchenskie Kazarmy, near the Volkhov). He was thirteen when his father, Fyodor, died in 1925, and the only son in a family of girls. Who was to manage his father's holding? Shurka took it on. The girls and his mother accepted him as head of the family. A working peasant and an adult now, he exchanged bows with other adults in the street. He was a worthy successor to his hard-working father, and when 1929 came his bins were full of grain. Obviously a kulak! The whole family was driven out!

Adamova-Sliozberg has a moving story about meeting a girl

called Motya, who was jailed in 1936 for leaving her place of banishment without permission to go to her native village, Svet-lovidovo near Tarussa, *two thousand kilometers on foot!* Sports-men are given medals for that sort of thing. She had been exiled with her parents in 1929 when she was a little schoolgirl, and deprived of schooling forever. Her teacher's pet name for her was "Motya, our little Edison": the child was not only an excellent pupil, but had an inventive turn of mind, had rigged up a sort of turbine worked by a stream, and invented other things for the school. After seven years she felt an urge to look just once more at the log walls of her unattainable school—and for that "little Edison" went to prison and then to a camp.

Did any child suffer such a fate in the nineteenth century?

Every *miller* was automatically a candidate for dekulakization —and what were millers and blacksmiths but the Russian village's best technicians? Take the miller Prokop Ivanovich Laktyunkin from the Ryazan region. No sooner was he dekulakized than they ground the millstones together too hard and burned the mill down. After the war he was pardoned and returned to his native village, but he could not reconcile himself to the fact that there was no mill. Laktyunkin obtained permission, cast the grinders himself, and set up a mill on the same spot (it had to be the same spot!), not for his own profit, but for the kolkhoz, or rather because the neighborhood was incomplete and less beautiful without it.

Now let us look at that other kulak, the village blacksmith. In fact, we'll start with his father, as Personnel Departments like to do. His father, Gordei Vasilyevich, served for twenty-five years in the Warsaw garrison, and earned enough silver to make a tin button: this soldier with twenty-five years' service was denied a plot of land. He had married a soldier's daughter while he was in the garrison, and after his discharge he went to his wife's native place, the village of Barsuki in the Krasnensky district. The village got him tipsy, and he paid off its tax arrears with half of his savings. With the other half he leased a mill from a landowner, but quickly lost the rest of his money in this venture. He spent his long old age as a herdsman and watchman. He had six daughters, all of whom he gave in marriage to poor men, and an only son, Trifon (their family name was Tvardovsky). The boy was sent away to serve in a haberdasher's shop, but fled back to Barsuki and found employment with the Molchanovs, who had the forge. After a year as an unpaid laborer, and four years as an apprentice, he became a smith himself, built a wooden house in the village of Zagorye, and married. Seven children were born (among them Aleksandr, the poet), and no one is likely to get rich from a forge. The oldest son, Konstantin, helped his father. If they smelted and hammered from one dawn to the next they could make five excellent steel axes, but the smiths of Roslavl, with their presses and their hired workmen, undercut their price. In 1929 their forge was still wood-built, they had only one horse, sometimes they had a cow and a calf, sometimes neither cow nor calf, and besides all this they had eight apple trees—you can see what bloodsuckers they were. . . . The Peasant Land Bank used to sell mortgaged estates on deferred payments. Trifon Tvardovsky had taken eleven desyatins of wasteland, all overgrown with bushes, and the year of the Plague found them still sweating and straining to clear it: they had brought five desyatins into cultivation, and the rest they abandoned to the bushes. The collectivizers marked them down for dekulakization—there were only fifteen households in the village and somebody had to be found. They assessed the income from the forge at a fantastic figure, imposed a tax beyond the family's means, and when it was not paid on time: Get ready to move, you damned kulaks, you!

If a man had a brick house in a row of log cabins, or two stories in a row of one-story houses—there was your kulak: Get ready, you bastard, you've got sixty minutes! There aren't supposed to be any brick houses in the Russian village, there aren't supposed to be two-story houses! Back to the cave! You don't need a chimney for your fire! This is our great plan for transforming the country: history has never seen the like of it.

But we still have not reached the innermost secret. The better off were sometimes left where they were, provided they joined the kolkhoz quickly, while the obstinate poor peasant who failed to apply was deported.

This is very important, the most important thing. The point of it all was not to dekulakize, but to force the peasants into the kolkhoz. Without frightening them to death there was no way of taking back the land which the Revolution had given them, and planting them on that same land as serfs.

It was a second Civil War—this time against the peasants. It was indeed the Great Turning Point, or as the phrase had it, the Great Break. Only we are never told what it was that broke.

It was the backbone of Russia.

No, we have been unfair to socialist realist writers: they have described the dekulakization, described it very fluently, too, and with great feeling for its heroes, as though they were hunters of snarling wolves.

But there are no descriptions of the long village street with every house in the row boarded up. Or of how you could walk through a village and see on the steps of a peasant house a dead woman with a dead child in her lap. Or an old man sitting under a fence, who asked you for bread—and when you walked back he had collapsed and died.

Nor shall we read in their works scenes like this: The chairman of the village soviet, taking the schoolteacher as a witness, goes into a hut where an old man and an old woman are lying on the sleeping bench. (The old man used to keep a teahouse—obviously a bloodsucker: who says wayfarers are glad of hot tea?) He brandishes his revolver. "Get down, you Tambov wolf!" The old woman starts howling, and the chairman fires at the ceiling, to intimidate them still more (a gun makes a very loud noise in a peasant hut). Both these old people died on their journey.

Still less will you read about this method of dekulakization: All the Cossacks (we are in a village on the Don) were summoned to a "meeting"—and there surrounded with machine guns, arrested, and driven away. Deporting their women later on was simplicity itself.

We can find described in books, or even see in films, barns and pits in the ground, full of grain hoarded by bloodsuckers. What they won't show us is the handful of belongings earned in a lifetime of toil: the livestock, the utensils—things as close to the owner as her own skin—which a weeping peasant woman is ordered to leave forever. (If some of the family survive, and are clever and persistent enough to persuade Moscow to rehabilitate them as "middle peasants," they will not find a stick of their "medium" property left when they return. It will have been pillaged by the *activists* and their women.)

What they will not show us are the little bundles with which the family are allowed onto the state's cart. We shall not learn that in the Tvardovsky house, when the evil moment came there was neither suet nor bread; their neighbor Kuzma saved them: he had several children and was far from rich himself, but brought them food for the journey.

Those who were quick enough fled from that Plague to the towns, sometimes with a horse—but there were no customers for horses in those times: that peasant horse, sure sign of a kulak, became as bad as the Plague itself. Its master would tie it to a hitching rail at the horse market, give it one last pat on the muzzle, and go away before anyone noticed.

The years 1929–1930 are generally regarded as the years of the Plague. But its deathly stink hung over the countryside long afterward. In the Kuban in 1932, all the wheat and rye to the very last grain was taken straight from the threshing machine to the state procurement point, and when the collective farmers, who had been given nothing to eat except their harvesters' and threshers' rations, found that hot dinners ended with the threshing, and that there was not a grain to come for their labor—how could the mobs of howling women be silenced? *"How many kulaks are there left in this place?"* Who should be deported? (Skripnikova's testimony enables us to judge the condition of the kulak-free countryside in the early days of the kolkhoz: she remembers some peasant women in 1930 sending parcels of dried crusts home to their native village *from Solovki!!*)

Here is the story of Timofey Pavlovich Ovchinnikov, born in 1886, from the village of Kishkino in the Mikhnevsko area (not far from Leninskiye Gorki, near the great highroad). He fought in the German war, he fought in the Civil War. When he had finished fighting, he returned to land given by the Decree, and married. He was clever, literate, experienced, an excellent worker. He had also acquired, self-taught, some skill in veterinary medicine, and gave a helping hand to the whole district. By tireless work he built himself a good house, planted an orchard, and reared a colt to become a fine horse. But the NEP confused him, and Timofey Pavlovich was rash enough to believe in it as he had believed in the Land Decree. In partnership with another peasant he started a little business making cheap sausages. (Now that the village has been without sausage for forty years, you may scratch your head and wonder what was so bad about that.) They made the sausage themselves, used no hired labor, and indeed sold their products through a cooperative. When they had worked at it for just two years, 1925–1927, crippling tax demands were made on them, since they allegedly earned large sums (these were dreamed up by tax inspectors in the line of duty, but envious drones in the village, incapable of making anything of themselves except *activists,* also blew into the taxman's ear). So the partners closed down. In 1929 Timofey was one of the first to join the kolkhoz, taking with him his good horse, his cow, and all his implements. He worked hard in the common fields, and also reared two steers for the kolkhoz. The kolkhoz began to collapse, and many walked or ran away from it, but Timofey had five children, and was stuck. Since the tax inspector couldn't forget old scores, and Timofey was still thought to be well off (partly because of the veterinary help he had given people), even now that he was in the kolkhoz they pursued him with tax demands, which he could not meet. When he had no money to pay, they started coming to his house and seizing bits of clothing: once his eleven-year-old son managed to drive their last three ewes off and hide them from the inventory takers, but on another occasion they, too, were taken. When they came yet again to list his belongings, the impoverished family had nothing else left, so the shameless tax inspectors put down the rubber plants and their tubs. This was more than Timofey could stand, and he hacked the plants to pieces before their very eyes. Now just consider the significance of his action. (1) He had destroyed property no longer belonging to him but to the state. (2) He had made use of an ax in a demonstration against the Soviet state. (3) He had sought to discredit the kolkhoz system.

Just then the kolkhoz system in the village of Kishkino was going to pieces fast. Nobody had any faith in it or wanted to work, half the peasants had left, and it was time to make an example of somebody. The hardened Nepman Timofey Ovchinnikov, who had wormed his way into the kolkhoz to wreck it, was now expelled as a kulak by decision of Shokolov, chairman of the village soviet. This was 1932, and mass deportation was over, so his wife and six children (one at the breast) were not deported, but only turned out of house and home. (A year later they made their way to Timofey in Archangel at their own expense. All the Ovchinnikovs had lived to be eighty, but after a life like this Timofey folded up at fifty-three.)[1]

---

1. What follows is not relevant to our immediate theme, but tells us a lot about our epoch. After a time, Timofey found a job in Archangel—in another "closed" sausage factory—where he was again one of two skilled men, this time with a manager over them. His own sausage factory had been closed as a menace to the working classes, whereas this other was "closed" so that the workers would not know of it. They privately supplied the rulers of that northern clime with a variety of expensive sausages. Timofey himself was sometimes sent to deliver their wares to the home of the regional Party secretary, Comrade Austrin

Even in 1935, drunken kolkhoz bosses went around the poverty-stricken village at Easter demanding money for vodka from those peasants who still farmed their own holdings. Give, or "We'll dekulakize you. We'll deport you." They could, too. If you farmed by yourself. This was what the Great Break was all about.

The journey itself, the peasant's *Via Crucis,* is something which our socialist realists do not describe at all. Get them aboard, pack them off—and that's the end of the story. Episode concluded. Three asterisks, please.

They were loaded onto carts . . . if they were lucky enough to be taken in the warm months, but it might be onto sledges in a cruel frost, with children of all ages, babes in arms as well. In February, 1931, when hard frosts were interrupted only by blizzards, the strings of carts rolled endlessly through the village of Kochenevo (Novosibirsk oblast), flanked by convoy troops, emerging from the snowbound steppe and vanishing into the snowbound steppe again. Even going into a peasant hut for a warm-up required special permission from the convoy, which was given only for a few minutes, so as not to hold up the cart train. (Those GPU convoy troops—they're still alive, they're pensioners now! I daresay they remember it all! Or perhaps . . . they can't remember.) They all shuffled into the Narym marshes—and in those insatiable quagmires they all remained. Many of the children had already died a wretched death on the cruel journey.

This was the nub of the plan: the peasant's seed must perish together with the adults. Since Herod was no more, only the Vanguard Doctrine has shown us how to destroy utterly—down to the very babes. Hitler was a mere disciple, but he had all the luck: his murder camps have made him famous, whereas no one has any interest in ours at all.

The peasants knew what was in store for them. And if it was their good fortune to be transported through inhabited places, when they halted they would slip small children not too small to climb through windows. Kind people may help you! Beg your way in the world! It's better than dying with us.

(In Archangel in the famine years of 1932–1933, the destitute children of resettled peasants were not given free school lunches

---

—a detached one-story house behind a high fence where Liebknecht Street meets Chumbarov-Luchinsky Street—and to the NKVD chief in the oblast, Comrade Sheiron.

and clothing vouchers, as were others in need.)

In that convoy of Don Cossacks, when the men arrested at the "meeting" were carried separately from their women, one woman gave birth to a child on the journey. Their rations were one glass of water a day, and 300 grams of bread not every day. Was there a medical attendant? Need you ask? The mother had no milk, and the child died on the way. Where were they to bury it? Two soldiers climbed in for a short trip between two stations, opened the door while the train was moving, and threw the tiny body out.

(This transport was driven to the great Magnitogorsk building operation.* Their husbands were brought to join them. Dig away, house yourselves! From Magnitogorsk on, our bards have done their duty and *reflected* . . . reality?)

The Tvardovsky family were carted only as far as Yelnya, and luckily it was April. There they were loaded into boxcars. The boxcars were locked, and there were no pails, or holes in the floor, for them to relieve themselves. Risking punishment, perhaps even imprisonment, for attempted escape, Konstantin Trifonovich cut a hole in the floor with a kitchen knife, while the train was moving and there was a lot of noise. The feeding arrangements were simple: once every three days pails of soup were brought along at main stations. True, they were only traveling for ten days (to a station called Lyalya in the Northern Urals). It was still winter there, and the transport was met by hundreds of sledges, which carried them up the frozen river into the forest. There they found a hut for twenty loggers, but more than five hundred people had been brought, and it was evening. The Komsomol in charge of the place, a Permian called Sorokin, showed them where to knock pegs into the ground: there'll be a street here, there'll be houses there. This was how the settlement of Parcha was founded.

It is hard to believe in such cruelty: on a winter evening out in the taiga they were told: You've arrived! Can human beings really behave like this? Well, they're moved by day so they arrive at nightfall—that's all there is to it. Hundreds and hundreds of thousands were carried into the wilds and dumped down like this, old men, women, children, and all. On the Kola Peninsula (Appatity) people lived through the dark polar winter in thin tents under the snow. But was it so very much more merciful to take trainloads of Volga Germans in summer (summer, 1931, not 1941 —don't confuse the dates!) to waterless places in the Karaganda steppe, ration their water, and order them to make themselves earth houses? There, too, winter would come soon enough (by the spring of 1932 the children and the old had all died of dysentery and malnutrition). In Karaganda itself, and in Magnitogorsk, they built long, low communal buildings of earth like vegetable storehouses. On the White Sea Canal the new arrivals were housed in huts vacated by prisoners. Those who were sent to work on the Volga Canal, even just beyond Khimki, were unloaded *before* there was a camp, tipped out on the ground as soon as the hydrographic survey was completed, and told to start swinging picks and wheeling barrows. (The papers reported the "delivery of machinery for the canal.") There was no bread. They had to build their earth houses in their spare time. (Nowadays pleasure boats carry Moscow sightseers over this spot. There are bones on the bottom, bones in the ground, bones in the concrete.)

As the Plague approached in 1929, all the churches in Archangel were closed: they were due to be closed anyway, but the very real need for somewhere to put the dekulakized hurried things along. Great streams of deported peasants poured through Archangel, and for a time the whole town became one big transit prison. Many-tiered sleeping platforms were put up in the churches, but there was no heat. Consignment after consignment of human cattle was unloaded at the station, and with dogs barking around them, the bast-shod went sullenly to church and a bed of planks. (S., then a boy, would never forget one peasant walking along with a shaft bow around his neck: he had been hurried away before he could decide what would be most useful. Another man carried a gramophone with a horn. Cameramen—there's work for you in this! . . .) In the Church of the Presentation, an eight-tiered bed platform which was not fastened to the wall collapsed in the night and several families were crushed. Their cries brought troops rushing to the church.

This was how they lived in that plague-stricken winter. They could not wash. Their bodies were covered with festering sores. Spotted fever developed. People were dying. Strict orders were given to the people of Archangel not to help the *special resettlers* (as the deported peasants were now called)! Dying peasants roamed the town, but no one could take a single one of them into his home, feed him, or carry tea out to him: the militia seized local inhabitants who tried to do so and took away their passports. A starving man would stagger along the street, stumble, fall—and die. But even the dead could not be picked up (besides the militia, plainclothesmen went around on the lookout for acts of kindness). At the same time market gardeners and livestock breeders from areas near big towns were also being expelled, whole villages at a time (once again—what about the theory that they were supposed to arrest exploiters only?), and the residents of Archangel themselves dreaded deportation. They were afraid even to stop and look down at a dead body. (There was one lying near GPU headquarters, which no one would remove.)

They were buried in an *organized* fashion: by the sanitation department. Without coffins, of course, in common graves, next to the old city cemetery on Vologda Street—out in open country. No memorials were erected.

And this was while the tillers were still only in transit. There was also a great camp for them beyond the village of Talagi, where some of them were given jobs loading timber. But one man contrived to write a letter abroad on a log (see what happens if you teach peasants to read and write!), and they were all taken off the job. Their path was still a long one—to Onega, Pinega, and up the river Dvina.

We had a joke in the camp: "They can't send you farther than the sun." But these peasants were sent farther, to a place where there would long be no shelter for a tallow dip.

The plight of these peasants differed from that of all previous and subsequent Soviet exiles in that they were banished not to a center of population, a place made habitable, but to the haunt of wild beasts, into the wilderness, to man's primitive condition. No, worse: even in their primeval state our forebears at least chose places near water for their settlements. For as long as mankind has existed no one has ever made his home elsewhere. But for the *special settlements* the Cheka (not the peasants themselves—they had no right of choice) chose places on stony hillsides (100 meters up above the river Pinega, where it was impossible to dig down to water, and nothing would grow in the soil). Three or four kilometers off there might be convenient water meadows—but no, according to instructions no one was supposed to settle there. So the hayfields were dozens of kilometers away from the settlement, and the hay had to be brought in by boat. Sometimes settlers were bluntly *forbidden to sow grain crops.* (What they should grow was also determined by the Cheka!) Yet another thing we town folk do not understand—what it means to have lived from time immemorial with animals. A peasant's life is nothing without animals—and here he was condemned for many years never to hear neighing or lowing or bleating; never to saddle, never to milk, never to fill a trough.

On the river Chulym in Siberia, the special settlement of Kuban Cossacks was encircled with barbed wire and towers were put up, as though it were a prison camp.

Everything necessary seemed to have been done to ensure that these odious work fiends should die off quickly and rid our country of themselves and of bread. Indeed, many such special settlements died off to a man. Where they once stood, chance wayfarers are gradually burning what is left of the huts, and kicking the skulls out of sight.

No Genghis Khan ever destroyed so many peasants as our glorious Organs, under the leadership of the Party.

Take, for instance, the Vasyugan tragedy. In 1930, 10,000 families (60,000–70,000 people, as families then went) passed through Tomsk and from there were driven farther, at first on foot, down the Tom although it was winter, then along the Ob, then upstream along the Vasyugan—still over the ice. (The inhabitants of villages on the route were ordered out afterward to pick up the bodies of adults and children.) In the upper reaches of the Vasyugan and the Tara they were marooned on patches of firm ground in the marshes. *No food or tools were left for them.* The roads were impassable, and there was no way through to the world outside, except for two brushwood paths, one toward Tobolsk and one toward the Ob. Machine-gunners manned barriers on both paths and let no one through from the death camp. They started dying like flies. Desperate people came out to the barriers begging to be let through, and were shot on the spot. Rather late in the day, when the rivers unfroze, barges carrying flour and salt were sent from the Tomsk Integralsoyuz (Producers' and Consumers' Cooperative), but they could not get up the Vasyugan. (Stanislavov was the Integralsoyuz agent in charge of the shipment, and it is from him that we know this.)

They died off—every one of them.

We are told that there was at least an inquiry into this business, and even that one man was shot. I am not much inclined to believe it. But even if it is so—the ratio is an acceptable one! The ratio with which we are familiar from the Civil War. For one of ours —a thousand of yours! For sixty thousand of yours—one of ours!

There's no other way to build the New Society.

■

And yet—exiles survived! Under their conditions it seems incredible—but live they did.

In the settlement at Parcha the day was started by foremen (Komi-Zyrians) with sticks. All their lives these peasants had begun their day themselves, but now they were driven out with sticks to fell and raft timber. By giving them no chance to get dry for months on end, by cutting down the flour ration, the masters exacted their stint from them—and in the evenings they could get on with homemaking. Their clothes fell to pieces on their bodies, and they wore sacks like skirts, or else stitched trousers from them.

If they had all died off, a number of towns we know today would not exist. Igarka, for instance. The building of Igarka, from 1929 on, was carried out and completed by—whom? The Northern and Polar Timber Trust? I wonder. Or was it perhaps dekulakized peasants? They lived in tents at 50 degrees below—but they made possible the first timber exports from the area as early as 1930.

The former kulaks lived in their *special* settlements like zeks in maximum punishment camps. Although there was no boundary fence, there was usually a man with a rifle living in the settlement, and he alone said yea and nay, he had the right, on his own authority and with no beating about the bush, to shoot anyone deemed unruly.

The civic category to which the special settlements belonged, their blood ties with the Archipelago, will quickly become clearer if you remember the law governing a fluid in interconnected receptacles. If a shortage of labor was felt at Vorkuta, special settlers were transferred (without retrial! without relabeling!) from their settlements to the camps. And they lived behind the wire as meekly as you could wish, went to work behind more wire, ate dishwater soup, only they paid for it (and for their guards, and for their huts) out of their wages. And no one saw anything surprising in this.

The special settlers were also torn from their families, and shifted from settlement to settlement, just as zeks were shifted from camp to camp.

In one of those strange vagaries to which our legislation is subject, the U.S.S.R. Central Executive Committee promulgated a decision on July 3, 1931, permitting the restoration of civil rights to former kulaks after five years, if—in a settlement under police control, mind you!—"if they have engaged in socially useful work and shown their loyalty to the Soviet regime" (by, let us say, helping the rifleman, the settlement manager, or the security officer at his tasks). But this was mere foolishness, the whim of a moment. And anyway, as the five years were ending, so the Archipelago was hardening.

There seemed to be never a year in which it was possible to make conditions easier: first there was the time after Kirov's assassination; then 1937–1938; then from 1939 there was war in Europe; and in 1941 our own war began. So that a safer way was found: from 1937 on, many of these same hapless alleged kulaks and their sons were plucked out of the special settlements, labeled with a clause from Article 58, and shoved into the camps.

True, when during the war there was a shortage of reckless Russian fighting power at the front, they turned among others to the "kulaks": they must surely be Russians first and kulaks second! They were invited to leave the special settlements and the camps for the front to defend their sacred fatherland.

And—they went. . . .

Not all of them, however. N. Kh——v, a "kulak's" son—whose early years I used for Tyurin,* but whose subsequent biography I could not bring myself to recount—was given the chance, denied to Trotskyite and Communist prisoners, however much they yearned to go, of defending his fatherland. Without a moment's hesitation, Kh——v snapped back at the head of the Prisoner Registration and Distribution Section: "It's your fatherland—you defend it, you dung-eaters! *The proletariat has no fatherland!*"

Marx's exact words, I believe, and certainly any camp dweller was still poorer, still lower, still less privileged than any proletarian—but the camp disciplinary board had not mastered this fact, and it sentenced Kh——v to be shot. He sat in jail two weeks with a *topper* hanging over him, hating them too much to appeal for clemency. It was they who made a move, and commuted his sentence to a further *tenner.*

It sometimes happened that they transported ex-"kulaks" out into the tundra or the taiga, let them loose, and forgot about them. Why keep count when you'd taken them there to die? They didn't even leave a rifleman—the place was too remote, too inaccessible. Now that the mysteriously wise leaders had dismissed them—

without horses, without plows, without fishing tackle, without guns—this hard-working and stubborn race of men, armed perhaps with a few axes and shovels, began the hopeless fight for life in conditions scarcely easier than in the Stone Age. And in defiance of the economic laws of socialism, some of these settlements not only survived, but became rich and vigorous!

In one such settlement, somewhere on the Ob, but on a backwater, nowhere near the navigation channel, Burov had landed as a boy, and there he grew up. He tells the story that one day before the war a passing launch noticed them and stopped. The people in the launch turned out to be the district bosses. They interrogated the Burovs—where had they come from and how long ago? The bosses were amazed at their wealth and well-being, the like of which they'd never seen in their collectivized region. They went away. A few days later plenipotentiaries arrived with NKVD troops, and once again, as in the year of the Plague, they were ordered to abandon within an hour all that they had earned for themselves, all the warmth and comfort of their settlement, and dispatched with nothing but a few bundles deeper into the tundra.

Perhaps this story is enough in itself to explain the true meaning of "kulak" and of "dekulakization"?

The things that could have been done with such people if they had been allowed to live and develop freely!!!

The Old Believers—eternally persecuted, eternal exiles—they are the ones who three centuries earlier divined the ruthlessness at the heart of Authority! In 1950 a plane was flying over the vast basin of the Podkamennaya Tunguska. The training of airmen had improved greatly since the war, and the zealous aviator spotted something that no one before him had seen in twenty years: an unknown dwelling place in the taiga. He worked out its position. He reported it. It was far out in the wilds, but to the MVD all things are possible, and half a year later they had struggled through to it. What they had found were the Yaruyevo Old Believers. When the great and longed-for Plague began—I mean collectivization—they had fled from this blessing into the depths of the taiga, a whole village of them. And they lived there without ever poking their noses out, allowing only their headman to go to Yaruyevo for salt, metal fishing and hunting gear, and bits of iron for tools. Everything else they made themselves, and in lieu of money the headman no doubt came provided with pelts. When he had completed his business he would slink away from the marketplace like a hunted criminal. In this way the Yaruyevo Old Believers had won themselves twenty years of life! Twenty years of life as free human beings among the wild beasts, instead of twenty years of kolkhoz misery. They were all wearing homespun garments and homemade knee boots, and they were all exceptionally sturdy.

Well, these despicable deserters from the kolkhoz front were now all arrested, and the charge pinned on them was . . . guess what? Links with the international bourgeoisie? Sabotage? No, Articles 58-10, on Anti-Soviet Agitation (!?!?), and 58-11, on hostile organizations. (Many of them landed later on in the Dzhezkazgan group of Steplag, which is how I know about them.)

In 1946 some other Old Believers were stormed in a forgotten monastery somewhere in the backwoods by our valiant troops, dislodged (with the help of mortars, and the skills acquired in the Fatherland War), and floated on rafts down the Yenisei. Prisoners still, and still indomitable—the same under Stalin as they had been under Peter!—they jumped from the rafts into the waters of the Yenisei, where our Tommy-gunners finished them off.

Warriors of the Soviet Army! Tirelessly consolidate your combat training!

No, the doomed race did not all die out! In exile more children were born to them—and they, too, were attached by inheritance to the same special settlements. ("The son does not answer for the father"—remember?) If a girl from outside married a special settler, she passed into the serf class, and lost her rights as a citizen. If a man married *one of those,* he became an exile himself. If a daughter came to visit her father, they corrected their error in missing her before, and added her to the list of special settlers. These additions made good the deficit as settlers were transferred to the camps.

Special settlers were very conspicuous in Karaganda and round about. There were a lot of them there. They were attached *in perpetuity* to the mines of Karaganda, as their ancestors had been to the factories of the Urals and the Altai. The "mine owner" was free to work them as hard and pay them as little as he liked. We are told that they greatly envied prisoners in agricultural Camp Divisions.

Until the fifties, and in some places until the death of Stalin, special settlers had no passports. Only with the war did the Igarka exiles begin receiving the polar wage rate.

But now that they have survived twenty years of plague and exile, now that they are free at last from police supervision, now that they hold their proud Soviet passports—who and what are they, in their hearts and in their behavior? Why, what else but Soviet citizens, guaranteed in good condition! Exactly the same as those reared simultaneously by the workers' settlements, trade union meetings, and service in the Soviet Army! They, too, can muster the courage to slam down their dominoes almost boldly. They, too, nod agreement to every shadowy presence on the television screen. When required, they, too, will angrily stigmatize the Republic of South Africa, or collect their kopecks for the benefit of Cuba.

So let us lower our eyes in awe before the Great Butcher, bend our heads and bow our shoulders in the face of the intellectual puzzle he sets us: was he right after all, that reader of men's hearts, to stir up that frightening mixture of blood and mud, and to go on churning year in and year out?

He was right, morally. No one bears him a grudge! In his day, so ordinary folk say, it was "better than under Khrushchev": why, on April Fool's Day, year in and year out, cigarettes went down a kopeck and fancy goods ten. Eulogies and hymns rang in his ears till the day of his death, and even today we are not allowed to denounce him. Not only will any censor stay your pen, but anyone standing in a shop or sitting in a train will hasten to check the blasphemy on your lips.

We honor Great Evildoers. We venerate Great Murderers.

And he was even more right politically. This bloody mix was the cement for obedient kolkhozes. No matter that within a quarter of a century the village would be a desert and the people spiritually extinct. No matter; our rockets would be flying into space, and the enlightened West fawning on us, cringing before our achievements and our might.

Chapter 3

■

The Ranks of Exile Thicken

Only the peasants were deported so ferociously, to such desolate places, with such frankly murderous intent: no one had been exiled in this way before, and no one would be in the future. Yet in another sense and in its own steady way, the world of exiles grew denser and darker from year to year: more were banished, they were settled more thickly, the rules became more severe.

We could offer the following rough time scheme. In the twenties, exile was a sort of preparatory stage, a way station before imprisonment in a camp. For very few did it all end with exile; nearly all were later raked into the camps.

From the mid-thirties and especially from Beria's time, perhaps because the world of exile became so populous (think how many Leningrad alone contributed!), it acquired a completely independent significance as a totally satisfactory form of restriction and isolation. In the war and postwar years, the exile system steadily grew in capacity and importance together with the camps. It required no expenditure on the construction of huts and boundary fences, on guards and warders, and there was room in its capacious embrace for big batches, especially those including women and children. (At all major transit prisons cells were kept permanently available for women and children, and they were never empty.)[1] Exile made possible a speedy, reliable, and irreversible cleansing of any important region in the "mainland." The exile

---

1. Husbands who were also being deported did not travel with their wives: there was a standing order that members of condemned families should be sent to different places. Thus, when the Kishinev lawyer I. K. Gornik was exiled as a Zionist to the Krasnoyarsk region, his family were sent to Salekhard.

system established itself so firmly that from 1948 it acquired yet another function of importance to the state—that of rubbish dump or drainage pool, where the waste products of the Archipelago were tipped so that they would never make their way back to the mainland. In spring, 1948, this instruction was passed down to the camps: at the end of their sentences 58's, with minor exceptions, were to be *released into exile.* In other words, they were not to be thoughtlessly unleashed on a country which did not belong to them, but each individual was to be delivered under escort from the camp guardhouse to the commandant's office in an exile colony, from fish trap to fish trap. Since the exile system embraced only certain strictly defined areas, these together constituted yet another separate (though interlocking) country between the U.S.S.R. and the Archipelago—a sort of purgatory in reverse, from which a man could cross to the Archipelago, but not to the mainland.

The years 1944–1945 brought to the exile colonies unusually heavy reinforcements from the "liberated" (occupied) territories, and 1947–1949 yet others from the Western republics. All these streams together, even without the exiled peasants, exceeded many times over the figure of 500,000 exiles which was all that Tsarist Russia, the prison house of nations, could muster in the whole course of the nineteenth century.

For what crimes was a citizen of our country in the thirties and forties punishable by exile or banishment? (Bureaucrats must have derived some strange pleasure from this distinction, since though not observed, it was continually mentioned throughout those years. When M. I. Bordovsky, who was persecuted for his religion, expressed surprise that he had been exiled without trial, Lieutenant Colonel Ivanov graciously explained: "The reason there was no trial is because this is not exile but banishment. We do not regard you as a convicted person, and are not even depriving you of your electoral rights." These, of course, being the most important constituent part of civil freedom!)

The commonest crimes can easily be indicated:

1. Belonging to a criminal nationality (for this see the next chapter).
2. A previous term of imprisonment in the camps.
3. Residence in a criminal environment (seditious Leningrad,

or areas in which there was a partisan movement, such as the Western Ukraine or the Baltic States).

And then many of the tributaries enumerated at the very beginning of this book branched out to feed the exile system as well as the camps, continually casting up some of their burden on the shores of exile. Who were these people? Above all the families of men condemned to the camps. But families were by no means always drawn in, and it was by no means only the families of prisoners who poured into exile. Just as it requires extensive knowledge of hydrodynamics to explain the currents in a fluid, without which you can only observe in despair the chaotically swirling and howling element, so here: we lack the information to study all the differential impulses which in various years, for no apparent reason, sent various people not to camps but into exile. We can only observe the bewildering mixture of resettlers from Manchuria; individual foreign subjects (even in exile they were not allowed by Soviet law to intermarry with fellow exiles who were nonetheless Soviet subjects); certain Caucasians (no one remembers a single Georgian among them) and Central Asians, who for having been prisoners of war were sentenced not to the usual ten years in a camp but a mere six years of exile; and even some former prisoners of war, Siberians, who were sent back to their native districts and lived there as free men, without having to report to the police, but also without the right to leave the district.

We cannot go into the different types and cases of exile, because all our knowledge of it derives from casual stories or letters. If A. M. Ar——v had not written his letter, the reader would not have the following story. In 1943 news came to a village around Vyatka that one of its kolkhoz peasants, Kozhurin, a private in the infantry, had either been sent to a punitive unit or shot outright. His wife, who had six children (the oldest was ten years, the youngest six months old, and two sisters of hers, spinsters nearing fifty, also lived with her), was immediately visited by the *executants* (you already know the word, reader—it is a euphemism for "executioners"). They gave the family no time to sell anything (their house, cow, sheep, hay, wood, were all abandoned to the pilferers), threw all nine of them with their smaller possessions onto a sledge, and took them sixty kilometers in a hard frost to the town of Vyatka (Kirov). Why they did not freeze on the way God only knows. They were kept for six weeks at the Kirov Transit Prison, then sent to a small pottery near Ukhta. The spinster sisters ate from rubbish heaps, both went mad and both died. The mother and children stayed alive only thanks to the help (the politically ignorant, unpatriotic, in fact anti-Soviet help) of the local population. The sons all served in the army when they grew up and are said to have "completed their military and political training with distinction." Their mother returned to her native village in 1960—and found not a single log, not a single brick from the stove, where her house had been.

A little cameo like this can surely be threaded on the necklace of our Great Fatherland Victory? But nobody will touch it—it isn't *typical.*

To what necklace will you add, to what category of exiles will you assign soldiers disabled in the Fatherland War, and exiled because of it? We know almost nothing about them; in fact, hardly anyone else seems to. But refresh your memory. How many such cripples—some of them still not old—were there crawling around market tearooms and suburban trains at the end of the war? Then their ranks were swiftly and discreetly thinned. This was also a current, a campaign. They were exiled to a certain northern island —exiled *because* they had consented to be mutilated in war for the glory of the Fatherland and *in order to* improve the health of a nation, which had by now won such victories in all forms of athletics and ball games. These luckless war heroes are held there on their unknown island, naturally without the right to correspond with the mainland (a very few letters break through, and this is how we know about it), and naturally on meager rations, because they cannot work hard enough to warrant generosity.

I believe they are still living out their days there.

The great purgatory in reverse, the land of exile that lay between the U.S.S.R. and the Archipelago, included big towns, small towns, settlements, and completely uninhabited areas. Exiles tried to get permission to go to the towns, rightly considering that "it's easier there for the likes of us," particularly in the matter of work. And that their lives would be more like those of ordinary people.

Perhaps the main capital of the exile world, and certainly one of its pearls, was Karaganda. I saw something of it before the end of general exile, in 1955. (The police had allowed me, an exile, to go there for a short time. I was going to get married, to another exile.) At the entrance to what was then a hungry town, near the ramshackle station building which trams could not approach too closely (for fear of falling through into mine workings near the surface), and next to the tram turntable stood a truly symbolic brick house, its wall buttressed by wooden beams to prevent it from collapsing. In the center of the new section of town the words "Coal is bread" (for industry) were set in stone in a stone wall. And indeed, black bread was on sale in the shops there every day —this was what made exile in town so much easier. There was heavy work to be had, too, and some less heavy. For the rest, the food shops were pretty empty, while dizzy prices made the market stalls unapproachable. Two-thirds if not three-quarters of the town's inhabitants at that time had no passports and were required to report to the police. From time to time I was recognized and hailed in the streets by former zeks, particularly zeks from Ekibastuz. What sort of life did an exile have there? At work an inferior position and depressed wage, for not everybody after the catastrophe of arrest, prison, and camp had the means of proving his qualifications, and any allowance for length of service was still less likely. Or perhaps they were simply in the position of Negroes who do not get paid on the same level as whites: if you don't want the job you can please yourself. Then there was a great shortage of apartments. Exiles lived in unscreened corners of corridors, in dark box rooms, in little sheds—and always paid through the nose, because it was all rented privately. Women no longer young, worn to a frazzle by the camps, with metal teeth, dreamed of having one crepe de Chine "going-out" blouse and one pair of "going-out" shoes.

Then again, distances in Karaganda are great and many people had a long way to travel to work. It took the tram a whole hour to grind its way from the center to the industrial outskirts. In the tram opposite me sat a worn-out young woman in a dirty skirt and broken sandals. She was holding a child with a very dirty diaper, and she kept falling asleep. As her arms relaxed, the child would slide to the edge of her lap and almost fall. People shouted, "You'll drop him!" She managed to grab him in time, but a few minutes later she would be falling asleep again. She was on the night shift at the water tower, and had spent the day riding around town looking for shoes—and never finding them.

That was what exile in Karaganda was like.

As far as I can judge, it was much easier in Dzhambul: in the fertile southern belt of Kazakhstan food was very cheap. But the smaller the town, the harder it was to find work.

Take the little town of Yeniseisk. When G. S. Mitrovich was being taken there from the Krasnoyarsk Transit Prison in 1948, he asked the lieutenant escorting his party whether there would be work. "Of course there will," the lieutenant cheerfully replied. "And somewhere to live?" "Of course." But having handed them over to the command post, the escort went off, traveling light. While the new arrivals had to sleep beneath overturned boats on the riverbank, or under open-sided sheds in the marketplace. They could not buy bread: bread was sold to those with a fixed address; and the new arrivals could not register: to find lodgings you had to put down money. Mitrovich, who was entitled to a disablement pension, asked for work in his own profession (he was a livestock specialist). The post commander did a bit of quick thinking and telephoned the Agricultural Department of the district soviet. "Listen, if you give me a bottle I'll give you a livestock specialist."

This was a place of exile where the threat "For sabotage you'll get 58-14; we'll put you back in the camp" frightened nobody. Another piece of information about Yeniseisk dates from 1952. One day when they went in to report, the desperate exiles started asking the post commander to arrest them and send them back to the camp. Grown men, they could not earn their bread in that place! The post commander chased them away. "The MVD isn't an employment office, you know."[2]

Here is an even more Godforsaken place—Taseyevo, in Krasnoyarsk territory, 250 kilometers from Kansk. This was used as a place of exile for Germans, for Chechens and Ingush, and for former zeks. It is not a new place, not recently invented. The village of Khandaly, where in the old days they used to forge chains (*kandaly*) for convicts, is nearby. But what is new there is a whole town of earth houses, with earth floors. In 1949 a group of *repeaters* was brought there toward evening and unloaded at the school. A commission met late at night to take over the new work force: the commander of the district MVD post, someone from the Timber Cooperative, the kolkhoz chairmen. Before the commission passed in slow procession people old and sick, exhausted by *tenners* in the camps, and most of them women—it was they

2. He had no obligation to know, and the prisoners were not allowed to know, the laws of the land of the Soviets—Article 35 of its Criminal Code, for instance: "Exiles must be endowed with land or provided with paid work."

whom the state in its wisdom had excluded from the dangerous towns and thrown into this harsh region to tame the taiga. The employers were reluctant to take such workers, but the MVD left them no choice. The most hopeless rejects among the goners were foisted on the salt works, whose representative was late for the meeting. The salt works were on the river Usolka at the village of Troitsk (also an ancient place of exile—Old Believers were driven there in the reign of Aleksei Mikhailovich). In the middle of the twentieth century, the technique in use was this: they drove horses around in circles to pump brine onto filters, and then obtained salt by evaporation. (The firewood was brought from the logging sections—this was where they hurled the old women into action.) A successful and well-known naval architect had landed in this party: he was given a job as close as possible to his old profession —packing salt in boxes.

Knyazev, a sixty-year-old worker from Kolomna, fetched up in Taseyevo. He was past working, and had to beg. Sometimes people would take him home for the night, sometimes he slept in the street. There was no room for him in the old people's home, and the hospital wouldn't keep him for long at a time. One winter's day he crawled up the steps of the raikom—the local committee of the workers' party—and froze to death.

Zeks transported from a camp to exile in the taiga (in twenty degrees of frost, in the backs of open lorries, ill clad, with only the clothes they were released in, in superannuated tarpaulin shoes, while the escort troops were wearing sheepskin coats and felt boots) could not believe their senses: was their release to mean no more than this? In the camp there were heated huts—and here they had an earth house left by loggers, and unheated since last winter. The power saws had roared in the camps, and they roared here. These saws were the only means in either place of earning a ration of half-baked bread.

So that new exiles were apt to make mistakes, and when in 1953 the Deputy Director of the Timber Enterprise, Leibovich, arrived all clean and handsome at Kuzeyevo, in the Sukhobuzim district, Yenisei, they took one look at his leather overcoat and his sleek, pale face and greeted him incorrectly: "Good day, *citizen chief!*"

Leibovich wagged his head reproachfully: "No, no—what's all this 'citizen' stuff! You must call me *comrade* now; you aren't prisoners any more."

The exiles were assembled in the one and only earth house, and the Deputy Director harangued them by the somber light of a spluttering oil lamp, like a man hammering nails into a coffin:

"Don't imagine that you are here temporarily. You really will have to live here *forever*. So get to work as quickly as you can! If you have a family, send for them; if you haven't, marry among yourselves without delay. Build yourselves homes. Have children. You will be given a grant for a house and a cow. To work, comrades, to work! Our country needs our timber!"

With this their *comrade* rode off in his motorcar.

It was as a special privilege that they were allowed to marry. In the wretchedly poor settlements on the Kolyma, near Yagodnoye, for instance, Retz recalls that there were women who were not allowed to go to the mainland, yet the MVD forbade exiles to marry them: families would have to be given accommodations.

But not allowing exiles to marry could also be a concession. In Northern Kazakhstan in 1950–1952 some MVD posts tried to tie the exiles down by confronting new arrivals with this stipulation: Marry within two weeks or we'll send you into the interior, into the desert.

It is a curious fact that in many places of exile the camp term "general duties" was used, quite straightforwardly, not as a joke. Because that was what they were—exactly the same as in the camp: the necessary, heartbreaking jobs which ruin a man's health and do not pay a living wage. And though the exiles, as *free men*, were now supposed to work shorter hours, two hours there (to the pit or the logging section) and two back brought their working day up to the camp norm.

The old worker Berezovsky, a trade union leader in the twenties, who had endured ten weary years of exile from 1938, only to be sentenced to ten years in the camps in 1949, tearfully kissed his camp ration in my presence and said happily that he would not perish in the camp, where he had a right to bread. As an exile, even if you go into a shop with money in your pocket and see a loaf on the shelf, they may look you brazenly in the eye, say "No bread!" and sell some to a local while you watch. It is the same with fuel.

The old Petersburg worker Tsivilko (and there is nothing namby-pamby about any of these people) expressed very similar sentiments. He said (in 1951) that after exile he felt like a human being again in a Special Hard Labor Camp: he worked his twelve hours and went back to the huts, whereas in exile the merest nonentity among the free population could order him (he was a bookkeeper) to work unpaid overtime in the evening or on his day off, or could call on him if he needed any sort of job done privately, and no exile would dare refuse, for fear of dismissal next day.

Nor was life sweet for the ex-prisoner who became a "trusty" in exile. Mitrovich was transferred to Kok-Terek in Dzhambul oblast. (This is how his new life there began. He and a companion were quartered in a donkey stable, which had no windows and was full of dung. They raked the dung away from one wall, made themselves a bed of wormwood, and lay down to sleep.) He was given a job as livestock specialist in the District Agricultural Department. He tried to *serve honestly*—and at once fell foul of the free men in the district Party leadership. The local petty officials were in the habit of taking cows newly in milk from the kolkhoz herd, and replacing them with heifers. They expected Mitrovich to register two-year-old animals as four-year-olds. When he began taking stock carefully, he discovered whole herds which did not belong to the kolkhozes, but were fed and tended by them. It turned out that they were the personal property of the first secretary of the District Party Committee, the chairman of the Executive Committee of the District Soviet, the head of the local tax office, and the militia chief. (Kazakhstan had taken the comfortable road to socialism!) "Just don't put them on the list," they ordered him. But he did. With a zeal for Soviet legality quite bizarre in an exile and ex-prisoner, he even ventured to protest against the appropriation of a gray astrakhan by the chairman of the Executive Committee. He was dismissed—and that was only the beginning of hostilities.

Still, even a district center is not such a bad place of exile. The real hardships of exile began where there was no semblance of a free settlement, beyond the fringe of civilization.

A. Tsivilko, again, tells us of the Zhana Turmys (New Life) kolkhoz in the Western Kazakhstan oblast, where he was from 1937. Before the exiles arrived, the Political Section of the Machine and Tractor Station had warned and indoctrinated the locals: Trotskyites and Counter-Revolutionaries are on the way. The frightened natives would not lend the newcomers so much as a bit of salt, in case they were accused of associating with enemies of the people! During the war, exiles had no bread cards. Our informant worked in a kolkhoz smithy for eight months—and earned a pood of millet. . . . The exiles themselves milled the grain they received between grindstones made from a sawn-up Kazakh monument. Some went to the NKVD and asked either to be jailed or to be allowed to move to the district center. (Someone may ask: "What about the natives?" Well . . . they're used to it. . . . And if you have a sheep or two, a goat, a cow, a yurt, some crockery —it all helps.)

In the kolkhoz, exiles are always badly off—no regulation clothing, no camp ration. There is no more dreadful place of exile than the kolkhoz. It is a kind of field test: Where is life harder, in the camp or in the kolkhoz?

Here we are at a *sale* of new exiles, S. A. Lifshitz among them, at the Krasnoyarsk Transit Prison. The *buyers* ask for carpenters, and the prison authorities answer: Take a lawyer and a chemist (Lifshitz) as well, and we'll give you your carpenter. Some sick old women are added as makeweight. Then they are all taken on open lorries in a mild 25-degree frost to a village deep, deep in the wilds, a village of three dozen households in all. What can a lawyer do there, or a chemist? Here's an advance to be getting on with: a sack of potatoes, some onions, some flour. (A generous advance, at that!) Money you'll get next year, if you earn it. Your work for the time being is getting the hemp from under the snowdrifts. At first there isn't even a sack to stuff with straw and use for a mattress. Your first impulse is to ask to be released from the kolkhoz! No, it can't be done: the kolkhoz has paid the Administration of Prisons 120 rubles a head (this was in 1952).

If only you could go back to the camp again! . . .

But the reader will err if he concludes that exiles were much better off in state farms than in collective farms. Take the state farm at the village of Minderla in the Sukhobuzim district. There are rows of huts, without a boundary fence, it's true, so that it looks like an open prison camp. Although it is a state farm, money is unknown and there is none in circulation. Instead, they put down meaningless figures: nine rubles (Stalin rubles) per man per day. They also put down how much gruel the man has eaten, and how much should be deducted for his padded jacket and the roof over his head. Deduction after deduction is made until—surprise —when the final account is drawn up the exile has no wage to come, but on the contrary is in debt to the state farm. A. Stotik recalls that two people at this farm hanged themselves in desperation.

(Stotik himself, the visionary, had learned nothing from his ill-fated attempt to study English in Steplag.[3] When he had taken a good look at his place of exile, he thought he would exercise the constitutional right of a citizen of the U.S.S.R. to . . . education! So he applied for a leave of absence, to go to Krasnoyarsk and *study!* On this impudent application, the like of which the land of exile had probably never seen, the state farm manager [a former raikom secretary] penned something more than a negative decision: a solemn prohibition for the future. "At no time is Stotik to be allowed to study!" However, a chance came his way. The Krasnoyarsk transit point was recruiting carpenters from the exiles in various districts. Stotik, although he was no carpenter, volunteered, went to Krasnoyarsk, where he lived in a hostel with thieves and drunks, and set about preparing for the competitive entrance examination to the Medical Institute. He passed with high marks. He got as far as the Credentials Commission, and still no one had examined his documents thoroughly. "I served at the front," he told them, "then came back and . . ." He dried up. "And what?" Stotik came out with it. "And then . . . I was . . . put in jail." The commission looked black. "But I've *served my sentence! I'm a free man now!* I got high marks!" Stotik insisted. In vain. Yet this was the year of Beria's fall!)

The more remote the farm, the worse things were; the wilder the place, the fewer the exile's rights. A. F. Makeyev, in his previously mentioned notes on Kengir, cites the story told by Aleksandr Vladimirovich Polyakov, the "slave of Turgai," about his exile to a remote pasture in the Turgai wilderness in the interval between two spells in the camps. The only authority there was the kolkhoz chairman, a Kazakh; even the MVD paid no parental visits. Polyakov's living quarters were in the same shed as the sheep, on a litter of straw. His duties—to be the slave of the chairman's four wives, help each of them with her chores and even empty their chamber pots after them. What was Polyakov to do? Leave the grazing ground and complain? Apart from the fact that he had no mount, it would have meant *attempted escape* and twenty years of *katorga.* There was not a single Russian out there. Months went by before a Russian, a tax official, turned up. Polyakov's story astonished him and he offered to pass on a written complaint to the district center. This complaint, treated as a foul libel

3. Part V, Chapter 5.

against the Soviet state, earned Polyakov another stretch in the camps, and he was happily serving his time in Kengir in the fifties. To him it was almost as though he had been released. . . .

And we cannot be sure that the "slave of Turgai" was the most miserable of all exiles.

Nor can we say without qualification that exile, as compared with imprisonment in a camp, has the advantage of providing a settled existence: this is where you live and where you will live, with no fear of transportation. Transports or no transports, an inexplicable and inexorable police order transferring you elsewhere, or the unexpected closure of a particular center or of a whole district to exiles, is an ever-present threat; our informants recall such cases in various localities over the years. Especially during the war—when vigilance was the watchword. All exiles in Taipak district must be ready within twelve hours! And off you go to Dzhembetinsk! Your wretched home, your goods and chattels, so pathetic and to you so precious, the leaky roof you've just mended—leave it all! Forget it! Quick-march, you sturdy beggars! You'll scrape it all together again if you live long enough!

Although life looked so free and easy (they didn't march about in columns, but went each his own way; they did not line up for work assignments; they did not doff their caps; they were not locked in for the night), exile had its own disciplinary code. It was more severe in some places than in others, but it made itself felt everywhere until 1953, when the general relaxation began.

In many places, for instance, exiles had no right to address complaints on civil matters to local government bodies except through the MVD command post, which alone decided whether to let them go forward or snuff them on the spot.

Whenever an MVD officer summoned him, the exile had to leave his work undone, drop whatever he was doing, and report. If you know the ways of the world, you will not need telling that no exile would refuse to carry out a (blackmailing) personal request from a post officer.

The officers in MVD command posts enjoyed a position and rights scarcely inferior to those of officers in camps. Indeed, they had much less to worry about: no restricted area, no guard roster, no hunting runaways, no escorting prisoners to work, none of the business of feeding and clothing a crowd of people. It was enough for them to tick off each name twice a month, and occasionally initiate a case against those who had broken some rule, as required by the Law. They were despotic, they were lazy, they were self-indulgent (a second lieutenant's pay was 2,000 rubles a month), and so for the most part malevolent creatures.

Escape, properly so called, from Soviet places of exile was almost unknown: the successful escaper would not gain very much in terms of civil freedom—the local free population, living all around him, had after all much the same rights as himself. These were not Tsarist times, when flight from exile easily passed into emigration. Besides, the punishment for running away was no light matter. Escapers were dealt with by a Special Board. Before 1937 it handed out its maximum sentence of five years in prison camp, and from 1937, ten years. After the war, however, a new law—unpublished but universally known—was invariably applied: the inordinately cruel penalty for running away from a place of exile was now *twenty years of katorga!*

Local MVD posts introduced their own interpretations of what should or should not be considered attempted escape, drew for themselves the forbidden line which the exile must not cross, decided whether or not he could go off to gather firewood or pick mushrooms. In Khakassiya, for instance, in the Ordzhonikidzevsky mining settlement, the ruling was that "uphill" absenteeism (into mountain country) counted as a mere breach of discipline punishable with five years in a camp, while "downhill" absenteeism (in the direction of the railway) was attempted escape, punishable with twenty years of *katorga.* This unpardonable leniency took such a hold on the settlement that when a group of Armenian exiles, driven to despair by the high-handedness of the mine management, went to the district center to complain—naturally without permission from the MVD post to absent themselves for this purpose—they got a mere six years for their *attempted escape.*

What were classified as attempts to escape were more often than not misinterpreted absences of this sort. These, and the ingenuous mistakes of older people who could not get the hang of our savage system and adapt themselves to it.

One Greek woman, more than eighty years old, was banished from Simferopol to the Urals toward the end of the war. When the war ended her son returned to Simferopol and she naturally went to live secretly with him. In 1949, now eighty-seven (!) years old, she was arrested, sentenced to twenty years' hard labor

(87 + 20 = ?), and transported to Ozerlag. Another old woman, also Greek, was well known in the Dzhambul oblast. When the Greeks were deported from the Kuban she was taken, together with two grown-up daughters, while her third daughter, who was married to a Russian, remained behind. The old woman lived awhile, and awhile longer, in exile, and decided to go home to this daughter to die. This was an "escape," punishable with hard labor for twenty years! In Kok-Terek we had a physiologist called Aleksei Ivanovich Bogoslovsky. He had benefited from the "Adenauer amnesty"* of 1955—but not in full: the period of exile was left in force, although it should not have been. He started sending in appeals and petitions, but this was a lengthy business, and in the meantime his mother, in Perm, who had not seen him in fourteen years, since he had gone off to the war and a prisoner-of-war camp, was going blind and longed to see him before her eyesight failed altogether. Risking hard labor, Bogoslovsky decided to visit her and be back within a week. He invented an official trip for himself to grazing lands out in the wilds, and boarded a train for Novosibirsk. No one in the district had noticed his absence, but in Novosibirsk a vigilant taximan reported him to the secret police, who asked to see his papers, and since he had none he had to make a clean breast. He was sent back to our clay-walled prison in Kok-Terek, and interrogation was under way when suddenly a ruling arrived that he was not to be treated as an exile. As soon as he was released he went to his mother. But he was too late.

We should paint a very inadequate picture of the Soviet exile system if we did not recall that in every district to which exiles were sent an Operations Department kept unsleeping watch, pulled in exiles for *little talks,* recruited informers, collected denunciations, and used them to pin *fresh sentences* on people. For the time was never far off when the isolated individual exile would exchange his monotonous and static existence for the animated congestion of the camps. A *second extension*—reinvestigation and a new sentence—was for many the natural *end of exile.*

If Pyotr Viksne had not deserted from the reactionary bourgeois Latvian army in 1922 and run away to the free Soviet Union, had he not been exiled in 1934 to Kazakhstan for corresponding with relatives still in Latvia (who came to no harm at all), yet refused to be downhearted, had he not worked indefatigably in exile as an engine driver at the Ayaguza depot and earned Stakhanovite status—on December 3, 1937, posters would not have been put up at the depot saying: "Model yourselves on Comrade Viksne!" and on December 4 Comrade Viksne would not have been put inside for an "extension of sentence," from which he was never to return.

In exile as in the camps, resentencing went on continually, to show them *up there* that the Operations Departments never slept. There, as elsewhere, *intensive methods* were used, to help the prisoner understand his fate more quickly and submit more unreservedly. (Tsivilko was put in the hole at Uralsk in 1937 for thirty-two days, and had six teeth knocked out.) But there were also special periods, as for instance in 1948, when a close-meshed net was cast in all places of exile, and either all exiles without exception were fished out for transportation to the camps, as at Vorkuta ("Vorkuta is becoming an industrial center and Comrade Stalin has given instructions to clean it up"), or else, in some other places, all males.

But even for those who did not land in jail for an extended sentence, the "end of exile" was a nebulous idea. Thus, on the Kolyma, where "release" from a camp meant no more than transfer from the care of the camp guardhouse to that of the special MVD post, there was, strictly speaking, no "end of exile" either, because there was no exit from the area. Those who did manage to break away and go to the mainland in the brief periods when it was permitted probably never stopped cursing their fate: on the mainland they were all sentenced to fresh spells in the camps.

The sky of exile, troubled enough without it, was continually darkened by the shadow of the Operations Department. Under the eye of the secret police, at the mercy of informers, continually working himself to the verge of collapse in the struggle to earn bread for his children—the exile lived a very isolated life, the life of a timorous recluse. There were none of those long intimate conversations, those confessions of things past, usual in prisons and camps.

That is why it is difficult to collect stories about life in exile.

The Soviet exile system has also left almost nothing in the way of photographs: the only photographs taken were meant for documents—for personnel departments and Special Sections. If a group of exiles had their photographs taken together . . . what could it mean? What it would certainly mean was immediate denunciation to the security authorities: There you are—our local underground anti-Soviet organization. They would use the snapshot to arrest the lot.

Exile in our day has left behind none of those rather jolly group photographs—you know the sort: third from the left Ulyanov, second on the right Krzhizhanovsky. All well fed, all neatly dressed, knowing neither toil nor want, every last beard tidily trimmed, every single cap of good fur.

Those, my children, were very dark times. . . .

Chapter 4

∎

Nations in Exile

Historians may correct us, but no instance from the nineteenth century, or the eighteenth, or the seventeenth, of forcible resettlement of whole peoples has lodged in the average man's memory. There were colonial conquests—on the South Sea Islands, in Africa, in Asia, in the Caucasus, the conquerors obtained power over the indigenous population—but somehow it did not enter the immature minds of the colonizers to sever the natives from the land which had been theirs of old, from their ancestral homes. Only the export of Negroes to the American plantations gives us perhaps some semblance, some anticipation of it, but there was no developed state system at work here: only individual Christian slave traders, in whose breasts the sudden revelation of huge gains lit a roaring fire of greed, so that they rushed to hunt down, to inveigle, to buy Negroes, singly or by the dozen, each on his own account.

Only when the twentieth century—on which all civilized mankind had put its hopes—arrived, only when the National Question had reached the summit of its development thanks to the One and Only True Doctrine, could the supreme authority on that Question patent the wholesale extirpation of peoples by banishment within forty-eight hours, within twenty-four hours, or even within an hour and a half.

Even to *Him*, of course, the answer did not become clear quite so suddenly. He once even committed himself to the incautious view that "there never has been and never can be an instance of anyone in the U.S.S.R. becoming an object of persecution because of his national origin."[1] In the twenties all those minority languages were encouraged; it was endlessly dinned into the Crimea that it was Tatar, Tatar, and nothing but Tatar; it even had the Arabic alphabet, and all the signs were in Tatar.

Then it turned out that this was . . . all a mistake.

Even when he had finished compressing the exiled peasant mass, the Great Helmsman did not immediately realize how conveniently this method could be applied to nations. His sovereign brother Hitler's experiment in the extirpation of Jews and Gypsies came late, when the Second World War had already begun, but Father Stalin had given thought to the problem earlier.

After the peasant Plague, and until the banishment of peoples, the land of exile could not begin to compare with the camps; although it handled hundreds of thousands, it was not so glorious and populous that the highroad of history lay through it. There were *exile settlers* (sentenced by the courts) and there were *administrative exiles* (untried), but both these groups consisted of persons individually registered, each with his own name, year of birth, articles of indictment, photographs full face and in profile; and only the Organs with their miraculous patience and their readiness for anything could weave a rope from these particles of sand, build a monolithic colony in each of their districts from the wreckage of so many families.

The business of banishment was immeasurably improved and speeded up when they drove the first *special settlers* into exile. The two earlier terms (exile settler and administrative exile) were from the Tsar's times, but *spetspereselenets* (special settler) was Soviet, our very own. *Spets*— so many of our favorite, our most precious words begin with this little prefix (special section, special assignment, special communications system, special rations, special sanatorium). In the year of the Great Break they designated the dekulakized as "special settlers"—and this made for much greater flexibility and efficiency; it left no grounds for appeal since it was not only kulaks who were dekulakized. Call them "special settlers," and no one can wriggle free.

Then the Great Father gave orders that this word be applied to banished nations.

Even *He* was slow to realize the value of his discovery. His first experiment was very cautious. In 1937 some tens of thousands of

---

1. Stalin, *Sochineniya (Works)*, Moscow, 1951, Vol. 13, p. 258.

those suspicious Koreans—with Khalkhin-Gol in mind, face to face with Japanese imperialism, who could trust those slant-eyed heathens?—from palsied old men to puling infants, with some portion of their beggarly belongings, were swiftly and quietly transferred from the Far East to Kazakhstan. So swiftly that they spent the first winter in mud-brick houses without windows (where would all that glass have come from!). And so quietly that nobody except the neighboring Kazakhs learned of this resettlement, no one who counted let slip a word about it, no foreign correspondent uttered a squeak. (Now you see why the whole press must be in the hands of the proletariat.)

He liked it. He remembered it. And in 1940 the same method was applied on the outskirts of Leningrad, cradle of the Revolution. But this time the banished were not taken at night and at bayonet point. Instead, it was called a "triumphal send-off" to the (newly conquered) Karelo-Finnish Republic. At high noon, with red flags flapping and brass bands braying, the Leningrad Finns and Estonians were dispatched to settle their new native soil. When they had been taken a bit farther from civilization (V.A.M. tells us what befell a party of some six hundred people), they were all relieved of their passports, put under guard, and carried forward, first in red prison boxcars, then by barge. At the harbor of their destination deep in Karelia, they were broken up into small groups and sent to "reinforce the collective farms." And these completely free citizens, fresh from their triumphal send-off . . . submitted. Only twenty-six rebels, our narrator among them, refused to go, and what is more, would not surrender their passports! A representative of Soviet power—in this case, the Council of People's Commissars of the Karelo-Finnish Republic—had also arrived and he warned them: "There will be casualties." "Will you turn machine guns on us?" they shouted back. Silly fellows —why machine *guns?* There they were, surrounded by guards, all in a bunch; a single barrel would have been enough for them (and nobody would have written poems about those *twenty-six* Finns!).* But a strange spinelessness, sluggishness, or reluctance to take responsibility prevented the carrying out of this sensible measure. In an attempt to separate them, they were told to report to the security officer singly—but all twenty-six answered the summons together. And their senseless obstinacy and courage prevailed! They were allowed to keep their passports and the cordon was removed. In this way they resisted falling to the level of collective farmers or exiles. But theirs was an exceptional case, and the great majority handed over their passports.

These were mere trial runs. Only in July, 1941, did the time come to test the method at full power: the autonomous and of course traitorous republic of the Volga Germans (with its twin capitals, Engels and Marxstadt) had to be expunged and its population hurled somewhere well to the East in a matter of days. Here for the first time the dynamic method of exiling whole peoples was applied in all its purity, and how much easier, how much more rewarding it proved to use a single criterion—that of nationality —rather than all those individual interrogations, and decrees each naming a single person. As for the Germans seized in other parts of Russia (and every last one was gathered in), local NKVD officers had no need of higher education to determine whether a man was an enemy or not. If the name's German—grab him.

The system had been proved and perfected, and henceforward would fasten its pitiless talons on any nation pointed out to it, designated and doomed as treacherous—and more adroitly every time: the Chechens; the Ingush; the Karachai; the Balkars; the Kalmyks; the Kurds; the Crimean Tatars; and finally, the Caucasian Greeks. What made the system particularly effective was that the decision taken by the Father of the Peoples was made known to a particular people not in the form of verbose legal proceedings, but by means of a military operation carried out by modern motorized infantry. Armed divisions enter the doomed people's locality by night and occupy key positions. The criminal nation wakes up and sees every settlement ringed with machine guns and automatic rifles. And they are given twelve hours (but that is a long time for the wheels of motorized infantry units to stop turning, and in the Crimea it was sometimes only two or even one and a half hours) to get ready whatever each of them can carry in his hands. Then each of them is made to sit cross-legged in the back of a lorry, like a prisoner (old women, mothers with babies at the breast: sit down, all of you; you heard the order!), and the lorries travel under escort to the railway station. From there prison trains take them to a new place. From which they may still have to make their way like Volga boatmen (as the Crimean Tatars did up the river Unzha—what more suitable place for them than those northern marshes?), towing rafts on which gray-bearded old men lie motionless, 150 to 200 kilometers against the current, into the wild forest (above Kologrivo).

From the air or from high up in the mountains it was probably a magnificent sight. The whole Crimean peninsula (newly liberated in April, 1944) echoed with the hum of engines and hundreds of motorized columns crawled snakelike, on and on along roads straight and crooked. The trees were just in full bloom. Tatar women were lugging boxes of spring onions from hothouses to bed them out in the gardens. The tobacco planting was just beginning. (And that was where it ended. Tobacco vanished from the Crimea for many years to come.) The motorized columns did not go right up to the settlements, but stayed at the road junctions while detachments of special troops encircled villages. Their orders were to allow the inhabitants an hour and a half to get ready, but political officers cut this down, sometimes to as little as forty minutes, to get it over with more quickly and be on time at the assembly point—and so that richer pickings would be lying around for the detachment of the task force to be left behind in the village. Hardened villages like Ozenbash, near Lake Biyuk, had to be burned to the ground. The motorized columns took the Tatars to the stations, and there they went on waiting in their trains for days on end, wailing, and singing mournful songs of farewell.[2]

Neatness and uniformity! That is the advantage of exiling whole nations at once! No special cases! No exceptions, no individual protests! They all go quietly, because . . . they're all in it together. All ages and both sexes go, and that still leaves something to be said. Those still in the womb go, too, and are exiled unborn, by the same decree. Yes, children not yet conceived go into exile, for it is their lot to be conceived under the high hand of the same decree; and from the very day of their birth, whatever that obsolete and tiresome Article 35 of the Criminal Code may say ("Sentence of exile cannot be passed on persons under 16 years of age"), from the moment they thrust their heads out into the light they will be special settlers, exiles in perpetuity. Their coming of age, their sixteenth birthdays, will be marked only by the first of their regular outings to report at the MVD post.

All that the exiles have left behind them—their houses, wide open and still warm, their belongings lying in disorder, the home

---

2. In the 1860s the landowners and the administration of Tavrida Province petitioned the government to expel all the Crimean Tatars to Turkey. Alexander II refused. In 1943 the Gauleiter of the Crimea made the same request. Hitler refused.

put together and improved by ten or even twenty generations—passes without differentiation to the agents of the punitive organs, then some of it to the state, some to neighbors belonging to more fortunate nations, and nobody will write to complain about the loss of a cow, a piece of furniture, or some crockery.

One final thing made the principle of uniformity absolute, raised it to the height of perfection—the secret decree did not spare even members of the Communist Party in the ranks of these worthless nations. No need then to check Party cards—another relief. Besides, the Communists could be made to work twice as hard as the rest in their new place of exile, and everybody would be satisfied.[3]

The only crack in the principle of uniformity was made by mixed marriages (not for nothing has our socialist state always been against them). When the Germans, and later the Greeks, were exiled, spouses belonging to other nationalities were not sent with them. But this caused a great deal of confusion, and left foci of infection in places supposedly sterilized. (Like those old Greek women who came home to their children to die.)

Where were the exiled nations sent? Kazakhstan was much favored—and there, together with the ordinary exiles, they formed more than half the republic's population, so that it could aptly be called Ka-zek-stan. But Central Asia, Siberia (where very many Kalmyks perished along the Yenisei), the Northern Urals, and the Northern European areas of the U.S.S.R. all received their fair share.

Should we, or should we not, regard the expulsions from the Baltic States as "deportation of nations"? They do not satisfy the formal requirements. The Balts were not deported wholesale: as nations they appeared to remain in their old homes. (It would have been so nice to move them all—but they were a little too close to Europe!) They appeared to be where they had been—but they were thinned out, their best people were removed.

The purge started early: back in 1940, as soon as our troops marched in, and even before those overjoyed peoples had voted

3. Of course, not even the Great Helmsman could foresee all the strange twists of history. In 1929 the Tatar princes and other high personages were expelled from the Crimea. This was done less harshly than in Russia: they were not arrested, but allowed to make their own way to Central Asia. There, among the local inhabitants, Moslems themselves, and kinsmen, they gradually settled down and made themselves comfortable. Then fifteen years later all the Tatar toilers came under the nit comb and were sent to the same place! Old acquaintances met again. Only the toilers were traitors and exiles, whereas their former princes had safe jobs in the local government apparatus, and many were in the Party.

unanimously in favor of joining the Soviet Union. Culling began with the officers. We must try to imagine what this first (and last) generation of native officers meant to these young states. They were not Baltic barons, not arrogant drones, but all that was most serious, most responsible, most energetic in these nations. While they were still schoolboys they had learned in the snows of Narva to shield a still infant country with their still childish frames. Now all this rich experience was mowed down with one sweep of the scythe. This was a very important part of the preparations for the plebiscite. The recipe was of course well tried—had not the very same thing been done in the Soviet Union proper? Quietly and speedily destroy those who might take the lead in resistance, and also those who might awaken resistance with their thoughts, their speeches, their books—and it will seem that the people is whole and in place, yet the people will be no more. Externally, a dead tooth looks for a while exactly like a live one.

But for the Baltic States in 1940, it was not exile, but the camps —or for some people, death by shooting in stone-walled prison yards. In 1941, again, as the Soviet armies retreated, they seized as many well-to-do, influential, and prominent people as they could, and carried or drove them off like precious trophies, and then tipped them like dung onto the frostbound soil of the Archipelago. (The arrests were invariably made at night, only 100 kilograms of baggage was allowed for a whole family, and heads of families were segregated as they boarded the train, for imprisonment and destruction.) Thereafter, the Baltic States were threatened (over Leningrad radio) with ruthless punishment and vengeance throughout the war. When they returned in 1944 the victors carried out their threats, and imprisoned people in droves. But even this was not deportation of whole nations.

The main epidemics of banishment hit the Baltic States in 1948 (the recalcitrant Lithuanians), in 1949 (all three nations), and in 1951 (the Lithuanians again). In these same years the Western Ukraine, too, was being scraped clean, and there, too, the last deportations took place in 1951.

Was the Generalissimo preparing to exile some national group in 1953? The Jews, perhaps? And who else besides? Perhaps the whole of the Right Bank Ukraine? We shall never know what his great scheme was. I suspect, for instance, that Stalin suffered from an unquenchable longing to exile all Finland to the wilderness on the Chinese border—but he had no luck either in 1940 or in 1947

(Leino's attempted coup). He could have found just the spot for the Serbs—say, beyond the Urals—or for the Greeks of the Peloponnese.

If this Fourth Pillar of the Vanguard Doctrine had stood another ten years, we should not recognize the ethnic maps of Eurasia. There would have been a great countermigration of the peoples.

For every nation exiled, an epic will someday be written—on its separation from its native land, and its destruction in Siberia. Only the nations themselves can voice their feelings about all they have lived through: we have no words to speak for them, and we must not get under their feet.

But to help the reader recognize that this is still the land of exile, which he has visited before, the same place of pollution adjoining the same Archipelago, let us look a little further into the deportation of the Baltic peoples.

The deportation of the Balts, far from being a violation of the sovereign will of the people, was carried out purely and simply in execution of it. In each of the three republics, its very own Council of Ministers freely reached the decision (in Estonia it was dated November 25, 1948) to deport certain specified categories of its fellow countrymen to distant and alien Siberia—and what is more, in perpetuity, never to return to their native land. (In this we see distinctly both the independence of the Baltic governments, and the exasperation to which their worthless and deplorable fellow countrymen had brought them.) These categories were: (*a*) the families of persons previously condemned (it was not enough that the fathers were perishing in prison camps; their whole stock had to be extirpated); (*b*) prosperous peasants (this greatly speeded up the now essential process of collectivization in the Baltic States) and all members of their families (students in Riga, and their parents on the farm, were picked up on the same night); (*c*) people who were in any way conspicuous, important in their own right, yet had somehow jumped over the nit comb in 1940, 1941, and 1944; (*d*) families who were simply hostile to the regime, but had not been quick enough to escape to Scandinavia, or were personally disliked by local activists.

So as not to injure the dignity of our great common Motherland, and gratify our Western *enemies*, this decree was not published in the newspapers, was not promulgated in the republics, was revealed even to the exiles themselves not at the moment of deportation but only on arrival at MVD posts in Siberia.

In the years which had passed since the deportation of the Koreans, or even that of the Crimean Tatars, the organization of such operations had improved to such a degree, the precious experience gained had been so widely spread and thoroughly assimilated, that they no longer counted in days or even hours, but in minutes. Practice proved that twenty or thirty minutes was time enough between the first bang on the door at night and the last scrape of the householder's heel on his threshold as he walked into the darkness and toward the lorry. Those few minutes gave the awakened family time to dress, take in the news that they were being exiled for life, sign a document waiving all property claims, collect their old women and children, get their bundles together, and leave their homes when the order was given. (Property left behind was dealt with in an orderly fashion. After the escort troops had left, representatives of the Tax Office arrived to draw up a list of confiscated items, which were then sold through commission shops for the benefit of the state. We have no right to reproach them if they stuffed some things under their coats, and "offloaded" others, while they were about it. They had no real need to do so. It was only necessary to get an extra receipt from the commission shop, and any representative of the people's power could quite legally carry home the article he had bought for a song.)

How could anyone think clearly in those twenty or thirty minutes? How could they decide what it would be most useful to take with them? The lieutenant who was evicting one family (a seventy-five-year-old grandmother, a fifty-year-old mother, a daughter of eighteen, and a son of twenty) advised them to "take your sewing machine, whatever you do!" Who would ever have thought of it! But later on that sewing machine fed the family—and without it they would have starved.[4]

Sometimes, however, the speed of the operation worked to the advantage of the victims. A whirlwind blew up—and was gone. Even the best broom leaves specks behind. If some woman managed to hang on for three days, spent the nights away from

4. These MVD troops—how much did they understand of what they were doing, and what did they think of it? Mariya Sumberg was deported by a Siberian soldier from the river Chulym. He was demobilized shortly after, went home, saw her there, grinned delightedly, and hailed her effusively: "Hello, there! Remember me? . . ."

home, then went to the Tax Office and asked them to unseal her apartment—well, sometimes they would. All right, damn you— live there till next time.

In small cattle cars, intended for the transport of eight horses, or thirty-two soldiers, or forty prisoners, they carried fifty or more exiled Talinners. They were in too much of a hurry to equip the cattle cars, and did not give immediate permission to hack holes in the floors. The old bucket in which the exiles re- lieved themselves was soon brimful, running over and splashing their belongings. From the first minute these two-legged mam- mals were made to forget that men and women are different. They were shut up for a day and a half without food and with- out water. A child died. (But of course, we read all this not so very long ago, didn't we? Two chapters ago, twenty years back —but nothing has changed. . . .) They stood for a long time in the station at Julemiste, with people running up and down out- side, banging on the sides of the cars, asking for friends and relatives by name, unsuccessfully trying to pass provisions and comforts to one or another of them. These people were chased away. While those locked in the boxcars went hungry. And Si- beria awaited its lightly clad guests.

The authorities began issuing bread on the journey, and at certain stations soup. All the trains had a long haul before them: to the provinces of Novosibirsk, Irkutsk, Krasnoyarsk. Barabinsk alone was the destination of fifty-two carloads of Estonians. Exiles to Achinsk were fourteen days on the way.

What can sustain people on such a desperate journey? The hope which is brought not by faith but by hatred: "Their end is near! There will be war this year, and we shall go home in the autumn."

No one who has not experienced such misfortune, either in the Western or in the Eastern world, can be expected to understand or sympathize with or perhaps even forgive the mood of those behind bars at that time. I have said already that we, too, had the same beliefs, the same yearnings, in those years, 1949–1950. These were the years in which the iniquity of the system, with its twenty- five-year sentences, and its *return trips* to the Archipelago, reached a new, explosive level, became so glaringly intolerable that its guardians could no longer defend it. (Let us put it gener- ally: if a regime is immoral, its subjects are free from all obligations to it.) Only by savagely mutilating their lives could you make thousands of thousands in cells, in prison vans and prison trains, pray for a devastating atomic war as their only way out!

But no one wept—no one. Hatred is dry-eyed.

Another thing the Estonians thought about on the journey was the reception they could expect from the people of Siberia. In 1940 the Siberians had stripped exiled Balts bare, bullied them into handing over their belongings, paid half a bucket of potatoes for a fur coat. (We were, of course, all so ill-clad in those days that the Balts really did look like bourgeois. . . .)

Now, in 1949, the word was put around in Siberia that those who were being brought were incorrigible kulaks. But these kulaks were dumped out of their cattle cars in rags and near the end of their endurance. At medical examinations Russian nurses were amazed that the women were so thin and so shabby, that they hadn't a clean rag for their babes. The new arrivals were sent to kolkhozes short of people—and there the peasant women of collectivized Siberia, keeping it a secret from their bosses, brought them whatever they could spare: half a liter of milk from one, a griddlecake made of sugar-beet pulp or very bad flour from another.

Now, at last, the Estonian women wept.

But there were also, of course, the Komsomol activists. They took the arrival of this Fascist rabble very much to heart ("They should drown the lot of you!" such people shouted), and greatly resented their reluctance to work—ingrates!—for the country which had liberated them from bourgeois slavery. These Komsomols were given the task of supervising the exiles and their work. And they were warned: at the first shot they should organize a roundup.

At Achinsk station there was an amusing mix-up. The Birilyussi district bosses *bought* from the convoy ten wagonloads of exiles, five hundred people or so, for their collective farms on the river Chulym, and briskly shifted them 150 kilometers to the north. They had in fact been assigned to the Saralinsk mining administration, in Khakassiya (but of course did not know it). The mine managers were awaiting their *contingent,* but the contingent had been sprinkled about collective farms in which the year before peasants had received 200 grams of grain for a workday. By that spring there was neither grain nor potatoes, the villages were loud with the bellowing of hungry cattle, and the cows flung themselves like mad things on half-rotted straw. So it was not out of malice or to keep the exiles on a tight rein that the collective farms gave these newcomers one kilogram of flour per person per week—that was a very respectable advance, almost equivalent to their total future earnings! For the Estonians it was an appalling change from their homeland. . . . (There were, in fact, big barns full of grain in a nearby settlement called Polevoy: stocks had mounted up from year to year because nobody had made arrangements to remove them. But this was now state grain, and the kolkhoz had no claim to it. The people all around were dying, but no grain from those barns was given to them: it belonged to the state. On one occasion, kolkhoz chairman Pashkov took matters into his own hands and issued five kilograms to each kolkhoznik still living— and was sentenced to the camps as a result. The grain belonged to the state; the kolkhoz's troubles were its own affair—and this is not the book in which to discuss them.)

There on the Chulym the Estonians lived a life of desperation, trying to master an astounding new law: *Steal or die.* They had begun to think that they were there *forever,* when suddenly they were all plucked out and driven off to the Saralinsk district of Khakassiya (the owners had found their missing contingent). There were no actual Khakassians to be seen there; every settlement was a place of exile, and in every settlement there was an MVD post. There were gold mines, new shafts being sunk, silicosis everywhere. (Indeed, these broad expanses were not so much part of Khakassiya, or the Krasnoyarsk region, as the territory of Khakzoloto [the Khakassian Gold Mining Trust] or Yeniseistroi [the Yenisei Development Authority]; they belonged not to the District Soviets and District Party Committees but to the generals of MVD troops, and the secretaries of District Party Committees truckled to the local MVD commanders.)

But those who were simply sent to the mines did not have the worst of it. Much worse off were those who were enrolled in "prospecting artels." *Prospectors!* It has a romantic ring. The word glistens as though lightly dusted with gold. But any idea you like to think of can be given an ugly twist in our country. Special settlers were forced into these artels because they dared not object. They were sent to work mines which the state had abandoned as unprofitable. There were no longer any safety measures in these mines, and water ran in continually as though it were raining heavily. The yield was low, however hard you worked, and it was impossible to earn a decent wage; these dying people were simply sent in to lick out the residual traces of gold which the state was too miserly to abandon. The teams came under the "Prospecting Sector" of the Mining Administration, which thought of nothing, recognized no obligation, except to hand down the plan and exact fulfillment. The artels were "free" not from the state, but only from the benefits of its legislation: they were not entitled to paid leave, nor as a matter of course to Sundays off (as even zeks in the camps were), since any month might be declared a "Stakhanov" month, with never a Sunday in it. The state's rights were preserved: a man who did not turn up for work was put on trial. Once every two months a people's judge arrived and condemned several exiles to 25 percent compulsory labor—there were always plenty of excuses. These "prospectors" earned 3–4 "gold" rubles (150–200 Stalin rubles) *per month.*

At certain mines near Kopyov the exiles were paid not in money but in *vouchers;* what need had they, in fact, of ordinary Soviet currency when they could not move around anyway, and the shop at the mine would accept payment in coupons as well?

Elsewhere in this book we have developed in detail the comparison between prisoners in the camps and peasants in the days of serfdom. But if we remember our Russian history we know that the hardest conditions were those not of the peasants, but of workers tied to factories. These vouchers expendable only in the mine shops bring memories of the gold mines and factories of the Altai flooding into our minds. The miners assigned to these enterprises in the eighteenth and nineteenth centuries deliberately committed crimes in order to get into *katorga,* and *have an easier life of it.* Even at the end of the last century workers in the Altai goldfields "had no right to refuse work even on Sunday" (!), they paid fines (cf. the compulsory labor system), and there were shops with poor-quality provisions, cheap drink, and short measure. "These shops, and not the incompetently conducted mining operations, were the main source of income for the gold mine owners"[5] —or, in modern terms, the Trust.

Strange that everything on the Archipelago should be so unoriginal! . . .

In 1952, Kh. S., a frail little woman, did not go to work in a hard frost because she had no felt boots. For this, the head of the woodworking artel sent her logging for three months—still without felt boots. Three months before childbirth she asked to be

5. Semyonov-Tyan-Shansky, *Rossiya (Russia)* (1899–1914), Vol. XVI.

given some lighter work than heaving logs, and the answer was: If you don't want the job, give it up. Then a benighted woman doctor got the date of her confinement wrong by a month, and did not send her on leave until two or three days before she had her child. Argument will get you nowhere out in the MVD's taiga.

But even there life was not wrecked beyond repair. Total ruin was an experience reserved for those special settlers who were sent to collective farms. There are people nowadays who debate (and it is not a foolish argument) whether life was really any easier in the kolkhoz than in the camps. But what, we may answer, if kolkhoz and camp are combined? Well, that was the position of the special settler in the kolkhoz. It was a kolkhoz to the extent that there were no regular rations: only at sowing time did they issue 700 grams of grain a day, and then it was half-rotten, mixed with sand, earth-colored (no doubt swept up from the floor of the barns). It was like a prison camp in that the settlers could be put in detention cells: if a foreman complained about one of the exiles in his gang, the kolkhoz management would telephone the MVD post, and the MVD would take the man inside. There was no way of picking up extra earnings—you couldn't possibly make ends meet: for her first year of work in the kolkhoz Mariya Sumberg got *twenty grams* of grain per workday (a little bird can find more hopping along the roadside) and fifteen Stalin kopecks (one and a half Khrushchev kopecks). With her earnings for a whole year she bought herself . . . an aluminum bowl.

What, you may ask, did they live on?! Why, on parcels from the Baltic States. Their people had been banished—but not the whole people.

But who was there to send the Kalmyks parcels? Or the Crimean Tatars? . . .

Walk among their graves and ask them.

Whether this, too, was part of the decision made by their own Baltic governments, or an example of Siberian correctness, special instructions concerning the Baltic exiles were observed until 1953, until the Father of the Peoples was no more: no work for them except the heaviest! Only the pick, the shovel, and the saw! *"You are here to become real human beings!"* If a management promoted someone, the MVD post would intervene and remove him for *general duties.* They would not even allow special settlers to dig the gardens at the Mining Administration's Rest Home—for fear of offending the Stakhanovites who were recuperating there.

The MVD post commander even had M. Sumberg dismissed from the post of calf-herd: "You haven't been sent here for your summer holidays—go and stack hay!" The chairman had the greatest difficulty in keeping her. (She had saved his calves from brucellosis. She had become very fond of Siberian cattle, which she found better-natured than Estonian beasts, and cows who were unused to kindness licked her hands.)

Grain suddenly had to be loaded onto barges in a hurry—and special settlers worked without pay or reward for thirty-six hours on end (Chulym). In that whole period there were two breaks of twenty minutes each for food and one rest period of three hours. "Either you do it or we'll send you farther north!" If an old man fell down under the weight of a sack, the Komsomol overseers kicked him up again.

Settlers had to report weekly. The MVD post is several kilometers away, you say? And the old woman is eighty? Get a horse and carry her in! Every time they reported, they were reminded that attempted escape meant twenty years hard labor.

The security officer's room is next door. Exiles are called in there, too. The bait of a better job is dangled. And they threaten to deport an only daughter beyond the Arctic Circle, separating her from her family.

Is there anything they can not do? At what forbidden limit was their hand ever stayed by conscience? . . .

They gave the exiles tasks to perform. Keep an eye on such-and-such. Gather the evidence which will imprison so-and-so.

If the merest sergeant from the command post entered a hut, all the exiles, even aged women, had to rise and remain standing unless permission to sit was given.

. . . I hope that the reader has not misunderstood me to mean that exiles were deprived of their civil rights . . .

Oh, no, no! Their civil rights were preserved intact! Their passports were not taken from them. They were not deprived of the right to participate in elections on a universal, equal, secret, and direct ballot. That supreme, that glorious moment—when you strike out all the candidates on a list except the one of your choice —was jealously preserved for them. Nor were they forbidden to subscribe to the state loan (remember what torments the Communist Dyakov suffered in the camp!). When *free* kolkhozniki, cursing and grumbling, grudgingly gave 50 rubles, 400 each were wrung out of the Estonians. "You're all rich. If anybody doesn't sign we won't let him have his parcels. And we'll move him farther north."

They would do it, too. What was to stop them?

The tedium of it all! Nothing but the same thing over and over again. At the beginning of this Part VI we appeared to be discussing something new: not the camps, but the exile system. And this chapter made a fresh start: our theme was no longer the administrative exiles, but the special settlers.

Yet we are back where we started.

Must we—and if so, how often must we repeat ourselves again and again and again—tell the story of other, and different, exile colonies? In other places? At other periods? Peopled by other exiled nations?

And if so, which? . . .

■

Groups of exiles belonging to different nationalities, interspersed and clearly visible to each other, displayed their own national characteristics, their own ways of life, their own special tastes and inclinations.

Far and away the most industrious were the Germans. They had hacked themselves free of their past lives more resolutely than any of the others (and what sort of homeland had they had on the Volga or the Manych?). As once they had rooted themselves in Catherine's fecund allotments, so now they put down roots in the harsh and barren soil Stalin had given, abandoned themselves to this new land of exile as their final home. They began settling in, not temporarily, until the next amnesty, the first act of clemency by the Tsar, but forever. They had been exiled in 1941 with not a stick or a stitch, but they were good husbandmen and indefatigable, they did not fall into despondency, and even in this place set to work as methodically and sensibly as ever. Is there any wilderness on earth which Germans could not turn into a land of plenty? Not for nothing did Russians say in the old days that "a German is like a willow tree—stick it in anywhere and it will take." In the mines, in Machine and Tractor Stations, in state farms, wherever it might be, the bosses could not find words enough to praise the Germans—they had never had better workers. By the fifties the

Germans—in comparison with other exiles and even with the locals—had the stoutest, roomiest, and neatest houses, the biggest pigs, the best milch cows. Their daughters grew up to be much-sought-after brides, not only because their parents were well off, but—in the depraved world around the camps—because of their purity and strict morals.

The Greeks, too, ardently embraced their work. Although they never stopped dreaming about the Kuban, they grudged no effort in this new place either. Compared with the Germans, they lived in rather cramped quarters, but they soon caught up with them in the number of their cows and the richness of their gardens. In the little marketplaces of Kazakhstan it was the Greeks who had the best cream cheese, the best butter, the best vegetables.

The Koreans prospered even more in Kazakhstan—but of course they had been exiled earlier, and by the fifties were already in large measure emancipated from serfdom: they were no longer required to report, and they traveled freely from oblast to oblast, provided they did not cross the borders of the republic. They did not excel as good home builders or husbandmen (their homes and steadings were uncomfortable and primitive until the younger people became Europeanized); but they responded very well to education, quickly filled the educational institutions of Kazakhstan (no one put obstacles in their way during the war), and became the main component of the educated stratum in the republic.

Other nations, secretly cherishing dreams of return, were incapable of such single-mindedness, of living wholly in the present. But as a rule they submitted to discipline, and gave the MVD little trouble.

The Kalmyks could not stand up to it, and grieved themselves to death. (Here, however, I cannot speak from observation.)

But there was one nation which would not give in, would not acquire the mental habits of submission—and not just individual rebels among them, but the whole nation to a man. These were the Chechens.

We have already seen how they behaved toward runaways from the camps. And how they alone among the exiles at Dzhezkazgan tried to support the Kengir rising.

I would say that of all the special settlers, the Chechens alone showed themselves zeks in spirit. They had been treacherously snatched from their home, and from that day they believed in nothing. They built themselves sakli—low, dark, miserable huts that looked as if you could kick them over. Their husbandry in exile was all of this sort—all just for a day, a month, a year, with nothing put by, no reserves, no thought for the future. They ate and drank, and the young people even dressed up. The years went by—and they owned just as little as they had to begin with. The Chechens never sought to please, to ingratiate themselves with the bosses; their attitude was always haughty and indeed openly hostile. They treated the laws on universal education and the state curriculum with contempt, and to save them from corruption would not send their little girls to school, nor indeed all of their boys. They would not allow their women to work in the kolkhoz. Nor did they believe in slaving in the kolkhoz fields themselves. They tried whenever possible to find themselves jobs as drivers: looking after an engine was not degrading, their passion for rough riding found an outlet in the constant movement of a motor vehicle, and their passion for thieving in the opportunities drivers enjoy. This last passion, however, they also gratified directly. "We've been robbed," "We've been cleaned out," were concepts which they introduced to peaceful, honest, sleepy Kazakhstan. They were capable of rustling cattle, robbing a house, or sometimes simply taking what they wanted by force. As far as they were concerned, the local inhabitants, and those exiles who submitted so readily, belonged more or less to the same breed as the bosses. They respected only rebels.

And here is an extraordinary thing—everyone was afraid of them. No one could stop them from living as they did. The regime which had ruled the land for thirty years could not force them to respect its laws.

How did this come about? Here is an episode in which the explanation is perhaps epitomized. One of the pupils in the Kok-Terek school when I was teaching there was the young Chechen Abdul Khudayev. He inspired no warm feelings and did not try to do so; he seemed to be afraid of demeaning himself by making himself pleasant, he was always ostentatiously cold, he was very arrogant, and he could be cruel. But you could not help admiring his clear, precise mind. In mathematics and physics he never remained on the surface, with his schoolmates, but always went deeper, asked questions, tirelessly searched for the heart of the matter. Like all the children of settlers, he was inevitably drawn into the so-called "social activities" of the school—first the Young

Pioneers organization, then the Komsomol, the school committees, wall newspapers, character training courses, political discussion groups—the spiritual price for education which the Chechens paid so unwillingly.

Abdul lived with his old mother. None of their close relations had survived, except for Abdul's older brother, who had long ago taken to crime, had served more than one spell in the camps for theft and murder, but had always emerged before his time either under an amnesty or because he got remission for good conduct. One day he appeared in Kok-Terek, drank for two days without pausing for breath, quarreled with a local Chechen, seized a knife, and rushed at him. An old Chechen woman who had nothing to do with either of them barred his way, flinging her arms wide to stop him. If he had obeyed Chechen law he should have thrown down his knife and given up the chase. But he was by now a thief first and a Chechen second: he brought down his knife and fatally stabbed the innocent old woman. The thought of what awaited him according to the law of the Chechens now entered his drunken head. He rushed to the MVD to make a clean breast of the murder, and they gladly put him in jail.

He had found a hiding place, but that left his younger brother Abdul, his mother, and also an old Chechen of their clan, Abdul's uncle. News of the murder went around the Chechen community of Kok-Terek in a flash, and all three surviving members of the Khudayev clan gathered in their house, laid in stocks of food and water, blocked the window, nailed up the door, and lay low in this fortress. The Chechens of the murdered woman's clan now had to take vengeance on some member of the Khudayev clan. Until Khudayev blood was spilled in payment for their blood, they would be unworthy to be called human beings.

So the siege of the Khudayev house began. Abdul did not go to school—all Kok-Terek and the whole school knew why. A member of one of the top classes in our school, a Komsomol, an outstanding pupil, was threatened with murder by the knife at any moment—perhaps this very minute, as the bell rings and the others take their places at their desks, perhaps now while the literature teacher is talking about socialist humanism. Everybody knew; no one forgot it for a moment. Between classes they talked about nothing else—and could not bring themselves to look at each other. Neither the Party organization in the school, nor the Komsomol, nor the directors of studies, nor the headmaster, nor the District Education Department—nobody went to try and save Khudayev, no one even approached his besieged home in Chechen territory, which was buzzing like a hornet's nest. And it wasn't only they: at the first breath of bloody vengeance, the District Party Committee, the Executive Committee of the District Soviet, the MVD post, and the militia behind their adobe walls, all of them so awesome to us until now, also froze in craven inactivity. A savage and ancient law had breathed on them—and all at once there was no Soviet power in Kok-Terek. Nor was its heavy hand quick to reach out from the provincial capital, Dzhambul, for three days went by and no plane flew in with troops, no firm instructions arrived, except an order to defend the jail with all forces to hand.

This helped the Chechens, and us, too, to see clearly the difference between real power on this earth and the mirage of power.

The Chechen elders were the only ones to show sense! They went first to the MVD and asked them to hand over the elder Khudayev for summary punishment. The MVD nervously refused. They came back to the MVD a second time, asking them to have Khudayev tried in public and to shoot him while they watched. If this was done, they promised that the vendetta against the Khudayevs would be at an end. No more reasonable compromise could have been devised. But think of the difficulties. A trial in open court! An execution promised in advance and carried out in public! Khudayev, after all, was not a political, he was a thief, a "class ally." The rights of 58's could be trampled underfoot, but not those of a multi-murderer. The MVD referred it to the oblast —and the request was turned down. "In that case," the old men informed them, "the younger Khudayev will be killed within the hour!" The bureaucrats of the MVD shrugged their shoulders: it was none of their business. They were not there to think about crimes as yet uncommitted.

But some faint awareness of the twentieth century touched . . . not the MVD—oh, no—but those hardened old Chechen hearts. In spite of everything, they forbade the avengers to exact vengeance! They sent a telegram to Alma-Ata. Some other old men, those most respected by the whole people, hurried down. They convened a council of elders, which anathematized the older Khudayev and sentenced him to death, where and whenever he came within reach of a Chechen knife. The other Khudayevs they summoned and told: "Go in peace. No one will touch you."

So Abdul took his books and went to school. There he was met with hypocritical smiles by the Party organizer and the Komsomol organizer. From the moment of his return he heard Communist social conscience extolled in lessons and political instruction periods, with never a mention of the vexatious incident. Not a muscle twitched in Abdul's somber face. He had learned all over again that the greatest force on earth is the law of vendetta.

We Europeans, at home and at school, read and pronounce only words of lofty disdain for this savage law, this cruel and senseless butchery. But the butchery is perhaps not so senseless after all. It does not sap the mountain peoples, but strengthens them. Not so very many fall victim to the law of vendetta—but what power the dread of it has over all around! With this law in mind, no highlander will casually insult another, as we insult each other in drink, from lack of self-control or just for the hell of it. Still less will any *non*-Chechen look for trouble with a Chechen, call him a thief or a ruffian, or accuse him of jumping the queue. For the answer may be given not in words, not in abuse, but with a knife in the ribs! And even if you grab a knife yourself (but you won't have one on you—you're civilized), you cannot give blow for blow, or your whole family will fall under the knife! The Chechens walk the Kazakh land with insolence in their eyes, shouldering people aside—and the "masters of the land" and non-masters alike respectfully make way for them. The law of vendetta creates a force field of fear—and so gives strength to its small mountain people.

"Strike your neighbors, that strangers may fear you!" The ancestors of the highlanders in remote antiquity could have found no stronger hoop to gird their people.

Has the socialist state offered them anything better?

Chapter 5

■

End of Sentence

In eight years of prison and prison camp I had never heard anyone who had experienced exile say a good word about it. But from his first days in jail under investigation and in transit, simply because the six flat stone surfaces of a cell press in on him too closely, the dream of exile burns like a secret light in the prisoner's mind, a flickering iridescent mirage, and the wasted breasts of prisoners on their dark bunks heave in sighs of longing: "If only they would sentence me to exile!"

I did not escape the common lot; far from it—the dream of exile had me more powerfully than most in its grip. At the Iyerusalim clay pit I listened to the cocks crowing in the village nearby, and dreamed of exile. From the roof of the checkpoint on the Kaluga road I looked at the unbroken mass of the alien capital and silently begged: Let me get as far from it as possible, to some out-of-the-way place of exile! I even sent a naïve appeal to the Supreme Soviet: for commutation of my eight years in the camps to exile for life, in however remote and wild a place. The elephant did not even sneeze in reply. (I had not yet realized that lifelong exile would always be waiting for me, but that it would come *after*, not *instead of*, the camp.)

In 1952 a dozen prisoners were "released" from the 3,000-strong "Russian" Camp Division at Ekibastuz. It looked very strange at the time: 58's, let out through the gates! Ekibastuz had been in existence for three years by then, and not a single man had been released, nor had anyone reached the end of his sentence. Evidently, for the few who had lived to see the day, the first wartime *tenners* had just ended.

We impatiently awaited letters from them. A few came, directly or indirectly. And we learned that nearly all of them had been taken from the camp to places of exile, although their sentences had not included exile. But this surprised no one. It was clear to our jailers and to us that justice, length of sentence, formal documentation, had nothing to do with it; the point was that once we had been declared enemies, the state would ever after assert the right of the stronger and trample us, crush us, squash us, until the day we died. And we were so used to it, it had become so much part of us, that no other state of affairs would have seemed normal either to the regime or to us.

In Stalin's last years it was not the fate of the exiles that caused alarm, but that of the nominally liberated, those who to all appearances were now safely beyond the gates, and unguarded, those from whom the tutelary gray wing of the MVD had apparently been withdrawn. Exile, which the powers that be obtusely regarded as an additional punishment, was a prolongation of the prisoner's irresponsible existence, the fatalistic routine in which he feels so secure. Exile relieved us of the need to choose a place of residence for ourselves, and so from troublesome uncertainties and errors. No place would have been right, except that to which they had sent us. This was the one and only place in the whole Soviet Union where no one could reproach us as intruders. Only there had we an assured and undeniable right to three square arshins of land.* And if, like me, you were alone in the world when you left the camp, with no one, anywhere, waiting for you, exile was perhaps the only place where you could hope to meet a kindred spirit.

Our masters are quick enough to arrest people, but less quick to release them. If some democratic Greek or socialist Turk had been kept in jail a single day longer than he should have been, the world press would have choked with indignation. I was happy enough to be held only a few days too long and then . . . released? No; after that I was in transit under guard. And they kept me on the road another month in what was now my *own* time.

Nonetheless, as we left the camp under guard we were still careful to respect the final prison superstitions: on no account must you look back at your last prison (or else you will return), and you must do the right thing with your spoon. (What was the right thing, though? Some said take it with you, or you would return for it; others said fling it at the prison, or else the prison would pursue you. I had molded my spoon myself in the foundry, and I took it with me.)

The transit prisons flashed by again—Pavlodar, Omsk, Novosibirsk. Although our sentences had expired, they searched us, took prohibited articles from us; herded us into cramped, overcrowded cells, prison vans, Stolypin wagons; mixed us up with the criminals; the guard dogs growled at us as of old, the lads with the automatic rifles yelled "eyes front," all just as before.

Then at the Omsk Transit Prison a good-natured warder called out our names from our dossiers, and asked the five of us from Ekibastuz: "Which god have you got working for you?" "Why? Where are we going?" We were all ears: obviously it was somewhere nice. "Why, to the south," said the warder, marveling at it.

From Novosibirsk we were in fact diverted southward. We were going where it was warm! Where there was rice, where there were grapes and apples. What could it mean? Surely somewhere in the Soviet Union Comrade Beria could find us a worse place? Could exile really be like this? (I already had plans, which I kept to myself, to write a cycle of poems and call it "Lines on the Beauties of Exile.")

At Dzhambul station we were transferred from the Stolypin car with the usual harsh treatment, taken through a living corridor of escort troops to a lorry, and made to sit on the bare floor of the vehicle, as though now that we had served our time we might be tempted to run away. It was the dead of night, and only the waning moon dimly illuminated the dark avenue along which we were being carried, but there was light enough for us to see that it really was an avenue—of Lombardy poplars! So this was exile! We might almost be in the Crimea. It was the end of February, and on the Irtysh, where we had come from, it was cruelly cold, but here a spring breeze caressed us.

They took us to the jail—and the jail admitted us without the usual body search and bath. The accursed walls were losing some of their harshness! In the morning the block superintendent unlocked the door and said almost in a whisper, "Come out and bring all your belongings."

The devil was unclenching his claws. . . .

We stepped out into the arms of a red spring morning. The dawn light was warming the brick walls of the jail. A lorry was waiting for us in the middle of the yard, with two zeks who were joining our party already sitting in the back. This was the time to breathe deeply, to look around, to steep ourselves in the uniqueness of that moment—but we simply could not waste the chance to strike up an acquaintance. One of our new companions, a skinny, gray-haired man with pale, watery eyes, was sitting on his crumpled belongings as upright, as majestic as a Tsar about to give audience to ambassadors. You might have thought that he was deaf or a foreigner, with no hope that we should find a common language. As soon as I was up on the lorry I decided to get into conversation with him—and he introduced himself in pure Russian, in a voice with no hint of a quaver in it: "I'm Vladimir Aleksandrovich Vasilyev."

And a spark of sympathy jumped between us. The heart senses who is a friend and who is no friend. This was a friend. In prison you must find out about people quickly—for all you know, you will be separated in a minute's time. No, of course we were no longer in prison, but still . . . Shouting above the noise of the engine, I interviewed him, and did not notice that the lorry had moved from the asphalted prison yard onto the cobbled street, forgot that I must not look back at my last jail (how many last jails would there yet be), did not even spare a glance for the bit of the outside world through which we were traveling—and soon we were back in the broad inner yard of a provincial MVD post, and once again we were forbidden to leave it for the town.

For the first minute you might have taken Vladimir Aleksandrovich for a man of ninety—his eyes that looked beyond time, his gaunt features, his lock of white hair, all told the same story. He was in fact seventy-three. He turned out to be one of the oldest surviving Russian engineers, one of our outstanding hydrotechnicians and hydrographers in the "Union of Russian Engineers." (What was that? I had never heard of it before. It was a high-powered public body created by the technical intelligentsia, one of those hundred-year leaps into the future, perhaps, of which Russia made several in the first two decades of the century, but all of which miscarried.) Vasilyev had been a prominent member, and he still recalled with solid satisfaction how "we refused to pretend that dates can be grown on dry sticks."

Which of course was why they were disbanded.

Half a century back, he had covered on foot or on horseback every inch of the Semirechye region, in which we had now arrived. Long ago, before the First World War, he had drawn up plans for flooding the Chuisk Valley, for the Naryn cascade, and for boring a tunnel through the Chu-Ili mountains, and, still before the First World War, had begun carrying them out himself. He had sent abroad for six "electric excavators" and put them to work in this part of the world as early as 1912. (All six survived the Revolution and were passed off as a new Soviet invention at Chirchikstroi in the thirties.) After serving a sentence of fifteen years for "sabotage" (the last three in the Verkhne-Uralsk isolator), he had obtained a special act of indulgence: permission to spend his exile and die right there in Semirechye, where he had begun his career. (Even this favor would never have been granted if Beria had not remembered him from the twenties, when engineer Vasilyev had divided the waters of the Caucasus between its three republics.)

And that was why he was now sitting, sphinx-like, absorbed in his thoughts, on a sack in the back of a lorry. For him it was not just his first day of freedom, but his homecoming to the land of his youth, the land of his inspiration. No, human life is not so short as all that if you leave memorials along the way.

Not so long ago V.A.'s daughter had stopped at a newspaper window on the Arbat in Moscow to look at *Trud.* A devil-may-care correspondent was lavishing well-paid words on a rousing account of his journey through the Chuisk Valley, which had been irrigated and brought to life by creative Bolsheviks, with descriptions of the Naryn cascade, the ingenious hydrotechnical installations, the happy collective farmers. And suddenly—who can have whispered in his ear?—he ended with this: "But very few people know that all these transformations are the realization of the dream of the talented Russian engineer Vasilyev, who found no support in old bureaucratic Russia.[1] How sad that the young enthusiast did not live to see the triumph of his noble ideas!" The precious lines in the newspaper blurred, swam, Vasilyev's daughter tore the newspaper out of its case, pressed it to her breast, and carried it off, with a militiaman blowing his whistle after her.

While this was going on, the "young enthusiast" was sitting in a damp cell in the Verkhne-Uralsk isolator. Rheumatism, or some sort of bone disease, had bent the old man double, and he could no longer straighten his spine. Luckily, he was not alone in the cell, but shared it with some Swede or other who

1. At the end of 1917 Vasilyev was for all practical purposes the head of the Department of Land Improvement.

had cured his spinal trouble by massage.

Swedes are not very often found in Soviet jails. I remembered that I, too, had been in with one of them. His name was Erik . . .

"Erik Arvid Andersen?"* V.A. eagerly interrupted. (He was very quick in his speech and movements.)

Of course, it had to be! It was Arvid who had massaged him back to health! A reminder from the Archipelago, by way of farewell, that it's a small world after all. So that was where they had taken Arvid three years ago—to the Uralsk isolation prison. And somehow or other neither NATO nor his multimillionaire papa had made much of an effort to save the dear lad.[2]

Meanwhile, they had started calling us into the oblast command post, which was right there on the MVD yard, and consisted of a colonel, a major, and several lieutenants who were in charge of all exiles in Dzhambul oblast. We, however, had no access to the colonel, while the major only scanned our faces as though they were newspaper headlines, and it was the lieutenants who exercised their beautiful penmanship on the task of *processing* us.

My camp experience gave me a sharp nudge in the ribs. Look out! In these few short minutes your whole future is being decided! Don't waste any time! Demand, insist, protest! Rack your brains, turn yourself inside out, invent some reason, any reason, why you must at all costs remain in the oblast capital, or be sent to the nearest and most convenient district. (There was, in fact, a good reason, although I did not know it: secondary growths had been developing in me for two years

2. Pavel Veselov (Stockholm), who is now studying other cases of Swedish citizens seized by the Soviet authorities, puts forward the following hypothesis after analyzing E. A. Andersen's stories about himself: Both his appearance and the form of the name which he gave make it more probable that E. A. Andersen was a Norwegian, but for reasons of his own preferred to pass himself off as a Swede. It was much commoner for Norwegians to escape from their country after 1940 and serve in the British army, though a very few Swedes may have done so. E. A. might have been related to some Robertsons in Britain, but invented a relationship with General Robertson to make himself more valuable in the eyes of the MGB. It is not impossible that he had served in West Berlin after the war in Allied military intelligence, which was what made him interesting to the MGB. He had probably visited Moscow as a member of a British or Norwegian, not a Swedish, delegation (there was, I believe, no such Swedish visit at the time), but was probably a person of minor importance in it. Perhaps the MGB invited him to become a double agent, and perhaps it was for refusing this offer that he got his twenty years. Erik's father may have been a businessman, but not on such a large scale as he claimed. However, Erik often exaggerated —among other things his father's acquaintance with Gromyko (which was why the MGB showed him to Gromyko), to interest the MGB in the idea of demanding ransom, and so letting the West know of his plight.

now since my incomplete operation in the camp.)

No-o-o . . . I was not the man I had been. No longer the man I had been when I started serving my sentence. A sort of inspired immobility came over me, and I pleasurably abandoned myself to it. I enjoyed not making use of my importunate camp experience. I loathed the thought of improvising some wretched poverty-stricken excuse. No human being can know the future. The greatest of disasters may overtake a man in the best of places, and the greatest happiness may seek him out in the worst. Anyway, I had not even had time to ask questions and find out which were the good and which the bad districts in the oblast, because I had been preoccupied with the fate of the old engineer.

There was some sort of saving clause in his papers, because they allowed him to go into the town on his own two feet, walk to the oblast irrigation construction office and ask for work. For the rest of us there was only one destination: the Kok-Terek district. This is a patch of desert in the north of the oblast, the beginning of the lifeless Bet-Pak-Dala, which occupies the whole center of Kazakhstan. So much for the grapes we had dreamed of! . . .

A form printed on coarse brown paper was put before each of us for signature, after his name had been entered in a flowing hand and the date stamp applied.

Where had I met something similar? Of course—when they informed me of the Special Board's decision. Then, too, nothing had been asked of me but to take the pen and sign. Only then it was smooth Moscow paper. The pen and ink, though, were just as cheap and nasty.

Of what, then, had I been "this day informed"? That I, the person herein mentioned, was exiled *in perpetuity* to such and such a district, under the open surveillance (that old Tsarist terminology!) of the district MGB, and that in case of unauthorized departure beyond the borders of the district I should be charged under a Decree of the Presidium of the Supreme Soviet providing for a penalty of 20 (twenty) years of imprisonment with hard labor.

What was there to say? It was all perfectly legal. There were no surprises for us here. We signed with alacrity.[3]

3. Years later I would get hold of the Criminal Code of the R.S.F.S.R. and read with great satisfaction under Article 35 that sentence of exile was passed for terms from three to ten years, or by way of supplementary punishment after imprisonment for *up to five* years. This is a source of pride to Soviet jurists: that with the 1922 Criminal Code, loss

It also says in Article 35 that no one is exiled except by special order of the *court*. Or of a Special Board, at least? But it was not even a Special Board; it was merely the lieutenant on duty who prescribed perpetual exile for us.

An epigram gradually took shape in my mind—rather a lengthy one, I must admit:

> Mere paper—with a hammer's speed
> It shatters frail hopes of a kinder fate.
> "Exiled eternally," I read,
> "With the MGB to keep watch on the gate."
> Yet I sign with a flourish, my heart is light.
> Like the Alps, the basalts, or the firmament,
> Like the stars (no, not those on your shoulders so bright!)—
> Oh, enviable lot, I am permanent!
> But can it, I wonder, be every word true?
> Can the MGB really be permanent, too?

When Vladimir Aleksandrovich got back from town, I read him my epigram, and we laughed, laughed like children, like prisoners, like innocents. V.A. had a very happy laugh—like that of K.I. Strakhovich. There was, in fact, a profound similarity between them: they were people who had withdrawn so deeply into the life of the mind that no bodily suffering could upset their spiritual equilibrium.

Not that he had much to laugh about even now. They had made one of the mistakes expected of them, and exiled him, of course, to the wrong place. Only Frunze could assign him to the Chuisk Valley, to the sites of his earlier work. Here the Hydraulic Engineering Authority was concerned only with irrigation channels. Its head, a smug, semiliterate Kazakh, graciously permitted the creator of the Chuisk irrigation system to stand in the doorway of his office, telephoned regional Party headquarters, and consented to take V.A. on as a junior hydraulic engineer, as though he were a little girl fresh from technical school. But to Frunze he could not go: it was in a different republic.

Shall we sum up the whole history of Russia in a single phrase? It is the land of smothered opportunities.

---

of civil rights, and all other penalties, for an indefinite term, ceased to exist in Soviet law —except for the most terrifying penalty of all—expulsion forever from the territory of the U.S.S.R. There is here "an important difference of principle between Soviet and bourgeois law" (see the collection of articles *From the Prisons . . .*). This is as may be, but to save the MVD labor, the simplest thing is to write out a *perpetual* sentence: there is then no need to watch out for sentences near expiry and worry your head looking for a way to renew them.

Still, the gray old man rubs his hands: his name is known to scientists; perhaps they will get him transferred. He, too, signs the document stating that he is exiled in perpetuity, and that if he absconds he will be sentenced to hard labor until he is ninety-three. I carry his things as far as the gate—the boundary line which I am forbidden to cross. Now he will look for someone kind enough to rent him one corner of a room, and he threatens to send for his old woman from Moscow. His children? . . . His children wouldn't come. They say that they cannot give up their Moscow apartments. Any other relatives? He has a brother. But his brother's lot has been a profoundly unhappy one: he is a historian who misunderstood the October Revolution, abandoned his homeland, and now, poor wretch, has the chair of Byzantine Studies at Columbia University. We laugh again, feel sorry for his brother, and embrace as we say goodbye. Yet another remarkable man has flashed by me and vanished forever.

The rest of us were for some reason kept for days in a tiny room, sleeping close-packed on a rough cracked floor, with scarcely room to stretch our legs out. It reminded me of the lockup in which I had begun my sentence eight years before. Discharged prisoners now, we were locked in for the night, and told to take the latrine pail inside if we wanted it. It differed from jail only in that for those few days we were no longer fed free of charge, but had to pay for food to be brought from the market.

On the third day a regular escort party with carbines arrived, made us sign for our travel money and rations, promptly took the travel money from us again (allegedly to buy tickets, but in reality they bullied the train conductors into letting us travel for nothing, and kept the money, considering that they had earned it), lined us up two by two with our belongings, and led us once more between the rows of poplars to the station. Birds were singing, the hum of spring was in the air—and it was only March 2! We felt hot in our padded clothes but we were glad to be in the south. Others may have different views, but for prisoners and exiles the greatest of hardships is severe cold.

For a whole day we were carried slowly back in the direction from which we had come, then, from the station at Chu, we were hurried along on foot for ten kilometers. Our sacks and cases made us sweat profusely, weighed us down, caused us to stumble, but still we dragged them along: our miserable carcasses might yet be grateful for every last rag carried out through the camp guardhouse. I was wearing two padded vests (one I had pinched during stocktaking), plus my long-suffering army greatcoat, worn threadbare by the earth of the trenches and that of the camp—how could I throw the faded, grimy old thing away now?

Day was ending, and we had not reached our destination. This meant spending yet another night in jail, at Novotroitsk. We had been free for such a long time—and still we went from jail to jail. The cell, the bare floor, the peephole, use the bucket, hands behind your backs, here's your hot water; the only thing they leave out is our rations, because we are *free men* now.

In the morning they sent up a lorry, and the same escort, after a night out of barracks, came to fetch us. Sixty kilometers farther into the steppe. We got stuck in muddy hollows, and jumped down from the lorry (something we could not have done as zeks) to heave and push it out of the mire, to get the eventful journey over and arrive in perpetual exile more quickly. The escort troops stood in a half-circle and kept guard over us.

The steppe sped by, kilometer after kilometer. To right and to left, as far as the eye could see, there was nothing but harsh gray inedible grass, and only very occasionally a wretched Kazakh village framed with trees. At length the tops of a few poplars (Kok-Terek means "green poplar") appeared ahead of us, over the curve of the steppe.

We had arrived! The lorry sped between Chechen and Kazakh adobe huts, raising a cloud of dust and drawing a pack of indignant dogs in pursuit. Amiable donkeys with little carts made way for us, and from one yard a camel turned slowly and contemptuously to look at us. There were people, too, but we had eyes only for the women—those unfamiliar, forgotten creatures: look at that pretty dark girl in the doorway, shading her eyes with her hand to watch our lorry pass; look at those three walking together in flowery red dresses. Not one of them Russian. "This is all right —we shall find wives for ourselves yet!" This cheerful shout in my ear came from V. I. Vasilenko, a forty-year-old sea captain who had lived an untroubled life in Ekibastuz in charge of the laundry, and was now on his way to freedom, to spread his cramped wings and look for a ship.

Past the district stores, the tearoom, the clinic, the soviet offices, the district Party headquarters with its slated roof, the House of Culture under its reed thatch. Our lorry stopped near the MVD-MGB building. Covered with dust, we jumped down, went into its front yard, and, unembarrassed by passers-by in the main street, washed ourselves down to the waist.

Directly across the street from the MGB stood an amazing building, one story, yet quite high; four Doric columns solemnly upheld a false portico, at the foot of the columns were two steps faced with smooth stone, and over all this—there was a blackened straw roof. My heart could not help beating faster. It was a school! A ten-year school. Stop pounding, be quiet, you nuisance. That building is nothing to you.

Crossing the main street to the magic gate goes a girl with waved hair, neatly dressed in a little wasp-waisted jacket. Surely she is walking on air? She is a *teacher!* She is too young to have graduated from an institute; she must have attended a seven-year school and then a teachers' training college. How I envy her! What a gulf there is between her and a common laborer like me. We belong to different estates, and I would never dare to walk arm in arm with her.

Meanwhile, someone was yanking the new arrivals into his quiet office one by one and busying himself with them. Who, would you think? Why, godfather, of course, the secret police officer. You find him in the land of exile, too. There, too, he plays the leading role!

The first encounter is very important: we shall be playing cat and mouse with him not just for a month but *in perpetuity.* Now I shall cross his threshold and we will discreetly look each other over. He is a very young Kazakh; he wears the mask of polite reserve, and I wear that of artlessness. We both know that our insignificant phrases, such as "Have a sheet of paper" and "Is there a pen I can use?" are the beginning of a duel. But it is important for me to pretend that I haven't the slightest idea of this. It must be self-evident that I am always the same, unbuttoned and guileless. "Come on, you brown devil, get it firmly into your head: this one needs no special surveillance, he's come here for a quiet life; imprisonment has taught him something."

What's this I must fill in? A questionnaire, of course. And a curriculum vitae. This will open the new file, lying ready on the table. Afterward all tales told against me, all character reports from official persons, will be filed there. And as soon as the outlines of a new *case* can be roughed out, and the mainland gives the signal to put a few more inside—inside they will put me (there's an adobe jailhouse right here, in the backyard), and slap another *tenner* on me.

I hand over the initial documents; the secret police officer peruses them and files them in his loose-leaf folder.

"Could you please tell me where the District Education Department is?" I suddenly ask him, politely and casually.

And he politely tells me. His eyebrows do not shoot up in surprise. From this I deduce that I can go and ask for work, with no objection from the MGB. (As an old prisoner, of course, I did not cheapen myself by asking him outright whether *I was allowed* to work in the school network.)

"Can you tell me when I shall be allowed to go there without escort?"

He shrugs his shoulders.

"Just for today, while you're being . . . it would be as well if you didn't go much beyond the gates. But you can pop out on business."

And off I *walk!* I wonder whether everybody knows the meaning of this great free word. I am walking along *by myself!* With no automatic rifles threatening me, from either flank or from the rear. I look behind me: no one there! If I like, I can take the right-hand side, past the school fence, where a big pig is rooting in a puddle. And if I like, I can walk on the left, where hens are strutting and scratching immediately in front of the District Education Department.

I walk the two hundred meters to the Department, and my spine, which seemed bent for eternity, is already just a little straighter, my manner already a little more relaxed. In the course of those two hundred meters I have graduated to the next higher civil estate.

I go in, wearing an old woolen tunic from my days at the front, and my old, my terribly old twill trousers. My shoes are camp issue, pigskin, and the obtrusive ends of my underpants are not entirely hidden in them.

Two fat Kazakhs sit there—inspectors of the District Education Department, according to their nameplates.

"I should like a job in a school," I say, with growing confidence and even a sort of offhandedness, as though I were asking where the water jug was.

They prick up their ears. After all, in a Kazakh village out in the desert, new teachers looking for jobs do not arrive every half hour. And although the Kok-Terek district covers a bigger area than Belgium, they know by sight everyone in it with seven years of schooling.

"What did you study?" they ask in fairly good Russian.

"Physics and mathematics, at the university."

They are quite startled. They exchange glances. They start gabbling in rapid Kazakh.

"And . . . where have you come from?"

As though there could be any mystery about it, I have to spell it out for them. What sort of idiot would come to this place looking for a job, and in March at that?

"I got here an hour ago; I'm an exile."

They assumed a knowing look and vanished one after the other into the director's office. When they had left I noticed that the typist, a Russian woman of fifty or so, was looking at me. An instantaneous spark of sympathy—we were compatriots. She, too, was from the Archipelago! Where was her home, why had she been put away, in which year? Nadezhda Nikolayevna Grekova, daughter of a Cossack family in Novocherkassk, arrested in '37, was a simple typist, but the Organs had used their whole arsenal to persuade her that she was a member of some fantastic terrorist organization. She had done ten years, and now she was a *repeater*, exiled in perpetuity.

Lowering her voice, and keeping one eye on the director's door, which was ajar, she briefed me succinctly: two ten-year schools, several seven-year schools, the district was gasping for math teachers, had not a single one with higher education, had never seen a physicist close enough to know what sort of animal he was. A ring from the inner office. In spite of her plumpness, the typist sprang up and hurried in, all eagerness to serve, and on her return summoned me in a loud official voice.

There was a red cloth on the table. Both inspectors were sitting very comfortably on a sofa. On a big armchair under a portrait of Stalin sat the director: a Kazakh woman, small, lithe, attractive, with something feline and something serpentine in her manner. Stalin grinned at me malevolently from his frame.

They made me sit over by the door, keeping me at a distance as though they were interrogating a prisoner, and began an excruciatingly pointless conversation, which might have been much shorter had they not followed every few sentences to me with a ten-minute conference in Kazakh, while I sat by like a halfwit. They questioned me in detail about where I had taught and when, and expressed their misgivings that I might have forgotten my subject or my teaching technique. Then, after endlessly stalling, endlessly sighing that there were no vacancies, that the schools in the district were chock-full of mathematicians and physicists, that it was difficult to squeeze out even half an extra stipend, that rearing the young in our epoch was a responsible task, they got around to the really important point. *What had I done* to land in jail? What crime, exactly, had I committed? The cat-snake screwed up her sly eyes in anticipation, as though the blood-red glare of my crime was already painfully beating on her Bolshevik face. I looked over her head at the sinister features of the Satan who had wrecked my whole life. With his portrait watching me, what could I tell them about our relationship?

I used a prison trick to frighten these educators off: what they wanted me to tell them was a state secret, and I had no right to do so. What I wanted to know, without further waste of time, was whether they would take me on.

The discussions in Kazakh went on and on. Who would be so bold as to employ a state criminal at his own risk? But they found a way out: making me write a curriculum vitae and complete a questionnaire in two copies. Here we go again! Paper can take anything. Surely it was no more than an hour ago that I had filled in these forms. I filled them in yet again and returned to the MGB.

I inspected their yard, and the homemade prison inside, with interest, saw how even they had imitated the grownups by quite unnecessarily knocking a little window for parcels in the adobe wall, although it was so low that baskets could be passed over it. Ah, but how can you have an MGB post without a parcels window? I wandered around their yard and found myself breathing much more freely than in the musty District Education Department. From there, the MGB was an enigma, and the inspectors' blood ran cold at the thought of it. But over here, I was at home, with *my own ministry*. The three dolts of commandants (two of them officers) were quite frankly there to keep watch on us—we were their bread and butter. There was no mystery about it.

They were easygoing and allowed us to spend the night not in a locked room but out in the yard, on hay.

A night under the open sky! We had forgotten what it was like. . . . There had always been locks, and bars, always walls and ceilings. I had no thought of sleep. I walked and walked and walked about the prison service yard, which was bathed in soft, warm light. A cart left where it had been unhitched, a well, a drinking trough, a small hayrick, the black shadows of horses under an open-sided shed—it was all so peaceful, so ancient, so free from the cruel imprint of the MVD. It was only the third of March, but there was not the slightest chill in the night air; it was still almost summery, as it had been in the daytime. Again and again the braying of donkeys rose over the sprawling town of Kok-Terek, long-drawn-out and passionate, telling the she-asses of their love, of the ungovernable strength flooding their bodies. Some of the braying was probably the she-asses answering. I found it difficult to distinguish one voice from another, but that powerful bass bellowing was perhaps the noise of camels. I felt that if I only had a voice I, too, would start baying at the moon: I shall be able to breathe here! I shall be able to move around!

Surely I should break through that paper curtain of forms! With that trumpeting night around me, I felt superior to all those timorous bureaucrats. To teach! To feel myself a man again! To sweep into the classroom, and run my burning eyes over childish faces! My finger points to a drawing—they all hold their breath! Given, to prove, construction, proof—they all breathe freely again.

I cannot sleep! I walk and walk and walk in the moonlight. The donkeys sing their song. The camels sing. Every fiber in me sings: I am free! I am free!

In the end I lie down beside my comrades, on some hay under the open-sided shelter. Two steps away from us, horses stand at their mangers peacefully champing hay all night long. Surely there could be no sweeter, no more friendly sound on this our first night of freedom.

Champ away, you mild, inoffensive creatures!

Next day we were allowed to move into private lodgings. I found myself a henhouse to suit my pocket, with a single bleary window and such a low roof that even where it was highest, in the middle, I could not stand upright. "Give me a low-roofed cottage," I once wrote in prison, dreaming of exile. It was not very pleasant, all the same, not being able to raise my head. Still, it was a little house of my own! The floor was earthen. I put my padded camp vest on it, and there was my bed! But Aleksandr Klimentiyevich Zdanyukevich, an exiled engineer, who had formerly taught at the Bauman Institute, quickly lent me a couple of wooden boxes, on which I managed to make myself comfortable. I had no oil lamp as yet—I had nothing!! an exile must select and buy every single thing he needs, as though he has just landed on this earth—but I did not feel the want of it. All those years, in our cells and our huts, the state's electricity had seared our souls, and now darkness was bliss. Even darkness can be an element of freedom! In the darkness and the silence (the noise of the loud-speaker on the town square might have reached me, but this was Kok-Terek, and it had been out of action for three days), I simply lay on my boxes and enjoyed it!

What more could I desire? . . .

But the morning of March 6 surpassed anything that I could have wished for! Chadova, my elderly landlady, an exile from Novgorod, whispered, because she dared not say it out loud: "Go and listen to the radio. I'm afraid to repeat what I've just heard."

Something told me to do as she said: I went over to the central square. A crowd of perhaps two hundred people—a lot for Kok-Terek—huddled around the post under the loudspeaker and the sullen sky. There were many Kazakhs, most of them old men, among the crowd. Their bald heads were bare, and they held their red-brown muskrat-fur hats in their hands. They were grief-stricken. The younger people seemed less concerned. Two or three tractor drivers had not removed their caps. Nor, of course, would I. Before I could make out what the announcer was saying (he spoke with a histrionic catch in his voice), understanding dawned on me.

This was the moment my friends and I had looked forward to even in our student days. The moment for which every zek in Gulag (except the orthodox Communists) had prayed! He's dead, the Asiatic dictator is dead! The villain has curled up and died! What unconcealed rejoicing there would be back home in the Special Camp! But where I was, Russian girls, schoolteachers, stood sobbing their hearts out. "What is to become of us now?" They had lost a beloved parent. . . . I wanted to yell at them across the square: "Nothing will become of you now! Your fathers will not be shot! Your husbands-to-be will not be jailed! And you will never be stigmatized as relatives of prisoners!"

I could have howled with joy there by the loudspeaker; I could even have danced a wild jig! But alas, the rivers of history flow slowly. My face, trained to meet all occasions, assumed a frown of mournful attention. For the present I must pretend, go on pretending as before.

All the same, my exile had begun with magnificent auguries!

Once again, a whole day was devoted to writing a poem: "The Fifth of March."

Ten days passed—and the rival barons, wrangling over portfolios and eyeing each other nervously, abolished the MGB altogether! So that I had been right to doubt whether the MGB was there *forever.* [4]

Injustice, inequality, and slavery apart—is anything "forever" in this world of ours?

---

4. Six months later, of course, it was back in the form of the KGB, staffed as before.

Chapter 6

■

The Good Life in Exile

1. Bicycle nails .... ½ kilo
2. Shoo 5
3. Ash-pan 2
4. Glashes 10
5. Fencil case, child's 1
6. Glope 1
7. Match .... 50 boxes
8. "Bat" lamp 2
9. Tooth past 8
10. Gingerbread .... 34 kilos
11. Vodka .... 156 half-liters

This was the list of all goods in stock and due for repricing in the general store at the village of Aidarly. Inspectors and stocktakers of the Kok-Terek District Consumer Cooperative (RaiPO) had drawn up the list, and I was now turning the handle of a calculating machine and reducing the prices of some items by 7.5 and of others by 1.5 percent. Prices were plunging alarmingly and it was to be expected that both "fencil case" and "glope" would be sold before the new school year began, that the nails would find a home in bicycles, and that only the great gingerbread pile, probably prewar, was on its way down into the category of unsalable stocks. As for the vodka, even if the price went up it wouldn't outlast May Day.

The price reductions, which according to Stalinist custom took place in time for April 1, and by which the toilers gained so many million rubles (the full benefit was calculated and published in advance), hit me hard.

By then I had spent a whole month in exile, going through all I had earned in the "self-financing" foundry at Dzhezkazgan—a free man, supporting myself on Gulag money—and calling regularly at the District Education Department to try and find out whether they would take me on. But the snakelike director had stopped receiving me, the two fat inspectors could scarcely find time to growl at me, and toward the end of the month I was shown a ruling from the Oblast Education Department that all schools in Kok-Terek district were fully staffed with math teachers and that there was no possibility of finding me work.

Meanwhile I had been writing a play *(Decembrists Without December)*, without having to go through the morning and evening body searches, and with no need to destroy what I had written at short intervals, as I used to. I had no other occupation, and after the camp I liked it that way. Once a day I went to the "tearoom" and ate some hot broth for two rubles—the same thin soup they sent out in a bucket for the prisoners in the local jail. Coarse black bread was sold freely in the local shop. I had bought potatoes, and even a lump of pork fat. I myself had brought a donkeyload of underbrush from the thickets, so that I could light my cooking stove if I liked. My happiness was not far from complete, and I thought to myself: If they won't give me a job, they needn't; while my money lasts I'll go on writing my play. I may never have so much freedom again!

When I was least expecting it, one of the three commandants crooked a finger at me in the street. He took me to the RaiPO, into the office of the director, a Kazakh as round as a cannonball, and said with solemn emphasis: "The mathematician."

What miracle was this? No one asked me why I had been inside, or gave me forms and curricula vitae to fill in! There and then, his secretary, an exiled Greek girl of cinematic beauty, tapped out with one finger an order appointing me a Planning Officer with a salary of 450 rubles a month. That same day two other unplaced exiles were assigned to the RaiPO with just as little formality, with no forms to fill in for leisurely study: Vasilenko, the oceangoing ship's captain, and someone I did not yet know, the very reserved Grigory Samoilovich M——z. Vasilenko was already nursing a plan to deepen the river Chu (cows could ford it in the summer months) and introduce a motorboat service: he had asked the command post to let him go and survey the channel. Captain

Mann, his classmate at navigation school and on the training brig *Tovarishch,* was just at that time fitting out the *Ob* for a voyage to the Antarctic—but Vasilenko was packed off to the RaiPO as a storeman.

But we were not wanted as planners, or storemen, or ledger clerks—all three of us were thrown in to deal with an emergency: the repricing of goods. On the night of March 31–April 1, year in and year out, the RaiPO was in the throes of despair, and there never were and never could be enough people. They had to make an inventory of all the goods (and expose all the thieving salespeople, though not with a view to prosecution), reprice them—and begin trading the very next day at the new prices, which were so very advantageous to the toilers. But the total length of all railways and highroads in our desert district was . . . zero kilometers, and shops out in the wilds just never could bring these new prices, so advantageous to the toiler, into operation before May 1: for a solid month all shops stopped trading altogether, while the lists were being reckoned up and confirmed by the RaiPO, and delivered by camel. But in the district center itself, trade could simply not be disrupted just before May Day!

By the time we arrived in the RaiPO, fifteen people—permanent staff and temporaries—were already working on it. Every desk was sheeted with inventories on rough paper, and nothing could be heard but the clicking of abacuses—on which the expert bookkeepers were both multiplying and dividing—and businesslike exchanges of abuse. They sat us down to work at once. I was soon fed up with multiplying and dividing on paper, and I asked for a calculating machine. There wasn't a single one in the RaiPO, and anyway none of them knew how to use such a thing, but someone remembered seeing some sort of gadget with figures in a cupboard at the District Statistics Administration, only nobody there used it either. They telephoned, went over, and came back with it. I started whirring, and quickly jotting down rows of figures, while the senior bookkeepers glowered at me, wondering whether I was a rival.

I just turned my handle and thought to myself: How quickly a zek gets cheeky—or, putting it in literary language, how quickly a man's requirements grow! I was dissatisfied because they had torn me away from the play I was writing in my dark hovel; dissatisfied because they had not given me a job in a school; dissatisfied because they had *forced* me—to what? to dig in frozen soil? to mix mud for bricks with my bare feet in icy water? No, they had forcibly put me at a clean desk to turn the handle of a calculating machine and enter figures in columns. At the beginning of my time in the camps, if they had ordered me to do this blissful work twelve hours a day, without pay, for as long as I was inside, I should have been beside myself with joy! As it was, they were paying me 450 rubles for it, I should be able to drink a liter of milk every day, and I was turning my nose up and wanting more.

The RaiPO had been bogged down in the repricing exercise for a week (each item had to be put in the right category for the overall price reduction, then reclassified for the rural price differential) and still not a single shop could begin trading. Then the obese chairman, who beat all the rest for idleness, assembled us all in his purely ceremonial office, and said:

"Right, listen. According to the latest findings of medical science, nobody needs eight hours' sleep. Four hours are absolutely adequate! So these are my orders: you'll start work at seven A.M., finish at two A.M.; one hour's break for dinner and one for supper."

As far as I could tell, not one of us was amused rather than awed by this stupefying outburst. We all shrank into ourselves, said nothing, plucked up our courage only to discuss the best time for the supper break.

This was it—the exile's lot, of which I had been warned. It was made up of orders like this. All those sitting there were exiles, trembling for their jobs: if they were dismissed, it would be a long time before they found anything else in Kok-Terek. And anyway, it was not for the director personally, it was for the country, *it had to be done*. They even thought the latest findings of medical science quite reasonable. Oh, how I longed to get up and jeer at that self-satisfied hog! To relieve my feelings just for once! But that would have been "Anti-Soviet Agitation," pure and simple—inciting people to sabotage an operation of major importance. You go through life from stage to stage—schoolboy, student, citizen, soldier, prisoner, exile—and it is always the same: the bosses are always too heavy, too strong for you, and you must bow down and keep silent.

If he had said until ten in the evening I would have stayed. But he was ordering us to face bloodless execution; ordering me, in this place, where I was free, to stop writing! Oh, no! To hell with you and with the price reductions. My camp experience suggested a way out: not to answer back, but quietly to do the opposite. I tamely listened to the order with the rest of them, but at five o'clock I rose from my desk and left. And I did not return until nine in the morning. All my colleagues were already sitting there, counting, or pretending to count. They looked at me as though I were crazy. M——z, who secretly approved of my behavior but dared not imitate me, informed me privately that the boss had stood over my empty desk screaming that he would drive me a hundred kilometers into the desert.

I admit that my heart was in my boots. The MVD could of course do anything. They could easily chase me out there. A hundred kilometers—why not?—and I would have seen the last of the district center. But I was born lucky: I had landed on the Archipelago after the war, missing the most lethal period; and now I had arrived in exile after Stalin's death. In the past month something new had crept through even to our local MVD command.

Almost imperceptibly, a new time was beginning—the mildest three years in the history of the Archipelago.

The chairman did not summon me, or come to see me. I finished the day's work still fresh, with dozing and delirious people all around me, and decided to leave at five in the evening again. I didn't care how it ended, as long as it was quick.

How often in my life have I observed that a man can safely sacrifice a great deal as long as he clings to the essential. The play which I had been carrying within me even in the Special Camp, at hard labor, I refused to sacrifice—and I triumphed. For a whole week they all worked nights—and they got used to my empty desk. Even the chairman just looked the other way when he passed me in the corridor.

But it was not my destiny to bring order into the rural cooperatives of Ka-zek-stan. A young Kazakh, head of a teaching department, suddenly appeared in the RaiPO. Until I appeared he had been the only university graduate in Kok-Terek, and very proud of it. My arrival, however, did not arouse his envy. Whether he wanted to reinforce his school before the first batch of pupils took their final exam, or to vex the snakelike director of the Education Department, I don't know, but he ordered me to "bring your diploma along quickly." I ran home like a schoolboy to fetch it. He put it in his pocket and went off to a trade union conference in Dzhambul. Three days later he looked in again and laid before me an extract from an order made by the Oblast Education Department. Over the same shameless signature which had certified in March that the schools in the district were fully staffed, I was now, in April, appointed to teach both math and physics, to both graduating classes, and just three weeks before their final examination! He was taking a chance, this director of studies. Not so much politically; what he had to fear was that I might have forgotten all my mathematics during my years in the camps. When the day of the written examination in geometry and trigonometry arrived, he did not allow me to open the envelope in the presence of my pupils, but took all the teachers into the headmaster's office and stood over me while I solved the problems. The fact that my answers coincided with those in the envelope put him, and the other math teachers, in a festive mood. It was easy enough to pass for a second Descartes in that place! I had still to learn that every year, during the final examinations, there were periodic calls from the villages to the district center: We can't work it out! It wasn't formulated properly! The teachers themselves had only seven years of schooling behind them. . . .

Shall I describe the happiness it gave me to go into the classroom and pick up the chalk? This was really the day of my release, the restoration of my citizenship: I stopped noticing all the other things which made up the life of an exile.

When I was in Ekibastuz our column was often marched past the local school. I would look upon it as at some inaccessible paradise, at the children running about the yard, at the teachers in bright dresses, and the tinkle of the bell from the front steps cut me to the heart. I had been reduced to such desperate longing by my hopeless prison years, my years of general labor in the camps. It seemed to me the supreme, heartbreaking happiness to enter a classroom carrying a register as that bell rang, and start a lesson with the mysterious air of one about to unfold wonders. (This was, of course, my teacher's gift craving satisfaction, but partly perhaps my hunger for self-esteem. I needed the contrast after years of humiliation, years of knowing that my talents were unwanted.)

But while my gaze was fixed on the life of the Archipelago and the state at large, I had lost sight of something very elementary: that sometime during or since the war the Soviet school had died; it no longer existed; there remained only a bloated corpse, a bag of wind. In the capital and in the hamlet the schools were dead.

When spiritual death creeps through the land like poison gas, the school and its pupils are of course among the first to suffocate.

Yet I only discovered this some years later, when I returned from the land of exile to metropolitan Russia. In Kok-Terek I had no inkling of it: in the deadly fog of obscuration all around us, the exile children had not yet choked, they still lived.

Those were very special children. They were growing up in the consciousness of their depressed status. In school council meetings and other waffling sessions, they were described and sometimes heard themselves described as Soviet children, growing up to live under Communism, whose freedom of movement was only temporarily restricted—no more than that. But they felt, every one of them felt the collar around his neck, had felt it from early childhood, as long as he could remember. The whole world which they knew from magazines and films—so varied, so rich, bubbling with life—was inaccessible to them, and there was no hope of entering it even for the boys, through the army. There was a very faint chance that a very few would obtain permission from the MVD post to go to a city, to be allowed when they got there to take an entrance examination, to be admitted to an institute, if they passed, and once in, to complete their studies successfully. So that all the discoveries they would ever make about the vast, inexhaustible world must be made there, in the school, which, for many years, was the beginning and end of their education. Moreover, life in the wilderness was so starkly simple that they were free from the distractions and dissipations which spoil twentieth-century urban youth from London to Alma-Ata. In the urban centers, children had lost the habit and the taste for study, studied as though discharging an irksome obligation, just to stay on the books till they were old enough to leave. But for the children in our exile colonies, if they were well taught, there was nothing more important in life, nothing else mattered. Studying avidly, they felt that they were rising above their second-class status, competing on equal terms with first-class children. Only in earnest study could they slake their ambitions.

(No, there were other ways: by holding elective office in school; in the Komsomol; and, from the age of sixteen, at the polls, in general *elections*. How they longed, poor things, for the illusion of equal rights, if nothing more. Many proudly joined the Komsomol and made sincere political declarations in their five-minute speeches. I tried to instill into one German girl, Victoria Nuss, who had won a place in a two-year teachers' training college, the idea that an exile should be proud of his position, not distressed by it. It was hopeless. She looked at me as though I had gone mad. Of course, there were others who did not hurry into the Komsomol. They were hauled in forcibly. You still haven't joined, although you're allowed to—now why is that? In Kok-Terek some young girls, Germans, members of a clandestine religious sect, were compelled to join to save their parents from being driven farther out into the desert. Whoso shall offend one of these little ones . . . it were better for him that a millstone were hanged about his neck.)

I have been speaking all this time about the "Russian" classes in the Kok-Terek school (there were scarcely any Russians proper among them—they were Germans, Greeks, Koreans, a few Kurds and Chechens, some Ukrainians from families which had settled in the region at the beginning of the century, and Kazakhs whose parents had "responsible posts" and wanted their children educated in Russian). But most of the Kazakh children were in "Kazakh classes." They were in very truth still savages, and most of them (those who were not corrupted by the high standing of their families) were very straightforward, sincere, with a sound sense of good and evil until false or conceited teachers perverted it. In fact, nearly all teaching in Kazakh was merely the propagation of ignorance: the first generation, dragged with difficulty through their diploma course, half-educated and hugely self-important, dispersed to instruct the rising generation, while Kazakh girls left schools and teachers' training colleges with "satisfactory" marks in spite of their utter and impenetrable ignorance. So that when these barely civilized children caught a glimpse of real teaching, they drank it in not just with their eyes and ears but with open mouths.

With such receptive children, I reveled in my teaching duties at Kok-Terek, and for three years this sufficed to keep me happy (and perhaps it would have done so for many years longer). There were not hours enough in the timetable for me to correct or make up for the mistakes and omissions of the past, so I prescribed additional evening classes, group discussions, field work, astronomical observations, and they turned up in greater numbers and higher spirits than if they had been going to the cinema. I was also put in charge of a class—a purely Kazakh class at that—but even this was almost enjoyable.

However, the bright side was bounded by the classroom door and the lesson bell. In the staff common room, the director's office, and the District Education Department, the atmosphere was smirched not just with the petty tyranny universal in our state, but with a more pernicious variety peculiar to the land of exile. When I arrived there were already some Germans, and some administrative exiles, among the teachers. We were all in the same oppressed situation: every opportunity was taken to remind us that we were allowed to teach on sufferance, and that this favor could always be withdrawn. The exiles on the staff, even more than other teachers (though they, too, were just as dependent), dreaded angering high officials in the district by failing to give their children good marks. They dreaded, too, angering the authorities with poor examination results—and deliberately marked too high, making their own contribution to the propagation of ignorance in Kazakhstan at large. Apart from this, special dues and duties fell upon exile teachers (and also their young Kazakh colleagues): 25 rubles were deducted every payday, for whose benefit nobody knew. The headmaster, Berdenov, might suddenly announce that it was his little daughter's birthday, and the teachers would have to contribute 50 rubles each for a present. Apart from this, one or another of the teachers would be called to the headmaster's office or the Education Department and asked for a "loan" of 300 to 500 rubles. (These things, however, were typical of the local style or system. Kazakh pupils were forced to give a sheep or half a sheep each for the graduation ceremony; those who did were assured of their certificates even if they were completely ignorant. The graduation party would turn into a great booze-up for the district Party activists.) In addition, all the district bosses were taking correspondence courses, and teachers in our school were forced to complete all their written tests for them (the orders were transmitted as though from feudal lords, through the directors of studies, and the teacher serfs were not even vouchsafed a look at *their* external students).

I don't know whether it was firmness on my part, made possible by my "irreplaceability," which became immediately obvious, or whether it was the milder climate of the times, or perhaps both, that helped me to keep my neck out of this harness. My pupils would be eager to learn only as long as I marked honestly, and I did so, with no thought for district secretaries. Nor did I pay any levies, or make "loans" to the bosses (the snaky head of the

Education Department had the impudence to ask!). I thought it quite enough that the needy state fleeced us of a month's salary every May (exile had restored to us the free man's privilege, denied us in the camps, of *subscribing to the state loan*). But my concern for principle stopped there.

I worked side by side with the biology and chemistry teacher Georgi Stepanovich Mitrovich, a Serb who had done a *tenner* on the Kolyma for "counter-revolutionary Trotskyite activity," and who, though now old and sick, fought doggedly for justice in the local affairs of Kok-Terek. Dismissed from the District Health Department, he had nonetheless been taken on by the school, and had transferred his efforts there. Indeed, wherever you looked in Kok-Terek there was lawlessness aggravated by ignorance, barbaric conceit, and smug clannishness. This lawlessness was a dark and tangled thicket, but Mitrovich fought selflessly and disinterestedly against it (with Lenin's name on his lips, it is true), exposed the corrupt at teachers' meetings and district teachers' conferences, failed ignorant high-ranking external students and "onesheep" graduating students, wrote complaints to the oblast center, to Alma-Ata, sent telegrams to Khrushchev in person (seventy parents' signatures were collected for appeals on his behalf, and such telegrams were dispatched in another district, because they would not have got through from ours). He demanded checkups, inspections, then, when inspectors arrived and turned against him, he would start writing again; he was analyzed at special teachers' meetings, accused of filling children with anti-Soviet propaganda (and came within a hairbreadth of arrest!), accused no less seriously of ill-treating the goats that browsed on the Young Pioneers' garden plots; he was dismissed and reinstated, he tried to get compensation for enforced absence from work, he was transferred to another school, refused to go, was dismissed again—he put up a splendid fight! If only I had joined him, what a drubbing we would have given them!

Yet I gave him no help at all. I held my peace. I always avoided taking part in the final vote (so as not to be against him) by slipping away to a club meeting or a tutorial. In this way I did nothing to prevent the external students with Party cards from obtaining pass marks: they *were* the regime—let them cheat the regime of which they were part. I had my own concerns to keep secret: I was writing and writing. I was saving myself for a different struggle later on. But there is a larger question to be answered:

was Mitrovich's struggle right? Was it necessary?

His battle was utterly hopeless, and he knew it: no one could unravel that tangled skein. And if he had won hands down, it would have done nothing to improve the *social order*, the system. It would have been no more than a brief, vague gleam of hope in one narrow little spot, quickly swallowed by the clouds. Nothing that victory might bring could balance the risk of rearrest—which was the price he might pay (only the Khrushchev era saved Mitrovich). Yes, his battle was hopeless, but it is human to be outraged by injustice, even to the point of courting destruction! His struggle could end only in defeat—but no one could possibly call it useless. If we had not all been so sensible, not all been forever whining to each other: "It won't help! It can't do any good!" our land would have been quite different! Mitrovich was not even a citizen—he was only an exile—but the district authorities feared the flash of his spectacles.

They feared him, yes, but when election time came around—the bright day on which we *elected* our beloved democratic rulers—the difference disappeared between the intrepid warrior Mitrovich (where was his fighting spirit now?), my noncommittal self, and M——z, who was even more reserved, and on the face of it the most pliant of the three; we all alike concealed our suffering and our disgust and took part in that festival of fools. Nearly all exiles had *permission* to take part in elections, they cost so little, and even those *deprived of rights* suddenly discovered themselves on the list and were hurried off to vote at the double. We did not even have voting booths in Kok-Terek. There was one box, with undrawn curtains, somewhere up a corner, so out of the way that it would have been embarrassing to make for it. Voting consisted in carrying the ballot forms to an urn as quickly as you could and tossing them in. Even stopping to scrutinize the candidates' names was enough to rouse suspicion: why read them—maybe you think the Party organs don't know whom to nominate? When he had cast his vote, everyone was entitled to go boozing (drink his wages, or an advance which would always be given at election time). Dressed in their best suits, they all (exiles included) exchanged solemn greetings, wishing each other a *happy holiday*. . . .

That's one good thing you can say for the camps—there were none of these elections!

Once, Kok-Terek elected a people's judge, a Kazakh—unanimously, of course. As usual, they congratulated each other on the great day. But a few months later this judge was accused of a criminal offense, by the district in which he had previously dealt justice (unanimously elected there, too). It turned out that in Kok-Terek, also, he had already had time to line his pockets comfortably with bribes from private persons. Alas, they had to remove him and hold a by-election. Once more the candidate came from outside, a Kazakh whom nobody knew. On Sunday the whole town dressed in its best, voted early in the morning and unanimously. The same happy faces exchanged felicitations in the streets, without a twinkle in the eye.

In hard-labor camp we had openly mocked at the whole farce, but in exile you must not be too ready to share your thoughts: the exile's life is like that of the free—and the first habit of free men he adopts is the worst: their reticence. M——z was one of the few with whom I could talk about such things.

They had sent him there from Dzhezkazgan—and without a kopeck: his money had been kept back somewhere along the line. This did not worry the MVD in the least—they simply took him off the prison dole and turned him onto the streets of Kok-Terek, to steal or die, just as he pleased. That was when I lent him ten rubles and earned his gratitude forever. It was a long time before he stopped reminding me how I had saved him. One of his fixed characteristics was never forgetting a kindness. Nor an injury either. (He bore a grudge against Khudayev, for instance—the Chechen boy who had nearly fallen victim to a blood feud. But nothing stands still in this world: Khudayev, after his narrow escape, viciously and unjustly took out his spite on M——z's son.)

As an exile with no professional qualifications, M——z could not find a decent job in Kok-Terek. The best that came his way was the post of assistant in the school labs, which he greatly prized. But his post required him to be at everyone's beck and call, never to answer back, never to have a mind of his own. He did keep his thoughts to himself, he was unreachable under his outward politeness, and no one knew the first thing about him, not even why at nearly fifty he had no profession. He and I somehow became friendly, we never clashed, and quite often helped each other; our reactions and our way of expressing them, acquired in the camps, were identical. So that although he kept quiet about it for a long time, I finally learned the carefully concealed story of his public and his private past. It is instructive.

Before the war he had been secretary of the District Party Committee at Zh——, and during the war was officer in charge of the cipher section of a division. He had always held high positions, always been a person of consequence, never experienced the petty troubles of lesser folk. Then one day in 1942 one regiment in the division did not receive the order to retreat in time, and the cipher section was to blame. The mistake had to be corrected, but it also happened that all M——z's subordinates had been killed or disappeared somewhere, and the general sent M——z himself to the forward sector, into the jaws of the pincers which were already closing on the regiment. Order them to retreat! Save them! M——z went on horseback, deeply distressed and fearing for his life. On the way he found himself in such danger that he decided to go no farther, and doubted whether he would survive even where he was. He deliberately stopped, abandoned the regiment to its fate, betrayed it, dismounted, threw his arms around a tree (or hid behind it from bursting shells), and . . . and solemnly swore to Jehovah that if he lived he would be zealous in the faith and observe the holy law punctiliously. It all ended happily: every man in the regiment was killed or taken prisoner, but M——z came out alive, was sentenced to ten years in a camp under Article 58, did his time—and here he was now in Kok-Terek, with me. How rigorously he fulfilled his vow! His Communist past had left no trace in his heart or his mind. Only when his wife tricked him into it would he eat unclean fish, fish without scales. He could not avoid coming to work on Saturdays, but endeavored to do nothing while there. At home he rigorously observed all the rituals, and prayed—secretly, as Soviet circumstances demand.

Understandably, he revealed his story to very few people.

To me, it doesn't seem so very simple. The only simple thing about it is something which people in our country are particularly reluctant to accept: that the innermost core of our being is religion and not Party ideology.

How are we to judge him? Under whatever code of law you like —criminal law, military law, the laws of honor, the laws of patriotism, and the laws of the Party—this man deserved death or obloquy; he had destroyed a whole regiment to save his own life, not to mention his failure at that moment to hate as he should the most terrible enemy the Jews had ever had.

But M——z could have appealed to some higher code of law and retorted: Are not all your wars caused by the imbecility of politicians? What made Hitler cut his way into Russia, if not imbecility—his own, and Stalin's, and Chamberlain's imbecility? And now you want to send *me* to my death? Was it *you* who brought me into the world? Some will object that he should have said this (and so then should all those in the doomed regiment!) on enlistment, when they were giving him a handsome uniform to wear, not out there with his arms around a tree. Logically I have no intention of defending him; logically I ought to have hated him, despised him, felt sick when I shook hands with him.

But *I had no such feelings* toward him. Because I had not belonged to that regiment, not felt what it was like to be in their situation? Because I suspected that a hundred other factors had combined to decide their fate? Because I had never seen M——z in his pride, but only when he was vanquished? Whatever the reason, we shook hands warmly and sincerely every day, and never once did I feel that there was anything disgraceful in it. One man can be bent into so many shapes in a lifetime! How different he may become, for himself as well as for others. And one of these different selves we readily, eagerly stone to death, obeying an order, the law, an impulse, or our blind misconception.

But what if the stone slips from your hand? What if you yourself are deep in trouble, and begin to look at things with different eyes? At the crime. At the criminal. At him *and* at yourself.

In this thick volume we have pronounced absolution so often. I hear cries of astonishment and indignation. Where do you draw the line? Must we forgive everyone?

No, I have no intention of forgiving everyone. Only those who have fallen. While the idol towers over us on his commanding eminence, his brow creased imperiously, smug and insensate, mutilating our lives—just let me have the heaviest stone! Or let a dozen of us seize a battering ram and knock him off his perch.

But once he is overthrown, once the first furrow of self-awareness runs over his face as he crashes to the ground—lay down your stones!

He is returning to humanity unaided.

Do not deny him this God-given way.

■

After the places of exile described earlier, I have to admit that we in Kok-Terek, and exiles in Southern Kazakhstan and Kirghizia generally, were privileged. They settled us in inhabited places, that is to say, where there was water and the soil was not altogether barren (in the Chu Valley and the Kurdai district it was indeed highly fertile). Very many of us landed in towns (Dzhambul, Chimkent, Talass, even Alma-Ata or Frunze), where lack of rights made no palpable difference between us and other townspeople who had rights. Food was cheap, and work easy to find in these towns, and especially in industrial settlements, because the local population lacked all enthusiasm for industry, skilled trades, and intellectual occupations. But even those who ended in rural areas were not invariably driven into the kolkhoz. There were four thousand people in our Kok-Terek, most of them exiles, but only some Kazakh sections of the town came under a kolkhoz. The rest all managed to find jobs at the Machine and Tractor Station, or took a nominal job somewhere, even at a derisory salary, and lived on a quarter of a hectare of irrigated garden, with a cow, some pigs, and some sheep. Significantly, a group of Western Ukrainians living among us (in administrative exile after five years in the camps), and working hard building adobe houses for the local construction agency, found life on the clay soil of Kok-Terek (which burned to dust unless it was heavily irrigated, but was at least not kolkhoz land) so much more comfortable than life on the kolkhoz in their beloved and flourishing Ukraine that when the release order came they all stayed in Kok-Terek for good.

The security officers in Kok-Terek were lazy, too—one instance in which the universal Kazakh laziness was beneficial. There were a few informers among us, but we hardly noticed and came to no harm from them.

The main reason for their inactivity, however, and for the steady relaxation of discipline was the onset of the Khrushchev era. Its impact, muffled by the jolts and wobbles along the way, at last reached even us.

There was a disappointment to begin with: the "Voroshilov amnesty" (as the Archipelago called it, although it was promulgated by the Seven Boyars).* Stalin's cruel joke with the politicals on July 7, 1945, was a lesson which had not sunk in and was forgotten. Both in the camps and in exile, whispered rumors of an amnesty flourished. People have a remarkable capacity for pigheaded credulity. N. N. Grekova, for instance, a repeater with fifteen years of hell behind her, kept a picture of clear-eyed Voroshilov on the wall of her little hut, and believed that a miracle would come from him. Well, the miracle came: it was in a decree signed by Voroshilov that the government enjoyed another laugh at our expense, on March 27, 1953.

Strictly speaking, it was impossible to invent any obvious and reasonable excuse for the grief-stricken rulers of a grief-stricken country to set criminals free just at that moment in March, 1953 —unless perhaps they were suddenly overwhelmed by a sense of the transience of all that is. Though in ancient Russia, too, so Kotoshikhin tells us, it was the custom to release criminals on the day of a Tsar's interment, which incidentally was the signal for an orgy of looting ("the Muscovites are not God-fearing by nature, they rob the male and the female sex alike of their garments in the street, and beat them to death").[1] It was just the same on this occasion. With Stalin in his grave, his successors were anxious for popularity, though their explanation connected the amnesty with "the eradication of crime in our country." (Who, then, were all those people *inside?* If it had been true, there would have been no one to release!) Since, however, they were still wearing Stalinist blinkers, still slavishly thinking along the same old lines, they amnestied thieves and gangsters, but only those 58's with "up to five years inclusive." The uninitiated, with the ways of decent governments in mind, might suppose that "up to five years" would enable three-quarters of the politicals to go home. In fact, only 1 or 2 percent of our kind had such childish sentences. (The thieves were set loose upon the local inhabitants like locusts, and it was only some time later and with considerable strain that the militia reinstalled the amnestied bandits in their old reserve.)

The amnesty had interesting repercussions in our place of exile. There were some among us who had served a "kid's sentence" (up to five years) in their time, but then instead of being allowed to go home had been exiled without trial. In Kok-Terek this was true of certain very lonely old men and women from the Western Ukraine and Western Byelorussia—the meekest and unhappiest people in the world. They cheered up greatly after the amnesty, and waited to be sent home. But some two months later came the usual heartless clarification: inasmuch as they had been exiled (after having served their time in the camps and without trial) not for five years but in perpetuity, the previous five-year term of

---

1. I quote from Plekhanov, *Istoriya Russkoi Obshchestvennoi Mysli (A History of Russian Social Thought),* Moscow, 1919, Vol. I, Part 2, Chap. 9.

imprisonment which had led to their exile did not count, and they were not covered by the amnesty. . . . Then there was Tonya Kazachuk, a completely free woman who had come from the Ukraine to join her exiled husband, and had been registered as an exile settler for the sake of tidiness. When the amnesty came she rushed to the MVD post. "Ah," they very reasonably retorted, "you didn't get five years like your husband; you're here indefinitely so the amnesty doesn't affect you."

Draco, Solon, and Justinian the lawgivers would have burst with indignation! . . .

So no one got anything out of the amnesty. But as the months went by, and especially after the fall of Beria, by slow degrees, without fanfares, genuine relaxation began creeping into the land of exile. They allowed the five-year people to go home. They began allowing the children of exiles to attend higher-educational institutions in the vicinity. At work they stopped addressing exiles with rude familiarity. Life was easier all the time! Exiles began to rise in their professions.

Vacant desks were seen in the MVD post. "That other officer —where is he?" "He doesn't work here now." The staff was being drastically reduced, thinned out. They started handling us more gently. The sacred duty to *report* ceased to be quite so sacred. If a man had not turned up by dinnertime—"Never mind, next time will do!" One national group after another had certain rights restored. Travel within the district was now unrestricted, and trips to other oblasts much freer. Rumors flew thicker and faster: "They're going to let us go home." Sure enough, they let the Turkmen go (those who were exiled for having been prisoners of war). Then the Kurds. Houses were put up for sale, and house prices tottered.

They also released some elderly administrative exiles: people had pleaded their cause in Moscow, and lo and behold, they were *rehabilitated.* Excitement rippled through the exiles, leaving them feverish and confused. "Shall we be on the move soon? Is it really possible? . . ."

Ridiculous! As though that regime could ever become any kinder. The camp had taught me to be consistent in my disbelief. And anyway, there was no special need for me to believe: here, in the great Russian heartland, I had neither family nor friends, whereas here, in exile, I was experiencing something like happiness. I doubt whether I had ever lived so comfortably.

True, during my first year of exile a deadly disease was tormenting me, as though it was in league with my jailers. And for a whole
year no one in Kok-Terek could even determine what it was. I
could hardly stand on my feet in front of my class. I slept badly,
and ate very little. All that I had previously written in the camp
and stored in my memory, and all that I had composed in exile
since, I had to write down hastily and bury in the ground. (I
remember clearly that night before I left for Tashkent, the last
night of 1953: it seemed as though for me life, and literature, was
ending right there. I felt cheated.)

But my illness abated, and my two years of truly Beautiful Exile
began, with only one hardship, one sacrifice to cast a shadow: I
dared not marry, because there was no woman whom I could trust
with the secrets of my lonely life, my writing, my hiding places.
But all my days were lived in a state of constant blissfully height-
ened awareness, and I felt no constraint on my freedom. At school
I could give as many lessons as I wanted, in both shifts—and every
lesson brought a throbbing happiness, never weariness or bore-
dom. And every day I had a little time left for writing—and there
was never any need for me to attune my thoughts: as soon as I sat
down the lines raced from under my pen. On those Sundays when
we were not turned out to thin beet in the kolkhoz, I wrote without
pause—the whole Sunday through! While I was there I also began
on a novel (impounded ten years later), and I had writing enough
for a long time ahead. As for publication—that was not to be
expected until after I was dead.

By now I had some money, so I bought a little clay house for
myself, and ordered a firm table to write on, but I went on sleeping
on the same old bare wooden boxes. I also bought a shortwave
radio set, covered my windows at night, glued my ear to the silk
over the speaker, and through the cascading crash of jamming
tried to catch some of the forbidden news we longed for, and to
reconstruct from the general sense the parts I could not hear.

We were so worn out by decades of lying nonsense, we yearned
for any scrap of truth, however tattered—and yet this work was
not worth the time I wasted on it: the infantile West had no riches
of wisdom or courage to bestow on those of us who were nurtured
by the Archipelago.

My little house stood on the extreme eastern edge of the settlement. Beyond my gate there was an irrigation channel, the steppe,
and the sunrise each morning. Whenever there was a puff of wind

from the steppe, my lungs drank it in greedily. In the dusk and at night, whether it was dark or moonlit, I strolled about alone out there, inhaling and exhaling like a lunatic. There was no other dwelling less than a hundred meters off, to left, to right, or to the rear of me.

I was fully resigned to living there, if not "in perpetuity," then for twenty years at least (I did not believe that conditions of general freedom would come about sooner and I was not far wrong). I seemed to have lost all desire to go elsewhere (although my heart stood still when I looked at a map of Central Russia). I was aware of the whole world not as something beckoning to me from outside, but as something experienced and assimilated, entirely within myself, so that nothing remained for me to do but write about it.

I was replete.

When Radishchev was in exile, his friend Kutuzov wrote to him as follows: "It grieves me to tell you this, my friend, but . . . your position has its advantages. Cut off from all men, remote from all the objects that dazzle us—you can all the more profitably voyage within yourself; you can gaze upon yourself dispassionately, and consequently form less biased judgments about things at which you previously looked through a veil of ambition and worldly cares. Many things will perhaps appear to you in a completely new aspect."

Precisely so. And because I cherished the purer vision it gave me, I was fully conscious that exile was a blessing to me.

Meanwhile, the shifts and stirrings in the exile world were more and more noticeable. The MVD officers became positively kindly, and their numbers were further reduced. The nominal punishment for running away was now only five years—and even this was not imposed. One after another, national groups ceased reporting to the MVD, and then were granted the right to leave. Tremors of joy and hope disturbed our quiet exile. Suddenly, without hint or warning, yet another amnesty crept up on us—the "Adenauer amnesty" of September, 1955. This was after Adenauer had visited Moscow and stipulated that Khrushchev should free all the Germans. No sooner had Nikita ordered their release than the absurdity of the situation—releasing the Germans and holding on to their Russian collaborators with twenty-year sentences—was realized. But since these were all Polizei, headmen appointed by the Germans, or Vlasovites, no one was anxious to draw much public attention to this amnesty. And anyway, there is a general law on the dissemination of information in our country: trivialities are shouted from the rooftops, important news stealthily leaked. So that the biggest of all political amnesties since the October Revolution marked no special day, and was proclaimed on September 9 in a single newspaper—*Izvestiya*—and even there on an inside page, with no comment whatsoever, not a single article to keep it company.

How could I help being agitated? I read it: "Amnesty for persons who collaborated with the Germans." Where did that leave me? Apparently it did not apply to me: I had served in the Red Army throughout. To hell with you, then—so much the less to worry about. Then my friend L. Z. Kopelev wrote from Moscow: he had flourished this amnesty in a Moscow militia station and talked them into giving him a temporary residence permit. Shortly after, they had sent for him. "What's the idea, trying to bamboozle us? You weren't a collaborator, were you?" "No." "And you did serve in the Soviet Army?" "Yes." "Right; get the hell out of Moscow within twenty-four hours." He stayed where he was, of course, but "after nine in the evening I'm on tenterhooks—every ring of the bell, I think they've come for me!"

I thought with pleasure how much better off I was! Once I'd hidden my manuscripts (I did this every evening), I slept like a babe.

From my clean desert I imagined the teeming, fretful, vainglorious capital—and felt no urge at all to go there.

My Moscow friends, however, insisted. "Why have you taken it into your head to stay there? . . . Ask for a judicial review. They have started reconsidering cases!"

Why should I? . . . Where I was, I could spend a whole hour watching the ants who had bored a hole in the mud-brick foundations of my house without foremen, warders, or commanders of Camp Divisions, carrying loads of husks for their winter store. One morning they suddenly failed to appear, although the ground outside the house was strewn with husks. As it turned out, they had anticipated long before, they *knew,* that there would be rain that day, although the cheerful sunny sky gave no warning. After the rain the clouds were still heavy and black, but they crept out to work: they knew for sure that it would not rain again.

There, in the silence of exile, I could see with perfect certainty the true course of Pushkin's life. His first piece of good fortune was his banishment to the south, his second and greatest his banishment to Mikhailovskoye. There he should have lived, and gone on living, instead of hankering for other places. What fatal compulsion drew him to Petersburg? What fatality prompted him to marry? . . .

Still, it is difficult for a human heart to follow where reason leads. Difficult for a wood chip not to sail with the pouring stream. The Twentieth Congress arrived. For a long time we knew nothing about Khrushchev's speech. (They started reading it to some people in Kok-Terek, but kept it secret from the exiles, and we learned about it from the BBC.) But Mikoyan's words in an ordinary open newspaper were enough for me: "This is the first Leninist Congress" . . . since such and such a year. I knew that my enemy Stalin had fallen, which meant that I was on the way up.

And so . . . I applied for a review of my case.

Then, in spring, they lifted sentences of exile from all 58's.

In my weakness I abandoned my crystalline exile. And went into the turbid outside world.

The feelings of a former zek as he crosses the Volga from east to west to travel all day in a clanking train over the wooded Russian plain do not form part of this chapter.

In Moscow that summer, I phoned the Public Prosecutor's Office to ask how my appeal was going. They asked me to call another number—and the cordial, unassuming voice of an interrogator invited me to look in at the Lubyanka for a chat. In the famous office on Kuznetsky Most where passes were issued, they told me to wait. Suspecting that somebody's eyes were already on me, studying my face, in spite of my inner tension I put on a look of good-natured weariness and pretended to be watching a child who was playing, with no enjoyment at all, in the middle of the waiting room. Just as I thought! My new investigating officer was standing there in civilian clothes, observing me! When he had satisfied himself that I was not a white-hot enemy, he came up and very amiably took me to the Big Lubyanka. While we were still on the way, he was full of regrets that my life had been so horribly messed up (by whom?), that I had been denied wife and children. But the stuffy, electric-lighted corridors of the Lubyanka had not changed since I had been taken through them with shaven skull, hungry, sleepless, buttonless, hands behind my back. "What a brute of an investigator you got, that Yezepov. I remember his name; he's been demoted since." (He was probably sitting in the next room and calling my man names. . . .) "I was in counterespionage, the naval branch of SMERSH. We didn't have people like that!" (Ryumin was one of yours. You had Levshin and Libin.) But I nodded innocently: of course you didn't. He even laughed at my 1944 witticisms about Stalin: "You put it very neatly!" He praised the stories I had written at the front, which were on the file as incriminating evidence. "There's nothing anti-Soviet in them! Take them if you like, try and get them published." But I refused in a feeble, almost expiring, voice. "Heavens, no, I've given up literature long ago. If I live another few years, my dream is to study physics." (Seasonal camouflage! This is the game we shall play from now on.)

Spare the rod and spoil the . . . man. Prison was bound to teach us something. If only how to behave in front of the Cheka-GB.*

Chapter 7

■

Zeks at Liberty

We have had a chapter in this book on "Arrest." Do we need one now called "Release"?

Of those on whom the thunderbolt of arrest at one time or another fell (I shall speak only of 58's), I doubt whether a fifth, I should like to think that an eighth lived to experience this "release."

And anyway, release is surely something everybody understands. It has been described so often in world literature, shown in so many films: unlock my dungeon—out into the sunshine—the crowd goes wild—open-armed relatives.

But there is a curse on those "released" under the joyless sky of the Archipelago, and as they move into freedom the clouds will grow darker.

Only in its long-windedness, its leisureliness, its otiose flourishes (what need has the law to hurry now?), does release differ from the lightning stroke of arrest. In all other respects, release is arrest all over again, the same sort of punishing transition from state to state, shattering your breast, the structure of your life and your ideas, and promising nothing in return.

If arrest is like the swift touch of frost on a liquid surface, release is the uncertain half-thaw between two frosts.

Between two arrests.

Because in this country, whenever someone is released, somewhere an arrest must follow.

The space between two arrests—that is what release meant throughout the forty pre-Khrushchev years.

A life belt thrown between two islands—splash your way from camp to camp! . . .

The wait between the very first day and the very last—that's what a *stretch* means. The walk from one camp boundary to the next—that's what is meant by release.

Your muddy green passport, which the poet so insistently bade all men envy,* is smirched with Article 39 of the law on passports, in black indelible ink. This means that you will not be given a residence permit in a town, even a small one, nor will you ever get a decent job. In the camp at least you received your rations, but here you do not.

Moreover, your freedom of movement is illusory. . . .

Not "released," but "deprived of exile" would be the best description of these unfortunates. Denied the blessings of an exile decreed by fate, they cannot force themselves to go into the Krasnoyarsk taiga or the Kazakh desert, where there are so many of their own kind, so many *ex*es, all around. No, they plunge deep into the tormented world of *freedom,* where everyone recoils from them, and where they are marked men, candidates for a new spell inside.

Natalya Ivanovna Stolyarova was discharged from Karlag on April 27, 1945. She could not go away immediately: she had to obtain a passport, she had no bread card, nowhere to live, the only work offered was splitting logs for firewood. When she had gone through the few rubles collected by friends inside, Stolyarova returned to the camp gates, told the guards some tale about coming to fetch her things (the local customs were patriarchal), and . . . went home to her hut! There never was such happiness! Her friends flocked around, brought her gruel from the kitchen (oh, how nice it tasted!), laughed and listened to her stories about how lost and homeless she had been outside. No, thanks—we're more comfortable here. Roll call. One too many! . . . The duty officer scolded, but let her stay in camp overnight (May Day eve), as long as she pushed off in the morning!

Stolyarova had worked hard in the camp, never eased up. (She had come to the Soviet Union from Paris when she was quite young, had been put inside shortly afterward, and had been longing to be freed so that she could see something of her Motherland!) As a reward for her "good work" she had been discharged on privileged terms: she was not assigned to any specified locality. Those who were sent to a particular place found jobs somehow: the militia could not chase them off elsewhere. But Stolyarova, with her "clean" certificate of discharge, was like a dog turned out of doors. Wherever she went, the militia refused to register her. Moscow families she knew well would make tea for her, but no one offered her a bed. She spent her nights at the stations. (It was bad enough that militiamen walked around all night to prevent people from sleeping, and turned them out before dawn so that the floors could be swept—but that was not all: any discharged zek whose path has ever lain through a big station will remember how his heart sank at the approach of a militiaman. How stern he looks! His nose, of course, tells him that you are a former zek! Any moment now he'll ask for your papers! He will take away your discharge certificate—and that's it, you're a zek again. For us rights, laws, even the human being himself, are nothing: the document is everything! Any minute now he'll lay hands on your certificate—and that's that. . . . These are our feelings. . . .) Stolyarova wanted to take a job in Luga, knitting gloves—not for toilers generally, mind you, just for German prisoners of war! Not only would they not take her, but the boss insulted her, with everyone listening. "Trying to worm her way into our organization! We know their sly tricks! We've read our Sheinin!" (Oh, that fat Sheinin! If he had any shame, he'd hang himself!)

It's a vicious circle: no job without a residence permit, no residence permit unless you have a job. And without a job you have no bread card either. Former zeks did not know the rule that the MVD is required to find them work. And those who did know were afraid to apply in case they were *put back inside*. . . .

You may be free, but your troubles are only beginning.

In my student days, there was a strange professor at Rostov University called N. A. Trifonov. His head was always pulled into his shoulders, he was always tense and jumpy, he didn't like people calling to him in the corridor. We learned later that he had *been inside*—and if anybody called out his name in the corridor he thought perhaps the security officers had sent for him.

At the Rostov Medical Institute after the war, a doctor discharged from a camp preferred not to wait for the second arrest he thought inevitable, and committed suicide. Anyone who has sampled the camps, who *knows* them, may very well make this choice. It is no harder.

Really unlucky were those who were released *too soon!* Avenir

Borisov's turn came in 1946. He went not to some big town but to his native settlement. All his old friends, people he had grown up with, tried to avoid him in the street, or to pass by without speaking (and these were yesterday's fearless front-line soldiers!); if conversation was quite unavoidable they made a few carefully noncommittal remarks as they edged away. *Nobody* ever asked him how life had treated him all those years (although most of us probably know less about the Archipelago than about Central Africa!). (Will our descendants ever realize how well trained our *free* citizens were!) However, one old friend of his student days did invite him to tea in the evening, after dark. Such friendliness! Such warmth! There can be no thaw without latent heat! Avenir asked to see some old photographs and his friend got out the albums for him. The friend had *forgotten*—and was surprised when Avenir suddenly rose and left without waiting for the samovar. Imagine Avenir's feelings when he saw his own face inked out in all the photographs![1]

Later on, Avenir moved up in the world a little, and became superintendent of an orphanage. Those growing up under his care were the children of soldiers, and they shed tears of outrage when the children of well-off parents called their superintendent "the jailer." (There is never anyone around to explain: their parents ought rather to be called the jailers, and Avenir the jail*ee*. In the last century the Russian people would never have shown such insensitivity to its own language!)

Although he was in under Article 58, Kartel was *administratively discharged* from a camp in 1943 with pulmonary consumption. He had a black mark in his passport—he could not live in any town, could not get work, he could only die slowly—and everyone shunned him. Then . . . the recruiting commission arrived in a hurry, looking for men to fight. Although he had tuberculosis in an advanced form, Kartel declared himself fit: he would perish among his peers! He served at the front almost till the end of the war. Only in the hospital did the eye of the Third Section detect an enemy of the people in this self-sacrificing soldier! In 1949 he was marked down for arrest as a repeater, but good people in the military registration department came to his help.

---

1. Five years later the friend put the blame for this on his wife: she had inked over the photographs. Ten years after that (in 1961) the wife came to Avenir in the district trade union committee to ask for a pass to go to Sochi. He gave it to her, and she reminisced effusively about their former friendship.

In the Stalin years the best way to be *released* was to go through the gates and stay right there. Such people were already known at work and were given jobs. Even the MVD men, when they met them in the street, looked on them as people who had passed the test.

Not always, though. When Prokhorov-Pustover was released in 1938 he stayed on at Bamlag as a free engineer. Rosenblit, the head of the Security Section, told him: "You've been released, but remember you'll be walking a tightrope. One little slip, and you're a zek again. *We shouldn't even have to put you on trial.* So watch your step—and don't start imagining that you're a free citizen."

Sensible zeks of this sort, who stayed on near the camps, voluntarily choosing prison as a variety of freedom, can still be found in hundreds of thousands out in the backwoods, in such places as the Nyrob or Narym districts. And if they have to go inside again —it's less trouble; it's just next door.

On the Kolyma there was really not much choice: they *hung on to* people. The discharged zek immediately signed a *voluntary* undertaking to go on working for Dalstroi. (Permission to leave for the mainland was even harder to obtain than your discharge.) N. V. Surovtseva, for instance, was unlucky enough to reach the end of her sentence. Only yesterday she had worked in the children's settlement, where she was warm and well fed, but now she was turned out to work in the fields, because there was nothing else. Only yesterday she was assured of her bed and her rations, and now she had neither rations nor a roof over her head; she found her way to a ruined house with rotten floors (this on the Kolyma!). She was thankful for her friends in the children's settlement: for a long time after her release they went on "slipping" her food. "The oppressive burden of freedom"—that is how she described her new experiences. Only gradually did she get on her feet, to become in the end . . . a house owner! See her proudly standing outside a shack which some dogs would not wish to own (Plate No. 6).

(In case the reader thinks that all this is true only of the unspeakable Kolyma, let us move over to Vorkuta and look at a typical VGS [Temporary Civil Construction] hut, of the sort in which comfortable members of the free population—former zeks, of course—are housed [Plate No. 7].)

So that release can easily mean something much worse than it did for M. P. Yakubovich. A prison just outside Karaganda was fitted out as a home for invalids—the Tikhonov home—and he was "released" into it, under surveillance and without the right to leave.

Rudkovsky, when no one would have him ("I suffered just as much as I had in the camps"), went to the virgin lands, near Kustanai ("You meet all sorts of people there"). . . . I. V. Shved lost his hearing, assembling trains at Norilsk in the fiercest blizzards; later he worked as a stoker twelve hours a day. But he lacked the necessary certificates. The social security officials shrugged their shoulders. "Can you produce witnesses?" Well, there were the walruses. . . . I. S. Karpunich had done twenty years on the Kolyma, and was worn out and sick. But at the age of sixty he still hadn't "twenty-five years of paid employment" behind him —so no pension. The longer a man had been in the camps, the sicker he was, the shorter his service record, and the slimmer his chances of a pension.

We have, of course, no Prisoners Aid Society, as they do in England. It is frightening even to think of anything so heretical.[2]

People write to me that "camp was one day in the life of Ivan Denisovich—and being back outside was another."

But wait a minute! Surely, since then, the sun of freedom has risen? Surely hands have reached out to the hapless: *"It will never happen again."* Surely tears have even dripped onto the platforms of Party Congresses?

Zhukov (from Kovrov): "Though I'm still not on my feet, I can more or less kneel now." But "we are still labeled convicts, and when staffs are reduced we're the first to go. . . ." P. G. Tikhonov: "I'm rehabilitated and work in a research institute, but it still feels like a continuation of the camp. Blockheads just like our bosses in the camps" still have power over him. . . . G. F. Popov: "Whatever anybody says, or writes, my colleagues only have to learn that I've been inside, and they start 'accidentally' turning their backs on me."

---

2. Nowadays the same sort of thing happens to *bytoviki* (nonpolitical offenders). A. I. Burlake went to the Ananiev district Party headquarters: "This is not an appointments bureau"; to the District Attorney's Office: "That's not our job"; and to the town soviet: "You'll have to wait." He was without work for five months (1964). At Novorossiisk (in 1965) they immediately made P. K. Yegorov sign an undertaking to leave the town in twenty-four hours. In the City Executive Committee's office he showed them a camp commendation for "excellent work"—and they laughed. The secretary of the City Party Committee simply threw him out. He went away, bribed somebody, and stayed in Novorossiisk.

Yes, the devil is strong! Our Fatherland is like this: to shove it a yard or two along the road to tyranny, a frown, a little cough, is enough. But to drag it an inch along the road to freedom you must harness a hundred oxen and keep after each of them with a cudgel: "Watch where you're pulling! Watch where you're going!"

What about the formalities of rehabilitation? Ch——, an old woman, receives a brusque summons: "Report to the militia at 10 A.M. tomorrow." Nothing more! Her daughter hurries there with the summons that evening. "I'm afraid for her; this could kill her. What's it all about? What should I prepare her for?" "Don't be afraid—it's something *that will please her:* her late husband has been rehabilitated." (But perhaps the news will be like wormwood to her? That would never occur to her benefactors.)

If these are the forms which our *mercy* takes, just imagine what our cruelty is like!

What an avalanche of rehabilitations there was! But even this could not crack the stony brow of our infallible society! For the avalanche fell not on the road along which you need only frown, but on that for which you must harness a thousand oxen.

"Rehabilitation was just a stunt!" Party bosses will tell you so frankly. *"Far too many were rehabilitated!"*

Voldemar Zarin (Rostov-on-the-Don) did fifteen years and meekly held his tongue for another eight. Then in 1960 he thought it right to tell his fellow employees how bad things were in the camps. He was put under investigation, and a KGB major told him: *"Rehabilitation does not mean that you were innocent,* only that your crimes were not all that serious. *But there's always a bit left over!"*

In Riga, also in 1960, Petropavlovsky's colleagues to a man joined in making his life a misery for three months on end, because he had concealed the fact that his father had been shot . . . in 1937!

Komogor is at a loss. "Who is supposed to be in the right nowadays, and who is guilty? Where do I put myself when some plugugly suddenly starts talking about equality and fraternity?"

After his rehabilitation Markelov became—of all things—chairman of an industrial insurance council, or putting it simply, of the trade union branch in a cooperative enterprise. The chairman of the enterprise would not risk leaving this chosen of the people alone in his office for a single minute! And Bayev, secretary of the Party Bureau, who was simultaneously *in charge of personnel,* to be on the safe side intercepted all mail intended for Markelov as head of the trade union committee. "There's a piece of paper about midterm elections to local committees—has it come your way?" "Yes, there was something a month ago." "Well, I need it!" "Here you are, then—read it, but be quick; it's nearly time to go home!" "Look, it's addressed to me! I'll let you have it back tomorrow!" "What are you talking about—it's an *official document.*" Put yourself in Markelov's shoes, with a thug like Bayev over you; imagine that your salary and residence permit depend entirely on the said Bayev—and fill your lungs with the air of freedom!

Deyeva, a schoolteacher, was dismissed for "immorality": she had lowered the dignity of her calling by marrying . . . a discharged convict (a pupil of hers in the camp)!

And this was under Khrushchev, not Stalin.

You have nothing concrete to show for your past except the *Certificate.*

A small sheet of paper, 12 centimeters by 18. Declaring the living "rehabilitated" and the dead "dead." The date of decease there is no way of checking. Death occurred at—a big capital "Z" (meaning "in prison"). Cause of death—you can leaf through a hundred of them and find the same "disease of the day."[3] Sometimes the names of (bogus) witnesses are added.

While the real witnesses all remain silent.

We . . . are silent.

From whom can future generations learn anything? The door is locked, nailed up; the cracks papered over. . . .

"Even the young," Verbovsky complains, "look upon the rehabilitated with suspicion and contempt."

Not all of them, though. The majority of young people could not care less whether we have been rehabilitated or not, whether twelve million people are still inside or are inside no longer; they do not see that it affects them. Just so long as they themselves are at liberty, with their tape recorders and their disheveled girl friends.

A fish does not campaign against fisheries—it only tries to slip through the mesh.

---

3. Young Ch——na asked an unsuspecting girl to show her all forty cards in a batch. On every one the same liver complaint had been entered in the same handwriting! . . . And then there was this sort of thing: "Your husband [Aleksandr Petrovich Malyavko-Vysotsky] died before he could be brought to trial, and cannot therefore be rehabilitated."

■

Just as a common illness develops differently in different people, the effects of freedom upon us varied greatly, as closer inspection will show.

Its physical effects, to begin with . . . Some had overstrained themselves in the fight to end their time in the camps alive. They had endured it all like men of steel, consuming for ten whole years a fraction of what the body requires; working and slaving; breaking stones half-naked in freezing weather—and never catching cold. But once their sentence was served, once the inhuman pressure from outside was lifted, the tension inside them also slackened. Such people are destroyed by a sudden drop in pressure. The giant Chulpenyov, who had never caught cold in seven years as a lumberjack, contracted a variety of illnesses once he was freed. . . . G. A. Sorokin: "After rehabilitation my mental health, which had been the envy of my comrades in the camp, steadily deteriorated. A succession of neuroses and psychoses . . ." Igor Kaminov: "After my release I weakened and went to pieces, and I seem to find things much harder outside."

There used to be a saying: The hard times brace you, and the soft times drive you to drink. Sometimes a man's teeth would all fall out in a year. Sometimes he would grow old overnight. Another man's strength would give out as soon as he got home, and he would die burned out.

Yet there were others who took heart when they were released. For them, it was time to grow younger and spread their wings. (I, for one, still look younger than I did in the first photograph I had taken in exile.) It comes as a sudden revelation: life after all is so *easy* when you're free! There, on the Archipelago, the force of gravity is quite different, your legs are as heavy as an elephant's, but here they move as nimbly as a sparrow's. All the problems which tease and torment men who have always been free we solve with a single click of the tongue. We have our own cheerful standards: "Things have been worse!" Things used to be worse—so now everything is quite easy. We never get tired of repeating it: Things have been worse! Things have been worse!

But the pattern of a man's future may be even more firmly drawn by the emotional crisis which he undergoes at the moment of release. This crisis can take very different forms. Only on the threshold of the guardhouse do you begin to feel that what you are leaving behind you is both your prison and your homeland. This was your spiritual birthplace, and a secret part of your soul will remain here forever—while your feet trudge on into the dumb and unwelcoming expanse of *freedom.*

The camps bring out a man's character—but so does release! This is how Vera Alekseyevna Korneyeva, whom we have met before in our story, took leave of a Special Camp in 1951. "The five-meter gates closed behind me, and although I could hardly believe it myself, I was weeping as I walked out to freedom. Weeping for what? . . . I felt as though I had torn my heart away from what was dearest and most precious to it, from my comrades in misfortune. The gates closed—and it was all finished. I should never see those people again, never receive any news from them. *It was as though I had passed on to the next world.* . . ."

To the next world! . . . Release as a form of death. Perhaps we had not been released? Perhaps we had died, to begin a completely new life beyond the grave? A somewhat ghostly existence, in which we cautiously felt the objects about us, trying to identify them.

Release into *the world of the living* had of course been thought of quite differently. We had pictured it as Pushkin did: "And your brothers shall return your sword to you."* But such happiness is the lot of very few generations of prisoners.

It was as though our freedom was stolen, not authentic. Those who felt like this seized their scrap of stolen freedom and ran with it to some lonely place. "While in the camp almost all my closest comrades thought, as I did, that if ever God allowed us to leave the camp alive, we would not live in towns, or even in villages, but somewhere in the depths of the forest. We would find work as foresters, rangers, or failing that, as herdsmen, and stay as far away as we could from people, politics, and all the snares and delusions of the world." (V. V. Pospelov) For some time after he was discharged Avenir Borisov shunned other people and took refuge in the countryside. "I felt like hugging and kissing every birch tree, every poplar. The rustle of fallen leaves (I was released in autumn) was like music to me, and tears came to my eyes. I didn't give a damn that I only got 500 grams of bread—I could listen to the silence for hours on end, and read books, too. Any sort of work seemed easy and simple now that I was free, the days flew by like hours, my thirst for life was unquenchable. If there is any happiness in the world at all, it is certainly that which comes to any zek in the first year of his life as a free man!"

It is a long time before people like this want to *own* anything: they remember that property is easily lost, vanishes into thin air. They have an almost superstitious aversion to new things, go on wearing the same old clothes, sitting on the same old broken chairs. One friend of mine had furniture so rickety that there was nothing you could safely sit or lean on. They made a joke of it. "This is the way to live—between camps." (His wife had also been inside.)

L. Kopelev came back to Moscow in 1955 and made a discovery: "I am ill-at-ease with successful people! I keep up an acquaintance only with those of my former friends who are in some way unlucky."

But then, only those who decline to scramble up the career ladder are interesting as human beings. Nothing is more boring than a man with a career.

But people vary. And many crossed the line to freedom with quite different feelings (especially in the days when the Cheka-KGB seemed to be closing its eyes a little). Hurrah! I'm free! One thing I solemnly swear: Never to land inside again! Now I'm going all out to make up for what I've missed.

Some want to make up for the appointments they should have held, some for the titles (learned or military) they might have won, some for the money they could have earned and salted away (it is considered bad taste among us to talk about salaries or savings accounts, but this does not mean people don't keep track). Others want to make up for their unborn children. Others, well . . . Valentin M. swore to us in jail that when he got out he would *make up* for all the girls he had missed, and true enough, for some years on end he spent his days at work and all his evenings, even in midweek, with girls, different girls all the time; he slept only four or five hours a night, became wizened and old. Others want to catch up on meals, or furniture or clothes (to forget how their buttons were cut off, how their best things were damaged beyond repair in the sterilization rooms of bathhouses). Once again buying becomes their favorite occupation.

How can you blame them when they have indeed missed so much? When such a large slice has been cut out of their lives?

There are these two reactions to freedom and, corresponding to them, two attitudes to the past.

You have lived through years of horror. You are not a foul

murderer, not a dirty swindler—so why should you try to forget prison and labor camp? Why need you be ashamed of them? Wouldn't it pay you to think yourself the richer for them? Wouldn't it be more fitting to take pride in them?

Yet so many (some of them neither weak nor stupid, people of whom you would not expect it) do their best *to forget!* To forget as quickly as they can! To forget utterly and finally! To forget as though it had never happened!

Y. G. Vendelstein: "As a rule you try not to remember—it's a defensive reflex." Pronman: "To be honest, I didn't want to meet former inhabitants of the camps and be reminded of things." S. A. Lesovik: "When I returned from the camp I tried not to remember the past. And you know, I very nearly succeeded" (until she read *One Day*). S. A. Bondarin (I had known for a long time that in 1945 he had been in the same cell in the Lubyanka before me; I offered to give him the names not only of our cellmates, but of those with whom he had previously shared a cell, people I had never known at all—and this is the answer I got): "I've tried to forget all those who were inside with me." (After that I did not, of course, even trouble to answer him.)

I can understand the orthodox avoiding old acquaintances from the camps: they are sick of barking one against a hundred; their memories are too painful. In any case, what would they want with those unclean philistines? And how can they call themselves loyal Communists, unless they forget and forgive and return to their former condition? After all, they had humbly petitioned for it four times a year: Take me back! Take me back! I've been good and I will be good![4] What did *going back* mean to them? Above all, recovering their Party cards. Their service records. Their seniority. Their merit awards.

> Vindicated—and as I take back my Party card
> A sun-warmed breeze caresses me.

Camp experience is a scummy contagion of which they must quickly swab themselves clean. Sift it, sluice it as long as you like —you will find not the tiniest grain of precious metal in your camp experience.

Take the old Leningrad Bolshevik Vasilyev. He had done two

---

4. This was their cry when they came pouring out in 1956: they brought with them the stale air of the thirties, just as though they had opened a stuffy chest, and hoped to start just where they had left off when they were arrested.

tenners, each with a five-year "muzzle" (deprivation of civil rights). He was given a special pension by the republic. "I am fully provided for. All praise to my Party and my people." (Wonderful words: there has been nothing like this since Job glorified God: for his sores, the loss of his cattle, the famine, the deaths of his children, his humiliations—blessed be Thy name!) But he is no loafer, this Vasilyev, no mere passenger: "I am a member of a commission for combating parasitism." In other words, he putters away as far as his aging powers permit, contributing to one of the worst legal abuses of the day. There you see it—the face of Righteousness! . . .

It is equally understandable that stoolies have no wish to remember, or to meet people: they fear exposure and reproach.

But what about the rest? Aren't they taking servility too far? Is it a voluntary pledge they make, for fear of landing inside again? Nastenka V., who landed in jail with a bullet wound, presses her fists to her temples and exclaims: "I want to forget it all like a hallucination, to escape from nightmarish visions of my past in those hellish camps." How could the classical scholar A.D., whose normal occupation was pondering over episodes in ancient history —how could even he command himself to "forget it all"? If he can do that, will he ever understand anything in the whole history of mankind?

Yevgeniya D., talking to me in 1965 about her spell in the Lubyanka in 1921, before her marriage, added: "But I never said a thing about it to my late husband—I *forgot.* " Forgot? Forgot to tell the person closest to her, the person with whom she shared her life? Perhaps we need to be jailed more often!

Or do you think I judge too harshly? Perhaps this is just the normal behavior of average human beings? Proverbs, after all, are made up about people: "If there's good luck tomorrow you'll forget all your sorrow." "As the paunch grows, the memory goes." Yes, men lose their shape—and their memories!

My friend and co-defendant Nikolai V.—the puerile efforts that put us behind bars were as much his as mine—thought of his whole past as a curse, as the shameful failure of a foolish man. He plunged into scholarship, the safest of all occupations, hoping to mend his fortunes. In 1959 I started talking to him about Pasternak, who was still alive but hemmed in by his persecutors. He didn't want to hear about it. "Never mind about all that rot. Just let me tell you about the *fight I'm having* in my department!" (He was always locking horns with somebody, looking for promotion.) Yet the tribunal had thought him worth ten years in the camps. Perhaps one good flogging was all he deserved?

Then there was Grigory M——z: he was discharged, his conviction was declared null and void, he was rehabilitated, they gave him his Party card back. (Nobody, of course, stops to ask whether you have started believing in Jehovah or Mohammed in the meantime; nobody entertains the possibility that not one particle of your old beliefs has survived all that time—here, take your Party card!) He passed through my town on his way back from Kazakhstan to his native Zh——, and I met his train. What was he thinking of now? Hold on—could he be hoping to get back into some Secret or Special or Specialist Section? Our conversation was rather disjointed. He never wrote me another line. . . .

Or take F. Retz. Nowadays he is the head of a housing bureau, and also a member of a volunteer patrol for assisting the militia. He talks self-importantly about his present life. And although he hasn't forgotten his old life—who could forget eighteen years on the Kolyma?—he talks about the Kolyma with much less feeling, wonderingly, as though not sure that it all really happened. Surely it couldn't have happened. . . . He has shed his past. He is sleek and satisfied with the world at large.

The thief who *goes straight* and the ersatz political who *forgets* are much alike. Now that they are back in the fold, the world becomes comfortable again; it neither prickles nor pinches. Just as they used to think that *everyone was inside,* they now think that no one is inside. May Day and the anniversary of the Revolution regain their old, comfortable meaning—they are no longer those grim days on which we stood in the cold to be searched with more than the ordinary sadistic thoroughness, and were packed in greater numbers than usual into the congested cells of the camp jail. But never mind these grand occasions. If during his day's work the head of the family is praised by his bosses—it's a red-letter day, and dinner will be a feast.

Only at home with his family does the sometime martyr permit himself an occasional grumble. Only there does he occasionally *remember,* to make them fuss over him and appreciate him the more. Once through the door—he has *forgotten* again.

We must, however, not be so unbending. It is of course a universal human characteristic: a man returns after hurtful experiences to his old self, to many of his old (not necessarily admirable) ways and habits. This is a manifestation of the stability of his personality, his genetic pattern. If it were not so, man would probably not be man. The same Taras Shevchenko whose despairing verses were quoted earlier[5] wrote joyfully ten years later that "not a single lineament in my inner self has changed. I thank my almighty Creator with all my heart and soul that He has not allowed my horrible experience to touch my beliefs with its iron claws."

But *how* do they manage to *forget?* Where can the art be learned? . . .

"No!" writes M. I. Kalinina. "Nothing is forgotten, and nothing in my life will come right. It gives me no pleasure to be as I am. I may be doing well at work, everything can be going smoothly in my home life—yet something endlessly gnaws at my heart, and I feel a boundless weariness. You will not, I hope, write that people discharged from camps have forgotten it all and are happy?"

Raisa Lazutina: "They tell us we shouldn't remember the bad things. But what if there is *nothing* good to remember? . . ."

Tamara Prytkova: "I was inside for twelve years, and although I've been free for eleven years since then I *simply cannot see the point of living.* I cannot see that there is any justice."

For two centuries Europe has been prating about equality—but how very different we all are! How unlike are the furrows life leaves on our souls. We can forget nothing in eleven years—or forget everything the day after. . . .

Ivan Dobryak: "It's all behind me—but not quite all. I have been rehabilitated, but there's no peace for me. There are not many weeks when I sleep peacefully—I keep dreaming of the camp. I jump up in tears, or somebody takes fright and wakes me up."

Ans Bernshtein dreams of nothing but the camp after eleven years. For five years I myself was always a prisoner in my dreams, never a free man. L. Kopelev fell ill fourteen years after his discharge—and in his delirium he was back in jail immediately. "Cabin" and "ward" are words our tongues cannot pronounce at all—it's always "cell."

Shavirin: "To this day I cannot look calmly at German shepherd dogs."

Chulpenyov walks through a wood, but he cannot simply

5. Part III, Chapter 19.

breathe deeply and enjoy himself: "I look at the pine trees; oh, they're good ones, these, not too knotty, hardly any trimmings to burn; they'll make neat blocks. . . ."

How can you forget if you settle in the village of Miltsevo, where nearly half the inhabitants have passed through the camps, most of them, it's true, for thieving. As you approach Ryazan station, you see that the fence is broken and three stakes are missing. Nobody ever repairs the hole—they seem to want it like that. Because the Stolypin cars stop exactly opposite this stop—they still do, they stop there to this day! The Black Marias are backed up to the gap and the zeks are pushed through it. (This is less awkward than taking zeks across a crowded platform.) You are asked to give a lecture (1957) by the All-Union Society for the Dissemination of Ignorance, and find that the order is made out for Corrective Labor Colony No. 2—a women's division attached to the prison. You go to the guardhouse, and a familiar cap appears at the judas window. You walk with the education officer across the prison yard, and shabby, dejected women keep greeting you obsequiously before you can greet them. Now you are closeted with the head of the Political Section, and while he entertains you, you know that out there they are driving women out of their cells, rousing the sleepers, snatching saucepans from those using the individual kitchen: Off you go, to the lecture, and look sharp about it! They round up enough to fill the hall. The hall is damp, the corridors are damp, the cells are probably damper still—and the unfortunate women laborers cough all through my lecture: loud, heavy, chronic coughing, short barking coughs, tearing gasping coughs. They are dressed not like women but like caricatures of women; the young ones are as angular and bony as old women, and they are all worn out and wishing my lecture would end. I feel ashamed. How I wish I could dissolve in smoke and vanish. Instead of these "achievements of science and technology," I want to cry out to them: "Women! How long can this go on? . . ." My eye immediately picks out some fresh-looking, well-dressed women, some of them even wearing jumpers. These are the trusties. If I let my gaze rest on them and do not listen to the coughing, I can get through the lecture quite smoothly. They listen eagerly, never taking their eyes off me . . . but I know that they are paying no heed to what I say, that it isn't the cosmos they are interested in: it's simply that they rarely see a man and so are looking me over carefully. . . . I start imagining things: any minute now my pass will be taken away and I shall have to stay here. And these walls, only a few meters away from a street I know well, from a familiar trolley-bus stop, will shut me off from life forever, will become, not walls, but years. . . . No, no, I shall be leaving shortly! For forty kopecks I shall ride home on the trolley bus and there I shall eat a tasty supper. But I must not forget: they will all be staying here. They will go on coughing like that. Coughing for endless years.

Each year on the anniversary of my arrest I organize myself a "zek's day": in the morning I cut off 650 grams of bread, put two lumps of sugar in a cup and pour hot water on them. For lunch I ask them to make me some broth and a ladleful of thin mush. And how quickly I get back to my old form: by the end of the day I am already picking up crumbs to put in my mouth, and licking the bowl. The old sensations start up vividly.

I had also brought out with me, and still keep, my number patches. Am I the only one? In some homes they will be shown to you like holy relics.

Once I was walking down Novoslobodskaya Street. The Butyrki Prison! "Parcels Reception Room." I went in. It was full, mostly of women, but there were men, too. Some were handing over parcels, others standing and talking. So this was where our parcels had come from! How interesting! With an air of complete innocence, I went over to read the rules. But a fat-faced sergeant major marked me with his eagle eye and strode toward me. "What do you want, citizen?" His nose told him that this was not a man with a parcel, but a man up to tricks. Obviously I still had a convict smell about me!

Should we visit the dead? Your own dead, lying where you, too, should be lying, run through with a bayonet? A. Y. Olenyev, an old man by then, went in 1965. With rucksack and stick, he made his way to the former Medical Section, and from there up the hill (near the settlement of Kerki), where the burial ground was. The hillside was full of skulls and bones, and nowadays the inhabitants call it "Bone Hill."

In a remote northern town, where there is half a year of daylight and a half year of night, lives Galya V. She has nobody in all the world, and what she calls home is one corner of a nasty, noisy room. Her recreation is to take a book to a restaurant, order wine, and spend her time drinking, smoking, and "grieving for Russia." Her best-loved friends are café musicians and doormen. "Many people who come back from *you know where* conceal their past. I am proud of my life story."

Associations of former zeks gather once a year, varying the place from time to time, to drink and reminisce. "And strangely enough," says V. P. Golitsyn, "the pictures of the past conjured up are by no means all dark and harrowing; we have many warm and pleasant memories."

Another normal human characteristic. And not the worst.

"My identification number in camp began with *yery*, V. L. Ginzburg rapturously informs me. "And the passport they issued to me was in the 'Zk' series!"

You read it—and feel a warm glow. No, honestly—however many letters you receive, those from zeks stand out unmistakably. Such extraordinary toughness they show! Such clarity of purpose combined with such vigor and determination! In our day, if you get a letter completely free from self-pity, genuinely optimistic— it can only be from a former zek. They are used to the worst the world can do, and nothing can depress them.

I am proud to belong to this mighty race! We were not a race, but they made us one! They forged bonds between us, which we, in our timid and uncertain twilight, where every man is afraid of every other, could never have forged for ourselves. The orthodox and the stoolies automatically removed themselves from our midst when we were freed. We need no explicit agreement to support each other. We no longer need to test each other. We meet, look into each other's eyes, exchange a couple of words—and what need for further explanation? We are ready to help each other out. Our kind has friends everywhere. And there are millions of us!

Life behind bars has given us a new measure for men and for things. It has wiped from our eyes the gummy film of habit which always clogs the vision of the man who has escaped shocks. And we reach such unexpected conclusions!

N. Stolyarova, who came from Paris of her own free will in 1934, into the trap which bit off the whole middle part of her life, far from torturing herself and cursing the day she came, says: "I was right to defy those around me, to defy the voice of reason, and come to Russia! Without knowing a thing about Russia, I understood intuitively what it would be like."

At one time I. S. Karpunich-Braven, then a fiery, successful, and impatient brigade commander in the Civil War, did not bother to examine the lists put before him by the head of the

Special Section, but simply took a blunt pencil and jotted down at the bottom, not the top of the page, "h m"—small letters, not capitals, and without periods, as though it was a triviality. ("H.M." meant "highest measure of punishment"—for all of them!) After that he was promoted, and after that he spent twenty years and a half on the Kolyma—and next we find him living on a lonely piece of land in the middle of the forest, where he waters his garden, feeds his hens, makes things in his carpenter's shop, does not petition for rehabilitation, uses foul language about Voroshilov, fills exercise books with endless angry answers to every broadcast he hears and every newspaper article he reads. But a few more years pass—and the farmyard philosopher thoughtfully copies this aphorism out of a book:

"It is not enough to love mankind—you must be able to stand people." And just before he died he wrote, in words of his own but words so startling that you wonder whether they belong perhaps to some mystic, or perhaps to the aged Tolstoi:

"I have lived and judged all things by their effect on me. But now I am a different man, and I have stopped judging things in this way."

The remarkable V. P. Tarnovsky simply stayed on the Kolyma after serving his sentence. He writes verses, but sends them to nobody. Verses like these:

> My life is set in this far place
> And lived in silence, by God's will:
> For I have seen Cain face to face
> And could not bring myself to kill.[6]

The only cause for regret is that we shall all die off one by one, without achieving anything worthwhile.

■

Freedom has something else in store for former convicts—reunion with family and friends. Reunion of fathers with sons. Of husbands with wives. And it is not often that good comes out of these reunions. In the ten or fifteen years lived apart from us, how could our sons grow in harmony with us: sometimes they are simply

---

6. Truth demands that I add what happened later: he left the Kolyma and married unhappily. He lost his spiritual equilibrium, and does not know how to get his neck out of the noose.

strangers, sometimes they are enemies. Nor are women who wait faithfully for their husbands often rewarded: they have lived so long apart, long enough for a person to change completely, so that only his name is the same. His experience and hers are too different—and it is no longer possible for them to come together again.

This is a subject which others can make into films and novels, but there is no room for it in this book.

Let me tell just one story, that of Mariya Kadatskaya. (Plate No. 8 shows Kadatskaya and her husband when they were young; Plate No. 9, Kadatskaya as she is now.)

"In the first ten years my husband wrote me six hundred letters. In the next ten—just one, and that one made me wish I was dead. When he got leave for the first time, after nineteen years, he went to stay with relatives and only came to see me and my son for four days on his way through. The train by which we expected him was taken off that day. After a sleepless night I lay down to rest. I heard a ring. A strange voice said: 'I want to see Mariya Venediktovna.' I opened the door. In came a stout elderly man in a raincoat and a hat. He walked right in without saying a word. Still half asleep, I had forgotten for the moment that I was expecting my husband. We stood there. 'Don't you recognize me?' 'No.' I could only think that he was some relative of mine—I had a great number of them, whom I had not seen for many years. Then I looked at his compressed lips, remembered that I was expecting my husband, and fainted. My son arrived, and to make matters worse, he was unwell. So there we stayed, the three of us, for four days, never once leaving my one and only room. Father and son were very reserved with each other, and my husband and I hardly had a chance to talk; the conversation was general. He told us about his own life, and was not a bit interested to know how we had got on in his absence. As he went back to Siberia, he said goodbye without looking me in the eye. I told him that my husband had died in the Alps (he had been in Italy, where the allies had liberated him)."

There are, however, reunions of a different and more cheerful kind.

You may meet a warder or a camp commander. At the Teberdinsk tourist center you suddenly recognize physical training instructor Slava as a screw from Norilsk. Or Misha Bakst suddenly sees a familiar face in the Leningrad Gastronom, and the other man notices him, too. Captain Gusak, commander of a Camp

Division, is now out of uniform. "Wait a bit now, wait a bit. Didn't I see you inside somewhere? . . . Oh, yes, I remember, we took your parcels from you for bad work!" (He remembers! But all this seems natural to them, as though they have been put in charge of us forever, and the present stage is just a short interval.)

You may (as Belsky did) meet your unit commander, Colonel Rudyko, who hastily agreed to your arrest to spare himself unpleasantness. Also in civilian clothes and a neat fur cap, with an air of scholarly respectability.

You may also meet your interrogator—the one who used to strike you or put you in to feed the bugs. Nowadays he is drawing a good pension, like Khvat for instance, who interrogated and murdered the great Vavilov and now lives on Gorky Street. God save you from such a meeting—it is you again who will suffer a blow to the heart, not he.

Then again, you might meet the man who denounced you—the one who put you inside and is obviously flourishing. For some reason heaven does not rain thunderbolts on him. A prisoner who returns to his native place is bound to see those who informed on him. "Listen—why don't you bring a case against them?" some hotheaded friend will urge him. "At least it would mean public exposure!" (And at most, as everybody knows . . .) "No, I don't think so. . . . Let's leave things as they are," the rehabilitated prisoner will answer.

Because to bring such a case you would have to begin by harnessing those hundred oxen.

"Let life punish them!" says Avenir Borisov, with a shrug.

It's all he can say.

"That lady over there, L., a member of our union, once put me inside," said K. the composer to Shostakovich. "Make a written statement," Shostakovich angrily suggested, "and we'll expel her from the union!" (Don't bet on it, though!) K. waved the thought away. "No, thank you. I've been dragged around the floor by this beard, and I don't want any more."

But is it just a matter of retribution? G. Polev complains that "when I came out, the dirty swine who landed me in jail the first time very nearly got me put away again—and would have pulled it off if I hadn't abandoned my family and left my home town."

Now that's how it should be, that's *our Soviet way of doing things!*

Which is the dream, which is the ignis fatuus—the past? or the present? . . .

In 1955 Efroimson took Deputy Prosecutor General Salin a whole volume of criminal charges against Lysenko. "We are not competent to deal with that," Salin told him. "Address yourself to the Central Committee."

Since when were they not competent? Or why had they not admitted as much thirty years before?

The two false witnesses who put Chulpenyov in the Mongolian hole—Lozovsky and Seryogin—are both flourishing. Chulpenyev went to see Seryogin in the Consumer Services bureau of the Moscow Soviet, together with someone they had both known in his unit. "Let me introduce you. He was with us at Khalkhin-Gol—do you remember?" "No, I don't." "A fellow by the name of Chulpenyov—do you remember him?" "No. I don't. The war scattered us in all directions." "So you don't know what became of him?" "I haven't the slightest idea." "Oh, you lousy bastard, you!"

There's no more to be said. In the District Party Committee which has Seryogin on its books: "It can't be true! He's such a conscientious worker!"

Conscientious worker . . .

Nothing, and nobody, has budged. The thunder growled once or twice—and passed over with hardly a drop of rain.

No, things are just where they were—so much so that Y. A. Kreinovich, the expert on languages of the Far North,[7] went back to work in the same institute, the same sector, with the very people who turned him in and who still hated him: with these very same people he hangs up his overcoat and sits around the conference table daily.

It's rather as though the victims of Auschwitz and their former overseers set up a fancy-goods shop together.

There are Obergruppen stoolies in the literary world, too. How many lives has Y. Elsberg destroyed? Or Lesyuchevsky? Everybody knows about them—and nobody dares touch them. Efforts to have them expelled from the Writers' Union came to nothing. There was even less hope of getting them dismissed from their jobs. Or, needless to say, from the Party.

7. It has been aptly said of him that whereas some members of the People's Will Party achieved fame as philologists thanks to the freedom they enjoyed in exile, Kreinovich preserved his fame in spite of Stalin's camps: even on the Kolyma he did his best to study the Yukagir language.

When our Criminal Code was being drawn up (1926), it was calculated that murder by slander was five times less serious and blameworthy than murder by the knife. (And anyway it was unimaginable that under the dictatorship of the proletariat anyone would resort to such a bourgeois weapon as slander!) Under Article 95 a wittingly false denunciation or deposition aggravated by (*a*) accusation of a serious crime, (*b*) motives of personal gain, or (*c*) the manufacture of evidence is punishable by deprivation of freedom for *not more than . . . two years.* And it can be as little as six months.

The drafters of this article were either complete idiots, or only too farsighted.

My guess is that they were farsighted.

At every amnesty since then (the Stalin amnesty in 1945, the "Voroshilov amnesty" of 1953), they have remembered to include those sentenced under this little article, to look after their most dedicated helpers.

Then, of course, there are *statutory limitations.* If you are falsely accused (under Article 58), there are no time limits. If you are the false accuser—there is a time limit, and you'll be protected.

The case against Anna Chebotar-Tkach and her family was a patchwork of false depositions. In 1944 she, her father, and her two brothers were arrested for the alleged murder, allegedly for political reasons, of her sister-in-law. All three men were done to death in jail. (They wouldn't confess.) Anna did ten years. The sister-in-law turned out to be unharmed! But another ten years went by with Anna vainly pleading for rehabilitation! As late as 1964 the reply from the Prosecutor's Office was: "You were convicted by due process of law and there are no grounds for review." When in spite of this they did rehabilitate her, the indefatigable Skripnikova wrote a petition for Anna, asking that the perjurers be brought to trial. U.S.S.R. Public Prosecutor G. Terekhov[8] answered: "Impossible because of the *time limit.* . . ."

In the twenties they dug up, dragged in, and shot the ignorant peasants who *forty* years earlier had carried out the Tsarist court's sentence of execution on the Narodovoltsy.* But those muzhiks were "not ours." Whereas these informers are flesh of our flesh.

8. The same Terekhov who would conduct the case against Galanskov and Ginzburg.

Such is the *freedom* into which former zeks are released. Look where you will, is there any historical parallel? When did so much generally recognized villainy escape judgment and punishment?

Why should we expect anything good to come of it? What can grow out of this stinking corruption?

How magnificently the wicked scheme of the Archipelago has succeeded!

END OF PART VI

PART VII

Stalin Is No More

■

"Neither repented they of their murders . . ."

REVELATION 9:21

NOTE: Passages in Part VII which were set in smaller type in the Russian edition have been omitted from the English translation at the author's request as being of limited interest to the non-Russian reader.

Chapter 1

■

Looking Back on It All

We never, of course, lost hope that our story *would* be told: since sooner or later the truth is told about all that has happened in history. But in our imagining this would come in the rather distant future—after most of us were dead. And in a completely changed situation. I thought of myself as the chronicler of the Archipelago, I wrote and wrote, but I, too, had little hope of seeing it in print in my lifetime.

History is forever springing surprises even on the most perspicacious of us. We could not foresee what it would be like: how for no visible compelling reason the earth would shudder and give, how the gates of the abyss would briefly, grudgingly part so that two or three birds of truth would fly out before they slammed to, to stay shut for a long time to come.

So many of my predecessors had not been able to finish writing, or to preserve what they had written, or to crawl or scramble to safety—but I had this good fortune: to thrust the first handful of truth through the open jaws of the iron gates before they slammed shut again.

Like matter enveloped by antimatter, it exploded instantaneously!

Its explosion touched off in turn an explosion of letters—that was to be expected. But also an explosion of newspaper articles—written with gritted teeth, with ill-concealed hatred and resentment: an explosion of official praise that left a sour taste in my mouth.

When former zeks heard this fanfare from all the newspapers in unison, learned that some sort of story about the camps had come out and that the journalists were slavering over it, their unanimous conclusion was: "More lying nonsense! Nothing's safe from those crafty liars!" That our newspapers, with their habitual immoderation, might suddenly start falling over each other to praise the truth was something no one could possibly imagine! Some of them were reluctant to risk soiling their hands on my story.

But when they started reading it, a single groan broke from all those thousands—a groan of joy and of pain. Letters poured in.

I treasure those letters. Only too rarely do our fellow country-men have a chance to speak their mind on matters of public concern—and former prisoners still more rarely. Their faith had proved false, their hopes had been cheated so often—yet now they believed that the era of truth was really beginning, that at last it was possible to speak and write boldly!

And they were disappointed, of course, for the hundredth time. . . .

"Truth has prevailed, but too late!" they wrote.

Even later than they thought, because it had still by no means prevailed.

There were also some sober people who did not put their names at the end of their letters ("I must think of my health for the little time left to me"), or who asked point-blank, while the journalistic adulation was at white heat: "Why, I wonder, did Volkovoy* allow you to publish this story? Please answer; I'm worried in case you're in the punishment cells." Or: "Why haven't you and Tvar-dovsky both been put away?"

Because their trap has jammed and failed to work, that's why. What, then, must the Volkovoys do? Take up their pens like the rest! Write letters themselves. Or refutations in the press. And indeed they proved to be remarkably literate.

From this second stream of letters we learn their names, and how they describe themselves. We had been looking for the right term for so long—"camp bosses," "camp personnel"—no, no: "practical workers," that was it! Golden words, these! "Chekists" was somehow not quite right; practical workers—that's the term they prefer.

And they write:

"Ivan Denisovich is a toady." (V. V. Oleinik, Aktyubinsk)

"One feels for Shukhov neither compassion nor respect." (Y. Matveyev, Moscow)

"Shukhov was rightly convicted. . . . What should a zek be

doing outside anyway?" (V. I. Silin, Sverdlovsk)

"These submen with their shabby little souls were *dealt with too leniently by the courts*. I feel no pity for people whose behavior during the Fatherland War was dubious." (E. A. Ignatovich, Kimovsk)

Shukhov is "a highly skilled, resourceful, ruthless scavenger. The consummate egoist, living only for his belly." (V. D. Uspensky, Moscow)[1]

"Instead of portraying the destruction of the most loyal citizens in 1937, the author has chosen 1941, when it was mainly self-seekers who landed in jail.[2] In '37 there were no Shukhovs[3] and people went to their deaths in grim silence, wondering *why anyone needed all this?*"[4] (P. A. Pankov, Kramatorsk)

On conditions in the camps:

"Why give a lot of food to those who do not work? Their energy remains unexpended. . . . I say the criminal world is being treated far too gently." (S. I. Golovin, Akmolinsk)

"Where rations are concerned we shouldn't forget one thing—that *they are not at a holiday resort*. They must atone for their guilt with honest toil." (Sergeant Major Bazunov of Oimyakon, age 55, grown gray in the camp service)

"There are fewer abuses of authority in the camps than in any other Soviet institution [!!]. I can affirm that things *are now stricter* in the camps." (V. Karakhanov, Moscow area)

"This story is an insult to the soldiers, sergeants, and officers of the MOOP.* The people are the makers of history, but how are the people portrayed here . . . ? As 'screws,' 'blockheads,' 'idiots.' " (Bazunov)

"We, the executors, were also human, we were also capable of heroism: we did not shoot every fallen prisoner, and by not doing so risked our posts." (Grigory Trofimovich Zheleznyak)[5]

"In the story the whole day is filled brimful with the negative behavior of prisoners, and the role of the administration is not shown. . . . But the detention of prisoners in camps *was not the*

---

1. Can this pensioner be the Uspensky who murdered his father, the priest, and made a career in the camps on the strength of it?

2. What he means is simple people, non-Party members, prisoners of war.

3. You'd be surprised! . . . There were more of them than *of your sort!*

4. What deep thinkers they were! Incidentally, they weren't as quiet as all that: they died with endless professions of regret and pleas for mercy.

5. Zheleznyak claims to remember me, too: "He arrived in irons, and he was a notorious troublemaker. Later he was sent to Dzhezkazgan and it was he, together with Kuznetsov, who headed the rising."

cause of all that happened in the period of the personality cult, but is simplyconnectedwithexecutionofthesentence."(A.I.Grigoryev)

"The guards did not know why any particular prisoner was inside."[6] (Karakhanov)

"Solzhenitsyn describes the whole *work* of the camp as though there was no Party leadership there at all. But Party organizations existed then, *just as they do now,* and *guided all our work according to their conscience.*"

Practical workers "only carried out what standing orders, periodic instructions, and operative decisions from above demanded of them. *The same people who were working there then work there still* [!!]'[7] (perhaps some 10 percent of the personnel are new), they have been repeatedly commended for their work, enjoy a good professional reputation. . . .

"All on the staff of MOOP are fired with outraged indignation. . . . The spiteful bitterness of this work is simply astounding. . . . He is deliberately trying to turn the people against the MVD! . . . Why do our Organs allow people to lampoon MOOP personnel? . . . *This is dishonorable!*" (Anna Filippovna Zakharova, Irkutsk Province. In the MVD since 1950, Party member since 1956!)

Just listen to it! Listen to it! *"This is dishonorable!"* A cry from the heart! For forty-five years they tortured the natives—and that was honorable. But someone publishes a story about it—and that's dishonorable!

"I've never before had to swallow such trash. . . . And this is not just my opinion. Many of us feel the same, our name is *Legion.*"[8]

In short: "Solzhenitsyn's story should be withdrawn immediately from all libraries and reading rooms." (A. Kuzmin, Orel)

Withdrawn it was, but by easy stages.

"This book should not have been published, the material should have been handed over to the Organs of the KGB instead." (Anon.,[9] a coeval of October)

October's contemporary shows insight: it very nearly happened that way.

---

6. What, us? "We were only carrying out orders"; "We didn't know."
7. A very important piece of information!
8. Quite right—their name is Legion. Only they were in too much of a hurry to check their reference to the Gospel. It was of course a Legion of *devils.*
9. Another one who conceals his identity, just in case: who the hell knows which way the wind will blow next!

And here's another "Anon.," a poet this time:

> Hear me, O Russia,
> Our souls are unspotted—
> Our conscience unblemished! . . .

That "accursed incognito"* again! It would be nice to know whether he had shot people himself, or merely sent them to their deaths, or whether he's just an ordinary orthodox citizen—but alas, he's anonymous! Anon., spotless Anon. . . .

Finally, we have the broad philosophical view:

"History has never had need of the past [!!], and the history of socialist culture needs it least of all." (A. Kuzmin)

History has no need of the past! That's the conclusion our loyalists are left with! What, then, does it need? The future, perhaps? And *these* are the people who write our history! . . .

What retort can we make to them all, faced with their massive ignorance? How can we now make them understand? . . .

Truth, it seems, is always bashful, easily reduced to silence by the too blatant encroachment of falsehood.

The prolonged absence of any free exchange of information within a country opens up a gulf of incomprehension between whole groups of the population, between millions and millions.

We simply *cease to be a single people,* for we speak, indeed, different languages.

■

Nonetheless, a breakthrough had been made! Oh, it was stout, the wall of lies, it looked so secure, looked built to last forever—but a breach yawned, and news broke through. Only yesterday we had had no camps, no Archipelago, and today there they were for the whole people, the whole world to see—prison camps! Camps, what is more, of the Fascist type!

What was to be done? You, whose skill in turning things inside out is of many years' standing! You venerable panegyrists! Surely you won't put up with it? You . . . surely you won't quail? Men like you . . . give way to this?

Of course they wouldn't! The distortion experts rushed unbidden into the breach! They might have been waiting all those years for just this: to cover the breach with their gray-winged bodies and with the joyous—yes, joyous!—flapping of their wings to hide the

Archipelago in its nakedness from astonished spectators.

Their first cry, which came to them instinctively, in a flash, was: *It will never happen again!* All praise to the Party! It will never happen again!

Such clever fellows, such expert gap-stoppers! Obviously, if it will never happen again, this automatically implies that it is not happening *today.* There will be nothing of the sort in the future —so of course it cannot exist today!

So cleverly did they flap their wings before the breach that the Archipelago became a mirage almost as soon as it rose into view: it does not exist, it will not exist, if you really must, you can just about say that it did exist once upon a time. . . . But of course— that was in the days of the *personality cult.* (Very convenient, this "personality cult": trot it out boldly and pretend you've explained something.) What manifestly does exist, what is left to us, what fills the breach, what will endure forever, is—"All praise to the Party!" (Praise is apparently due in the first place because "it can't happen again," but very shortly it begins to sound like praise for the Archipelago—the two things merge and cannot be separated: on every side we heard it, even before they got hold of the magazine with my story in it: "All praise to the Party!" Even before they reached the passage about how Volkovoy used to lay on with the lash, cries of "All praise to the Party" thundered all around!)

In this way the cherubim of the lie, the guardians of the Wall, dealt with the first moment of danger—admirably.

■

When Khrushchev, wiping the tear from his eye, gave permission for the publication of *Ivan Denisovich,* he was quite sure that it was about Stalin's camps, and that he had none of his own.

Tvardovsky, too, when he was worrying the highest in the land to give their imprimatur, sincerely believed that it was all in the past, that it was over and done with. . . .

Tvardovsky can be forgiven. Around him, everybody in public life in Moscow was sustained by this one thought: "There's a thaw, they've stopped *snatching* people, we've had two cathartic congresses, people are returning from nowhere, and in large numbers!" The Archipelago was lost in a beautiful pink mist of rehabilitations, and became altogether invisible.

But I (even I!) succumbed, and I do not deserve forgiveness. I did not mean to deceive Tvardovsky! I, too, genuinely believed that the story I had brought him was about the past! Could *my* tongue have forgotten the taste of gruel? I had sworn never to forget. Had I somehow never learned the mentality of dog handlers? Could I have failed, as I schooled myself to be the chronicler of the Archipelago, to understand how closely akin it is to the state, and how necessary to it? I was so sure that I at least was not in the power of the law I have mentioned:

"As the paunch grows, the memory goes."

I did get fatter. I fell for it and . . . believed. I let myself be persuaded by the complacent mainland. Believed what my own new-found prosperity would have me believe. And the stories told by the last of my friends to come from *you know where.* Conditions were easier! Discipline had relaxed! They were letting people out all the time! Closing whole camp areas! Dismissing NKVD men!

No—we are creatures of mortal clay! Subject to its laws. No measure of grief, however great, can leave us forever sensitive to the general suffering. And until we transcend our clay there will be no just social system on this earth—whether democratic or authoritarian.

So that I was taken by surprise when I received yet a third stream of letters—from *present-day* zeks, although this was the most natural of all, and the first thing I should have expected.

On crumpled scraps of paper, in a blurred pencil scrawl, in stray envelopes often addressed and posted by free employees, in other words, *on the sly,* today's Archipelago sent me its criticisms, and sometimes its angry protests.

These letters, too, were a single many-throated cry. But a cry that said: *"What about us!!??"*

In the uproar over my story the press had nimbly avoided all that free citizens and foreigners did not need to know and the theme of its trumpetings was: "Yes, it happened, but it will never happen again."

And the zeks set up a howl: What do you mean, never happen again? We're *here inside now,* and our conditions are just the same!

"Nothing has changed since Ivan Denisovich's time"—the message was the same in letters from many different places.

"Any zek who reads your book will feel bitterness and disgust because everything is just as it was."

"What has changed, if all the laws providing for twenty-five years' imprisonment issued under Stalin are still in force?"

"Once more we're being put inside for nothing at all. Whose *personality cult* is to blame this time?"

"A black mist has covered us, and no one can see us."

"Why have people like Volkovoy gone unpunished? . . . They are still in charge of our re-education."

"From the shabbiest warder to the head of the camps administration, the existence of the camps is of vital concern to them all. The warders fabricate charges against us on the most trivial excuses; the security officers blacken our personal files. . . . Prisoners like me with twenty-five years are their favorite fare, and they gorge themselves on it, these corrupt creatures whose mission it is to exhort us to virtue. They are just like the colonizers who used to pretend that Indians and Negroes were inferior human beings. It takes no effort at all to set public opinion against us; all you need do is write an article called 'The Man Behind Bars'[10] . . . and next day the people will be holding mass meetings to get us burned in ovens."

True. Every word of it true.

"You have taken up your position with the rear guard!" says Vanya Alekseyev, to my dismay.

After reading all these letters, I who had been thinking myself a hero saw that I hadn't a leg to stand on: in ten years I had lost my vital link with the Archipelago.

For *them*, for *today*'s zeks my book is no book, my truth is no truth unless there is a continuation, unless I go on to speak of them, too. Truth must be told—and things must change! If words are not about real things and do not cause things to happen, what is the good of them? Are they anything more than the barking of village dogs at night?

(I should like to commend this thought to our modernists: this

---

10. Kasyukov and Monchanskaya, "The Man Behind Bars," *Sovetskaya Rossiya,* August 27, 1960. An article inspired by government circles, which put an end to the brief period (1955–1960) of mildness on the Archipelago. In the view of the authors, the conditions created in the camps had turned them into "charitable institutions"; "punishment is forgotten"; "the prisoners do not want to know about their obligations," while "the administration has many fewer rights than the prisoners." (?) They assure us that the camps are "free boardinghouses" (for some reason no charge is made for laundry, haircuts, and the use of visiting rooms). They are outraged to find that there is only a forty-hour working week in the camps, and indeed that—so they say—"prisoners are not obliged to work." (??) They call for "hard and strict conditions," so that criminals will be *afraid* of jail (hard labor, bare planks without mattresses to sleep on, civilian clothes to be prohibited, "no more stalls selling sweets," etc.); they also call for the abolition of early discharge ("and if a man is guilty of an offense against discipline, *let him stay in longer!*"). And they further recommend that "when he has served his time, the prisoner should not count on charity."

is how our people usually think of literature. They will not soon lose the habit. Should they, do you think?)

And so I came to my senses. And through the pink perfumed clouds of rehabilitation I could make out the familiar rocky piles of the Archipelago, its gray outlines broken by watchtowers.

The state of our society can be well described in terms of an electromagnetic field. All the lines of force in it point away from freedom and toward tyranny. These lines are very stable, they have etched their way in and set hard in the grooves; it is almost impossible to perturb them, deflect them, twist them about. Any charge, any mass introduced into the field is blown effortlessly in the direction of tyranny, and can never break through toward freedom. For that ten thousand oxen must be yoked.

Now that my book has been openly declared harmful, its publication recognized as a mistake (one of the "consequences of voluntarism in literature"), now that it is being removed even from libraries used by the free population—the mere mention of Ivan Denisovich's name, or my own, has become an irreparably seditious act on the Archipelago. But even in those days! even then—when Khrushchev shook my hand and to the accompaniment of applause presented me to the three hundred who considered themselves the artistic elite; when I was getting a "big press" in Moscow, and reporters waited in suspense outside my hotel room; when it was explicitly announced that *such books had the support of* the Party and the government; when the Military Collegium of the Supreme Court was proud of having rehabilitated me (just as it now no doubt regrets it) and the lawyer colonels declared from its bench that the book *should be read* in the camps!—even then the mute, the secret, the unnamed forces were invisibly resisting . . . and the book was stopped! *Even then* it was stopped! Only rarely did it reach a camp legally, so that readers could take it out of the Culture and Education Section library. It was removed from camp libraries. If it was sent from outside by book post, it was confiscated. Free employees sneaked it in under their coats, and charged zeks five rubles a time, or even, so we are told, as much as twenty (Khrushchev rubles! from zeks! but no one who knows the unscrupulous world about the camps will be surprised). The zeks carried it in like a knife past the search point; hid it in the daytime and read it at night. In a camp somewhere in the Northern Urals they made a metal binding for it so that it would last longer.

But why talk about zeks when the same tacit but generally accepted prohibition was extended to the world about the camps, too. At Vis station on the Northern Railway, Mariya Aseyeva, *a free woman,* wrote to the *Literary Gazette* expressing her approval of the story—and whether she posted the letter or imprudently left it on the table, five hours after she had put her opinion on paper the secretary of the local Party organization, V. G. Shishkin, was accusing her of political provocation (the words they think up!)—and she was arrested on the spot.[11]

In Corrective Labor Colony No. 2, at Tiraspol, the sculptor G. Nedov, himself a prisoner, modeled the figure of a prisoner (Plate No. 10) in his trusty's workshop, in Plasticine to begin with. The disciplinary officer, Captain Solodyankin, discovered it: "So you're making a prisoner, are you? Who gave you the right? This is *counter-revolution!*" He seized the figure by its legs, tore it apart, and hurled the halves on the floor. "Been reading too many of those Ivan Denisoviches!" (But he didn't go on to stamp on it, and Nedov hid the halves.) On Solodyankin's complaint, Nedov was called in to see the camp chief, Bakayev, but in the meantime he had managed to come upon a few newspapers in the Culture and Education Section. "We'll put you on trial!" thundered Bakayev. "You're trying to turn people against the Soviet regime!" (So they understand what the sight of a zek can do!) "Permit me to tell you, Citizen Chief... Nikita Sergeyevich says here... And here's what Comrade Ilyichev says..." "He talks to us like equals," gasped Bakayev. Only six months later did Nedov dare retrieve the halves; he glued them together, made a Babbit metal cast and sent the figure out of the camp with the help of a free man.

Repeated searches for the story began in Corrective Labor Colony No. 2. A thorough inspection was made of the living area. They did not find it. One day Nedov decided to get his own back: he settled down after work with Tevekelyan's *Granite Does Not Melt,* behaving as though he was hiding something from the rest of the room (with stoolies in earshot, he asked the lads to screen him), but making sure that he could be seen through the window. The stoolies did their work quickly. Three warders ran in (while a fourth stayed outside and watched through the window, in case Nedov passed the book to someone). They seized it! They carried it off to the warders' room and put it away in the safe.

---

11. How this incident ended I just don't know.

Looking Back on It All | 481

Warder Chizhik, with his huge bunch of keys, arms akimbo: "We've found the book! You'll be put in the hole now!" But next morning an officer took a look. "Oh, you idiots! . . . Give it back to him."

This was how the zeks read a book "approved by the Party and the people"!

■

In a declaration by the Soviet government dated December, 1964, we read: "The perpetrators of monstrous crimes must never and in no circumstances escape just retribution. . . . The crimes of the Fascist murderers, who aimed at the destruction of whole peoples, have no precedent in history."

This was to prevent the Federal German Republic from introducing a statute of limitations for war criminals after twenty years had elapsed.

But they show no desire to face judgment *themselves,* although they, too, "aimed at the destruction of whole peoples."

A lot of articles appear in our press on the importance of punishing fugitive West German criminals. There are even people who specialize in such articles—Lev Ginzburg, for instance. He writes as follows (some say he intends us to see an analogy): What moral training did the Nazis have to undergo for mass murder to seem natural and right to them? Now the lawgivers try to defend themselves by saying that they were not the ones to carry out the sentences! And those who did carry them out, by saying that it was not they who enacted the laws!

It's all so familiar. We have just read what our *practical workers* have to say: "The custody of prisoners is a matter of carrying out the sentence of the courts. The guard did not know who was inside for what."

So you should have *made it your business to know,* if you are human beings! That is what makes you villains—that you looked upon the people in your custody neither as fellow citizens nor as fellow men. Did not the Nazis have their *instructions,* too? Did not the Nazis believe that they were saving the Aryan race?

Nor will our interrogators stumble over their answer (they have it pat already): Why, they ask, did prisoners make statements against themselves? In other words, "They should have stood fast when we tortured them!" And why did informers pass on false reports? We relied on these as though they were sworn evidence!

For a short time they became uneasy. V. N. Ilyin, the former lieutenant general of the MGB of whom we spoke earlier, said of Stolbunovsky (General Gorbatov's interrogator, and mentioned by him): "Oh, dear, dear, what an awful business it is! All these troubles he's having. And the poor fellow draws such a good pension." This is also why A. F. Zakharova took up her pen—she was worried that they might all come under attack soon. Of Captain Likhosherstov (!),* whom Dyakov had "denigrated," she wrote emotionally: "He is still a captain, is secretary of a Party organization [!], and devotes his *labor* to agriculture. You can imagine how difficult it is for him to *work* now, with people writing such things about him! There is talk of *looking into* Likhosherstov, and perhaps even taking him to court.[12] *What for?* If it stops at talk, well and good, but it is not impossible that they will get around to it. That really will cause uproar among the MOOP personnel. Look into him, for *carrying out fully the instructions he was given from above?* Is he now supposed to answer for those who gave the instructions? How very clever! When there's an accident —blame the switchman!"

But the commotion was soon over. No, no one would have to *answer.* No one would be *looked into.*

Staffs, perhaps, were somewhat reduced in places—but have a little patience and they will expand again! Meanwhile, the security boys, those who have not yet reached pensionable age, and those who need to top up their pensions, have become writers, journalists, editors, antireligious lecturers, ideological workers, or, some of them, industrial managers. They will just change their gloves and go on leading us as before. It's safer that way. (And if someone is content to live on his pension—let him enjoy his ease. Take Lieutenant Colonel [ret.] Khurdenko, for example. Lieutenant colonel—that's quite a rank! A former battalion commander, no doubt? No, he began in 1938 as a simple screw; he used to hold the force-feeding pipe.)

While in the records office they carry out a leisurely inspection and destroy all unwanted documents: lists of people shot, orders committing prisoners to solitary confinement or the Dis-

12. "Putting him on trial" is unthinkable, and she cannot bring herself to use the words.

ciplinary Barracks, files on investigations in the camps, denunciations from stoolies, superfluous information about practical workers and convoy guards. In Medical Sections, accounts offices—everywhere, in fact, there are superfluous papers, unnecessary clues to be found. . . .

> We take our seats at your feast like silent ghosts.
> And you who hated us living shall be our hosts.
> Living we could not move you, but speechless and dead
> We are the vengeful presence you cannot but dread.
>
>   Victoria G.—a woman graduate of the Kolyma camps

A word in passing. Why, indeed, is it always the switchmen? What about the traffic managers? What about those a little higher up than the screws—the practical workers, the interrogators? Those who only pointed a finger? Those who only spoke a few words from a platform? . . .

How does it go again? "The perpetrators of monstrous crimes . . . in no circumstances . . . righteous retribution . . . have no precedent in history . . . aimed at the destruction of whole peoples . . ."

Shh! Shh! Now we see why in August, 1965, from the platform of the Ideological Conference (a closed conference on the Direction Our Minds Should Take), the following proclamation was made: *"It is time to rehabilitate* the sound and useful concept of *enemy of the people!"*

Chapter 2

∎

Rulers Change, the Archipelago Remains

The Special Camps must have been among the best-loved brain-children of Stalin's old age. After so many experiments in punishment and re-education, this ripe perfection was finally born: a compact, faceless organization of numbers, not people, psychologically divorced from the Motherland that bore it, having an entrance but no exit, devouring only enemies and producing only industrial goods and corpses. It is difficult to imagine the paternal pain which the Visionary Architect would have felt if he had witnessed in turn the bankruptcy of this great system of his. While he yet lived it was shaken, it was giving off sparks, it was covered with cracks—but probably caution prevailed and these things were not reported to him. When the Special Camp system began it was inert, sluggish, unalarming—but it underwent a rapid process of overheating, and within a few years its state was that of a boiling volcano. If the Great Coryphaeus had lived a year or eighteen months longer, it would have been impossible to conceal these explosions from him, and his weary senile brain would have been burdened with a new decision: either to abandon his pet scheme and mix the camps again, or, on the contrary, to crown it by systematically shooting all the index-lettered thousands.

But, amid weeping and wailing, the Thinker died somewhat too soon for this.[1] He died, and soon afterward his frozen hand brought crashing down his still rosy-cheeked, still hale and vigor-

1.

484

ous comrade in arms—the Minister of those extraordinarily extensive, intricate, and irresolvable Internal Affairs.

The fall of the Archipelago's Boss tragically accelerated the breakdown of the Special Camps. (What an irreparable historic mistake it was! What sense does it make to disembowel the Minister of Arcane Affairs!, lay oily paws on sky-blue shoulder boards?!)

Number patches—the supreme discovery of twentieth-century prison-camp science—were hurriedly ripped off, thrown away, and forgotten! This alone was enough to rob the Special Camps of their austere uniformity. It hardly mattered, when the bars had also been removed from hut windows and locks from hut doors, so that the Special Camps had lost the pleasant jail-like peculiarities which distinguished them from Corrective Labor Camps. (Perhaps they needn't have hurried with the bars—but they couldn't afford to be late either; in times like those it's best to show where you stand!) Sad though it was, the stone jailhouse at Ekibastuz, which had held out against the rebels, was now pulled down, razed quite officially. . . .[2] But what could you expect, if they could suddenly release to a man the Austrian, Hungarian, Polish, and Rumanian prisoners in Special Camps, showing scant concern for their black crimes and their fifteen- or twenty-year terms of imprisonment, and so undermine altogether the prisoner's awe of heavy sentences. They also lifted restrictions on correspondence, which more than anything had made prisoners in Special Camps really feel buried alive. They even allowed visits—dreadful thought! Visits! . . . (Even in mutinous Kengir they began building separate little houses for this purpose.) The tide of liberalism swept on so irresistibly over the erstwhile Special Camps that prisoners were allowed to choose their own hair styles (and aluminum dishes started vanishing from the kitchens for conversion into aluminum combs). Instead of credit accounts, instead of Special Camp coupons, the natives were allowed to handle ordinary Soviet currency and settle their bills with cash like people outside.

Carelessly, recklessly they demolished the system which had fed them—the system which they had spent decades weaving and binding and lashing together.

And were those hardened criminals at all mollified by this pampering? They were not! On the contrary! They showed their depravity and ingratitude by adopting the profoundly inappropri-

2. And we were denied the possibility of opening a museum there in the eighties.

ate, offensive, and nonsensical word "Beria-ites"—and now whenever something upset them they would yell this insult at conscientious convoy guards, long-suffering warders, and their solicitous guardians, the camp chiefs. Not only did the word pain the tenderhearted practical workers, it could even be dangerous so soon after Beria's fall, because someone might make it the starting point of an accusation.

For this reason the head of one of the Kengir Camp Divisions (by then purged of mutineers and replenished with prisoners from Ekibastuz) was compelled to deliver the following address from the platform: "Men!" (In those few short years, from 1954 to 1956, they found it possible to call the prisoners "men.") "You hurt the feelings of the supervisory staff and the convoy troops by shouting 'Beria-ites' at them! Please stop it." To which the diminutive V. G. Vlasov replied: "Your feelings have been hurt in the last few months. But I've heard nothing but 'Fascist' from your guards for eighteen years. Do you think we have no feelings?" And so the major promised to cut out the abusive word "Fascist." A fair trade.

After all these pernicious and destructive reforms we may consider the separate history of the Special Camps concluded in 1954, and need no longer distinguish them from Corrective Labor Camps.

Throughout the topsy-turvy Archipelago easier times set in from 1954 and lasted till 1956—an era of unprecedented indulgences, perhaps the period of greatest freedom in its history, if we disregard the BDZ's (detention centers for nonprofessional criminals) in the mid-twenties.

Instruction capped instruction, inspector vied with inspector, encouraging ever wilder displays of liberalism in the camps. Permission for the use of female labor in the forests was withdrawn —yes, it was acknowledged that lumbering was perhaps excessively heavy work for women (although thirty years of continual use were the proof that it was not too heavy at all). Parole was reintroduced for those who had served two-thirds of their sentence. They began paying money wages in all camps, and prisoners flooded the shops, which were subject to no prudent disciplinary restrictions—and how could they be when so many prisoners were unguarded so much of the time? Indeed, they could spend their money in the settlement, too, if they wished. All huts were wired for radio, prisoners had more than their fill of newspapers and wall newspapers, agitators were assigned to each work team. Comrade lecturers (colonels, even!) came to address the camps' population on various themes—even on the perversion of history by Aleksei Tolstoi—but the administration did not find it so easy to collect an audience (they could no longer drive them in with sticks, and subtler methods of pressure and persuasion were necessary). There was a continual buzz of private conversation in the hall, and nobody listened to the lecturers. Prisoners were permitted to subscribe to the loan, but no one except the loyalists was moved by this, and their mentors had to tug each prisoner by the hand toward the subscription list and squeeze the odd ten (one ruble, Khrushchev style) out of him. They started organizing joint shows for men's and women's camps on Sundays—people flocked to these willingly, and men even bought themselves ties in the camp shops.

Much of the Archipelago's gold reserve was put into circulation again. There was a revival of the selfless voluntary activities by which it had lived in the days of the Great Canals. "Councils of Activists" were set up, with sectors like those of a local trade union committee (for Industrial Training, Culture and Recreation, and Services), with the struggle for higher productivity and better discipline as their main task. "Comrades' courts" were recreated, with the right to censure offenders, to request disciplinary measures or the nonapplication of the "two-thirds" rule.

These measures had once served our Leaders well—but that was in camps which had not gone through the Special Camp course in murder and mutiny. Now things were simpler: one chairman of an Activists' Council was murdered (at Kengir), a second beaten up—and suddenly nobody wanted to join. (Captain Second Class Burkovsky worked at this time in a Council of Activists, worked honestly and conscientiously, but because he was always being threatened with the knife he was also very cautious, and attended meetings of the Banderist brigade to listen to criticism of his actions.)

And still the pitiless blows of liberalism staggered and rocked the camp system. "Light Discipline Camp Divisions" were set up (there was even one in Kengir!): in effect, you need only sleep in the camp area, because you went to work without escort, by any route and at any time you pleased (everybody, in fact, tried to set out early and return late). On Sunday a third of the prisoners were let out on the town before dinner, a third of them after dinner, and only one-third were not allowed an outing.[3]

Let the reader put himself in the position of the camp authorities and tell me: Could they operate in such conditions? Could anyone expect good results?

One MVD officer, my traveling companion on a Siberian train journey in 1962, described this whole era in the life of the camps until 1954 as follows: "Things were completely out of hand! Those who didn't want to didn't even go to work. They bought television sets with their own money."[4] He had retained very dark memories of that short but unpleasant period.

For no good can come of it when his mentor stands before a prisoner like a suppliant, with neither whips nor punishment cells nor graded short rations to fall back on!

But this was apparently still not enough: the next assault on the Archipelago was with a battering ram called the *off-camp* (living-out) system. Prisoners were allowed to move out of camp altogether, to acquire houses and families; they were paid wages like free men, and paid in full (with no more deductions for the upkeep of the camp, the guards, and the administration). Their only remaining link with the camp was that once a fortnight they went in to report.

This was the end—the end of the world, or of the Archipelago, or of both at once!—but the organs of the law applauded the off-camp system as an outstandingly humane, a pioneering discovery of the Communist order![5]

After these blows there seemed to be nothing left but to disband the camps and be done with it. To destroy the great Archipelago; ruin, scatter, and demoralize hundreds of thousands of practical workers with their wives, children, and domestic animals; make faithful service, rank, and an impeccable record things of no account.

It seemed to have begun already: something called "Commis-

3. This does not mean that such leniency was universal. Punitive Camp Divisions were also preserved, like the "All-Union Punitive Camp" at Andzyoba, near Bratsk, with our old friend from Ozerlag, the bloodstained Captain Mishin in charge. In the summer of 1955 there were about four hundred prisoners there for special punishment (Tenno among them). But even there the prisoners, not the warders, were masters of the camp.

4. If they didn't work, where did the money come from? If this was in the North, and in 1955 at that, where did the television sets come from? Still, I never dreamed of interrupting him. I enjoyed listening.

5. Described, we may add (together with "remission for good conduct" and "conditional early release"), by Chekhov in *Sakhalin:* convicts classified as "expected to reform" had the right to build themselves houses and to marry.

sions of the Supreme Soviet," or more simply "unloading parties," began arriving at the camps, cold-shouldering the camp authorities, taking over the staff hut for their sessions, and writing out orders of release as light-heartedly and irresponsibly as though they were warrants for arrest.

A deadly threat hung over the whole caste of practical workers. They had to think of something! They had to *fight!*

■

One of two fates awaits any important social event in the U.S.S.R.: either it will be hushed up or it will be the subject of calumny. I can think of no significant event in our country which has slipped past the roadblock.

So it was throughout the existence of the Archipelago: most of the time it was hushed up, and whatever was written about it was lies: whether in the days of the Great Canals, or about the unloading commissions of 1956.

As far as these commissions were concerned, we ourselves, with no insidious prompting from the newspapers, under no pressure from outside, assisted the work of sentimental falsification. Who would not be deeply moved: we were used to attack from our own lawyers, and now we saw public prosecutors taking our side! We pined for *freedom,* we felt that out there a new life was beginning, we could see as much even from the changes in the camps—and suddenly there came a miracle-working plenipotentiary commission which talked to each prisoner for five or ten minutes, then handed him a rail ticket and a passport (some of them even registered for Moscow). What, except praise, could burst from the emaciated, chronically bronchitic, wheezy breasts of us prisoners?

But what if we had risen for a moment above the happiness that set our hearts painfully pounding, and ourselves running to stuff our rags into our duffel bags—and asked ourselves whether *this* was the proper ending to all Stalin's crimes? Should not the commission have stood before a general line-up of prisoners, bared their heads, and said:

"Brothers! We have been sent by the Supreme Soviet to beg your forgiveness. For years and decades you have languished here, though you are guilty of nothing, while we gathered in ceremonial halls under cut-glass chandeliers with never a thought for you. We submissively confirmed the cannibal's inhuman decrees, every one of them, we are accomplices in his murders. Accept our belated repentance, if you can. The gates are open, and you are free. Out there on the airstrip, planes are landing with medicines, food, and warm clothing. There are doctors on board."

Either way, they obtained their freedom—but this was the wrong way to confer it, a denial of its true meaning. The unloading commission was like a careful janitor following the trail of Stalin's vomit, and diligently mopping it up—that was all. This was no way to lay new moral foundations for our society.

I am quoting in what follows the judgment of A. Skripnikova, with which I entirely agree. Prisoners are summoned one by one (as usual, to keep them disunited) into the office where the commission sits. A few factual questions are asked about each man's case. The questions are perfectly polite, and apparently well meant, but their drift is that *the prisoner must admit his guilt* (not the Supreme Soviet, but the unhappy prisoner again!). He must be silent, he must bow his head, he must be put in the position of one forgiven, not one who forgives! In other words, they want to coax out of him with the promise of freedom what previously they could not wring from him even by torture. Why? you may ask. It is most important: he must return to freedom a timid soul! And at the same time the commission's records will make it appear to History that those inside were, most of them, guilty, that the pictures of brutal lawlessness have been greatly overdrawn. (There may have been a handsome financial benefit, too: without rehabilitation, no compensation need be paid.)[6] If it meant no more than this, the release of prisoners held no dangers: it would not blow the whole camp system sky-high, it created no obstacle to *new admissions* (which went on uninterruptedly even in 1956–1957) and no obligation to release the newcomers in their turn.

What of those who out of incomprehensible pride refused to acknowledge their guilt to the commission? *They were left inside.* There were quite a few of them. (Unrepentant women prisoners in Dubrovlag in 1956 were rounded up and dispatched to the Kemerovo camps.)

Skripnikova tells us of the following incident. One Western

6. Incidentally, there was a proposal at the beginning of 1955 to pay compensation for *every year spent inside.* This was only natural, and such payments were in fact being made in Eastern Europe. But not to so many people and not for so many years! They totted it all up, and were horrified: "We shall ruin the state!" So they opted for two months' compensation.

Ukrainian woman had been given ten years because her husband was a supporter of Bandera. She was now called upon to admit that she was in because he was a *bandit.* "No, I won't say it!" "Say it, and you'll go free!" "No, I won't say it. He's no bandit, he's in the OUN."* "All right, if you don't want to go—you can stay!" said Solovyov, chairman of the commission. A few days later her husband visited her, on his way home from the North. He had been sentenced to twenty-five years, but he readily admitted that he was a bandit, and was pardoned. He showed no appreciation of his wife's staunchness and heaped reproaches on her. "Why didn't you say that I was the devil himself, that you'd seen my tail and my cloven hoof? How am I going to manage the farm and the children now?"

I should mention that Skripnikova herself refused to acknowledge her guilt, and remained inside another three years.

So even the era of freedom came to the Archipelago in a public prosecutor's gown.

All the same, the alarm of the practical workers was not baseless. In 1955–1956 the stars over the Archipelago were in a conjunction never seen before. These were fateful years for it, and might have been its last!

If the people who were invested with supreme power and weighed down with the fullness of their knowledge about their country had also been steeped in that Doctrine of theirs, but believed in it genuinely and wholeheartedly, surely those years were the time for them to look back in horror and to sob aloud. How can they gain admittance to the "kingdom of Communism" with that bloody sack at their backs? It oozes blood; their backs are one great crimson stain! They have let out the politicals—but who or what has produced all those millions of nonpolitical (and nonprofessional) criminals? Our "relations of production"? *The social milieu? You,* perhaps?

They should have let their space program go to hell! Let Sukarno's navy and Kwame Nkrumah's guard regiments look after themselves! They might at least have sat down and scratched their heads: Where do we go next? Why are our laws, the best in the world, rejected by millions of our citizens? What makes them so ready to crawl under that murderous yoke—and the more intolerable it is, the more densely they flock to shoulder it? What must we do to stem the stream? Perhaps our laws are not what they should be? (And here it would be worth thinking a little about the harassed schools, the neglected countryside, and all those things which we can only call "injustices," with no class label.) How are we to bring the fallen back to life? Not by cheap and facile gestures like the Voroshilov amnesty, but by a sympathetic effort to understand each of them, his case, and his character.

Should we *put an end* to the Archipelago or not? Or is it there *forever?* For forty years it has been an ulcer in our flesh—isn't that enough?

Evidently not! It is not enough! Nobody wants to tire the convolutions of his brain, there is no answering ring in the soul. Let the Archipelago stand for another fifty years while we get on with the Aswan dam and the unification of the Arabs!

Historians attracted to the ten-year reign of Nikita Khrushchev —when certain physical laws to which we had grown accustomed suddenly seemed to stop operating, when objects miraculously began defying the forces in the electromagnetic field, defying the pull of gravity—will inevitably be astounded to see how many opportunities were briefly concentrated in those hands, and how playfully, how frivolously they were used before they were nonchalantly tossed aside. Endowed with greater power than anyone in our history except Stalin, a power which though impaired was still enormous, he used it like Krylov's Mishka in the forest clearing, rolling his log first this way, then that, and all to no purpose. He was given the chance to draw the lines of freedom three times, five times more firmly, and he failed to understand his duty, abandoned it as though it were a game—for space, for maize, for rockets in Cuba, for Berlin ultimatums, for persecution of the church, for the splitting of Oblast Party Committees, for the battle with abstract art.

He never carried anything through to its conclusion—least of all the fight for freedom! Stir him up against the intelligentsia? Nothing could have been simpler. Use his hands, the hands that wrecked Stalin's camps, to reinforce the camps now? That was easily achieved! And just think *when!*

In 1956, the year of the Twentieth Congress, the first orders limiting relaxation of the camp regime were promulgated! They were extended in 1957—the year when Khrushchev achieved undivided power.

But the caste of practical workers was still not satisfied. Scenting victory, they went over to the offensive. We can't go on like this! The camp system is the main prop of the Soviet regime and it is collapsing!

For the most part, their influence was of course brought to bear discreetly—at official banquets, in the passenger cabins of aircraft, at dacha boating parties—but their activities sometimes came out into the open, as for instance in B. I. Samsonov's speech at a session of the Supreme Soviet (December, 1958): Prisoners, said he, live *too well*, they are satisfied (!) with their food (whereas they should be permanently dissatisfied . . .), they are treated *too well*. (In a parliament which had never acknowledged its earlier guilt, no one of course rebuked Samsonov.) Or in the article about "The Man Behind Bars" (1960).

Yielding to this pressure, without examining anything closely, without pausing to reflect that crime had not increased in those last five years (or that if it had, the causes must be sought in the political system), without considering how these new measures could be squared with his faith in the triumphal advance of Communism, or attempting to study the matter in detail, or even to look at it with his own eyes—this Tsar who had spent "all his life on the road" light-heartedly signed the order for nails to knock the scaffold together again quickly, in its old shape and as sturdy as ever.

And all this happened in the very year—1961—when Nikita made his last, expiring effort to tug the cart of freedom up into the clouds. It was in 1961—the year of the Twenty-second Congress —that a decree was promulgated on the death penalty in the camps for "terrorist acts against reformed prisoners [in other words, stoolies] and against supervisory staff" (something which had never happened), and the plenum of the Supreme Court confirmed (in June, 1961) *regulations for four disciplinary categories in camps*—Khrushchev's camps now, not Stalin's.

When he climbed onto the Congress platform for another attack on Stalin's tyranny, Nikita had only just allowed the screws of his very own system to be turned no less tight. And he sincerely believed that all this could be fitted together and made consistent!

The camps today are as approved by the Party before the Twenty-second Congress. Six years later they are just as they were then.

They differ from Stalin's camps not in regime, but in the composition of their population: there are no longer millions and millions of 58's. But there are still millions inside, and just as before, many of them are helpless victims of perverted justice: swept in simply to keep the system operating and well fed.

Rulers change, the Archipelago remains.

It remains because *that particular* political regime could not survive without it. If it disbanded the Archipelago, it would cease to exist itself.

■

Every story must have an end. It must be broken off somewhere. To the best of our modest and inadequate ability we have followed the history of the Archipelago from the crimson volleys which greeted its birth to the pink mists of rehabilitation. In the glorious period of leniency and disarray on the eve of Khrushchev's measures to make the camps harsher again, on the eve of a new Criminal Code, let us consider our story ended. Other historians will appear—historians who to their sorrow know the Khrushchev and post-Khrushchev camps better than we do.

Two have in fact appeared already: S. Karavansky[7] and Anatoly Marchenko.[8] And they will float to the surface in great numbers, because soon, very soon, the era of publicity will arrive in Russia!

Marchenko's book, for instance, fills with pain and horror even the heart of an old camp hand, inured to suffering as it is. In its description of prison conditions today it gives us a jail of a Still Newer Type than the one of which my own witnesses speak. We learn that the horn, the second horn of imprisonment (see Part I, Chapter 12), juts more boldly, sticks in the prisoner's neck more sharply than ever. By comparing the buildings of the Vladimir Central Prison—the Tsarist and the Soviet buildings—Marchenko shows concretely where the analogy with the Tsarist period of Russian history breaks down: the Tsarist building is dry and warm, the Soviet building damp and cold (your ears may get frostbitten in your cell! padded jackets are never taken off); the windows of the Tsarist building are blocked with four layers of Soviet bricks—and don't forget the muzzles!

■

7. S. Karavansky, "Petition," samizdat, 1966.
8. A. Marchenko, *My Testimony,* samizdat, 1967; New York: E. P. Dutton, 1969.

The NKVD men are a power in the land. And they will never give way of their own free will. If they stood their ground in 1956 they can certainly hold out a bit longer.

It isn't just the corrective labor organs. It isn't just the Ministry for the Protection of Public Order. We have seen already how eagerly newspapers and deputies to the Supreme Soviet support them.

Because they are the backbone. The backbone of so many things.

They have strength, but that is not all—they have arguments, too. Debate with them is not so easy.

I have tried it.

Not that I ever meant to. But those letters drove me to it— letters from today's natives which took me completely by surprise. The natives looked at me in hope and begged me to tell their story, to defend them, to make them human again!

But—tell whom? Supposing that anyone will listen to me . . . If we had a free press I would publish all this: There, it's all out in the open, now let's discuss it!

As things are at present (January, 1964) I wander around the corridors of institutions, a secret and timid suppliant, bow my head to the hatches through which passes are issued, feel upon myself the disapproving and suspicious stare of the soldiers on duty. How hard a writer and commentator on public affairs must work before busy government officials will do him the honor, will condescend to lend him an ear for half an hour!

But even this is not the greatest difficulty. My greatest difficulty is just what it was all that time ago at the foremen's meeting in Ekibastuz: what can I talk to them *about?* And *in what language?*

To speak any of my real thoughts, as set out in this book, would be both dangerous and completely hopeless. Why lose my head in the hushed privacy of an inner office, unheard by the public, unbeknown to all those who long to hear it, and without advancing the cause a single millimeter?

How, then, can I speak? As I cross their mirrorlike marble thresholds, go up their softly carpeted stairways, I must voluntarily trammel myself with silken threads drawn through my tongue, my ears, my eyelids, and then stitched to my shoulders, the skin of my back, and my belly. I must at the very least voluntarily accept two things:

1. All praise to the Party, for our whole past, present, and

future! (Which means that our general penal policy cannot be wrong.) I dare not express my doubts as to whether the Archipelago need ever have existed. And I must not maintain that "the majority are inside for no good reason."

2. The high-ranking personages with whom I shall be talking are dedicated to their work and concerned for the prisoners. They must not be accused of insincerity, coldness, or ignorance (if a man puts his whole heart into a job, how can he possibly not know all about it!).

Much more dubious are *my* motives in interfering. What am I up to? Why me, when I have no official duty to perform? Perhaps I have some dirty ends of my own? ... Why must I meddle, when the Party sees everything, and will get it all right with no assistance from me?

To make my position look a little stronger, I choose the month of my nomination for a Lenin Prize,* and I make my move, like a pawn of some importance: he may yet go up in the world and become a rook!

■

Supreme Soviet of the U.S.S.R. Legislative Proposals Commission. I discover that it has been engaged for some years in drafting a new Corrective Labor Code, a Code, that is, to govern the whole future life of the Archipelago, replacing the 1933 Code, which existed and yet never existed, which might as well never have been written. And now they have arranged to see me so that I, an alumnus of the Archipelago, can acquaint myself with their wisdom, and put before them my own trumpery notions.

They are eight in number. Four of them are surprisingly young, boys who may just about have had time to complete their higher education, and then again may not. How quickly they are rising to positions of power! They look so much at home in this marble-floored palace to which I was admitted with great precautions. The chairman of the commission is Ivan Andreyevich Badukhin, an elderly man, who seems infinitely good-natured. His looks seem to say that if it depended on him, he would disband the Archipelago tomorrow. But his role is this: to take a back seat and say nothing throughout our conversation. The real beasts of prey are two little old men, just like those old men in Griboyedov, who remembered "Ochakov taken and Tartary subdued,"* set fast in postures adopted long ago. I will take my oath that they have not even opened a newspaper since March 5, 1953, so sure is it that nothing capable of influencing their views can ever happen. One of them wears a blue coat, and I imagine that it is some sort of uniform worn at Catherine's court. I can even distinguish the mark left when he unpinned Catherine's silver star, which must have half covered his chest. Both old men totally disapprove of me and my visit from the moment I cross the threshold, but they are determined to make a show of tolerance.

It's never harder to speak than when you have too much to say. Besides, I have all these threads stitched onto me, and I feel them with every movement I make.

Still, I have my main harangue ready, and I anticipate no painful tugs at the strings. I speak as follows:

Where does the idea come from (I pretend to assume that it is not theirs) that the prison camps are in danger of becoming *health resorts*, and that unless a camp is garrisoned by cold and hunger, blessed ease will enthrone itself there? I ask them in spite of their defective personal experience to try and imagine the densely ringed stockade of privations and punishments which is the reality of imprisonment: a man is deprived of his native place; he lives with men with whom he has no wish to live; he wants to live with his family and friends, but cannot; he does not see his children growing up; he is deprived of his normal surroundings, his home, his belongings, right down to his wristwatch; his name is disgraced, and forgotten; he is deprived of freedom of movement; denied as a rule even the possibility of working at his own trade; he feels the constant pressure of strangers, some of them hostile to him, of other prisoners, whose background, outlook, and habits are different from his; denied the softening influence of the other sex (not to mention the physical deprivation); and even the medical attention he gets is incomparably poorer. In what way does all this resemble a Black Sea sanatorium? Why are they so much afraid of the "health resort" jibe?

No, this thought doesn't bowl them over. They aren't rocked in their chairs by it.

So I broaden my theme: *Do we or do we not want* to restore these people to society? If we do, why do we make them live like outcasts? Why is the whole content of our *disciplinary regimes* the systematic humiliation and physical exhaustion of prisoners? What advantage is it to the state to make cripples of them?

So, I've unburdened myself. And they start showing me where I am wrong. I have no real idea of what the present "contingent" is like, I judge from old impressions, I'm behind the times. (Yes, this is my weak spot: I cannot indeed *see* who is inside *now.*) For habitual criminals sentenced to close confinement, all the things in my list are no privation at all. The present disciplinary regimes are the only thing that can teach them sense. (Two painful jerks at my strings. They are the experts here. They know best *who is inside.*) Restore them to society? . . . Yes, of course, of course, the old men say in wooden voices, and what I hear is: "No, of course, let them all die; it'll make our lives easier—and yours, too."

The disciplinary regimes? One of the veterans of Ochakov, a public prosecutor—the one in blue, with the star on his chest, and sparse ringlets of gray hair, who even looks a little like Suvorov:

"We have already begun to see a *return* from the introduction of strict regimes. Instead of *two thousand murders a year*"—here it *can* be said—"there are only a few dozen."

An important figure. I make a discreet note of it. This appears to be the most useful result of my visit.

Who *is* inside? Of course, to argue about prison regimes you have to know who is inside. Dozens of psychologists and lawyers would have to go and talk to the prisoners without obstruction— then we should be in a position to argue. This is just the one thing my camp correspondents never put in their letters—why they and their comrades are inside.[9]

The general part of our discussion is at an end, and we turn to particulars. The commission, of course, has no doubts, and has already made up its mind about everything. I can be of no use to them; they were merely curious to know what I look like.

Parcels? Only five kilograms at a time, and at the same intervals as at present. I suggest that they should at least double the number allowed, and make the parcels eight kilograms each. "They're hungry in there! Starvation is no way to reform criminals!"

9. The great variety of these *habitual criminals* defeats the imagination. In the Tavdinsk colony, for instance, there is an eighty-seven-year-old, formerly an officer in the Tsar's army, and probably in one of the White armies, too. By 1962 he had served eighteen years of his *second twenty.* He has a beard like Father Christmas and works as a tally clerk in a glove factory. We can't help wondering whether forty years' imprisonment is not rather a high price to pay for the beliefs of your youth. And there are so many unfortunates of this sort—each of them unique. We should have to find out about *every one* of them before we could form a judgment on the regime imposed on them *all.*

"What do you mean, hungry?" The commission's indignation is unanimous. "We've *been there ourselves,* and seen left-over bread *carried out of a camp by the truckload!*" (For the warders' pigs, you mean?)

What can I do? Shout: "You're lying! That just can't be true" —but I feel a painful tug at my tongue, attached by a thread running over my shoulders to a place behind my back. I must not violate our working assumptions: they are well-informed, they are sincere, they care. Shall I show them the letters from my zeks? It would all be Greek to them, and the well-thumbed, crumpled scraps of paper would look absurd and contemptible on the red velvet tablecloth.

"But it costs the state nothing to allow more parcels!"

"Ah, but *who* will benefit from them?" they retort. "Mainly rich families. [They use the word "rich": realistic discussion of policy cannot do without it.] Those who have a lot of stolen goods hidden outside. So that by increasing the number of parcels, we should put *working families* at a disadvantage!"

Now the threads are cutting and tearing! This is an unchallengeable assumption: the interests of the toiling strata are above everything else. They are of course only sitting here for the good of the toiling strata.

I find myself lost for an answer. I don't know what retort I can make. I could say, "I'm not convinced"—and a fat lot they would care. What do I think I am—their boss or something?

I keep pushing. "What about the shops? Where does the socialist principle of remuneration come in? If a man's earned something, give it to him!"

They hit back. "He has to build up reserves for when he is released! Otherwise when he gets out he becomes a charge on the state."

The interests of the state come first—that's stitched on my back; I dare not tug at it. Nor can I suggest that prisoners' wages should be raised *at the state's expense.*

"Well, at least let their Sundays off be sacrosanct!"

"That's provided for—they are."

"But there are dozens of ways of ruining a Sunday in camp. Say expressly that no one must do so!"

"We can't include such minute regulations in the Code."

The camps work an eight-hour day. I half-heartedly put it to them that seven hours is enough, but in my own mind I know that this is impertinence: it isn't a twelve- or a ten-hour day—will you never be satisfied?

"Correspondence gives the prisoner a feeling of participation in the life of our socialist society. [Such were the arguments I had taught myself to use.] Why restrict it?"

But they cannot reconsider. The allowance is fixed, and not so harshly as in our day. . . . They show me also the schedule of visits —including "private" three-day visits—and we, of course, had none at all for years, so this seems tolerable. Indeed, the arrangements for visits seem quite generous to me, and I barely restrain myself from praising them.

I am tired. I am completely sewn up, I can't stir a muscle. I'm doing no good here. It's time to go.

And indeed, seen from this bright, festive room, from these comfortable armchairs, to the accompaniment of their smoothly flowing eloquence, the camps look not horrible but quite rational. You see—left-over bread by the truckload. Well, would *you* let these terrible people loose on the community? I remember the master thieves and their ugly mugs. . . . It's ten years since I was in myself; how can I begin to guess who is there now. My sort, the politicals, are supposed to have been freed. . . . The national groups have been released. . . .

The other disagreeable old man wants my views on hunger strikes: surely I cannot disapprove of forced feeding if the food given is more nourishing than gruel?[10]

I get up on my hind legs and bellow at them that a zek has every right not only to go on a hunger strike—his only means of self-defense—but to starve himself to death.

My arguments seem crazy to them. But I am all sewn up: I cannot talk about the connection between hunger strikes and public opinion in the country at large.

I leave, feeling tired and jaded. I even feel a little less sure of myself, whereas they are not the least bit shaken. They will do just as they please, and the Supreme Soviet will confirm it unanimously.

■

10. But from Marchenko we learn of a new practice: pouring hot water down the tube to damage the esophagus.

Vadim Stepanovich Tikunov, Minister for the Protection of Public Order. What wild fantasy is this? Can I, the miserable convict Shch 232, be on my way to teach the Minister of Internal Affairs how to run the Archipelago! . . .

On the approaches to the Minister, nothing but colonels, bullet-headed, sleekly pallid, but very agile. No door leads on beyond the chief secretary's room. Where it should be stands an enormous mirror-fronted cupboard with gathered silk curtains behind the glass, big enough to take two men on horseback—and this turns out to be the vestibule of the Minister's sanctum. His office would seat two hundred comfortably.

The Minister himself is unhealthily fat, with a heavy jaw; his face is a trapezium, broadening toward the chin. Throughout the conversation his manner is strictly official; he hears me out as a matter of duty, but with no sign of interest.

I launch at him the same old tirade about "health resorts." And once again, the same old questions: Is it our common aim (his and mine!) to *reform* the zeks? (My views on "reform" are behind us in Part IV.) Why the sharp change of course in 1961? Why those four camp regimes? I repeat boring things for him, all the things I have written in this chapter—about diet, camp shops, parcels, clothing, work, bullying warders, the mentality of the practical workers. (I have chosen not to bring the letters, in case someone here pounces on them, and have simply copied out excerpts, omitting the authors' names.) I go on talking to him for forty minutes or an hour—it seems a very long time anyway—and I am surprised myself that he is listening to me.

He interrupts now and then, but only to accept or reject some statement outright. He does not attempt a crushing refutation. I was expecting a blank wall of arrogance, but he is much softer. He agrees with much of what I say! He agrees that shopping money must be increased, and that there should be more parcels and that there is no need to define the contents of parcels as minutely as the Legislative Proposals Commission does (but this does not depend on him: the new Corrective Labor Code, not the Minister, will decide all these things). He agrees that prisoners should be allowed to boil or bake food they have acquired themselves (except that they never have any), and that there should be no limit on letters and printed matter sent through the mails (though this would put a great burden on the camp censorship). He is even against martinets who overdo drills and line-ups (but it wouldn't be tactful to interfere: it's easy to wreck discipline, difficult to impose it). He agrees that the grass in the camp area should not be weeded out. (What had happened in Dubrovlag around the engineering workshops was quite another matter. They had planted little kitchen gardens, the machine operators busied themselves there during their break, and each man had two or three square meters sown with tomatoes or cucumbers. But the Minister ordered them to dig it all under, to destroy it at once, and he's proud of it. I tell him that "man's ties with the earth have a moral importance," and he tells me that individual garden plots foster property-owning instincts.) The Minister even shudders at the thought that people had been sent back behind the wire after living outside. (I don't like to ask what his position was at the time and what he did to prevent it.) More than all this, the Minister acknowledges that *zeks are kept in harsher conditions now* than in the days of Ivan Denisovich.

This being so, I need not waste my time persuading him! There is nothing for us to discuss. (And it is pointless for him to take note of suggestions from someone with no official position.)

What can I suggest? Breaking up the whole Archipelago, and letting prisoners live without guards? I can't get the words out. It's utopian. And anyway, the solution of a big problem never depends on a single individual; it winds snakelike through many departments, and is at home in none.

On the other hand, the Minister is emphatically sure that the striped uniform for habitual criminals is necessary. ("If you only knew what sort of people they are!") My critical remarks about warders and convoy troops simply offend him: "You are confused, or else your own experiences have given you a peculiar way of looking at things." He assures me that you can no longer round up recruits for the prison service, because *the old privileges have been done away with.* ("It shows a healthy attitude on the part of ordinary people if they won't join up," I almost exclaim—but I feel the warning tug of the strings at my ears, my eyelids, my tongue. However, I am overlooking something: it's only sergeants and corporals who *won't join;* you can't keep the officers out.) They have to make use of conscripts. Contradicting me again, the Minister tells me that only prisoners use vulgar abuse, while warders are invariably correct in their manner of speaking to prisoners.

When there is such a discrepancy between the letters of insignificant zeks and the words of a minister, whom shall we believe? Clearly, the prisoners are lying.

He even quotes his own observation at first hand. Because he, of course, *does* see something of the camps, and I do not. Perhaps I would like to visit one? Kryukovo, or Dubrovlag? (These two names come to him so easily that they are obviously Potemkin structures.* Besides, *in what capacity* would I go? As a ministry inspector? If I did, I couldn't look the zeks in the face. . . . I refuse.)

The Minister, disagreeing with me, expresses the view that zeks lack feeling and do not respond to the efforts made for them. You go to the Magnitogorsk colony, and ask: "Any complaints about your treatment?" and—with the head of the Camp Division standing by—would you believe it, they shout in chorus, "None!"

Now for what the Minister sees as "splendid results of the re-education program in the camps":

A machine operator's pride when the head of his Camp Division commended him.

The pride of prisoners whose work (they made kettles) was destined for heroic Cuba.

The reports and elections of Internal Order Sections (that is, "A Bitch Went Walking").*

The abundance of flowers (provided by the state) in Dubrovlag.

His main concern is to create in every camp an industrial base of its own. The Minister reckons that by increasing the number of interesting jobs he can cut out escape attempts.[11] (When I object that all human beings long for freedom, he simply doesn't understand.)

I left with the weary conviction that *there was no end to it*. That I had not advanced my cause by a hairbreadth and that they would always take a sledge hammer to crack a nut. I left depressed by the realization that two human minds could think so differently. Zek will understand minister when he ensconces himself in a ministerial sanctum, and minister will understand zek when he, too, goes behind the wire, has his own garden plot trodden down,

11. All the more easily because, as we now know from Marchenko, they no longer try to catch escapers, but just shoot them down.

and in return for his freedom is offered the chance to master a machine.

Institute for the Study of the Origins of Criminality. This was an interesting discussion with two cultured deputy directors and several members of the research staff—lively people, each with his own opinions, given to arguing among themselves. Afterward, V. N. Kudryavtsev, one of the deputy directors, chided me as he led me along the corridor: "It's all very well, but you don't take all points of view into account. Now, Tolstoi would have done so. . . ." And suddenly I found that he had tricked me into taking a wrong turn. "We'll just look in and meet the director, Igor Ivanovich Karpets."

This visit was not part of the plan! We'd finished our discussion —what was the point of it? Well, all right, I'll drop in to shake hands! I need not have expected a polite exchange of greetings here. It was hard to believe that the deputy directors and heads of section I had seen worked for a boss like this, that he presided over all their research. (The most important thing about him I would learn much later: Karpets was vice-president of the International Association of *Democratic Lawyers!*)

The man who rose to meet me was hostile and disdainful (as I remember it, we remained on our feet throughout our five-minute conversation), as if he were reluctantly granting an interview for which I had begged and pleaded. In his face I saw well-fed prosperity, firmness, and distaste (for me). With no thought for his nice suit, he had pinned a large badge on his chest as though it were a medal: a vertical sword piercing something down below, over the legend "MVD." (This appears to be a very important badge. It shows that the wearer has had "clean hands, fire in his heart, and a cool head" much longer than most.)

"Let's have it, then—what've you been talking about?" he asks with a scowl.

I have no use for him at all, but out of politeness I repeat some of it.

"Oh, that," says the Democratic Lawyer, as though he had heard enough. "Liberalization, eh? Babying the zeks?"

Then, suddenly, out they come—full answers to all the questions which I had carried in vain over marble floors and between mirrored walls.

Raise the living standards of prisoners? *Can't be done!* Because the free people around the camp would be living *less well* than the zeks, which cannot be allowed.

Receive parcels frequently, and bigger ones? *Can't be done!* Because this would have a bad effect on the warders, who get no food from Moscow shops.

Reprimand warders and teach them to behave better? *Can't be done!* We're trying to hold on to them! Nobody wants the job, we can't pay much, and some of their privileges have been taken away.

We deny prisoners payment for their work according to socialist principles? Their own fault—they've cut themselves off from our socialist society!

But don't we want to reclaim them for normal life?! . . .

Reclaim them??? The sword-bearer is astonished. "That's not what the camps are about. A camp is a place of retribution!"

Retribution! The word fills the whole room.

Retribution!

Rrrretrrribution!!!

The sword stands upright—to smite, to pierce. You'll never ease it out again!

Rret-rrib-ution!!

The Archipelago was, the Archipelago remains, the Archipelago will stand forever!

Without it, who can be made to suffer for the errors of the Vanguard Doctrine? For the fact that people will not grow into the shapes devised for them?

Chapter 3

■

The Law Today

The reader has seen throughout this book that from the very beginning of the Stalin age there have been no *politicals* in our country. The crowds, the millions driven past while you watched, all those millions of 58's, were merely common *criminals.*

Besides, merry, mouthy Nikita Sergeyevich took so many bows from so many platforms: *Politicals?* Not a one!! We just don't have them!

And as grief grew forgetful, as distance softened craggy contours, as fat formed under the skin—we almost believed it! Even former zeks did. Millions of zeks were released for all to see—so perhaps there really were no politicals left? We had returned, others joined us, our friends and families were back. The gaps in our little world of urban intellectuals seemed to be filled, the ring closed. You could sleep undisturbed, and no one would have been taken from the house when you awoke. Friends would telephone —no one was missing. Not that we altogether believed it—but for practical purposes we accepted that there were no longer any politicals in jail. Well, yes, even today (1968) a few hundred Balts are not allowed to go home to their republics, and the curse has not been lifted from the Crimean Tatars—but very soon, no doubt . . . From outside, as always (as indeed under Stalin), all was clean and tidy, nothing showed.

And Nikita was there, glued to his platform. "There can be no return to deeds and occurrences such as these, either in the Party or in the country generally" (May 22, 1959—that was before Novocherkassk). "Now everyone in our country can breathe freely . . . with no need to worry about the present or the future"

(March 8, 1963, *after* Novocherkassk).

Novocherkassk! A town of fateful significance in Russia's history. As though the Civil War had not left scars enough, it thrust itself beneath the saber yet again.

Novocherkassk! A whole town rebels—and every trace is licked clean and hidden. Even under Khrushchev the fog of universal ignorance remained so thick that no one abroad got to know about Novocherkassk, there were no Western broadcasts to inform us of it, and even local rumor was stamped out before it could spread, so that the majority of our fellow citizens do not know what event is associated with the name Novocherkassk and the date June 2, 1962.

Let me then put down here all that I have been able to gather.

We can say without exaggeration that this was a turning point in the modern history of Russia. If we leave out the Ivanovo weavers at the beginning of the thirties (theirs was a large-scale strike, but it ended without violence), the flare-up at Novocherkassk was the first time the people had spoken out in forty-one years (since Kronstadt and Tambov): unorganized, leaderless, unpremeditated, it was a cry from the soul of a people who could no longer live as they had lived.

On Friday, June 1, one of those carefully considered enactments of which Khrushchev was so fond was published throughout the Union—raising the prices of meat and butter. On that very same day, as demanded by another and quite separate economic plan, piece rates at the huge Electric Locomotive Works in Novocherkassk (NEVZ) were lowered, in some cases by 30 percent. That morning the workers in two shops (the forge and the foundry), usually obedient creatures of habit, geared to their jobs, could not force themselves to work—so hot had things become for them. Their loud, excited discussions developed into a spontaneous mass meeting. An everyday event in the West, an extraordinary one for us. Neither the engineers nor the chief engineer himself could persuade them. Kurochkin, the works manager, arrived. When the workers asked him, "What are we going to live on now?" this well-fed parasite answered: "You're used to guzzling meat pies—put jam in them instead." He and his retinue barely escaped being torn to pieces. (Perhaps if he had answered differently it would all have blown over.)

By noon the strike had spread throughout the enormous locomotive works. (Runners were sent to other factories, where the workers wavered but did not come out in support.) The Moscow-Rostov railway line runs close to the works. Either to make sure that the news would reach Moscow more quickly, or to prevent troops and tanks from moving in, a large number of women sat down on the tracks to hold up trains, whereupon the men began pulling up the rails and building barriers. Strike action of such boldness is unusual in the history of the Russian workers' movement. Slogans appeared on the works building: "Down with Khrushchev!" "Use Khrushchev for sausage meat!"

While all this was happening, troops and police began converging on the works (which stands, with its settlement, three to four kilometers from Novocherkassk, across the river Tuzlov). Tanks took up position on the bridge over the Tuzlov. From evening until the following morning, movement inside the city or across the bridge was completely forbidden. Even during the night the workers' settlement did not quiet down for a moment. Overnight about thirty workers were arrested as "ringleaders" and carried off to the city police station.

On the morning of June 2, some other enterprises in the town struck (but by no means all of them). Another spontaneous mass meeting at NEVZ decided on a protest march into the town to demand the release of the arrested workers. The procession (only about three hundred strong to begin with—you had to be brave!), with women and children in its ranks, carrying portraits of Lenin and peaceful slogans, marched over the bridge past the tanks without obstruction, then uphill into the town. Here their numbers were quickly swelled by curious onlookers, individual workers from other enterprises, and little boys. At several places in the city people stopped lorries and used them as platforms for speechmaking. The whole town was seething. The NEVZ demonstrators marched along the main street (Moskovskaya) and some of them began trying to break down the locked doors of the town police station in the belief that their arrested comrades were inside. They were met with pistol shots. Further on, the street led to the Lenin monument[1] and by two narrow paths around a public garden to the headquarters of the town Party committee (formerly the ataman's palace, in which General Kaledin had shot himself in 1918). All the streets were choked with people and here, on the square,

1. It had replaced Klodt's statue of Ataman Platonov, which had been melted down for scrap.

the crowd was densest. Many little boys had climbed trees in the garden to get a better view.

The Party offices were found to be empty—the city authorities had fled to Rostov.[2] Inside the building there was broken glass and the floors were strewn with documents, as they must have been after a retreat in the Civil War. A couple of dozen workers walked through the palace, came out on its long balcony, and harangued the crowd in halting speeches.

It was about 11 A.M. There were no police to be seen in the town, but there were more and more troops. (A revealing picture: at the first slight shock the civil authorities hid behind the army.) Soldiers had occupied the post office, the radio station, the bank. By this time the whole of Novocherkassk was beleaguered, and every entry and exit barred. (For this task they had brought in, among others, cadets from the officers' training schools in Rostov, leaving some behind to patrol that city.) Tanks crawled slowly along Moskovskaya Street, following the route the demonstrators had taken toward Party headquarters. Boys started scrambling onto the tanks and obstructing the observation slits. The tanks fired a few blank shells, rattling the windows of shops and houses all along the street. The boys scattered and the tanks crawled on.

And the students? Novocherkassk is of course a town of students! Where were they all? . . . The students of some institutes, including the Polytechnic, and of some technical secondary schools, had been *locked in* their dormitories or in other school buildings from early morning. Their rectors had thought quickly. But we may as well say it: the students for their part showed little civic courage. They were presumably glad of this excuse to do nothing. It would take more than the turn of a key to hold back rebel students in the West today (and took more in Russia in days gone by).

A scuffle broke out inside the Party building, step by step the speakers were dragged back inside and soldiers emerged onto the balcony, more and more of them. (Remember how the military observed the Kengir mutiny from the balcony of the Steplag head

2. The First Secretary of the Rostov Oblast Party Committee, Basov, whose name, together with that of Pliev, commander of the North Caucasus Military District, will one day be inscribed over the site of the mass shooting, had arrived in Novocherkassk in the meantime, but had rushed back to Rostov in terror. (It is even said that he made his escape by jumping from a second-story balcony.) Immediately after the Novocherkassk events, he went with a delegation to heroic Cuba.

office?) A file of riflemen began forcing the crowd back from the small square immediately before the palace, toward the railings of the garden. (Several witnesses say unanimously that *these* soldiers were all non-Russians—Caucasians brought in from the other end of the oblast to replace the cordon from the local garrison previously posted there. But not all witnesses agree that the previous cordon had been ordered to open fire, and that the order was not carried out because the captain who received it killed himself in front of his men rather than pass it on.[3] That an officer committed suicide is beyond doubt, but accounts of the circumstances are vague and no one knows the name of this hero of conscience.) The crowd backed away, but no one expected the worst. It is not known who gave the order,[4] but *these* soldiers raised their rifles and fired a first volley over the heads of the crowd.

Perhaps General *Pliev** had no immediate intention of firing on the crowd; perhaps the situation got out of hand. The burst fired over the heads of the crowd found the trees in the little garden and the boys who had climbed into them, some of whom fell to the ground. The crowd, it seems, gave a roar, whereupon the soldiers, whether at a command, or because they saw red, or in panic, started firing freely into the crowd, and—yes—with dumdum bullets.[5] (Remember Kengir? The sixteen at the guardhouse?) The crowd fled in panic, jamming the narrow paths around the garden, but the troops *went on firing at their backs as they retreated*. They continued firing until the large square beyond the garden and the Lenin statue was completely empty—all along the former Platovsky Prospekt, and as far as Moskovskaya Street. (An eyewitness says that the area looked like one great mound of corpses. But many of those lying there were of course only wounded.) Information from a variety of sources is more or less unanimous that some seventy or eighty people were killed.[6] The soldiers looked around for lorries and buses, commandeered them, loaded them with the dead and the wounded, and dispatched them to the high-walled military hospital. (For a day or two afterward these buses went around with bloodstained seats.)

3. According to this version, the soldiers who refused to fire into the crowd were exiled to Yakutya.
4. Those who stood near enough know, but they were either killed or taken out of circulation.
5. There is reliable evidence that forty-seven were killed by dumdum bullets alone.
6. Rather fewer than before the Winter Palace, yet all Russia was outraged by January 9 and observed its anniversary yearly. When shall we begin commemorating June 2?

That day, just as in Kengir, movie cameras took pictures of the rebels on the streets.

The firing ceased, the terror passed, the crowd poured back onto the square, and *was fired upon again*.

All this happened between noon and 1 P.M.

This is what an observant witness saw at 2 P.M.: "There are about eight tanks of different types standing on the square in front of Party headquarters. A cordon of soldiers stands before them. The square is almost deserted, there are only small groups of people, mostly youngsters, standing about and shouting at the soldiers. On the square puddles of blood have formed in the depressions in the pavement. I am not exaggerating; I never suspected till now that there could be so much blood. The benches in the public garden are spattered with blood, there are bloodstains on its sanded paths and on the whitewashed tree trunks in the public garden. The whole square is scored with tank tracks. A red flag, which the demonstrators had been carrying, is propped against the wall of Party headquarters, and a gray cap splashed with red-brown blood has been slung over the top of its pole. Across the façade of the Party building hangs a red banner, there for some time past: 'The People and the Party are one.'

"People go up to the soldiers, to curse them or to appeal to their conscience. 'How could you do it?' 'Who did you think you were shooting at?' 'Your own people you were shooting at!' They make excuses: 'It wasn't us! We've only just been brought in and posted here. We had nothing to do with it.'

"That's how efficient our murderers are (and yet people talk about bureaucratic sluggishness). *Those* soldiers have already been taken away, and perplexed Russians put in their place. He knows his business, that General Pliev. . . ."

Toward five or six o'clock the square gradually filled with people again. (They *were* brave, the people of Novocherkassk! The town radio kept appealing to them: "Citizens, do not fall for provocation, go home quietly!" The riflemen still stood there, the blood had not been mopped up, and again they pressed forward.) Shouts from the crowd, more and more people, and another impromptu meeting. They knew by now that six senior members of the Central Committee had flown in (probably arriving before the first shootings?), among them, needless to say, Mikoyan (the expert on Budapest-type situations) and Frol Kozlov. (The names of the other four are not known for certain.) They stayed in the

KUKKS* building (formerly the headquarters of the Cadet Corps), as though it were a fortress. And a delegation of younger workers from NEVZ was sent to tell them what had happened. A buzz went through the crowd: "Let Mikoyan come down here! Let him see all this blood for himself!" Mikoyan wouldn't come down, thank you. But a reconnaissance helicopter flew low over the square around six o'clock. Inspected it. Flew off again.

Shortly afterward the workers' delegation came back from KUKKS. As agreed, the military cordon let the delegates through and officers escorted them to the balcony of the Party building. Silence. The delegates reported to the crowd that they had seen the Central Committee members and told them about this "bloody Saturday," and that *Kozlov had wept* when he heard about the children falling from the trees at the first volley. (You know Frol Kozlov, the Leningrad Party gang boss, the cruelest of Stalinists? He wept! . . .) The Central Committee members had promised to investigate these events and severely punish those responsible (the very promises made to us in Osoblag), but for the present everyone must go home to prevent the outbreak of fresh disorders in the town.

The meeting, however, did not disperse! The crowd grew ever denser toward the evening. The desperate courage of Novocherkassk! (There is a story that the Politburo team made the decision that evening to *deport the whole population of the town, every last one of them!* I can believe this; it would have been nothing extraordinary after the deportation of nations. Wasn't the same Mikoyan close to Stalin when that happened?)

Around nine in the evening they tried to drive the people away from the palace with tanks. But as soon as the drivers switched on their engines people clustered around the tanks, blocking the hatches and the observation slits. The tanks stalled. The riflemen stood by and made no effort to help the tank crews.

An hour later tanks and armored personnel carriers appeared from the opposite side of the square, with an escort of Tommygunners perched on top of them. (Our battle experience counts for something! We are the ones who defeated the Fascists!) Advancing at high speed (to the jeers of young people on the footpaths—the students had been released toward evening), they cleared the roadways of Moskovskaya Street and the former Platovsky Prospekt.

At last, toward midnight, the riflemen began firing tracer bullets into the air and the crowd slowly dispersed.

(What power there is in a popular disturbance! How quickly it changes the whole political situation! The night before there had been a curfew, and people had been frightened anyway, but now the whole town was strolling about and hooting at the soldiers. A people transformed—can it be so near to breaking through the crust of this half-century, into a completely different atmosphere?)

On June 3 the town radio broadcast speeches by Mikoyan and Kozlov. Kozlov did not weep. Nor did they any longer promise to find the culprits (those in higher places). What they now said was that *these events were the result of enemy provocation,* and that *these enemies would be severely punished.* (The people had of course gone from the square by now.) Mikoyan said further that *dumdum bullets had never been adopted as part of the equipment of Soviet troops, and that they must therefore have been used by enemies of the state.*

(But who were these enemies? How had they parachuted into the country? Where were they hiding? Show us just one! We are so used to being treated like fools: "Enemies," they say, and all is explained. In the Middle Ages it was "devils.")[7]

The shops were immediately the richer for butter, sausage, and many other things not seen in those parts for a long time, or anywhere outside the capitals.

The wounded all vanished without trace; not one of them went home. Instead, the *families* of the wounded and the killed (who of course wanted to know what had become of their kin) *were deported to Siberia.* So were many of those involved in the demonstration who had been noticed or photographed. Some participants were dealt with in a series of trials in camera. There were also two "public" trials (with entry by ticket for factory Party officials and for the town apparatchiki). At one of these, nine men were sentenced to be shot and two women to fifteen years' imprisonment.

The membership of the town Party committee remained as before.

---

7. This is a woman schoolteacher (!) from Novocherkassk holding forth in a train in 1968: "The military did not shoot anyone. They fired only one warning burst into the air. The shooting was done by saboteurs, with dumdum bullets. Where did they get them? Saboteurs can get absolutely anything. They shot at soldiers and workers alike. . . . The workers seemed to go mad, attacked the soldiers and beat them—but how were the soldiers to blame? Afterward Mikoyan walked around the streets and went into people's houses to see how they lived. The women offered him strawberries. . . ." *This* is all that history has preserved to date.

On the Saturday following "bloody Saturday," the town radio announced that the "workers of the Electric Locomotive Works have solemnly undertaken to fulfill their seven-year plan ahead of time."

. . . If the Tsar had not been such a ninny, he would have realized that all he needed to do on January 9 in Petersburg was hunt down the workers carrying banners and pin charges of banditry on them. After that there would have been no "revolutionary movement" worth mentioning.

At Alexandrovo in 1961, a year before Novocherkassk, the police beat a man to death while he was under arrest and then would not allow his body to be carried past their "precinct" to the cemetery. The crowd was furious and burned down the police station. Arrests followed immediately. (There was a similar incident about the same time in Murom.) What would the appropriate charge now be? Under Stalin, even a tailor who stuck a needle in a newspaper could get Article 58. Now a more sensible view was taken: wrecking a police station should not be regarded as a political act. It was ordinary banditry. *Instructions were handed down* to this effect: "mass disorders" should not be treated as political offenses. (If they are not political, what is?)

So all at once—there were no more *politicals*.

But one stream has never dried up in the U.S.S.R., and still flows. A stream of criminals untouched by the "beneficent wave summoned to life . . ." etc. A stream which flowed uninterruptedly through all those decades—whether "Leninist norms were infringed" or strictly observed—and flowed in Khrushchev's day more furiously than ever.

I mean the believers. Those who resisted the new wave of cruel persecution, the wholesale closing of churches. Monks who were slung out of their monasteries (Krasnov-Levitin has given us a great deal of information about this). Stubborn sectarians, especially those who refused to perform military service: there's nothing we can do about it, we're really very sorry, but you're directly aiding imperialism; we let you off lightly nowadays—it's five years first time around.

These are in no sense politicals, they are "religionists," but still they have to be *re-educated*. Believers must be dismissed from their jobs merely for their faith; Komsomols must be sent along to break the windows of believers; believers must be officially compelled to attend antireligious lectures, church doors must be cut down with blowtorches, domes pulled down with hawsers attached to tractors, gatherings of old women broken up with fire hoses. (Is this what you mean by *dialogue,* French comrades?)

As the monks of the Pochayev Monastery were told in the Soviet of Workers' Deputies: *"If we always observe Soviet laws, we shall have to wait a long time for Communism."*

Only in extreme cases, when *educational* methods do not help, is recourse to the *law* necessary.

Here we can dazzle the world with the diamond-pure nobility of our laws today. We no longer try people in closed courts, as under Stalin, we no longer try them in absentia, we try them semi-publicly (that is to say, in the presence of a semi-public).

I hold in my hand a record of the trial of some Baptists at Nikitovka in the Donbas, in January, 1964.

This is how it's done. On the pretense that their identity must be checked, the Baptists who arrived to attend the trial were held in jail for three days (until the trial was over, and to give them a fright). Someone (a free citizen!) who threw flowers to the defendants got ten days. So did a Baptist who kept a record of the trial, and his notes were taken away (but another record survived). A bunch of hand-picked Komsomols were let in before the general public by a side door, so that they could occupy the front rows. While the trial was in progress there were shouts from the spectators: "Pour kerosene over the lot and set fire to them!" The court did nothing to curb this righteous indignation. Typical of its procedures: it admitted the evidence of hostile neighbors and also that of terrorized minors; little girls of nine and eleven were brought before the court (who the hell cares what effect it has on them as long as we get our verdict). Their exercise books with texts from the Scriptures were introduced as exhibits.

One of the defendants, Bazbei, father of *nine* children, was a miner who had never received any support from the Union committee at his pit because he was a Baptist. But they managed to confuse his daughter Nina, a schoolgirl in the eighth grade, and to suborn her with fifty rubles from the Union committee and a promise to place her in an institute later on, so that during the investigation she made fantastic statements against her father: he had tried to poison her with a sour fruit drink; when the believers were hiding in the woods for their prayer meetings (because they were persecuted in the settlement) they had had a radio transmitter—"a tall tree with wire wound all around it." Afterward these lying statements began to prey on Nina's mind, she became mentally ill and was put in the violent ward of an asylum. Nonetheless, she was produced in court in the expectation that she would stick to her evidence. But she repudiated every word of it! "The interrogator dictated what I had to say himself." It made no difference. The shameless judge ignored her latest statements and regarded only her earlier evidence as valid. (Whenever depositions favorable to the prosecution come unstuck, this is the typical and regular dodge used by the courts: they ignore what is brought out in court and base themselves on faked evidence obtained in the preliminary investigation: "Now, what do you mean by that? It says here in your deposition . . . You testified during the investigation . . . What right have you to retract now? That's an offense, too, you know!")

The judge is not at all interested in the substance of the case, in the truth. The Baptists are persecuted because they do not accept preachers sent by an atheist plenipotentiary of the state, but prefer their own. (Under Baptist rules, any brother can preach the Gospel.) There is a directive from the Oblast Party Committee: put them on trial and forcibly take their children from them. And this will be carried out, although with its left hand the Presidium of the Supreme Soviet has just (July 2, 1962) signed the world convention on "the fight against discrimination in the sphere of education."[8] One of its points is that "parents must be allowed to provide for the religious and moral education of their children in accordance with their own convictions." But that is precisely what we cannot allow! Anyone who speaks in court on the substance of the case, anyone who tries to clarify the issue, is invariably interrupted, diverted from his train of thought, deliberately confused by the judge, who conducts the debate on this level: "How can you talk about the end of the world when we are committed to the building of Communism?"

This is from the closing statement made by one young girl, Zhenya Khloponina. "Instead of going to the cinema or to dances, I used to read the Bible and say my prayers—and just for that you are taking my freedom from me. Yes, to be free is a great happiness, but to be free from sin is a greater still. Lenin said that only in Turkey and Russia did such shameful phenomena as religious

8. But of course, we signed it for the sake of the American Negroes. How else could it concern us?

persecution still exist. I've never been in Turkey and know nothing about it, but how things are in Russia you can see for yourselves." She was cut short.

The sentences: Two of them got five years in the camps, two of them four years, and Bazbei, father of all those children, got three. The defendants accepted their sentences *joyfully,* and said a prayer. The "representatives from enterprises" shouted: "Not long enough! Make it more!" (Throw kerosene over them and put a match to it. . . .)

The long-suffering Baptists took note and kept count: and set up a "Council of Prisoners' Relatives," which began issuing manuscript bulletins about all the persecutions. From these bulletins we learn that from 1961 to June, 1964, 197 Baptists were condemned, 15 of them women.[9] (They are all listed by name. Prisoners' dependents, now left without means of support, have also been counted: 442, of whom 341 are under school age.) The majority get five years of exile, but some get five years in a *strict* regime camp (narrowly escaping the hardened criminals' motley!), with three to five years of exile in addition. B. M. Zdorovets from Olshany in Kharkov oblast got seven years of strict regime for his faith. A seventy-six-year-old, Y. V. Arend, was put inside, as were the whole Lozovoy family (father, mother, and son). Yevgeny M. Sirokhin, a (Group 1) disabled veteran of the Fatherland War, *blind in both eyes,* was condemned in the village of Sokolovo, Zmievski district, Kharkov oblast, to three years in a camp for bringing up his children Lyuba, Nadya, and Raya as Christians, and they were taken away from him by order of the court.

The court trying the Baptist M. I. Brodovsky (at Nikolayev, October 6, 1966) was not too squeamish to use crudely faked documents; when the defendant protested—"This is dishonest of you!"—they barked back at him: "The *law* will crush you, smash you, destroy you!"

The law, my friend. Not one of your acts of "extrajudicial vengeance," as practiced in the years when "norms were still observed."

We recently got to know S. Karavansky's soul-chilling "Petition," which was transmitted from a camp to the outside world.

9. One of the trials of Populists a hundred years ago was called "the trial of the 193." Lord, what a fuss there was! What emotions were stirred! It even found its way into textbooks.

The author had been sentenced to twenty-five years, had served sixteen of them (1944–1960), had been released (evidently under the "two-thirds" rule), had married, had begun a university course —but no! In 1965 they came for him again. Get yourself ready! You still have nine years to go.

Where else is this possible, under what other code of law on earth except ours? They had hung *quarters* around people's necks like iron collars. Sentences which would end sometime in the seventies! Suddenly a new Code is promulgated (1961)—with no sentence higher than fifteen years. Even a first-year law student can see that those twenty-five-year sentences are thereby rescinded. Only we do not agree that they are. Yell yourself hoarse, beat your head on the wall if you like—they are not rescinded. We feel, rather, that you should step back inside and finish your time!

There are quite a few people like this. People who were not affected by the epidemic of releases under Khrushchev, the teammates, cellmates, transit prison acquaintances whom we left behind. We have long ago forgotten them in our new lives, but they still shuffle hopelessly, drearily, numbly about the same little patches of trampled earth, with the same watchtowers and barbed-wire fences all around them. The faces in the papers change, the speeches from platforms change, people fight against the *cult* and then stop fighting—but the twenty-five-year prisoners, Stalin's godchildren, are still inside. . . .

Karavansky cites the blood-freezing prison careers of several such people.

All you freedom-loving "left-wing" thinkers in the West! You left laborites! You progressive American, German, and French students! As far as you are concerned, none of this amounts to much. As far as you are concerned, this whole book of mine is a waste of effort. You may suddenly understand it all someday—but only when you *yourselves* hear "hands behind your backs there!" and step ashore on our Archipelago.

■

Still, there really is no comparison between the numbers of political prisoners now and in Stalin's time; they are no longer counted in millions or in hundreds of thousands.

Is this because the *law* has been reformed?

No, it is just that the ship has changed course (for a time).

Courtroom epidemics flare up just as they used to, lightening the labors of the legal brain. Even the newspapers will keep you abreast if you know how to read them: when they start writing about hooligans, you know that the courts are jailing people wholesale on charges of hooliganism; if they write about theft from the state, you know that the fashionable charge is embezzlement.

Zeks writing from today's colonies tell us despondently that:

"It is useless trying to find justice. What you read in the press is one thing; real life is another." (V.I.D.)

"I'm sick of being an outcast from my society and my people. But where can I get justice? The interrogator's word carries more weight than mine. Yet what knowledge or insight can she—a young girl of twenty-three—have? How can she possibly imagine the fate they can send a man to?" (V.K.)

"The reason they never reopen cases is that if they did, some of them might become redundant." (L——n)

"Stalinist methods of investigation and trial have simply migrated from the political to the criminal sphere, and that's all there is to it." (G.S.)

Let us note carefully what these suffering people have told us.

> 1. Retrial is impossible (because the judicial caste might collapse).
> 2. Nowadays they use the criminal clauses to make mincemeat of people, just as they once used Article 58. (If they did not, what would they feed on? And what would become of the Archipelago?)

Briefly—suppose one citizen wishes to rid the world of another whom he dislikes (not, of course, straightforwardly, with a knife between the ribs, but legally). What is the surest way of doing it? Formerly, he would have had to write a denunciation under Article 58-10. But now he should begin by consulting the *professionals* (investigating officers, policemen, court officials)—*that* sort of citizen always has friends of *this* sort—to find what is in fashion this year. For what type of offender are the nets being laid? In which category are the courts required to increase their yield? Find the appropriate clause, and stick that in him—it's as good as any knife.

Thus, a storm of accusations under the *Rape* clause raged for a long time after Nikita in a heated moment ordered minimum sentences of twelve years. Thousands of hammers in every locality began busily riveting on twelve-year fetters—for the smiths must never stand idle! Now, this clause deals with delicate and very private matters. Weigh it carefully, and you will see that in some ways it resembles Article 58-10. The offenses covered by each are committed tête-à-tête, they are difficult to verify, they are shy of witnesses—and that is just what the courts require.

Take the S——v case. Two Leningrad women were summoned to the police station. Had they been at a party with some men? Yes. Had sexual intercourse taken place? (This had already been established with the aid of a reliable informer.) Er—yes. Right, then; which is it: did you take part in the sexual act voluntarily or against your will? If voluntarily, we shall have to regard you as prostitutes, you will hand over your passports and get out of Leningrad in forty-eight hours. If it was against your will, you must bring a charge of rape! The women were not a bit anxious to leave Leningrad! So the men got twelve years each.

Our obtuse, our blinkered, our hulking brute of a judicial system can live only if it is infallible. The brute is so strong and so sure of itself only because it never reconsiders its decisions, because every officer of the court can lay about him as he pleases in the certainty that no one will ever correct him. To this end there exists a tacit understanding that every complaint, whatever summit of summits you send it to, will be referred back to the very authority of which you are complaining. Let no officer of the court (prosecutor or investigator) be censured for abusing his office, for giving free rein to bad temper or a desire for personal vengeance, for making a mistake or for misconducting a case. We will cover up for him! Protect him! Form a wall around him! We are the Law —and that is what Law is for.

What is the good of beginning an investigation and then not bringing charges? Does this not mean that the interrogator's work is wasted? What is the good of a hearing without a conviction? Wouldn't the people's court be letting the investigating officer down and wasting his time? What does it mean when an oblast court overturns the decision of a people's court? It means that the higher court has added another botched job to the oblast's record. Think of the discomfort you would be causing your comrades in the profession—what's the point of it? *Once begun,* as the result of a denunciation, let's say, an *investigation* must end without fail in a conviction, *which cannot possibly be quashed.* Above all—don't let one another down! And don't let the raikom down—do what they tell you. In return they will see that you come to no harm.

Another very important thing about the courts today: there is no tape recorder, no stenographer, just a thick-fingered secretary with the leisurely penmanship of an eighteenth-century schoolgirl, laboriously recording some part of the proceedings in the transcript. This record is not read out during the session, and no one is allowed to see it until the judge has looked it over and approved it. Only what the judge confirms will remain on record, will have happened in court. While things that we have heard with our own ears vanish like smoke—they never happened at all!

In his mind's eye the judge can always see the shiny black visage of truth—the telephone in his chambers. This oracle will never fail you, as long as you do what it says.

Endure and flourish, O noble company of judges! We exist for you! Not you for us! May justice be a thick-piled carpet beneath your feet. If it goes well with you, then all is well!

The proven reliability of the judicial system makes the lives of the police much easier. It enables them to apply without misgivings the method known as the "trailer" or the "crime sack." Because of the slackness, the inefficiency, the boneheadedness of the local police, crime after crime after crime remains unsolved. But to keep the books straight, criminals must be "exposed" (and cases "closed"). So they wait for a suitable opportunity. A man lands in the police station—somebody pliant, easily bullied, not too bright—and they saddle him with all these unsolved crimes. He's the one! All this year! The elusive master criminal! Pummel and starve him till he *confesses* everything, puts his name to it all, earns himself a sentence commensurate with the grand total of his crimes—and so wipe a blot from the district.

The health of society is much improved, since no sin goes unpunished. And the police in charge of criminal investigations are given prizes.

The health of society has improved still further, and justice has been further reinforced in recent years, since the cry went up that *parasites* should be seized, tried, and deported. This decree was also a partial replacement for the elastic 58-10, now only a memory: accusations made under it proved just as insidious, just as flimsy—and just as irrefutable. (They managed to use it against I. Brodsky, the poet!)

The meaning of the word was skillfully distorted from the start. Real parasites, highly paid drones, sat on the bench or at their bureaucratic desks while sentences rained down on paupers with skills and an appetite for work who knocked themselves out trying to earn a bit extra when the working day was over. How viciously —with the undying hatred of the overfed for the hungry—they fell upon these "idlers." Two of Adzhubei's unscrupulous journalists[10] had the effrontery to declare that parasites were not being banished far enough from Moscow. They were allowed to receive parcels and money orders from relatives! Discipline was not strict enough! "They are not made to work from dawn to dusk." These are their very words: "from dawn to dusk." What Communist dawn, what constitutional order, we may wonder, can call for such drudgery?

We have listed several important *streams* which (together with the endless spate of embezzlers) ensure that the Archipelago is continually replenished.

Nor is it altogether wasted effort for the "people's brigades" *(druzhinniki)*—those freebooters or storm troopers commissioned by the militia, unmentioned in the Constitution, and free from responsibility before the law—to walk the streets, or stay comfortably in their command posts knocking out the teeth of prisoners.

Reinforcements flow in to the Archipelago. And although we have had a classless society for so long, although the glow of the Communist conflagration half-fills the sky, we are used to the idea that crime never ceases, never decreases, and indeed that no one now seems to promise any such thing. In the 1930s they assured us: We're almost there, just a few more years. They don't even make such statements any more.

The Law in our country, in its might and its flexibility, is unlike anything called "law" elsewhere on earth.

The stupid Romans had a formula: "The law has no retroactive force." With us—it has! An old reactionary proverb may mutter: "Laws aren't written for what's gone and done." In our country —they are! If a modish new Decree comes out and the Law itches to apply it to persons already in custody—why not, let it do so!

10. *Izvestiya,* June 23, 1964.

This is what happened to the currency speculators and bribe takers. Lists were sent from, say, Kiev to Moscow, where the names of those to whom the Law could be *retrospectively applied* were ticked off (and they were given a *longer stretch* or promoted to *nine grams of lead* accordingly).

Then again, in our country the Law is clairvoyant. You might suppose that before a trial takes place, the course of the hearing, and the verdict, would be unknown. But you may find *Socialist Legality* publishing all this *before* the trial takes place. How can it know? Just ask yourself.[11]

Then again, Soviet Law has *forgotten all about* the sin of bearing false witness—and simply does not regard it as a crime! A legion of false witnesses thrives in our midst, they go sedately on their way to an honorable old age, bask in a golden sunset at the end of their days. Ours is the only country in the world and in history to pamper perjurers!

Then again, Soviet Law does not punish *murdering* judges and *murdering* prosecutors. They all enjoy long and honorable careers, and live to be noble elders.

Then again, no one can deny that Soviet Law is capable of those abrupt changes of course, those sudden swerves characteristic of all anxious creative thought. At times, the Law veers toward "sharp reduction of crime in a single year!" Arrest fewer! Try fewer! Release convicted offenders on probation! At other times, it veers in the opposite direction. Evildoers endlessly multiply! No more probation! Send more to hard labor and special regime camps! Stiffer sentences! Execute the villains!

Whatever storms may buffet it, the vessel of the Law sails smoothly and majestically on. Our Supreme Courts, our Supreme Prosecutors, are old hands, and no gust will take them by surprise. They will conduct their Plenary Sessions, they will issue. their Instructions—and every insane change of course will be shown to be a long-felt need, a logical result of our whole historical develop-

11. See *Sotsialisticheskaya Zakonnost* (organ of the Public Prosecutor's Office of the U.S.S.R.), No. 1, January, 1962. Signed for the press on December 27, 1961. On pp. 73–74 there is an article by Grigoryev (Gruzda) (called "Fascist Hangmen"). It contains a report on the trial of some Estonian war criminals at Tartu. The writer describes the questioning of witnesses, the exhibits before the court, the cross-examination of one defendant ("the murderer cynically answered"), the reactions of the public, the prosecutor's speech. It further reports that sentence of death was passed. All these things, indeed, occurred *exactly as described*—but not till January 16, 1962 (see *Pravda* for January 17), by which time the journal was already in print and on sale. (The trial had been postponed, and the journal had not been warned. The journalist concerned got one year's forced labor.)

ment, prophetically envisaged in the One True Doctrine.

The vessel of Soviet Law is ready for the sharpest turn. If orders come tomorrow to put millions inside again for their way of thinking, or to deport whole peoples (the same peoples as before, or others) or rebellious towns, or to pin four numbers on prisoners again—its mighty hull will scarcely tremble, its stem will not buckle.

There remains—what Derzhavin tells us, what only those who have experienced it for themselves can feel in their hearts:

"An unjust court is worse than brigandage."

Yes, that remains true. As true as it was under Stalin, as it was all through the years described in this book. Many Fundamental Principles, Decrees, and Laws, contradictory or complementary, have been promulgated and printed—but it is not in accordance with them that our country lives, and that arrests are made, trials held, expert evidence given. Only in those few cases (15 percent, perhaps?) in which the subject of investigation and judicial proceedings affects neither the interests of the state, nor the reigning ideology, nor the personal interests or comfort of some officeholder—only very rarely can the officers of the court enjoy the privilege of trying a case without telephoning somebody to seek instructions; of trying it on its merits and as conscience dictates. All other cases—the overwhelming majority: criminal or civil, it makes no difference—inevitably affect in some important way the interests of the chairman of a kolkhoz or a village soviet, a shop foreman, a factory manager, the head of a Housing Bureau, a block sergeant, the investigating officer or commander of a police district, the medical superintendent of a hospital, a chief planning officer, the heads of administrations or ministries, special sections or personnel sections, the secretaries of district or oblast Party Committees—and upward, ever upward! In all such cases, calls are made from one discreet inner office to another; leisurely, lowered voices give friendly *advice*, steady and steer the decision to be reached in the trial of a wretched little man caught in the tangled schemes, which he would not understand even if he knew them, of those set in authority over him. The naïvely trusting little newspaper reader goes into the courtroom conscious that he is in the right. His reasonable arguments are carefully rehearsed, and he lays them before the somnolent, masklike faces on the bench, never suspecting that sentence has been passed on him already—that there are no courts of appeal, no proper channels and due procedures through which a malignant, a corrupt, a soul-searingly unjust verdict can be undone.

There is—only a wall. And its bricks are laid in a mortar of lies.

We called this chapter "The Law Today." It should rightly be called *"There Is No Law."*

The same treacherous secrecy, the same fog of injustice, still hangs in our air, worse than the smoke of city chimneys.

For half a century and more the enormous state has towered over us, girded with hoops of steel. The hoops are still there. There is no law.

Afterword

Instead of my writing this book alone, the chapters should have been shared among people with special knowledge, and we should then have met in editorial conference and helped each other to put the whole in true perspective.

But the time for this was not yet. Those whom I asked to take on particular chapters would not do so, but instead offered stories, written or oral, for me to use as I pleased. I suggested to Varlam Shalamov that we write the whole book together, but he also declined.

What was really needed was a well-staffed office. To advertise in the newspapers and on the radio ("Please reply!"), to carry on open correspondence, to do what was done with the story of the Brest fortress.*

Not only could I not spread myself like this; I had to conceal the project itself, my letters, my materials, to disperse them, to do everything in deepest secrecy. I even had to camouflage the time I spent working on the book with what looked like work on other things.

As soon as I began the book, I thought of abandoning it. I could not make up my mind: should I or should I not be writing such a book by myself? And would I have the stamina for it? But when, in addition to what I had collected, prisoners' letters converged on me from all over the country, I realized that since all this had been given to me, I had a duty.

I must explain that *never once* did this whole book, in all its parts, lie on the same desk at the same time! In September, 1965, when work on the Archipelago was at its most intensive, I suffered a setback: my archive was raided and my novel* impounded. At this point the parts of the Archipelago already written, and the materials for the other parts, were scattered, and never reassembled: I could not take the risk, especially when all the names were given correctly. I kept jotting down reminders to myself to check this and remove that, and traveled from place to place with these bits of paper. The jerkiness of the book, its imperfections, are the true mark of our persecuted literature. Take the book for what it is.

I have stopped work on the book not because I regard it as finished, but because I cannot spend any more of my life on it.

Besides begging for indulgence, I want to cry aloud: When the time and the opportunity come, gather together, all you friends who have survived and know the story well, write your own commentaries to go with my book, correct and add to it where necessary (but do not make it too unwieldy, do not duplicate what is there already). Only then will the book be definitive. God bless the work!

I am surprised to have finished it safely, even in this form. I have several times thought they would not let me.

I am finishing it in the year of a double anniversary (and the two anniversaries are connected): it is fifty years since the revolution which created Gulag, and a hundred since the invention of barbed wire (1867).

This second anniversary will no doubt pass unnoticed.

Ryazan—Ukryvishche
April 27, 1958–February 22, 1967

P.P.S.

I was in a hurry when I wrote what you have just read, because I expected that even if I did not perish in the explosion set off by my letter to the Writers' Congress I should lose my freedom to write and access to my manuscripts. But as things turned out, I was not only not arrested as a result of the letter, but found myself on a granite footing. I realized then that I must and could complete and correct this book.

A few friends have now read it. They have helped me to see the serious defects in it. I did not try it out on a wider circle, and if this ever becomes possible, it will be too late for me.

In this last year I have done what I could to improve it. Let no one blame me for its incompleteness; there is no end to the additions which could be made, and every single person who has had the slightest contact with the subject or thought seriously about it will always be able to add something—often something precious. But there are laws of proportion. In size my book has reached the utmost limit. Push in a few more little grains and the whole cliff will come tumbling down.

For sometimes expressing myself badly, for repetition in places and loose construction in others, I ask forgiveness. I was not granted a quiet year after all, and during the last few months the ground has been burning under my feet again, and the desk under my hand. Even while preparing this last version I have *never once* seen the whole book together, never once had it all on my desk at one time.

The full list of those without whom this book could not have been written, revised, or kept safe cannot yet be entrusted to paper. They know who they are. They have my homage.

Rozhdestvo-na-Iste
May, 1968

Notes

Page

7 **katorga:** This word also serves as the general title of Part V. The standard English translation is "hard labor" or "penal servitude," and the Russian term derives from the Greek word for the forced labor of a slave chained to the oar of a galley. It is important to note here that the word *katorga* (the first syllable is stressed) had come to stand for a specifically Tsarist type of punishment; it summoned to mind images of idealistic revolutionaries toiling in Siberian mines.

A person sentenced to *katorga* is a *katorzhanin* (masc.); the plural form is *katorzhane.*

7 **DOPR:** Acronym for *Dom Prinuditelnykh Rabot* (Institution of Compulsory Labor).

7 **ITL:** Acronym for *Ispravitelno-Trudovoi Lager* (Corrective Labor Camp).

7 **"officer," "general,"** etc.: These terms and a number of other items in military and administrative terminology were abandoned in 1917 as expressive of the bourgeois class system. The word "officer" was in special disrepute because the majority of the Imperial officer corps had sided with the Whites during the Civil War. In Solzhenitsyn's short story "Incident at Krechetovka Station," set in 1941, the protagonist feels "wounded as if by a bayonet" by the very thought that his interlocutor might be an "officer" in disguise.

"General" was reinstated in 1940, "officer" was brought back as a standard military term in 1943. "Director" and "supreme" had reappeared in common usage in the 1930's.

9 **zeks:** That is, "convicts" or "prisoners." For one theory about the origin of the term "zek," see Volume Two, page 506.

10 **twenty-eight letters:** The Russian (Cyrillic) alphabet has thirty-three letters, but five of these—*yo, i kratkoye,* the "hard sign", *yery,* and the "soft sign"—are not generally used in any serial notation that involves letters.

10 **Polizei, burgomasters:** Terms that describe, respectively, members of the police units recruited by the Germans from among the population of the occupied territories, and minor local officials appointed by the Germans.

11 **Organs:** That is, Organs of State Security, a Soviet designation for the

Page

political police. The term "Organs"—without any modifiers—was commonly used by the personnel of the internal security agencies.

11 **Kabanikha:** Sanctimonious and tyrannical woman in Aleksandr Ostrovsky's play *The Thunderstorm* (1860).

16 **Oktyabr:** Literary grouping of proletarian writers, organized in 1922, which began publishing a monthly periodical of the same name in 1924. Characterized by doctrinaire excesses in the 1920's, the journal has remained a mouthpiece for hard-line Party views.

21 **OGPU:** Acronym for *Obyedinyonnoye Gosudarstvennoye Politicheskoye Upravlenie* (Unified State Political Administration), the name of the Soviet internal security agency from 1924 to 1934. Between 1922 and 1924 the agency's name was simply GPU, and this shorter form was often used informally for the later period as well.

22 **Levitan:** Yuri Levitan is the best-known radio announcer in the Soviet Union. He has delivered the official news broadcasts on Radio Moscow for more than a generation.

27 **Vlasov movement:** Russian anti-Communist movement during World War II associated with the name of General Andrey Vlasov (see Glossary), which envisaged an armed overthrow of the Soviet regime with the help of the German military. In practice it mainly involved the recruiting of help (both armed and unarmed) for the German army from among the vast numbers of Soviet POW's. Viewed with considerable suspicion by the Nazi hierarchy, the Vlasov movement was not allowed to concentrate its forces on the Eastern front, nor did General Vlasov receive formal authority over Russian units in the German army until the very last phases of the war. See Solzhenitsyn's earlier comments on the movement in *The Gulag Archipelago*, Volume One, pages 251–262.

27 **Chekist:** Originally and narrowly, members of the Cheka, the first Soviet internal security agency. The name is often applied, by extension, to personnel of the succeeding security agencies.

28 **Politburo and Orgburo of the CPSU(b):** The "b" in parentheses stands for "Bolsheviks" and "Communist Party of the Soviet Union (Bolsheviks)" was the official title of the Party until 1952.

The Politburo is the chief policy-making body of the Communist Party. The Orgburo is a subcommittee of the Party's Central Committee and is concerned with organizational and procedural matters.

29 **raikoms and gorkoms:** Acronyms for *rayonny komitet* and *gorodskoy komitet*, local administrative bodies of the Communist Party at the *rayon* and the municipal level respectively. (A *rayon* is a territorial subdivision of an *oblast* or of a large municipality.)

31 **Pugachev rising:** Yemelyan Pugachev (Pugachov) headed a large popular revolt in 1773–1775. He promised a liberation of the serfs and received considerable support in the Volga and Ural areas; it required extraordinary efforts on the part of the government of Catherine II to quell the uprising.

32 **Cadets:** Usual designation of the Russian Constitutional Democrats, moderate liberals whose party was formed in 1905 and outlawed by the Bolsheviks. (The name "Cadets" is derived from the Russian abbreviation of the party's name, K.D., pronounced "kah-deh.")

34 **Gorlag, etc.:**
Gorlag = Gorny lager—Mountain Camp
Berlag = Beregovoi lager—Waterside Camp

Page

Minlag = Mineralny lager—Mineral Camp
Rechlag = Rechnoi lager—Riverside Camp
Dubrovlag = Dubrovny lager—Leafy Grove Camp
Ozerlag = Ozernoi lager—Lakeshore Camp
Steplag = Stepnoi lager—Steppe Camp
Peschanlag = Peschany lager—Sandy Camp
Luglag = Lugovoi lager—Meadow Camp
Kamyshlag = Kamyshovy lager—Reed Camp

34 **red-tabbed guards:** That is, guards drawn from regular army troops.

37 **"At all costs steer clear of general duties":** On this special-assignment prisoner and his advice, see *The Gulag Archipelago,* Volume One, pages 563–64.

38 **Stolypin car:** A railroad car designed for transporting prisoners.

42 **bitches:** Translation of *suki,* term for professional criminals who choose to collaborate with the authorities. In abusive force similar to English "scab" directed at a strikebreaker.

42 **"godfather":** Translation of *kum,* prison-camp slang for the chief security officer. In its literal sense, *kum* is the term used to designate the godfather of one's child. Since this type of relationship implies friendship, the word *kum* has also come to be used as an ironic expression for a person with "pull" who can influence one's career positively. The prison-camp term seems to be a further development of this meaning.

42 **SR:** Abbreviation for Socialist Revolutionary. This radical populist party enjoyed considerable support in Russia and was outlawed by the regime in 1922.

43 **"muzzles":** Louvers or shutters attached to the windows of prisons.

44 **They had sown the seed themselves:** Reference to the substantial role played by the Latvian Rifle Regiments in the establishment of the Soviet regime.

44 **Tambov peasants:** In 1920–1921, a major peasant uprising against the Soviet regime took place in Tambov Province under the leadership of Aleksandr Antonov, an adherent of the SR party.

46 **Vanguard Doctrine:** That is, Marxist-Leninist ideology, which claims to be the world's most progressive philosophy.

51 **Campaign on the Ice:** Also known as the First Kuban Campaign, this episode in the Russian Civil War refers to the withdrawal of the White Volunteer Army from Rostov-on-the-Don in February, 1918, its march across the frozen steppe to the Kuban river area, and its triumphant return to the Don region after almost three months of constant fighting.

58 **numbers beginning with yery:** In Russian serial notation that involves letters of the Cyrillic alphabet, *yery* (ы) is not generally used.

67 **it:** The "it" in the biblical passage refers to the "beast from the earth" that enforced emperor-worship.

70 **Finnish huts:** Prefabricated units imported from Finland or based on such a design.

75 **gophers:** A pun in Russian. Apart from the reference to the desert animals who would be the only ones to see the prison van anyway, the word "gopher" *(suslik)* is a slang term for "gullible fool."

79 **Assemblies of the Land:** Translation of *Zemsky Sobor,* a term for assemblies convoked in Muscovy in the sixteenth and seventeenth centuries. Played prominent role in reestablishing order during the turbulent period in the early seventeenth century.

Page
79 **"SK"**: Abbreviation for *ssylno-katorzhny*—i.e., "one exiled to hard labor."

79 **Decembrist rising**: An unsuccessful attempt to overthrow the Tsarist regime undertaken by liberal-minded aristocratic officers in 1825. Some three thousand rebel troops formed on Senate Square in St. Petersburg on December 14, 1825, but the mutiny collapsed by the end of the day. Five leaders were eventually hanged, several dozen more were sentenced to Siberian exile.

80 **"On the Senate Square"**: That is, with the Decembrist mutineers (see note above). The poet Aleksandr Pushkin had many close friends among the Decembrist leadership, but at the time of the uprising he was confined to his parents' country estate in Mikhailovskoye. Summoned to Moscow for a personal audience with Tsar Nicholas I, Pushkin openly admitted his sympathies.

83 **Iskra**: First Russian Marxist newspaper, founded by Lenin in 1900 and published abroad until 1905, *Iskra* carried many of Lenin's important early essays.

87 **"Stolypin reaction"**: The revolutionaries' term for the period following the suppression of the 1905 revolution, and associated with the policies of Minister of the Interior P. A. Stolypin (1862–1911).

88 **punitive operation at Novocherkassk**: Reference to the bloody suppression of the disturbances in Novocherkassk in June, 1962. This episode is described below, Part VII, Chapter 3.

88 **Livadia**: Site of the summer residence of the royal family in the vicinity of Yalta.

92 **Dal**: Vladimir Dal (Dahl) was a prominent Russian nineteenth-century lexicographer. His four-volume dictionary of the Russian language contains a great mass of material not included in any other dictionary and is one of Solzhenitsyn's favorite sources.

103 **Vasily Tyorkin**: Long narrative poem by Aleksandr Tvardovsky, published in installments during World War II, which describes the life and adventures of a cheerful and resourceful Russian front-line soldier. The poem has enjoyed enormous popularity.

114 **Kady trial**: See *The Gulag Archipelago,* Volume One, pages 419–431.

116 **Mtsyri**: Hero of Mikhail Lermontov's narrative poem of the same name (1840). Mtsyri is a novice who escapes from a Caucasian monastery in an attempt to return to the land of his birth. He fails, and his wanderings through the surrounding forests symbolize the inescapable tragedy of life.

119 **I was on the roads of East Prussia**: That is, engrossed in the long narrative poem *Prussian Nights,* in which Solzhenitsyn describes the tumultuous advance of the Soviet Army through East Prussia during the last phases of World War II.

121 **goners**: Translation of *dokhodyagi,* prison slang for zeks whose physical state indicates that their days are numbered.

123 **Ninth of January**: On Sunday, January 9, 1905, a large procession of workers attempted to present a petition to the Tsar. They were met with gunfire, which left over one hundred dead and several hundred wounded. This episode became known as "Bloody Sunday" and the date is commemorated yearly in the Soviet Union. (But see below, page 255*n*.)

127 **NEP**: Abbreviation for "New Economic Policy," a temporary relaxation of controls in agriculture, trade, and industry between 1921 and 1928. Private enterprise of all sorts flourished during this period.

Page

148 **"half-caste"**: Translation of *priblatnyonny*, prison slang for a zek who acts like a professional criminal.

149 **shock workers**: Soviet designation of workers whose productivity is considerably above the norm.

150 **makhorka**: Coarse, low-grade tobacco.

159 **"green prosecutor"**: Prison-camp expression for escape.

163 **beshbarmak**: A Kazakh meat dish.

163 **aksakal**: Kazakh term of respect, literally "white beard."

163 **yok**: Kazakh for "there is none."

171 **sovkhoz**: Soviet farm administered directly by the state and operated like an industrial enterprise. A sovkhoz lacks the cooperative structure characteristic of a collective farm (kolkhoz).

193 **"class ally" type of prisoners**: That is, common or professional criminals. The term "class allies" (*sotsialno-blizkie*, which might also be rendered as "social allies" or "those of a kindred class") is derived from Marxist-Leninist class theory, according to which felons are seen as potential allies in the building of Communism due to their proletarian background. See further in *The Gulag Archipelago*, Volume Two, especially pages 434 f. The antonym of "class ally" is "socially alien element," once again defined as such on the strength of social background.

219 **of Stalin fur**: That is, without fur of any kind. This ironic expression probably arose by analogy to the Russian phrase *na rybyem mekhu* ("lined with fish fur"), used to describe a flimsy garment.

226 **bluecaps**: That is, members of the Soviet security agencies. The reference is to the light-blue cap band that distinguished the uniform of the NKVD.

251 **"kulak sabotage"**: The word *kulak*, (literally "fist"), which in its figurative sense used to mean "village usurer," came to be applied in Soviet times to almost any peasant who was successful or well-to-do. (See below, Part VI, chapter 2.) Resistance to the policy of forced collectivization of agriculture initiated in 1929 was termed "kulak sabotage." .

251 **kutya**: A dish prepared from boiled wheat (or rice) sweetened with honey, and usually with an admixture of poppy seeds, raisins, or nuts. It is traditionally eaten on Christmas Eve and at certain other special occasions.

271 **isolator**: Special prison or camp for important political prisoners who are kept incommunicado.

279 **March 5**: The death of Iosif Stalin was officially announced on March 5, 1953.

289 **Yevtushenko's poem**: Bratsk, a town on the Angara River in southeastern Siberia, is the site of a huge hydroelectric station, completed in the 1960's. Yevgeny Yevtushenko has written a lengthy narrative poem celebrating this project (*Bratskaya GES*, 1965).

327 **that summer thirteen years before**: That is, the summer of 1941, the time of the surprise attack launched by Germany on Soviet Russia.

337 **"Witte" committee**: This appears to refer to a provincial branch of an organization that promoted Russian industrial development. Count Sergei Witte (1845–1915) served as Russia's Minister of Finance between 1892 and 1903 and did a great deal to stimulate the growth of industry.

339 **subbotnik**: According to the official definition, this is a voluntary contri-

Page

bution of one's time to perform "socially useful" labor. No compensation is given for such work, which was originally scheduled on Saturdays.

340 **War Communism:** The name applied to the policies and practices of the Soviet regime between 1918 and 1921, particularly in the sphere of economics. This included a large-scale requisitioning of produce from the peasants and the nationalization of industry and trade.

342 **Stakhanovite:** The Stakhanovite movement was launched by the regime in an attempt to increase the productivity of workers. The movement takes its name from Aleksei Stakhanov, a coal miner who was widely touted to have exceeded the production norm by a factor of 14 on one fine day in 1935. Workers who emulated such feats, and those who tried earnestly to do so, were called "Stakhanovites." The term has been supplanted by "shock workers."

343 **SD's:** That is, members of the Social Democratic party, the Russian Marxists who had not accepted Bolshevism.

360 **Magnitogorsk building operation:** One of the major industrial projects during the first Five-Year Plan was the construction of the Magnitogorsk Metal Works, a huge steel-producing plant in the Urals, built together with the town of Magnitogorsk in 1929–1931. Many writers visited the site and hymned the project in prose and verse. Among the better-known works in this style is Valentin Katayev's novel *Time, Forward* (1932).

365 **Tyurin:** A hardbitten and camp-wise brigade leader in Solzhenitsyn's *One Day in the Life of Ivan Denisovich.*

382 **"Adenauer amnesty":** Amnesty granted in 1955 by Khrushchev to persons accused of having collaborated with the Germans during World War II. It was a direct result of the repatriation of many thousands of German POW's that had been negotiated by Chancellor Konrad Adenauer in Moscow.

387 **twenty-six Finns:** For the sake of argument, the author is here suggesting a comparison with the twenty-six Bolshevik commissars executed in Baku by British troops in 1918. A formidable cult has been created in the Soviet Union to commemorate those twenty-six names and a large number of literary works treat this subject (the best known is "The Ballad of the Twenty-Six" by Sergei Esenin [1924]).

407 **three square arshins of land:** That is, enough land to dig a grave. An arshin is a Russian traditional unit of length equal to 28 inches; a grave plot would be three arshins long and one arshin wide.

411 **Eric Arvid Andersen:** For some biographical details about this mysterious prisoner, see *The Gulag Archipelago,* Volume one, pages 521–522, 551–554.

437 **Seven Boyars:** Solzhenitsyn's designation of the seven heirs of Stalin in the early period of "collective leadership" following Stalin's death. They are, in alphabetical order: Nikolai Bulganin, Lazar Kaganovich, Nikita Khrushchev, Georgi Malenkov, Anastas Mikoyan, Vyacheslav Molotov, and Kliment Voroshilov. The term "Seven Boyars" *(Semiboyarshchina)* is taken from Russian history, where it was used to refer to a boyar oligarchy.

444 **Cheka-GB:** A composite term that covers all the Soviet internal security agencies, from the earliest (the Cheka) to the present-day KGB.

446 **passport, which the poet . . . bade all men envy:** Reference to Vladimir Mayakovsky's poem entitled "Verses About a Soviet Passport" (1929),

Page

in which the poet describes with relish the effect produced by his Soviet documents on a foreign official. The poem ends:

Read this

and turn green with envy—

I am a citizen
of the Soviet Union.

"And your brothers shall return your sword to you": Quotation from Aleksandr Pushkin's "Message to Siberia" (1827), addressed to the Decembrists exiled to hard labor in Siberia. Pushkin ends his poetic epistle on the following hopeful note:

The heavy-hanging chains will fall,
The walls will crumble at the word;
And Freedom greet you in the light
And brothers give you back the sword.

(Trans. Max Eastman)

467   **Narodovoltsy:** Members of the clandestine terrorist organization called Narodnaya Volya (The People's Will), formed in 1879. Their major goal was the assassination of high-ranking government officials, with the Tsar as the principal target. After several attempts, the Narodovoltsy succeeded in murdering Alexander II in 1881. Five members of the organization were hanged as a result.

472   **Volkovoy:** Lieutenant Volkovoy is the sadistic disciplinary officer in the prison camp depicted in Solzhenitsyn's *One Day in the Life of Ivan Denisovich.*

473   **MOOP:** Acronym for *Ministerstvo Okhrany Obshchestvennogo Poriadka* (Ministry for the Protection of Public Order). The MVD was given this name in 1962, but the original title (MVD) was restored in 1968. A major function of the MOOP, like that of the MVD before and after it, was to administer the system of camps and prisons. It was also charged with a broad range of other security tasks, from riot control to border duty. The author of the letter quoted by Solzhenitsyn is, strictly speaking, guilty of an anachronism when he refers to the guards depicted in *One Day in the Life of Ivan Denisovich* as serving in the MOOP: the action of the novel is set in 1951, when the MVD was in charge. But as another outraged letter quoted below points out, the administrative personnel throughout these years remained largely the same.

475   **"accursed incognito":** A famous quote from Nikolai Gogol's play *The Inspector General* (1836).

482   **Likhosherstov (!):** The exclamation mark draws attention to the bizarre surname, which means something like "vicious fur" or "bad pelt."

491   **OUN:** *Organizatsiya Ukrainskikh Natsionalistov* (Organization of Ukrainian Nationalists), a militant organization that sporadically collaborated with the Germans during World War II. After the conclusion of the war, various Ukrainian nationalist groups (commonly known as Banderists) waged guerrilla warfare against Soviet forces for a number of years. It has always been Soviet practice to refer to armed resistance of this sort as "banditry" so as to deny that it possesses broad public support.

Page
496 **my nomination for a Lenin Prize:** Solzhenitsyn's novel *One Day in the Life of Ivan Denisovich* was nominated for the Lenin Prize in literature in December, 1963. (The prize was eventually awarded to a less "controversial" writer.)

496 **"Ochakov taken and Tartary subdued":** Quote from Griboyedov's comedy *Woe from Wit* (1824). The action of the play is set in the 1820's and this line characterizes a mentality hopelessly out of date: Ochakov was captured by the Russians in 1788, and the Tartar Khanate of Crimea annexed in 1783.

503 **Potemkin structures:** That is, showcase institutions entirely uncharacteristic of the usual conditions in prison camps. During Catherine the Great's 1787 tour of the newly annexed Crimean territories, her statesman and favorite Grigory Potemkin undertook elaborate measures to make this area appear to be wealthier and more populous than it was in reality. Among other things, he ordered several fake villages to be constructed along Catherine's route. The phrase "Potemkin village" has since then become a common designation of a fraud designed to deceive outsiders.

503 **"A Bitch Went Walking":** A complex pun in Russian. In his book *My Testimony*, Anatoly Marchenko relates that the acronym SVP, which stands for *Sektsiya Vnutrennego Poriadka* (Internal Order Section—a unit that operates in contemporary Soviet prison camps), has been deciphered by zek wits as *Suka Vyshla Pogulyat* (A Bitch Came Out for a Walk). The word "bitch" here stands for an active collaborator with the authorities, and the whole expression therefore suggests that the Internal Order Sections depend primarily on turncoats and informers.

510 **Pliev: "Pli!" ("Pali!")** is the word of command meaning "Fire!"

512 **KUKKS:** Acronym for *Kursy Usovershenstvovaniya Kavaleriiskogo Komandnogo Sostava* (Courses for the Upgrading of the Command Structure in the Cavalry).

526 **what was done with the story of the Brest fortress:** The city of Brest near the Polish border came under attack on the very first day of the German onslaught on Soviet Russia in 1941. The Germans met unexpectedly stiff resistance here, and the citadel of Brest held out for over a week against vastly superior forces. In the post-Stalin epoch, when the defenders of Brest were no longer classified as traitors to the Motherland (all who capitulated had automatically been branded as such during the war), the writer Sergei Smirnov received permission to collect documents and memoirs relating to this episode. To this end Smirnov published appeals in the newspapers and even made special radio broadcasts. The material gathered in this manner was incorporated in two books on the defense of Brest published by Smirnov in the late 1950's.

527 **my novel:** Solzhenitsyn's *The First Circle,* which had at that point not been published anywhere.

The Notes to this edition have been prepared by Alexis Klimoff.

Glossary

This Glossary is selective and the reader is referred to the corresponding sections of Volumes I and II for additional entries.

Aksakov, Ivan Sergeyevich (1823–1886). Slavophile essayist and thinker. Banned from Moscow in 1878 after giving a lecture on Slavic affairs.

Aleksei Mikhailovich (1629–1676). Tsar of Moscow from 1645.

Alexander I (1777–1825). Became Tsar in 1801.

Alexander II (1818–1881). Became Tsar in 1855. Assassinated by revolutionary terrorists.

Alexander III (1845–1894). Became Tsar in 1881. Under his rule, the campaign against revolutionary movements was intensified.

Arakcheyev, Aleksei Andreyevich (1769–1834). General and Minister of War under Alexander I. Inventor of the "military colonies," which were worked by soldier-farmers under strict discipline. His name became synonymous with reaction.

Arany, Janos (1817–1882). Hungarian poet, active supporter of the 1848–49 revolution.

Bandera, Stepan (1909–1959). Leader of a militant Ukrainian nationalist movement. Attempted to collaborate with Germans during World War II, but was arrested and interned as too unreliable and independent. His followers are referred to as Banderists.

Berdyayev, Nikolai Aleksandrovich (1874–1948). Philosopher, essayist, brilliant defender of human freedom against the encroachments of ideology. Lived in emigration after 1922.

Beria, Lavrenti Pavlovich (1899–1953). Sinister head of Stalin's internal security apparatus between 1938 and 1953.

Breshko-Breshkovsky, Nikolai Nikolayevich (1874–1943). Émigré author of several dozen low-grade works of a type known in Russian as "boulevard novels" (accurately rendered by the British expression "penny dreadfuls").

Bulgakov, Mikhail Afanasyevich (1891–1940). Writer of prose and drama.

Burtsev, Vladimir Lvovich (1862–1942). Populist revolutionary. Emigrated at the beginning of the twentieth century. Publisher of the magazine *Byloye,* which concerned itself with the history of the revolutionary movement. Returned to Russia in 1917, but emigrated again during the Civil War.

Catherine II (1729–1796). Became Empress in 1762.

Chang Tso-lin (1873–1928). Powerful Chinese military leader and war lord. Governor of Manchuria in 1911. Occupied Peking several times.

Chekhov, Anton Pavlovich (1860–1904). Major prose writer and greatest Russian playwright. In 1890, he undertook a journey to Sakhalin Island in order to see the penal settlements there; he published a book of his observations in 1895.

Cot, Pierre (1895–). French progressive political figure. Member of the World Council for Peace.

Denikin, Anton Ivanovich (1872–1947). Russian general who served 1918–1919 as commander of anti-Bolshevik forces in southern Russia during the Civil War.

Dmitri Tsarevich (1582–1591). Prince of Uglich. Son of Ivan the Terrible. Historians are divided as to whether his death was accidental or the result of a plot.

Dolgoruky, Yuri (1090–1157). Russian prince, considered the founder of Moscow in 1147.

Doroshevich, Vlas Mikhailovich (1864–1922). Russian journalist of radical sympathies, noted for his description of the penal settlements on Sakhalin Island, which he toured in the late 1890's.

Dutov, Aleksandr Ilyich (1864–1921). White Russian military leader during the Civil War. Led a Cossack revolt in the Orenburg region in November–December, 1917.

Dyakov, Boris Aleksandrovich (1902–). Soviet writer, arrested in 1949. Published two books of prison-camp memoirs in the 1960's, in which he proclaimed his unwavering faith in the Party.

Dzerzhinsky, Feliks Edmundovich (1877–1926). Organizer and head of the first Soviet internal security agency, the Cheka. Until his death served also as the head of the succeeding security agencies, the GPU and the OGPU.

Ehrenburg, Ilya Grigoryevich (1891–1967). Poet, prose writer, and journalist notorious for his ability to adjust smoothly to every fluctuation of the Party line. Played prominent role in the Soviet peace campaign after World War II.

Elizabeth I (Elizaveta Petrovna) (1709–1762). Daughter of Peter the Great. Became Empress in 1741.

Elsberg, Yakov Yefimovich (1901–). Critic and literary historian. After the Twentieth Party Congress, he was accused of having denounced and caused the arrest of many writers. Threatened with exclusion from the Writers' Union, he saved himself with a letter in which he explained that he was "mistaken, along with the Party."

Frunze, Mikhail Vasilyevich (1885–1925). Military and political figure. Trotsky's successor as Commissar for Military Affairs.

Galanskov, Yuri Timofeyevich (1939–1972). Dissident poet. Editor of samizdat journal. Arrested for "anti-Soviet activity," he died in prison.

Ginzburg, Aleksandr Ilyich (1937–   ). Dissident, compiler of "White Book" on Sinyavsky-Daniel trial. Codefendent with Galanskov in 1968 trial. Rearrested in 1977 for aid to families of political prisoners.

Gorbatov, Aleksandr Vasilyevich (1891–). A general, arrested in 1929 and deported to the Kolyma; freed in 1941. Author of memoirs of the camps in line with official policy.

Griboyedov, Aleksandr Sergeyevich (1795–1829). Writer and diplomat. His great play in verse, *Woe from Wit* (1824), depicts the collision of an idealistic aristocrat nurtured on liberal ideas with the representatives of a hidebound Moscow society.

Gumilyov, Nikolai Stepanovich (1886–1921). Major poet of the Acmeist school, executed by the Bolsheviks for alleged counterrevolutionary activities.

Herzen, Aleksandr Ivanovich (1812–1870). Liberal socialist writer and journalist. Lived abroad after 1847 and published the influential émigré Russian journal *Kolokol (The Bell)*. Herzen supported the Poles in their 1863 uprising against Russian domination.

Ilyichev, Leonid Fyodorovich (1906–). Communist leader; member of the Central Committee, 1961–1966. Appointed to deal with ideological problems during the time of Nikita Khrushchev.

Isakovsky, Mikhail Vasilyevich (1900–   ). Soviet pastoral poet.

Ivan IV (Ivan the Terrible) (1530–1584). Became first Tsar of Russia in 1547.

Jochelson (Iokelson), Vladimir Ilyich (1855–1943). Militant in the People's Will terrorist movement; deported to the Kolyma in 1888. An ethnographer and linguist, he lived in the United States after 1922.

Kaledin, Aleksei Maksimovich (1861–1918). General in the Tsarist army. Led a White Russian insurrection in the Don area, October 1917–February 1918. Committed suicide.

Kalyaev, Ivan Platonovich (1877–1905). Member of the Socialist Revolutionary Party, involved in an attempted assassination of Plehve in 1904. On February 4, 1905, he killed the governor general of Moscow with a bomb.

Kamo (Ter-Petrossyan, Semyon Arkakovich) (1882–1922). Georgian

Bolshevik and friend of Stalin; the two men successfully arranged several major robberies ("expropriations").

Karavansky, Svyatoslav (1920–). Ukrainian dissident. Has spent more than 25 years in camps and prison.

Karpenko-Kary (Tobilyevich, Ivan Karpovich) (1845–1907). Ukrainian dramatist and actor.

Karpov, Yevtikhi Pavlovich (1857–1926). Dramatist and theatrical producer. Participated in the Populist movement.

Karsavina, Tamara Platonovna (1885–). Prima ballerina of Serge Diaghilev's Ballets Russes from 1909 to 1929. After the Revolution she settled in London and taught ballet.

Kasatkin, Ivan Mikhailovich (1880–1938). Writer and landscape painter; member of the Social Democratic Party. Took care of children orphaned by the Revolution. Worked in the State Publishing House during the thirties. Victim of Stalin's purge.

Kautsky, Karl (1854–1938). German Social Democrat. Leader of the Second International. Editor in chief of the *Neue Zeit*, 1883–1917.

Khalturin, Stepan Nikolayevich (1856–1882). Russian revolutionary; set off bomb in the Winter Palace in February, 1880.

Kirillov, Vladimir Timofeyevich (1890–1943). A sailor, he participated in the 1905 revolution, and was sent into exile. Became an important proletarian poet. Victim of Stalin's purge.

Kochetov, Vsevolod Anisimovich (1912–1973). Typifies the Socialist Realism school of writing. Author of many novels.

Kon, Feliks Yakovlevich (1864–1941). Polish revolutionary; later a Bolshevik. Settled in the U.S.S.R. after the October Revolution. Member of the Third International.

Kopelev, Lev Zinovyevich (1912–). German scholar. Solzhenitsyn's prison companion. Author of memoirs.

Korneychuk, Aleksandr Yevdokimovich (1905–1972). Russian-Ukrainian dramatist and high public official. Director of the Writers' Union, and Peace Movement; member of the Central Committee; president of the Supreme Soviet of the Ukrainian S.S.R.; holder of five Stalin Prizes.

Kotoshikhin, Grigory Karpovich (c.1630–1667). Muscovite official who fled in 1666 to Sweden, where he wrote *Russia Under the Reign of Aleksei Mikhailovich*, a valuable historical, literary, and linguistic document.

Kozlov, Frol Romanovich (1908–1965). Communist leader. Member of the Presidium, 1957–1964. After the purge of Leningrad, was secretary of the City Committee (1949–1952), then of Leningrad oblast (1952–1957).

Krasin, Leonid Borisovich (1870–1926). Engineer and one of the very early Bolsheviks. First U.S.S.R. ambassador to France, 1925.

Krasnov, Pyotr Nikolayevich (1869–1947). Important White Russian military leader during the Civil War. In 1944–1945, led a Cossack unit which fought against the Bolsheviks. Delivered to Stalin by the British, and executed.

Kravchenko, Viktor Andreyevich (1905–1966). High Soviet official who defected to the West in 1944. Published *I Chose Freedom* (1946), an exposé of Stalinist crimes; sued the directors of the French Communist press when they accused him of publishing fraudulent data.

Krupskaya, Nadezhda Konstantinovna (1869–1939). Lenin's wife and collaborator.

Krzhizhanovsky, Gleb Maksimilianovich (1872–1959). One of Lenin's leading comrades in arms in the revolutionary movement before 1917; later an important official in the Soviet campaign for electrification and the development of energy resources. Spent 1897–1900 in Siberian exile.

Kurochkin, Vasily Stepanovich (1831–1875). Satiric poet and journalist.

Lavrov, Pyotr Lavrovich (1823–1900). Revolutionary; his *Historical Letters* (1868–1869) were the philosophical basis of the Populist movement.

Lermontov, Mikhail Yuryevich (1814–1841). Major Russian poet and novelist. Killed in duel during Caucasian exile.

Lesyuchevsky, Nikolai Vasilyevich (1908–). Director of the Sovetski Pisatel publishing house. Accused of having denounced several writers at the end of the 1930's, causing their arrest.

MacDonald, James Ramsay (1866–1937). English labor leader. Prime Minister, 1924, 1929–1935. Involved in several incidents with the U.S.S.R.

Malenkov, Georgi Maksimilianovich (1902–). Stalin's secretary, then his successor as head of state in 1953. Resigned in 1955; removed from political life in 1957.

Mandelstam, Nadezhda Yakovlevna (1899–). Widow of the poet Osip Mandelstam, who died in transit camp c. 1938. Author of memoirs which are essential reading for an understanding of Russian twentieth-century intellectual history.

Marchenko, Anatoly Tikhonovich (1938–). Spent 1960–67 in prison; described this experience in *My Testimony*. Rearrested several times since then.

Maslennikov, Ivan Ivanovich (1900–1954). General; commander of combined armies during World War II. Connected with GPU-NKVD-MVD before and after war.

Mayakovsky, Vladimir Vladimirovich (1893–1930). Called the "poet of the Russian Revolution." Committed suicide.

Meyerhold, Vsevolod Emilyevich (1874–c. 1942). Actor and avant-garde director. Disappeared in Stalin's purges.

Mikhailovsky, Nikolai Konstantinovich (1842–1904). Positivist philosopher, literary critic, and publicist. Leading theorist of the Populist movement.

Mikoyan, Anastas Ivanovich (1895–   ). Bolshevik; member of the Politburo, 1935–1966. Close collaborator of Stalin's; foreign policy counselor to Khrushchev. Specialized in "difficult situations": sent to Budapest during 1956 insurrection, to Cuba after Soviet missile withdrawal.

Mokrousov, Boris Andreyevich (1909–   ). Composer, songwriter, winner of a Stalin Prize.

Nadezhdin, Nikolai Ivanovich (1804–1856). Essayist and literary critic; professor at University of Moscow. Exiled 1836–1838.

Nicholas I (1796–1855). Became Tsar in 1825. Quelled Decembrist uprising; known for his hostility to liberalism.

Nicholas II (1868–1918). Tsar from 1894 to 1917. Shot by the Bolsheviks, together with his family.

Nogin, Viktor Pavlovich (1878–1924). Bolshevik; arrested several times. After October, 1917, became People's Commissar of Commerce and Industry.

Ostrovsky, Aleksandr Nikolayevich (1823–1886). Major Russian playwright.

Panin, Dmitri (1911–   ). Engineer; prison companion of Solzhenitsyn. Has written memoirs. Now lives in France.

Parvus (Helphand), Aleksandr Izrailovich (1867–1924). Active member of the St. Petersburg soviet during 1905 revolution. Sentenced to three years of exile, he escaped abroad. He amassed a large fortune, and sometimes contributed to Bolshevik causes.

Petlyura, Simon Vasilyevich (1879–1926). Ukrainian nationalist leader, 1917–1920. Followers referred to as Petlyurovites.

Petöfi, Sándor (1849–1923). Hungarian poet. Killed during battle of Segesvár.

Pisarev, Dmitri Ivanovich (1840–1868). Radical literary critic. Fierce opponent of "art for art's sake." Imprisoned in the Peter and Paul Fortress for four years.

Platov, Matvey Ivanovich (1751–1818). Hetman of the Don Cossacks. Hero of the 1812 war; buried in Novocherkassk.

Plehve, Vyacheslav Konstantinovich (1846–1904). Minister of the Interior from 1902. Assassinated.

Podbelsky, Vadim Nikolayevich (1887–1920). Bolshevik. Commissar of the Postal Service in 1920.

Pushkin, Aleksandr Sergeyevich (1799–1837). Greatest Russian poet. Spent 1820–26 in exile (Odessa, Moldavia, Mikhailovskoye).

Radishchev, Aleksandr Nikolayevich (1749–1802). Russian nobleman and writer; spent several years in Siberian exile for the attack on the

Russian serf-owning system contained in his book *A Journey from St. Petersburg to Moscow* (1790).

Rudenko, Roman Andreyevich (1907–). Public prosecutor for the U.S.S.R. at Nuremberg, 1945–1946. Attorney General of the U.S.S.R. from 1953.

Sazonov, Igor S. (1879–1910). Socialist Revolutionary. One of Plehve's assassins.

Semashko, Nikolai Aleksandrovich (1874–1949). Social Democrat, later Bolshevik. Emigrated in 1905, returned to Russia after February, 1917. People's Commissar for Health, 1918–1930; professor of medicine.

Semyonov-Tyan-Shansky, Pyotr Petrovich (1827–1914). Famous explorer and geographer. His nineteen-volume geography of Russia is a classic.

Shalamov, Varlam Tikhonovich (1907–). Writer; spent 17 years in Kolyma camps; author of *Kolyma Stories* (Paris, 1969) and *Essays on the Criminal World.*

Sheinin, Lev Romanovich (1905–1967). Soviet writer and prosecutor; served as interrogator during the purge years. Apart from publishing detective and spy stories with titles like *The Military Secret,* Sheinin has described some aspects of his career in *Notes of an Investigator* (1938).

Shevchenko, Taras Grigoryevich (1814–1861). Greatest Ukrainian poet. Spent ten years in exile for nationalistic activities.

Shostakovich, Dmitri Dmitriyevich (1906–1975). Composer. Criticized at various times for his "formalism."

Skoropadsky, Pavel Petrovich (1873–1945). General in the Tsarist army. Hetman of the Ukraine, April–December, 1918. Supported by the Germans; emigrated.

Solovyov, Aleksandr₍ Konstantinovich (1846–1879). Revolutionary. Made unsuccessful attempt to assassinate Alexander II in 1879. Hanged.

Spiridonova, Mariya Aleksandrovna (1884–1941). Born in Tambov. Sentenced to prison in 1906 for shooting a policeman who was quelling a peasant uprising. In 1917 she became a leader of the left wing of the Socialist Revolutionaries. Died in a Soviet camp.

Staroselsky, Vladimir Aleksandrovich (1860–1916). Agronomist. Governor of the province of Kutaisi (Georgia), 1905–1906. Helped the revolutionaries; later became a Bolshevik.

Strakhovich, Konstantin Ivanovich. Russian scientist, specialist in aerodynamics. Served as one of Solzhenitsyn's informants in compilation of *The Gulag Archipelago.*

Surkov, Aleksei Aleksandrovich (1899–). Soviet poet, winner of several

Stalin Prizes; member of the World Council for Peace; a director of the Writers' Union.

Suvorin, Aleksei Sergeyevich (1834–1912). Enterprising journalist and publicist, man of letters, theatre director.

Suvorov, Aleksandr Vasilyevich (1730–1800). Outstanding Russian military commander; led Russian armies to a series of victories over Turkish and French forces in the late eighteenth century.

Tan-Bogoraz, Vladimir Germanovich (1865–1936). Deported to the Kolyma at the end of the nineteenth century for his part in the People's Will terrorist movement. Ethnographer, linguist, folklorist, specialist in northern Siberian affairs.

Tevekelyan, Varktes Arutiunovich (1902–1967). Party member; director of a textile mill; writer. His novel *Granite Does Not Melt* (1962) glorifies the Chekists.

Tikhonov, Nikolai Semyonovich (1896–). Soviet poet and prose writer. As chairman of the Soviet Peace Committee after World War II, he made numerous trips abroad.

Tolstoi, Aleksei Nikolayevich (1882–1945). Novelist; well known before the Revolution. Emigrated after the Revolution, but returned to the Soviet Union in 1923. Winner of several Stalin Prizes.

Turgenev, Ivan Sergeyevich (1818–1883). Major Russian novelist. Banished to his estate in 1852 for having written an obituary of Gogol that was forbidden by the censors.

Tvardovsky, Aleksandr Trifonovich (1910–1971). Poet and journalist. Between 1958 and 1970 was Editor in chief of *Novy Mir,* the most prestigious—and at the time the most liberal—Soviet literary monthly. Tvardovsky's support was instrumental in allowing Solzhenitsyn's *One Day in the Life of Ivan Denisovich* to be published in the Soviet Union, and this novel first appeared on the pages of *Novy Mir.*

Vlasov, Andrei Andreyevich (1900–1946). General of the Soviet Army, captured by the Germans in 1942; agreed to lend his name to the Russian anti-Communist movement during the war. Surrendered to American forces in 1945, and was handed over to Soviet authorities and executed.

Yermak Timofeyevich (?–1585). Cossack hetman. Conqueror of eastern Siberia, 1581–1582.

Yevtushenko, Yevgeny Aleksandrovich (1933–). Poet of the "thaw"; alternately conformist and outspoken.

Zhukovsky, Vasily Andreyevich (1783–1852). Poet, tutor of future Alexander II. He helped Aleksandr Pushkin on many occasions due to his connections with the court.

Index

About the author

About the book

Read on

Insights,
Interviews
& More...

Meet Aleksandr I. Solzhenitsyn

© The Nobel Foundation

ALEKSANDR I. SOLZHENITSYN was born in Kislovodsk, Russia, on December 11, 1918. He earned a degree in mathematics and physics from Rostov University and studied literature through a correspondence course from the Moscow Institute of History, Philosophy, and Literature. A captain in the Soviet Army during World War II, he was arrested in 1945 for criticizing Stalin and the Soviet government in private letters. He was sentenced to eight years of incarceration, to be followed by "perpetual" internal exile, but was cleared of all charges in 1957 as part of Nikita Khrushchev's campaign of de-Stalinization. Solzhenitsyn vaulted from unknown schoolteacher to internationally famous writer in 1962 with the publication of his novella *One Day in the Life of Ivan Denisovich,* which Khrushchev himself authorized. The writer's increasingly vocal opposition to the regime resulted in another arrest, a charge of treason, and expulsion from the USSR in 1974. For eighteen years of his exile, he and his family lived in Vermont. In 1994 he returned to Russia, thus fulfilling his longstanding prediction. He died at his home in Moscow on August 3, 2008.

Solzhenitsyn's major works include the novels *In the First Circle* and *Cancer Ward,* the memoirs *The Oak and the Calf* and *Invisible Allies,* a cycle of historical novels with the series title *The Red Wheel,* and the monumental history of the Soviet prison system *The Gulag Archipelago,* which *Time* Magazine named the "Best Nonfiction Work of the Twentieth Century." In 1970 Solzhenitsyn received the Nobel Prize in Literature. ∽

Written in Secret
The Nobel Lecture

WHILE THE *Nobel Lecture traditionally is delivered at the annual award ceremony in Stockholm, Solzhenitsyn could not risk the trip in December 1970, for fear he would be barred re-entry into the Soviet Union. Instead, he stayed in Moscow and worked on the lecture in secret. The finished manuscript was photographed and transferred onto a film negative; Solzhenitsyn then arranged for it to be smuggled to Sweden via a network of supporters. Transported across the Soviet border inside a portable radio, Solzhenitsyn's lecture finally was presented to the Swedish Academy in 1972. In this lecture Solzhenitsyn introduced to the world the term "Gulag Archipelago." He was already at work on the monumental history of the same name.*

1

Just as that puzzled savage who has picked up—a strange cast-up from the ocean?—something unearthed from the sands?—or an obscure object fallen down from the sky?—intricate in curves, it gleams first dully and then with a bright thrust of light. Just as he turns it this way and that, turns it over, trying to discover what to do with it, trying to discover some mundane function within his own grasp, never dreaming of its higher function. So also we, holding

Art in our hands, confidently consider ourselves to be its masters; boldly we direct it, we renew, reform and manifest it; we sell it for money, use it to please those in power; turn to it at one moment for amusement—right down to popular songs and night-clubs, and at another—grabbing the nearest weapon, cork or cudgel— for the passing needs of politics and for narrow-minded social ends. But art is not defiled by our efforts, neither does it thereby depart from its true nature, but on each occasion and in each application it gives to us a part of its secret inner light. But shall we ever grasp the whole of that light? Who will dare to say that he has *defined* Art, enumerated all its facets? Perhaps once upon a time someone understood and told us, but we could not remain satisfied with that for long; we listened, and neglected, and threw it out there and then, hurrying as always to exchange even the very best—if only for something new! And when we are told again the old truth, we shall not even remember that we once possessed it. One artist sees himself as the creator of an independent spiritual world; he hoists onto his shoulders the task of creating this world, of peopling it and of bearing the all-embracing responsibility for it; but he crumples beneath it, for a mortal genius is not capable of bearing such a burden. Just as man in general, having declared himself the center of existence, has not succeeded in creating a balanced spiritual system. And if misfortune overtakes him, he casts the blame upon the age-long disharmony of the world, upon the complexity of today's ruptured soul, or upon the stupidity of the public. Another artist, recognizing a higher power above, gladly works as a humble apprentice beneath God's heaven; then, however, his responsibility for everything that is written or drawn, for the souls which perceive his work, is more exacting than ever. But, in return, it is not he who has created this world, not he who directs it, there is no doubt as to its foundations; the artist has merely to be more keenly aware than others of the harmony of the world, of the beauty and ugliness of the human contribution to it, and to communicate this acutely to his fellow-men. And in misfortune, and even at the depths of existence—in destitution, in prison, in sickness—his sense of ▶

Written in Secret *(continued)*

stable harmony never deserts him. But all the irrationality of art, its dazzling turns, its unpredictable discoveries, its shattering influence on human beings—they are too full of magic to be exhausted by this artist's vision of the world, by his artistic conception or by the work of his unworthy fingers. Archeologists have not discovered stages of human existence so early that they were without art. Right back in the early morning twilights of mankind we received it from Hands which we were too slow to discern. And we were too slow to ask: *for what purpose* have we been given this gift? What are we to do with it? And they were mistaken, and will always be mistaken, who prophesy that art will disintegrate, that it will outlive its forms and die. It is we who shall die—art will remain. And shall we comprehend, even on the day of our destruction, all its facets and all its possibilities? Not everything assumes a name. Some things lead beyond words. Art inflames even a frozen, darkened soul to a high spiritual experience. Through art we are sometimes visited—dimly, briefly—by revelations such as cannot be produced by rational thinking. Like that little looking-glass from the fairy-tales: look into it and you will see—not yourself—but for one second, the Inaccessible, whither no man can ride, no man fly. And only the soul gives a groan . . .

2

One day Dostoevsky threw out the enigmatic remark: "Beauty will save the world." What sort of a statement is that? For a long time I considered it mere words. How could that be possible? When in bloodthirsty history did beauty ever save anyone from anything? Ennobled, uplifted, yes—but whom has it saved? There is, however, a certain peculiarity in the essence of beauty, a peculiarity in the status of art: namely, the convincingness of a true work of art is completely irrefutable and it forces even an opposing heart to surrender. It is possible to compose an outwardly smooth and elegant political speech, a headstrong article, a social program, or a philosophical system on the basis of both a mistake and a lie. What is hidden, what distorted, will not immediately become obvious. Then a contradictory speech, article, program, a differently constructed philosophy rallies in opposition—and all just as elegant and smooth, and once again it works. Which is why such things are both trusted and mistrusted. In vain to reiterate what does not reach the heart. But a work of art bears within itself its own verification: conceptions which are devised or stretched do not stand being portrayed in images, they all come crashing down, appear sickly and pale, convince no one. But those works of art which have scooped up the truth and presented it to us as a living force—they take hold of us, compel us, and nobody ever, not even in ages to come, will appear to refute them. So perhaps that ancient trinity of Truth, Goodness and Beauty is not simply an empty, faded formula as we thought in the days of our self-confident, materialistic youth? If the tops of these three trees converge, as the scholars maintained, but the too blatant, too direct stems of Truth and Goodness are crushed, cut down, not allowed through—then perhaps the fantastic, unpredictable, unexpected stems of Beauty will push through and soar *to that very same place*, and in so doing will fulfill the work of all three? In that case Dostoevsky's remark, "Beauty will save the world," was not a careless phrase but a prophecy? After all *he* was granted to see much, a man of fantastic illumination. And in that case art, literature might really be able to help the world today? It is the small insight which, over the years, I have succeeded in gaining into this matter that I shall attempt to lay before you here today.

3

In order to mount this platform from which the Nobel lecture is read, a distant platform offered only once in a lifetime, I have climbed not three or four makeshift steps, but hundreds and even thousands of them; unyielding, precipitous, frozen steps, leading out of the darkness and cold where it was my fate to survive, while others—perhaps with a greater gift and stronger than I—have ▶

perished. Of them, I myself met but a few on the Archipelago of
GULAG, shattered into its fractionary multitude of islands; and
beneath the millstone of shadowing and mistrust I did not talk
to them all, of some I only heard, of others still I only guessed.
Those who fell into that abyss already bearing a literary name are
at least known, but how many were never recognized, never once
mentioned in public? And virtually no one managed to return.
A whole national literature remained there, cast into oblivion not
only without a grave, but without even underclothes, naked, with
a number tagged on to its toe. Russian literature did not cease
for a moment, but from the outside it appeared a wasteland!
Where a peaceful forest could have grown, there remained, after
all the felling, two or three trees overlooked by chance. And as
I stand here today, accompanied by the shadows of the fallen,
with bowed head allowing others who were worthy before to pass
ahead of me to this place, as I stand here, how am I to divine and
to express what *they* would have wished to say? This obligation has
long weighed upon us, and we have understood it. In the words of
Vladimir Solov'ev:

> Even in chains we ourselves must complete
> That circle which the gods have mapped out for us.

Frequently, in painful camp seethings, in a column of prisoners,
when chains of lanterns pierced the gloom of the evening frosts,
there would well up inside us the words that we should like to cry
out to the whole world, if the whole world could hear one of us.
Then it seemed so clear: what our successful ambassador would
say, and how the world would immediately respond with its
comment. Our horizon embraced quite distinctly both physical
things and spiritual movements, and it saw no lop-sidedness in
the indivisible world. These ideas did not come from books,
neither were they imported for the sake of coherence. They were
formed in conversations with people now dead, in prison cells and
by forest fires, they were tested against *that* life, they grew out of
that existence. When at last the outer pressure grew a little weaker,

my and our horizon broadened and gradually, albeit through a minute chink, we saw and knew "the whole world." And to our amazement the whole world was not at all as we had expected, as we had hoped—that is to say a world living "not by that," a world leading "not there"; a world which could exclaim at the sight of a muddy swamp, "what a delightful little puddle!" and at concrete neck stocks, "what an exquisite necklace!"; but instead a world where some weep disconsolate tears and others dance to a lighthearted musical. How could this happen? Why the yawning gap? Were we insensitive? Was the world insensitive? Or is it due to language differences? Why is it that people are not able to hear each other's every distinct utterance? Words cease to sound and run away like water—without taste, color, smell. Without trace. As I have come to understand this, so through the years has changed and changed again the structure, content and tone of my potential speech. The speech I give today.

And it has little in common with its original plan, conceived on frosty camp evenings.

4

From time immemorial man has been made in such a way that his vision of the world, so long as it has not been instilled under hypnosis, his motivations and scale of values, his actions and intentions are determined by his personal and group experience of life. As the Russian saying goes, "Do not believe your brother, believe your own crooked eye." And that is the most sound basis for an understanding of the world around us and of human conduct in it. And during the long epochs when our world lay spread out in mystery and wilderness, before it became encroached by common lines of communication, before it was transformed into a single, convulsively pulsating lump—men, relying on experience, ruled without mishap within their limited areas, within their communities, within their societies, and finally on their national territories. At that time it was possible for individual human beings to perceive and accept a general ▶

scale of values, to distinguish between what is considered normal, what incredible; what is cruel and what lies beyond the boundaries of wickedness; what is honesty, what deceit. And although the scattered peoples led extremely different lives and their social values were often strikingly at odds, just as their systems of weights and measures did not agree, still these discrepancies surprised only occasional travelers, were reported in journals under the name of wonders, and bore no danger to mankind which was not yet one. But now during the past few decades, imperceptibly, suddenly, mankind has become one—hopefully one and dangerously one—so that the concussions and inflammations of one of its parts are almost instantaneously passed on to others, sometimes lacking in any kind of necessary immunity. Mankind has become one, but not steadfastly one as communities or even nations used to be; not united through years of mutual experience, neither through possession of a single eye, affectionately called crooked, nor yet through a common native language, but, surpassing all barriers, through international broadcasting and print. An avalanche of events descends upon us—in one minute half the world hears of their splash. But the yardstick by which to measure those events and to evaluate them in accordance with the laws of unfamiliar parts of the world—this is not and cannot be conveyed via sound waves and in newspaper columns. For these yardsticks were matured and assimilated over too many years of too specific conditions in individual countries and societies; they cannot be exchanged in mid-air. In the various parts of the world men apply their own hard-earned values to events, and they judge stubbornly, confidently, only according to their own scales of values and never according to any others. And if there are not many such different scales of values in the world, there are at least several; one for evaluating events near at hand, another for events far away; aging societies possess one, young societies another; unsuccessful people one, successful people another. The divergent scales of values scream in discordance, they dazzle and daze us, and in order that it might not be painful we steer clear of all other values, as though from insanity, as though from illusion, and we confidently judge the whole world according to our own home values. Which is why we take for the greater, more painful and less bearable disaster not that which is in fact greater, more painful and less bearable, but that which lies closest to us. Everything which is further away, which does not threaten this very day to invade our threshold—with all its groans, its stifled cries, its destroyed lives, even if it involves millions of victims—this we consider on the whole to be perfectly bearable and of tolerable proportions. In one part of the world, not so long ago, under persecutions not inferior to those of the ancient Romans, hundreds of thousands of silent Christians gave up their lives for their belief in God. In the other hemisphere a certain madman, (and no doubt he is not alone), speeds across the ocean to *deliver* us from religion—with a thrust of steel into the high priest! He has calculated for each and every one of us according to his personal scale of values! That which from a distance, according to one scale of values, appears as enviable and flourishing freedom, at close quarters, and according to other values, is felt to be infuriating constraint calling for buses to be overthrown. That which in one part of the world might represent a dream of incredible prosperity, in another has the exasperating effect of wild exploitation demanding immediate strike. There are different scales of values for natural catastrophes: a flood craving two hundred thousand lives seems less significant than our local accident. There are different scales of values for personal insults: sometimes even an ironic smile or a dismissive gesture is humiliating, while for others cruel beatings are forgiven as an unfortunate joke. There are different scales of values for punishment and wickedness: according to one, a month's arrest, banishment to the country, or an isolation-cell where one is fed on white rolls and milk, shatters the imagination and fills the newspaper columns with rage. While according to another, prison sentences of twenty-five years, isolation-cells where the walls are covered with ice and the prisoners stripped to their ▶

Written in Secret *(continued)*

underclothes, lunatic asylums for the sane, and countless
unreasonable people who for some reason will keep running
away, shot on the frontiers—all this is common and accepted.
While the mind is especially at peace concerning that exotic
part of the world about which we know virtually nothing, from
which we do not even receive news of events, but only the trivial,
out-of-date guesses of a few correspondents. Yet we cannot
reproach human vision for this duality, for this dumbfounded
incomprehension of another man's distant grief, man is just
made that way. But for the whole of mankind, compressed into
a single lump, such mutual incomprehension presents the threat
of imminent and violent destruction. One world, one mankind
cannot exist in the face of six, four or even two scales of values:
we shall be torn apart by this disparity of rhythm, this disparity
of vibrations. A man with two hearts is not for this world, neither
shall we be able to live side by side on one Earth.

5

But who will co-ordinate these value scales, and how? Who will
create for mankind one system of interpretation, valid for good
and evil deeds, for the unbearable and the bearable, as they are
differentiated today? Who will make clear to mankind what is
really heavy and intolerable and what only grazes the skin
locally? Who will direct the anger to that which is most
terrible and not to that which is nearer? Who might succeed
in transferring such an understanding beyond the limits of his
own human experience? Who might succeed in impressing upon
a bigoted, stubborn human creature the distant joy and grief of
others, an understanding of dimensions and deceptions which he
himself has never experienced? Propaganda, constraint, scientific
proof—all are useless. But fortunately there does exist such a
means in our world! That means is art. That means is literature.
They can perform a miracle: they can overcome man's detrimental
peculiarity of learning only from personal experience so that the
experience of other people passes him by in vain. From man to

man, as he completes his brief spell on Earth, art transfers the whole weight of an unfamiliar, lifelong experience with all its burdens, its colors, its sap of life; it recreates in the flesh an unknown experience and allows us to possess it as our own. And even more, much more than that; both countries and whole continents repeat each other's mistakes with time lapses which can amount to centuries. Then, one would think, it would all be so obvious! But no; that which some nations have already experienced, considered and rejected, is suddenly discovered by others to be the latest word. And here again, the only substitute for an experience we ourselves have never lived through is art, literature. They possess a wonderful ability: beyond distinctions of language, custom, social structure, they can convey the life experience of one whole nation to another. To an inexperienced nation they can convey a harsh national trial lasting many decades, at best sparing an entire nation from a superfluous, or mistaken, or even disastrous course, thereby curtailing the meanderings of human history. It is this great and noble property of art that I urgently recall to you today from the Nobel tribune. And literature conveys irrefutable condensed experience in yet another invaluable direction; namely, from generation to generation. Thus it becomes the living memory of the nation. Thus it preserves and kindles within itself the flame of her spent history, in a form which is safe from deformation and slander. In this way literature, together with language, protects the soul of the nation. (In recent times it has been fashionable to talk of the leveling of nations, of the disappearance of different races in the melting-pot of contemporary civilization. I do not agree with this opinion, but its discussion remains another question. Here it is merely fitting to say that the disappearance of nations would have impoverished us no less than if all men had become alike, with one personality and one face. Nations are the wealth of mankind, its collective personalities; the very least of them wears its own special colors and bears within itself a special facet of divine intention.) But woe to that nation whose literature is ▶

disturbed by the intervention of power. Because that is not just a violation against "freedom of print," it is the closing down of the heart of the nation, a slashing to pieces of its memory. The nation ceases to be mindful of itself, it is deprived of its spiritual unity, and despite a supposedly common language, compatriots suddenly cease to understand one another. Silent generations grow old and die without ever having talked about themselves, either to each other or to their descendants. When writers such as Achmatova and Zamjatin—interred alive throughout their lives—are condemned to create in silence until they die, never hearing the echo of their written words, then that is not only their personal tragedy, but a sorrow to the whole nation, a danger to the whole nation. In some cases moreover—when as a result of such a silence the whole of history ceases to be understood in its entirety—it is a danger to the whole of mankind.

6

At various times and in various countries there have arisen heated, angry and exquisite debates as to whether art and the artist should be free to live for themselves, or whether they should be for ever mindful of their duty towards society and serve it albeit in an unprejudiced way. For me there is no dilemma, but I shall refrain from raising once again the train of arguments. One of the most brilliant addresses on this subject was actually Albert Camus' Nobel speech, and I would happily subscribe to his conclusions. Indeed, Russian literature has for several decades manifested an inclination not to become too lost in contemplation of itself, not to flutter about too frivolously. I am not ashamed to continue this tradition to the best of my ability. Russian literature has long been familiar with the notions that a writer can do much within his society, and that it is his duty to do so. Let us not violate the *right* of the artist to express exclusively his own experiences and introspections, disregarding everything that happens in the world beyond. Let us not *demand* of the artist, but—reproach, beg, urge and entice him—that we may be allowed to do. After all, only in part does he himself develop his talent; the greater part of it is blown into him at birth as a finished product, and the gift of talent imposes responsibility on his free will. Let us assume that the artist does not *owe* anybody anything: nevertheless, it is painful to see how, by retiring into his self-made worlds or the spaces of his subjective whims, he *can* surrender the real world into the hands of men who are mercenary, if not worthless, if not insane. Our Twentieth Century has proved to be crueler than preceding centuries, and the first fifty years have not erased all its horrors. Our world is rent asunder by those same old cave-age emotions of greed, envy, lack of control, mutual hostility which have picked up in passing respectable pseudonyms like class struggle, racial conflict, struggle of the masses, trade-union disputes. The primeval refusal to accept a compromise has been turned into a theoretical principle and is considered the virtue of orthodoxy. It demands millions of sacrifices in ceaseless civil wars, it drums into our souls that there is no such thing as unchanging, universal concepts of goodness and justice, that they are all fluctuating and inconstant. Therefore the rule—always do what's most profitable to your party. Any professional group no sooner sees a convenient opportunity to *break off a piece*, even if it be unearned, even if it be superfluous, than it breaks it off there and then and no matter if the whole of society comes tumbling down. As seen from the outside, the amplitude of the tossings of western society is approaching that point beyond which the system becomes metastable and must fall. Violence, less and less embarrassed by the limits imposed by centuries of lawfulness, is brazenly and victoriously striding across the whole world, unconcerned that its infertility has been demonstrated and proved many times in history. What is more, it is not simply crude power that triumphs abroad, but its exultant justification. The world is being inundated by the brazen conviction that power can do anything, justice nothing. Dostoevsky's *Devils*—apparently a provincial nightmare fantasy of the last century—are crawling across the whole world in front of our very eyes, infesting countries where they could not ▶

Written in Secret *(continued)*

have been dreamed of; and by means of the hijackings, kidnappings, explosions and fires of recent years they are announcing their determination to shake and destroy civilization! And they may well succeed. The young, at an age when they have not yet any experience other than sexual, when they do not yet have years of personal suffering and personal understanding behind them, are jubilantly repeating our depraved Russian blunders of the Nineteenth Century, under the impression that they are discovering something new. They acclaim the latest wretched degradation on the part of the Chinese Red Guards as a joyous example. In shallow lack of understanding of the age-old essence of mankind, in the naive confidence of inexperienced hearts they cry: let us drive away *those* cruel, greedy oppressors, governments, and the new ones (we!), having laid aside grenades and rifles, will be just and understanding. Far from it! . . . But of those who have lived more and understand, those who could oppose these young—many do not dare oppose, they even suck up, anything not to appear "conservative." Another Russian phenomenon of the Nineteenth Century which Dostoevsky called *slavery to progressive quirks.* The spirit of Munich has by no means retreated into the past; it was not merely a brief episode. I even venture to say that the spirit of Munich prevails in the Twentieth Century. The timid civilized world has found nothing with which to oppose the onslaught of a sudden revival of barefaced barbarity, other than concessions and smiles. The spirit of Munich is a sickness of the will of successful people, it is the daily condition of those who have given themselves up to the thirst after prosperity at any price, to material well-being as the chief goal of earthly existence. Such people—and there are many in today's world—elect passivity and retreat, just so as their accustomed life might drag on a bit longer, just so as not to step over the threshold of hardship today—and tomorrow, you'll see, it will all be all right. (But it will never be all right! The price of cowardice will only be evil; we shall reap courage and victory only when we dare to make sacrifices.) And on top of this we are threatened by destruction in the fact that the physically compressed, strained world is not allowed to blend spiritually; the molecules of knowledge and sympathy are not allowed to jump over from one half to the other. This presents a rampant danger: *the suppression of information* between the parts of the planet. Contemporary science knows that suppression of information leads to entropy and total destruction. Suppression of information renders international signatures and agreements illusory; within a muffled zone it costs nothing to reinterpret any agreement, even simpler—to forget it, as though it had never really existed. (Orwell understood this supremely.) A muffled zone is, as it were, populated not by inhabitants of the Earth, but by an expeditionary corps from Mars; the people know nothing intelligent about the rest of the Earth and are prepared to go and trample it down in the holy conviction that they come as "liberators." A quarter of a century ago, in the great hopes of mankind, the United Nations was born. Alas, in an immoral world, this too grew up to be immoral. It is not a United Nations organization but a United Governments organization where all governments stand equal; those which are freely elected, those imposed forcibly, and those which have seized power with weapons. Relying on the mercenary partiality of the majority, the UN jealously guards the freedom of some nations and neglects the freedom of others. As a result of an obedient vote it declined to undertake the investigation of private appeals—the groans, screams and beseechings of humble individual *plain people*— not large enough a catch for such a great organization. The UN made no effort to make the Declaration of Human Rights, its best document in twenty-five years, into an *obligatory* condition of membership confronting the governments. Thus it betrayed those humble people into the will of the governments which they had not chosen. It would seem that the appearance of the contemporary world rests solely in the hands of the scientists; all mankind's technical steps are determined by them. It would seem that it is precisely on the international goodwill of scientists, ▶

and not of politicians, that the direction of the world should depend. All the more so since the example of the few shows how much could be achieved were they all to pull together. But no; scientists have not manifested any clear attempt to become an important, independently active force of mankind. They spend entire congresses in renouncing the sufferings of others; better to stay safely within the precincts of science. That same spirit of Munich has spread above them its enfeebling wings. What then is the place and role of the writer in this cruel, dynamic, split world on the brink of its ten destructions? After all we have nothing to do with letting off rockets, we do not even push the lowliest of hand-carts, we are quite scorned by those who respect only material power. Is it not natural for us too to step back, to lose faith in the steadfastness of goodness, in the indivisibility of truth, and to just impart to the world our bitter, detached observations: how mankind has become hopelessly corrupt, how men have degenerated, and how difficult it is for the few beautiful and refined souls to live amongst them? But we have not even recourse to this flight. Anyone who has once taken up the *word* can never again evade it; a writer is not the detached judge of his compatriots and contemporaries, he is an accomplice to all the evil committed in his native land or by his countrymen. And if the tanks of his fatherland have flooded the asphalt of a foreign capital with blood, then the brown spots have slapped against the face of the writer forever. And if one fatal night they suffocated his sleeping, trusting Friend, then the palms of the writer bear the bruises from that rope. And if his young fellow citizens breezily declare the superiority of depravity over honest work, if they give themselves over to drugs or seize hostages, then their stink mingles with the breath of the writer. Shall we have the temerity to declare that we are not responsible for the sores of the present-day world?

7

However, I am cheered by a vital awareness of *world literature* as of a single huge heart, beating out the cares and troubles of our world, albeit presented and perceived differently in each of its corners. Apart from age-old national literatures there existed, even in past ages, the conception of world literature as an anthology skirting the heights of the national literatures, and as the sum total of mutual literary influences. But there occurred a lapse in time: readers and writers became acquainted with writers of other tongues only after a time lapse, sometimes lasting centuries, so that mutual influences were also delayed and the anthology of national literary heights was revealed only in the eyes of descendants, not of contemporaries. But today, between the writers of one country and the writers and readers of another, there is a reciprocity if not instantaneous then almost so. I experience this with myself. Those of my books which, alas, have not been printed in my own country have soon found a responsive, worldwide audience, despite hurried and often bad translations. Such distinguished western writers as Heinrich Böll have undertaken critical analysis of them. All these last years, when my work and freedom have not come crashing down, when contrary to the laws of gravity they have hung suspended as though on air, as though on *nothing*—on the invisible dumb tension of a sympathetic public membrane; then it was with grateful warmth, and quite unexpectedly for myself, that I learnt of the further support of the international brotherhood of writers. On my fiftieth birthday I was astonished to receive congratulations from well-known western writers. No pressure on me came to pass by unnoticed. During my dangerous weeks of exclusion from the Writers' Union, the *wall of defense* advanced by the world's prominent writers protected me from worse persecutions; and Norwegian writers and artists hospitably prepared a roof for me, in the event of my threatened exile being put into effect. Finally even the advancement of my name for the Nobel Prize was raised not in the country where I live and write, but by François Mauriac and his colleagues. And later still entire national writers' unions have expressed their support for me. Thus I have understood and felt that world literature is no longer an abstract anthology, nor ▶

Written in Secret *(continued)*

a generalization invented by literary historians; it is rather a certain common body and a common spirit, a living heartfelt unity reflecting the growing unity of mankind. State frontiers still turn crimson, heated by electric wire and bursts of machine fire; and various ministries of internal affairs still think that literature too is an "internal affair" falling under their jurisdiction; newspaper headlines still display: "No right to interfere in our internal affairs!" Whereas there are no *internal affairs* left on our crowded Earth! And mankind's sole salvation lies in everyone making everything his business; in the people of the East being vitally concerned with what is thought in the West, the people of the West vitally concerned with what goes on in the East. And literature, as one of the most sensitive, responsive instruments possessed by the human creature, has been one of the first to adopt, to assimilate, to catch hold of this feeling of a growing unity of mankind. And so I turn with confidence to the world literature of today—to hundreds of friends whom I have never met in the flesh and whom I may never see. Friends! Let us try to help if we are worth anything at all! Who from time immemorial has constituted the uniting, not the dividing, strength in your countries, lacerated by discordant parties, movements, castes and groups? There in its essence is the position of writers: expressers of their native language—the chief binding force of the nation, of the very earth its people occupy, and at best of its national spirit. I believe that world literature has it in its power to help mankind, in these its troubled hours, to see itself as it really is, notwithstanding the indoctrinations of prejudiced people and parties. World literature has it in its power to convey condensed experience from one land to another so that we might cease to be split and dazzled, that the different scales of values might be made to agree, and one nation learn correctly and concisely the true history of another with such strength of recognition and painful awareness as it had itself experienced the same, and thus might it be spared from repeating the same cruel mistakes. And perhaps under such conditions we artists will be able to cultivate within ourselves a field of vision to embrace the *whole world*: in the center observing like any other human being that which lies nearby, at the edges we shall begin to draw in that which is happening in the rest of the world. And we shall correlate, and we shall observe world proportions.

And who, if not writers, are to pass judgment—not only on their unsuccessful governments, (in some states this is the easiest way to earn one's bread, the occupation of any man who is not lazy), but also on the people themselves, in their cowardly humiliation or self-satisfied weakness? Who is to pass judgment on the light-weight sprints of youth, and on the young pirates brandishing their knives? We shall be told: what can literature possibly do against the ruthless onslaught of open violence? But let us not forget that violence does not live alone and is not capable of living alone: it is necessarily interwoven with falsehood. Between them lies the most intimate, the deepest of natural bonds. Violence finds its only refuge in falsehood, falsehood its only support in violence. Any man who has once acclaimed violence as his *method* must inexorably choose falsehood as his *principle*. At its birth violence acts openly and even with pride. But no sooner does it become strong, firmly established, than it senses the rarefaction of the air around it and it cannot continue to exist without descending into a fog of lies, clothing them in sweet talk. It does not always, not necessarily, openly throttle the throat, more often it demands from its subjects only an oath of allegiance to falsehood, only complicity in falsehood. And the simple step of a simple courageous man is not to partake in falsehood, not to support false actions! Let *that* enter the world, let it even reign in the world—but not with my help. But writers and artists can achieve more: they can *conquer falsehood*! In the struggle with falsehood art always did win and it always does win! Openly, irrefutably for everyone! Falsehood can hold out against much in this world, but not against art. And no sooner will falsehood be dispersed than the nakedness of violence will be revealed in all its ugliness—and violence, decrepit, will fall. That is why, my ▶

Written in Secret *(continued)*

friends, I believe that we are able to help the world in its white-hot hour. Not by making the excuse of possessing no weapons, and not by giving ourselves over to a frivolous life—but by going to war! Proverbs about truth are well-loved in Russian. They give steady and sometimes striking expression to the not inconsiderable harsh national experience: *one word of truth shall outweigh the whole world.* And it is here, on an imaginary fantasy, a breach of the principle of the conservation of mass and energy, that I base both my own activity and my appeal to the writers of the whole world.

Copyright © The Nobel Foundation ∾

More from Aleksandr I. Solzhenitsyn

IN THE FIRST CIRCLE

Moscow, Christmas Eve, 1949. The Soviet secret
police intercept a call made to the American
embassy by a Russian diplomat who promises to
deliver secrets about the nascent Soviet Atomic
Bomb program. On that same day, a brilliant
mathematician is locked away inside a Moscow
prison that houses the country's brightest minds.
He and his fellow prisoners are charged with
using their abilities to sleuth out the caller's
identity, and they must choose whether to aid
Joseph Stalin's repressive state—or refuse and
accept transfer to the Siberian Gulag camps . . .
and almost certain death.

First written between 1955 and 1958, *In the
First Circle* is Solzhenitsyn's fiction masterpiece.
In order to pass through Soviet censors, many
essential scenes—including nine full chapters—
were cut or altered before it was published in a
hastily translated English edition in 1968. Now
with the help of the author's most trusted
translator, Harry T. Willetts, here for the first
time is the complete, definitive English edition of
Solzhenitsyn's powerful and magnificent classic.

"Solzhenitsyn's best novel."
—*Washington Post*

"A classic. . . . Future generations will read it
with wonder and awe."
—*New York Times*

"So profound in its vision and its implications
that it transcends both its locale and the
specificities of its subject matter."
—*New Republic*

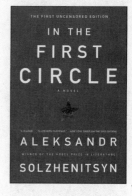

THE FIRST UNCENSORED EDITION

IN THE
FIRST
CIRCLE
A NOVEL

ALEKSANDR
SOLZHENITSYN
WINNER OF THE NOBEL PRIZE IN LITERATURE

The thrilling cold
war masterwork,
published in full
for the first time

More from Aleksandr I. Solzhenitsyn
(continued)

THE GULAG ARCHIPELAGO, ABRIDGED

"It is impossible to name a book that had a greater effect on the political and moral consciousness of the late twentieth century. Not only did Solzhenitsyn deliver the historical truth of the Gulag, he conveyed, as no one else did, its demonic atmosphere and the psychology of both the prisoners and the guards, as well as the mark it left on the entire society."

—David Remnick, *The New Yorker*

THE GULAG ARCHIPELAGO, VOLUME 2

"Volume Two is concerned with the daily life and death of the prisoners, among whom Solzhenitsyn spent eight years. . . . [P]assionate and sharply ironic. . . . Both a powerful chronicle of brutal abuses and at the same time a testament to the tensile strength of the human spirit."

—*Newsweek*

THE GULAG ARCHIPELAGO, VOLUME 3

"[An] enthralling record of camp uprisings, of escapes, of defiance by individuals and groups of victims. . . . In poignant closing chapters, [Solzhenitsyn] recalls his own resurrection from the house of the dead."

—*New Yorker* ∾

Don't miss the next book by your favorite author. Sign up now for AuthorTracker by visiting www.AuthorTracker.com.